Reproduction in Plants: Biology and Physiology

Reproduction in Plants: Biology and Physiology

Editor: Aimee Charlton

www.callistoreference.com

Callisto Reference,
118-35 Queens Blvd., Suite 400,
Forest Hills, NY 11375, USA

Visit us on the World Wide Web at:
www.callistoreference.com

ISBN: 978-1-64116-737-6 (Hardback)

Cataloging-in-Publication Data

Reproduction in plants : biology and physiology / edited by Aimee Charlton.
    p. cm.
Includes bibliographical references and index.
ISBN 978-1-64116-737-6
1. Plants--Reproduction. 2. Plant physiology. I. Charlton, Aimee.
QK825 .I58 2023
575.6--dc23

Table of Contents

Preface

Every book is a source of knowledge and this one is no exception. The idea that led to the conceptualization of this book was the fact that the world is advancing rapidly; which makes it crucial to document the progress in every field. I am aware that a lot of data is already available, yet, there is a lot more to learn. Hence, I accepted the responsibility of editing this book and contributing my knowledge to the community.

The process of reproduction in plants involves the production of new offspring. There are two modes of reproduction in plants, namely, sexual and asexual reproduction. The process of sexual reproduction involves the fusion of gametes that results in offspring being genetically different from the parent plants. Meiosis and fertilization are two fundamental processes involved in sexual reproduction. In asexual reproduction, offspring are produced without the fusion of gametes. Asexual reproduction results in the formation of genetically identical offspring. Budding, fragmentation, spore formation, regeneration and vegetative propagation are different ways of asexual reproduction in plants. The most common form of plant reproduction technique used by cultivators is the use of seeds. Asexual techniques such as cutting, grafting, budding, layering, division, and sectioning of rhizomes, roots, tubers and bulbs, are used to produce desirable characteristics that are usually not obtained from seeds. This book outlines the biology and physiology of plant reproduction in detail. It aims to equip students and experts with the advanced topics and upcoming concepts in this area of study.

While editing this book, I had multiple visions for it. Then I finally narrowed down to make every chapter a sole standing text explaining a particular topic, so that they can be used independently. However, the umbrella subject sinews them into a common theme. This makes the book a unique platform of knowledge.

I would like to give the major credit of this book to the experts from every corner of the world, who took the time to share their expertise with us. Also, I owe the completion of this book to the never-ending support of my family, who supported me throughout the project.

Editor

Unresolved issues in pre-meiotic anther development

Timothy Kelliher[1]*, Rachel L. Egger[2], Han Zhang[2] and Virginia Walbot[2]

[1] *Syngenta Biotechnology Inc., Research Triangle Park, NC, USA*
[2] *Department of Biology, Stanford University, Stanford, CA, USA*

Edited by:
Dazhong Dave Zhao, University of Wisconsin-Milwaukee, USA

Reviewed by:
Scott D. Russell, University of Oklahoma, USA
Hongchang Cui, Florida State University, USA

*Correspondence:
Timothy Kelliher, Syngenta Biotechnology Inc., 3054 East Cornwallis Road, Research Triangle Park, NC 27709, USA
e-mail: tim.kelliher@syngenta.com

Compared to the diversity of other floral organs, the steps in anther ontogeny, final cell types, and overall organ shape are remarkably conserved among Angiosperms. Defects in pre-meiotic anthers that alter cellular composition or function typically result in male-sterility. Given the ease of identifying male-sterile mutants, dozens of genes with key roles in early anther development have been identified and cloned in model species, ordered by time of action and spatiotemporal expression, and used to propose explanatory models for critical steps in cell fate specification. Despite rapid progress, fundamental issues in anther development remain unresolved, and it is unclear if insights from one species can be applied to others. Here we construct a comparison of Arabidopsis, rice, and maize immature anthers to pinpoint distinctions in developmental pace. We analyze the mechanisms by which archesporial (pre-meiotic) cells are specified distinct from the soma, discuss what constitutes meiotic preparation, and review what is known about the secondary parietal layer and its terminal periclinal division that generates the tapetal and middle layers. Finally, roles for small RNAs are examined, focusing on the grass-specific phasiRNAs.

Keywords: arabidopsis, rice, maize, cell fate specification, tapetum, meiosis, phased small RNA

INTRODUCTION

Successful anther development results in pollen dispersal. Steps required to achieve this are conveniently divided into three phases: organ patterning and initial cell differentiation, meiosis, and post-meiotic gametophyte development. Historically most studies have focused on meiosis and pollen biogenesis, with the assumption that a simple lineage model explained how the typical four somatic wall layers with central archesporial (AR) cells arose from a stamen primordium (**Figure 1A**). Starting with recovery and analysis of mutants defective in cell fate specification about 20 years ago (Sheridan et al., 1996; Canales et al., 2002; Zhao et al., 2002; reviewed by Ma, 2005), new theories and molecular insights into the first phase of anther development were proposed, disputed, and continue to be revised and elaborated.

EARLY STEPS IN ANTHER ONTOGENY

The most detailed description of cellular numbers, shapes, and volumes during fate acquisition is available for maize utilizing 3-D reconstruction from confocal microscopy (Kelliher and Walbot, 2011) rather than transverse sectioning. Patterning to achieve the four somatic wall layers and central pre-meiotic cells typical of anther lobes is summarized in **Figure 1B**, employing pale coloration to indicate initial specification, with cell types darkening as differentiated features emerge. First, note that these stages, numbered one through eight, are processes, not discrete events. For example, starting from a primordium full of pluripotent cells, AR cells are the first to differentiate in anther lobes. In maize, discrete differentiation events from different precursors generate a column of ~10 AR cells over the course of 1 day. The first molecular marker of differentiating AR cells is MAC1 secretion; this ligand triggers pluripotent subepidermal cells to become bipotent Primary Parietal Cells (PPC) (**Figure 1B**, stage 2). The PPC then divide once periclinally to generate Endothecium (EN) and Secondary Parietal Layer (SPL) cells (**Figure 1B**, stage 3). In a given transverse section these steps occur successively, but viewing the anther longitudinally it is clear that AR differentiation in the anther base and tip occurs simultaneously with PPC periclinal divisions in the middle of the lobe (**Figure 1C**). Although it was long assumed that L1 presumptive epidermal cells and L2 internal cells have distinct fates locked in by their positions within an apical meristem, maize AR cells can differentiate from L1 cells during stage 2 low oxygen treatments: thus it appears that every maize anther primordium cell is pluripotent (Kelliher and Walbot, 2012). These observations lay to rest the lineage model where germinal and somatic cell fates diverge from a single "hypodermal cell" division event within each lobe.

Periclinal divisions generate new anther cell types and add cell layers to the anther wall, but most anther cell division is anticlinal (within layers). In maize a rapid elongation phase featuring exclusively anticlinal divisions prolongs stage 4 (**Figure 1B**), prior to SPL periclinal division into the Middle Layer (ML) and Tapetum (TAP) (**Figure 1B**, stages 5–7). This is followed by a second phase of rapid anther growth, prior to differentiation of post-mitotic AR cells into meiotically competent Pollen Mother Cells (PMC) (**Figure 1B**, stage 8). Although rice anthers are similar in size to maize during AR specification, both periods of rapid anticlinal cell division and expansion are absent (**Figure 1B**). As a result rice anthers starting meiosis are about one-third the length of maize; in rice, there is substantial anticlinal division during and just after

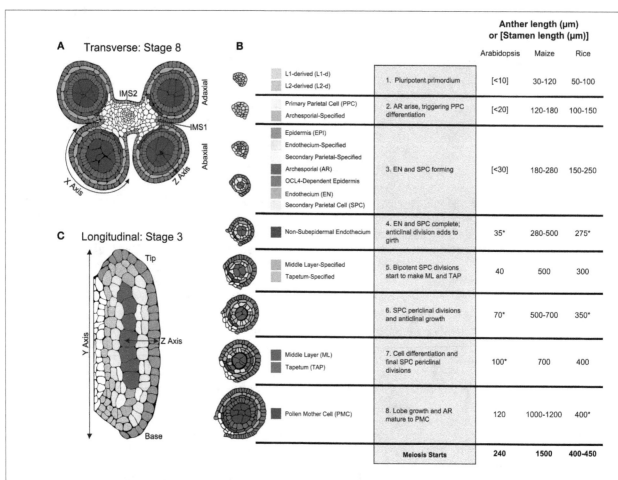

		Anther length (μm) or [Stamen length (μm)]		
			Arabidopsis	Maize	Rice
		1. Pluripotent primordium	[<10]	30-120	50-100
		2. AR arise, triggering PPC differentiation	[<20]	120-180	100-150
		3. EN and SPC forming	[<30]	180-280	150-250
		4. EN and SPC complete; anticlinal division adds to girth	35*	280-500	275*
		5. Bipotent SPC divisions start to make ML and TAP	40	500	300
		6. SPC periclinal divisions and anticlinal growth	70*	500-700	350*
		7. Cell differentiation and final SPC periclinal divisions	100*	700	400
		8. Lobe growth and AR mature to PMC	120	1000-1200	400*
		Meiosis Starts	240	1500	400-450

FIGURE 1 | Pre-meiotic anther development. (A) The four-lobed anther typical of flowering plants with a central column of vasculature that extends into the stamen filament surrounded by connective tissue [stage 8]. **(B)** Tracings of confocal images of single lobes of the W23 maize inbred are colorized to show the progression of cell fate specification and anther lobe patterning. At stage [1] the lobe consists of pluripotent Layer1- and Layer2-derived cells, colored in beige and light gray, respectively. For all cell types, just-specified cells are colorized in a pale shade, which gradually darkens as the cells acquire stereotyped differentiated shapes, volumes, and staining properties. The first specification event results in visible archesporial (AR) cells centrally within each lobe. In maize, the glutaredoxin encoded by *Msca1* responds to growth-generated hypoxia to initiate AR differentiation, marked by secretion of the MAC1 protein, which is required for cell specification of the subepidermal L2-d cells to Primary Parietal Cells (PPC) [stage 2]. PPC divide periclinally generating the subepidermal Endothecium (EN) and the bipotent Secondary Parietal Cells (SPC). In the same timeframe, Epidermal (EPI) cells differentiate; signals controlled by expression of the OCL4 epidermal-specific transcription factor suppress excess periclinal divisions in the EN (Vernoud et al., 2009) [stage 3]. Following these early patterning events that result in a three-layered wall surrounding the AR, there is a period of anticlinal division that expands anther cell number and organ size [stage 4]. Subsequently, each SPC divides once periclinally to generate the ML and TAP and the final four somatic walled architecture of the pre-meiotic anther lobe is achieved [stages 5–7]. Prior to meiosis, anticlinal divisions occur to increase anther size, and the individual cell types acquire differentiated properties [stages 6–8], including dramatic enlargement of AR as they mature into Pollen Mother Cells (PMC) capable of meiosis [stage 8]. Comparison of anther lengths at the 8 stages plus meiotic entry in Arabidopsis (Smyth et al., 1990; Sanders et al., 1999), maize (Kelliher and Walbot, 2014; Zhang et al., 2014), and rice (Zhang et al., 2011) is summarized in the table; lengths marked with an (*) are inferred assuming linear growth in length in between known stages. **(C)** Longitudinal view of an anther lobe [stage 3] illustrates how AR column formation occurs simultaneously within both the tip and base of the lobe and that periclinal division of the PPC is stochastic. Parts of the illustrations in 1A,B are based on figures published in Zhang et al. (2014).

meiosis. Arabidopsis anther primordia are considerably smaller than either grass—a combination of fewer, smaller cells—and like rice there is only modest pre-meiotic growth.

Given conserved internal anther anatomy, we expect similar regulatory processes among different flowering plants. Highlights of these common themes include: (1) many anther-specific mutants (Ma, 2005; Timofejeva et al., 2013); (2) a complex and dynamic transcriptome (Zhang et al., 2014); (3) communication between cell layers using secreted proteins (Wang et al., 2012); (4) presumptively locally produced hormones (Zhang et al., 2014), and (5) developmentally regulated small RNAs (Johnson et al., 2009). It is difficult to propose imposition of hormone or other gradients from materials delivered through the central vasculature (**Figure 1A**). Thus, unanswered questions include how the pace of anticlinal cell division is regulated autonomously within an anther lobe and what specific cues (activators and repressors) regulate periclinal divisions. It is striking that AR and later PMC development does not require normal somatic layers, as documented for the maize *mac1* mutant (Wang et al., 2012). Indeed, once specified, AR cells express a unique transcriptome, including precocious synthesis of transcripts for meiotic proteins, and robust production of transcripts for ribosomes and RNA binding proteins (Kelliher and Walbot, 2014). A major unanswered question is whether anther cell types express but then sequester mRNAs for use at later stages, a developmental mechanism widely employed in animal germlines (Zhang et al., 2014).

INITIAL EVENTS RESULTING IN AR COLUMN FORMATION

A major breakthrough in plant reproductive genetics was identification of the Arabidopsis transcription factor *SPOROCYTELESS (SPL)* which was shown to be essential to AR cell differentiation and meiotic entry in both anthers and ovules (Yang et al., 1999). The late onset of AR-specific expression (~stage 4), however, suggested that *SPL* is only involved after initial fate acquisition, which at the time was thought to depend upon the inheritance of reproductive determinants via asymmetric periclinal division of a single founding hypodermal cell in each lobe. This lineage model ruled the field for decades, largely because early-acting mutants were challenging to identify and characterize, especially in Arabidopsis and rice. Then, in 2003 Chaubal et al. reported on a maize mutant, *male sterile converted anther1 (msca1)*, in which AR cells failed to form and were replaced by vascular bundles. When *MSCA1* was identified as a glutaredoxin (Albertson et al., 2009), and homologs were shown to affect AR fate acquisition in rice (Hong et al., 2012) and Arabidopsis (Xing and Zachgo, 2008), the first connection between protein redox status and plant reproductive cell fate was made.

The application of confocal microscopy enabled a detailed morphometric analysis of AR column formation, which established that instead of just one founding germinal cell per lobe determined by inheritance, there were many initial AR cells specified from multiple progenitors. The lineage model was incorrect, and it was proposed that the internal position of AR cells within lobes determined their ultimate fate (Kelliher and Walbot, 2012). It was reasoned that such internal cells should be more hypoxic than their neighbors because of the high metabolic demand of rapid proliferation and the lack of air space in the tightly packed tissue (**Figure 2A**). To test the idea, hypoxia treatments were applied and rescued AR fate acquisition in *msca1*, dramatically increased AR cell counts in fertile anthers, and stimulated ectopic AR formation in both fertile and mutant anthers (Kelliher and Walbot, 2012). While redox manipulations were not attempted in other species, the mutant phenotypes of *Msca1* homologs in rice (*MIL1*) (Hong et al., 2012) and Arabidopsis (*ROXY1/ROXY2*) (Xing and Zachgo, 2008) indicate that glutaredoxin-based control of AR fate might be a conserved feature across flowering plants.

A major unanswered question is whether MSCA1 is expressed and active in the pluripotent L2-d cells surrounding the AR column, which experience mild hypoxia but will acquire a somatic fate. Glutaredoxins such as MSCA1 are known to bind and activate bZIP-type TGA transcription factors (Murmu et al., 2010); we hypothesize that among the MSCA1 targets are transcription factors that promote *Mac1* expression, the first molecular marker of AR fate acquisition. A secreted peptide, MAC1 signals pluripotent cells to differentiate as PPC (somatic) and divide periclinally forming the EN and SPL (**Figure 2B**). If MSCA1 is active

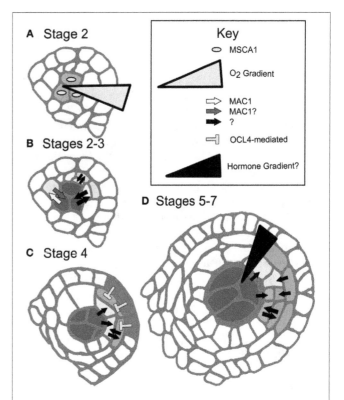

FIGURE 2 | Defined and proposed signaling networks in pre-meiotic maize development. (A) Specification of AR cells is dependent on oxygen status interaction with MSCA1; hypoxic conditions stimulate AR cell differentiation. **(B)** As AR cells differentiate, they secrete the ligand MAC1, which is putatively perceived by LRR-RLK-type receptors on the L2-d cells, specifying these cells as soma and likely stimulating them to become PPC. MAC1 is also inferred to negatively regulate proliferation of AR cells until an entire column is formed in W23 (Wang et al., 2012). At these and later stages there are likely other cell-to-cell communication networks, indicated as black arrows with questions marks. These signals may be other ligand-receptor pairs, siRNAs, or other as yet undiscovered factors. **(C)** The differentiated EPI expresses transcription factor OCL4, which indirectly represses periclinal division in the neighboring EN, possibly by assisting with EN differentiation (Vernoud et al., 2009). **(D)** The trigger for periclinal division of the SPL, differentiation of the ML and TAP cell fates and maintenance of these fates is completely unexplored. A variety of cell-to-cell communication pathways might be in use (black arrows), or there may be a locally produced hormone gradient(s) along the Z-axis.

in these presumptive PPCs, how does MAC1 signaling counteract or attenuate MSCA1 activity, overcoming mild hypoxia to enforce a somatic destiny? Likewise, does low oxygen tip the balance toward MSCA1 in the growing AR column, where newborn AR cells differentiate despite MAC1 secretion from existing AR cells (**Figure 1C**)? Ectopic AR cells arising from redox manipulation have column-forming abilities, even in *mac1*; therefore, column formation may be an emergent property of AR cells that is neither inhibited nor promoted by MAC1. Future work will continue to examine potential interactions between MSCA1 and MAC1 with respect to column formation.

Reproductive cell fate acquisition has put glutaredoxins at the forefront of the study of the genetic-environmental interface. These proteins appear to provide flexibility and responsiveness to hard-wired developmental programs and continue to be implicated in diverse plant developmental processes. In rice, the *mil1* glutaredoxin affects AR and SPL cell fate following AR differentiation (Hong et al., 2012) (stages 3–4), while the Arabidopsis glutaredoxin double mutant *roxy1/2* differentially affects ab- and ad-axial AR development (Xing and Zachgo, 2008), causing defects in stage 2 adaxial lobes and stage 7 abaxial lobes, implying a continued requirement for redox management during anther development and the involvement of multiple glutaredoxins in the differentiation of both germinal and somatic anther tissues. Mis-expression of MSCA1 in the shoot apical meristem gives opposite phyllotaxy in *Abph2* (Jackson et al., 2013), demonstrating the power of this protein to influence body plan elaboration. While genetic programming has a major role in early anther development—initial AR cell counts vary among inbred lines from ~10 in W23 to ~25 in A619—hypoxic treatments can double or triple initial AR counts without forcing a trade off in somatic cell populations. Therefore, the initial size of the AR population is controlled by both genetic factors and environmental conditions, and pollen production can be manipulated via genetic or extrinsic, redox-based treatments.

PATTERNING THE ANTHER SOMATIC NICHE

The somatic niche surrounding the developing AR cells is equally important for anther fertility. Contingent on MAC1-mediated signaling, L2-d cells are specified as somatic niche cells. The first morphological marker of somatic fate is PPC periclinal division that results in two somatic daughter cells (stage 3), that through subsequent periclinal divisions will differentiate as EN, ML, and TAP (Kelliher and Walbot, 2011; Timofejeva et al., 2013). Although EN, SPL, ML, and TAP cells are typically represented as differentiating immediately after a periclinal division, this is an over-simplification. Indeed, careful analysis of cell shape in maize indicates that these cells must expand in preferential dimensions before gradually achieving mature cell morphology (Kelliher and Walbot, 2011). Borrowing terminology from animal development, we consider somatic cells as passing through multiple steps, starting with cell fate specification, commitment, and finally differentiation. Periclinal division is the first marker of specification, and differentiation has occurred by the morphological end-point, but without deeper knowledge of cellular and molecular details, determining commitment timing is impossible. Does specification occur before a periclinal division, or after? When is commitment reached, and how are we to interpret the numerous mutants that persist in making periclinal divisions, i.e., *ocl4*, *ms23*, and *ms32* in maize (Chaubal et al., 2000; Vernoud et al., 2009; Moon et al., 2013, **Figure 2C**), and *tdf1* in Arabidopsis (Zhu et al., 2008)?

A second issue is within-layer stochasticity in periclinal division initiation. AR cells initiate meiosis synchronously, yet the periclinal divisions resulting in the EN/SPL and later the ML/TAP are both asynchronous and exhibit no discernible spatial pattern (Kelliher and Walbot, 2011; Zhang and Li, 2014). Each cell appears to divide periclinally after a variable number of anticlinal divisions. Could signals directing periclinal divisions be perceived during a restricted portion of the cell cycle, resulting in a large population of cells deaf to the signal when first available?

Asynchrony of anticlinal somatic division would then result in asynchrony of periclinal divisions.

The complexity and lack of information on signaling within the soma makes investigations into the regulation of somatic cell patterning and specification in the anther ripe for pursuit. Despite the relative dearth of information, there are some hints that allow predictions (**Figure 2**). First, the secreted peptide and receptor model may be a common theme. Mutants in Arabidopsis and rice homologs *AtTPD1* and *OsTDL1A* have excess AR cells but develop a normal EN and SPL (stage 5, **Figure 1B**) (Feng and Dickinson, 2010). Clues from these mutants suggest the MAC1 peptide family binds LRR-RLK receptors (*AtEXS/EMS1* and *OsMSP1*, respectively). As LRR-RLKs are known to dimerize, hetero-dimerization utilizing different isoforms or family members may contribute to differential signal interpretation in peripheral L2-d and presumptive AR cells during column formation, or between other cell types at later stages (Wang et al., 2012). Elucidating the suite of LRR-RLKs could uncover a distinct "combinatorial code" of dimeric receptors on different cell types. This idea is consistent with the presence of many putatively secreted peptides (Wang et al., 2012) and the expression of numerous LRR-RLKs in anthers (Kelliher and Walbot, 2014; Zhang et al., 2014). The anther is one of many organs where peptide signaling has emerged as a major developmental paradigm. In particular, tissues that lack direct access to vasculature, such as stem cells in the shoot apical meristem and meristemoid cells of the stomatal lineage (Li and Torii, 2012), often employ this mode of cell-cell communication.

The role of hormones or other chemical gradients within anther lobes is another area yet to be explored on a cell type-specific level. In a commonly cited example, periclinal division in Arabidopsis roots involves an auxin gradient delivered by polarized auxin flow (Cruz-Ramirez et al., 2012). Given the anatomy of the anther, it is unlikely that a similar flow exists because cell types are not a uniform distance from the vasculature. Local hormone production and perception would more likely be a feature of anther development (**Figure 2D**). Transcriptome analysis indicates that both gibberellin and brassinolide associated genes are differentially regulated in stage 7 pre-meiotic anther development; these would make strong initial candidates for further exploration as regulators of anther cell fates (Zhang et al., 2014).

Despite identification of many maize male-sterile mutants disrupting pre-meiotic somatic patterning (**Table 1**), this phase of early anther development is underrepresented in the rice and Arabidopsis literature. The small size of rice and Arabidopsis anthers at these stages (20–70 and 100–350 μm, respectively, compared to 120–700 μm for maize) along with the rapidity of development (in maize, events span nearly 3 days, roughly twice as long as Arabidopsis) makes isolation and precise analysis of discrete stages difficult (**Figure 1B**). Many mutants that fail to complete meiosis have been categorized as meiotic mutants, even when the defect is earlier and the primary defect is in the soma. There is no consensus concerning the specific roles of each somatic cell type. The tapetum has been characterized as a highly transcriptionally active, secretory cell type required for exine formation. But the ML can be present as either a

Table 1 | A comprehensive list of anther mutants sequentially organized from organ specification through meiosis.

Stage affected	A. thaliana	Rice	Maize	Annotation	Phenotype	doi
1	agamous	mads3, mads58	Zmm2, Mads2	MADS-box transcription factor	stamens converted to petals (Arabidopsis) or lodicules (grasses)	10.1105/tpc.3.8.749
1			ems71924, ems72032	not cloned	stamen adaxialization	10.1534/g3.112.004465
1			ems71990, ms-si*355	not cloned	absence of anthers in some florets	10.1534/g3.112.004465
1	rdr6	rol (SHOOTLESS2)	rdr6	RNA-directed RNA polymerase	stamen abaxialization (defect in tasi-ARF biosynthesis)	10.1105/tpc.110.075291
2	bam1 bam2 double mutant			LRR receptor like kinases	all internal lobe cells become AR; no somatic cells	10.1105/tpc.105.036871
2		mil1	msca1	glutaredoxin	AR fail to differentiate (Os) or differentiate as vasculature (Zm)	10.1007/s00425-002-0929-8
2 (ad), 8 (ab)	roxy1 roxy2 double mutant			glutaredoxin (thioreductase)	adaxial lobes: AR specification failure; abaxial: PMC degrade	10.1111/j.1365-313X.2007.03375.x
3	nzz = spl	no homology in grasses		MADS-box transcription factor	AR differentiation failure; somatic cell layer defects	10.1104/pp.109.145896
3 (Zm/Os), 5 (At)	tpd1	tdl1a = mil2	mac1	small secreted protein ligand	somatic cell specification failure; overproliferation of AR	10.1105/tpc.016618
3	exs = ems1	msp1		LRR receptor-like kinase	somatic cell specification failure; overproliferation of AR	10.1016/S0960-9822(02)01151-x
4			ocl4	HD-ZIP IV transcription factor	additional periclinal divisions in subepidermal cell layer	10.1111/j.1365-313X.2009.03916.x
4			ems63089, tcl1, mtm00-06	not cloned	undifferentiated somatic cell layers	10.1534/g3.112.004465
5		tip2		bHLH transcription factor	all three anther wall layers fail to differentiate properly	10.1105/tpc.114.123745
5	er erl1 erl2 triple mutant			LRR receptor-like kinases	missing anthers and somatic cell differentiation defects	10.1093/mp/ssn029
6	tdf1			R2R3 Myb transcription factor	early vacuolization in epidermis and endothecium, tapetal failure	10.1111/j.1365-313X.2008.03500.x
6	serk1 serk2 double mutant			LRR receptor-like kinases	SPL periclinal division failure	10.1105/tpc.105.036731
7			ms23, ms*6015	not cloned	additional periclinal divisions in the tapetal layer	10.1534/g3.112.004465

(Continued)

Table 1 | Continued

Stage affected	*A. thaliana*	Rice	Maize	Annotation	Phenotype	doi
7			*ems72063*	*not cloned*	undifferentiated soma; excess periclinal divisions in tapetum	10.1534/g3.112.004465
7			*ems72091*	*not cloned*	additional periclinal divisions in the middle layer	10.1534/g3.112.004465
7	*mpk3 and mpk6*			MAP kinases	somatic cell specification failure; overproliferation of AR	10.1093/mp/ssn029
8		*dtm1*		ER membrane protein	tapetal differentiation failure	10.1111/j.1365-313X.2011.04864.x
8	*dyt1*			bHLH transcription factor	tapetal differentiation failure	10.1111/j.1365-313X.2012.05104.x
8	*myb33 myb65 double mutant*			GAMYB-like transcription factor	tapetal differentiation failure	10.1105/tpc.104.027920
8			*ms9, ms11, ms13, ms14*	*not cloned*	tapetal differentiation failure	10.1534/g3.112.004465
8			*ms32*	bHLH transcription factor	excess periclinal divisions in tapetum after normal wall is built	10.1111/tpj.12318
8			*csmd1*	*not cloned*	excess pre-meiotic callose and slow dissolution of the tetrad	10.1007/s00497-011-0167-y
8			*ms8*	beta-1,3-galactosyl transferase	cell growth defects in epidermis and tapetum, meiotic arrest	10.1007/s00497-013-0230-y
8		*gamyb-4*		GAMYB transcription factor	tapetal differentiation failure; meiotic arrest	10.1111/j.1744-7909.2010.00959.x
meiosis		*udt1*		bHLH transcription factor	tapetal differentiation failure; meiotic arrest	10.1105/tpc.105.034090
meiosis		*mel1*		Argonaute	tapetal differentiation failure; meiotic arrest	10.1105/tpc.107.053199

The left hand column indicates the first developmental stage at which a mutant phenotype is observed, using the staging rubric outlined in **Figure 1**. *In cases where the onset of the phenotype differs among species or tissues, abbreviations are used ("Zm" = maize, "Os" = rice, "At" = Arabidopsis, "ab" = abaxial anther lobes, and "ad" = adaxial anther lobes). If a given mutant is phenocopied by homologs from the other two species, the gene names are given in the corresponding species' column. An exception to this rule was made for the first maize entry, "Zmm2, Mads2," because while mutants in these genes have not been found, the genes are clearly agamous orthologs by sequence comparison and expression pattern in the third floral whorl, which is characteristic of C class genes. The next two columns contain the phenotypic description of the mutant and protein annotation if a causative gene has been cloned. Uncloned mutants are indicated as "not cloned," and these clustered in a single row in cases where they roughly phenocopy each other (for example, maize ems71924 and ems72032 have nearly identical anther polarity phenotypes, and may be allelic). In the final column, a doi is provided for the founding mutant of each class. The high number of blank spaces in the species' columns reflects the challenge of comparisons between model species. The bHLH, MYB, and LRR-RLK genes are all found in large families making identification of orthologs between species problematic. Furthermore, mutations in a single gene that cause a clear phenotype in one plant species may not be available in others because of functional gene redundancy from lineage specific gene duplication. And there is already evidence that orthologs can regulate different steps reflecting evolutionary diversification of developmental pathways. For these reasons we do not anticipate a high degree of correspondence between Arabidopsis, rice, and maize.*

single layer (as in rice, maize, and Arabidopsis), or several layers, but is always consistent within a species (Esau, 1965; D'Arcy and Keating, 1996). Middle-layer like tissue and dartboard lobe architecture are relatively ancient, dating to the gymnosperm microsporangia (Esau, 1965; D'Arcy and Keating, 1996), but as of yet no specific function has been proposed for the ML. Secondary wall thickening of the EN is involved in anthesis at the end of anther development, but no attention has been given to earlier roles. Germ cell establishment and the subset of tapetal mutants that result in meiotic arrest have received the vast majority of attention.

ROLE OF SMALL RNAs IN ANTHER PATTERNING

Developing anthers depend on small RNAs for gene regulation, cell-to-cell communication, and epigenetic reprogramming. For example, in rice, anther adaxial-abaxial polarity is regulated by trans-acting siRNAs (tasiRNAs), which are secondary siRNAs derived from *TAS* transcripts (Toriba et al., 2010). These transcripts are first processed by miRNA-guided cleavage and then converted to double-stranded RNAs by RNA-dependent RNA polymerase6 (RDR6), followed by Dicer4 (DCL4) cleavage yielding 21-nt tasiRNAs (Allen et al., 2005). Mutants that disable components of the tasiRNA biogenesis pathway display severe defects in floral morphology and fertility. Rice *dcl4* exhibits a disruption in lemma abaxial-adaxial polarity and abnormal anther development (Liu et al., 2007). In the *rod-like lemma (rol/rdr6)* mutants, stamen, and lemma development are severely compromised (Toriba et al., 2010). A single nucleotide polymorphism in the same gene results in reduced fertility at high temperature (Song et al., 2012a). The phenotypes of these mutants are largely caused by loss of tasiRNAs, resulting in upregulation of target *Auxin Response Factor* genes required for abaxial identity (Pekker et al., 2005).

Recently two size classes of phased siRNAs (phasiRNAs) were found to be preferentially expressed in grass inflorescences, particularly in anthers (Johnson et al., 2009; Song et al., 2012b). Both phasiRNA types are derived from low copy intergenic regions, and each requires a specific miRNA trigger (miR2118 for the 21-nt class and miR2275 for 24-nt class) to initiate cleavage and RDR6-dependent second-strand synthesis. The resulting double-stranded RNAs are then cleaved by DCL4 to produce 21-nt phasiRNAs and by DCL5 for 24-nt phasiRNAs (Arikit et al., 2013). Despite similarities to tasiRNA biogenesis, the functions of these 21- and 24-nt phasiRNAs and their targets are largely unknown. The phasiRNAs lack complementarity to genes or transposons, suggesting novel roles rather than post-transcriptional gene silencing or RNA-directed DNA methylation.

Post-meiotically, small RNAs do provide germinal cell genome surveillance in microspores. Triggered by miRNAs and dependent on RDR6 and DCL4 (Creasey et al., 2014), the 21-nt epigenetically-activated siRNAs (easiRNAs) are derived from transposon transcripts exclusively expressed in the vegetative nucleus of Arabidopsis pollen grains; easiRNAs then move to sperm cells to direct transcriptional gene silencing (Slotkin et al., 2009). As a result, sperm cell chromatin is highly condensed and enriched in the repressive epigenetic modifications while chromatin in the vegetative cell nucleus is largely decondensed (Slotkin et al., 2009).

Despite sharing biogenesis factors including DCL4 and RDR6, 21-nt phasiRNAs are different from 21-nt easiRNAs in three ways. First, although 21-nt phasiRNAs have been found in dicots, they are mostly derived from a single *NB-LRR* defense gene family (Zhai et al., 2011); in contrast hundreds of grass loci encode phasiRNAs. Second, phasiRNAs are produced prior to meiosis or gametogenesis (Komiya et al., 2014) while easiRNAs are found specifically in gametophytes. Third, easiRNAs are derived from repetitive regions, while phasiRNAs are produced from unique or low copy sequences. These distinctions between phasiRNAs and easiRNAs suggest their functional divergence and imply a sporophytic role for phasiRNAs in anther development.

ARGONAUTE proteins have diverged rapidly in plants, with 10 members in Arabidopsis (Chen, 2009), 19 in rice (Kapoor et al., 2008), and 17 in maize (Zhai et al., 2014); many are preferentially expressed in germinal cells (Zhai et al., 2014). The continued diversification of ARGONAUTEs suggests that many new functions for small RNAs during plant reproduction await discovery. MEL1, a rice homolog of Arabidopsis AGO5, mainly localizes to the cytoplasm of pre-meiotic sporocytes. *mel1* loss-of-function mutants exhibit abnormal tapetal formation and contain aberrant PMC that arrest in early meiosis (Nonomura et al., 2007). Recently MEL1 was demonstrated to bind 21-nt phasiRNAs preferentially (Komiya et al., 2014). Histone methylation patterns are altered in *mel1* mutant meiocytes (Nonomura et al., 2007), hinting at a role for MEL1 and its associated 21-nt phasiRNAs in chromatin modification. Because the tapetal somatic defect precedes the documented meiocyte phenotypes in *mel1* anthers, it is not yet clear if MEL1/21-nt phasiRNAs act directly or indirectly to disrupt meiosis.

Maize *AGO104* is a homolog of Arabidopsis *AGO9* and its transcripts accumulate specifically during pre-meiosis just following germinal differentiation in anthers (Kelliher and Walbot, 2014) and during meiosis (Singh et al., 2011). Maize *ago104* mutants show defects in meiotic chromatin condensation and subsequent failure to properly segregate chromosomes (Singh et al., 2011). Small RNAs interacting with AGO104 are yet to be profiled, but the heterochromatic decondensation in maize *ago104* mutants suggests a role in germinal cell epigenetic reprogramming. With rapid progress in deep sequencing of small RNAs and RNA-IP-seq, we are likely to acquire a better understanding of the diverse ways small RNAs contribute to anther development and plant reproduction.

AUTHOR CONTRIBUTIONS

Each author wrote one section of the manuscript, edited all sections, and approved final submission.

ACKNOWLEDGMENTS

Support for our research on pre-meiotic maize anthers was from NSF 07-01880. As a graduate student at Stanford, Timothy Kelliher was supported by an NIH Biotechnology Training Grant (5-T32-GM008412-17). Rachel L. Egger was supported in part by an NIH Cell and Molecular Biology Training Grant (5T32GM007276-36), by the 2013 American Society of Plant

Biologists-Pioneer Hi-Bred Graduate Student Fellowship, and by an Agriculture and Food Research Initiative Competitive Grant from the USDA National Institute of Food and Agriculture (2013-67011-21096).

REFERENCES

Albertson, M. C., Fox, T., Trimnell, M., Wu, Y., Lowe, K., Li, B., et al. (2009). *Msca1 Nucleotide Sequences Impacting Plant Male Fertility and Method of Using Same. WO2009020458.A1, ed USTPO.* Alexandria, VA: E.I. Du Pont de Nemours and Company.

Allen, E., Xie, Z., Gustafson, A. M., and Carrington, J. C. (2005). microRNA-directed phasing during trans-acting siRNA biogenesis in plants. *Cell* 121, 207–221. doi: 10.1016/j.cell.2005.04.004

Arikit, S., Zhai, J., and Meyers, B. C. (2013). Biogenesis and function of rice small RNAs from non-coding RNA precursors. *Curr. Opin. Plant Biol.* 16, 170–179. doi: 10.1016/j.pbi.2013.01.006

Canales, C., Bhatt, A. M., Scott, R., and Dickinson, H. (2002). EXS, putative LRR receptor kinase, regulates male germline cell number and tapetal identity and promotes seed development in Arabidopsis. *Curr. Biol.* 20, 1718–1727. doi: 10.1016/S0960-9822(02)01151-X

Chaubal, R., Zanella, C., Trimnell, M. R., Fox, T. W., Albertsen, M. C., and Bedinger, P. (2000). Two male-sterile mutants of *Zea mays* (Poaceae) with an extra cell division in the anther wall. *Am. J. Bot.* 87, 1193–1201. doi: 10.2307/2656657

Chaubal, R., Zanella, C., Trimnell, M. R., Fox, T. W., Albertsen, M. C., and Bedinger, P. (2003). The transformation of anthers in the *msca1* mutant of maize. *Planta* 216, 778–788. doi: 10.1007/s00425-002-0929-8

Chen, X. (2009). Small RNAs and their roles in plant development. *Annu. Rev. Cell Dev. Biol.* 25, 21–44. doi: 10.1146/annurev.cellbio.042308.113417

Creasey, K. M., Zhai, J., Borges, F., Van Ex, F., Regulski, M., Meyers, B. C., et al. (2014). miRNAs trigger widespread epigenetically activated siRNAs from transposons in Arabidopsis. *Nature* 508, 411–415. doi: 10.1038/nature13069

Cruz-Ramirez, A., Diaz-Triviño, S., Blilou, I., Grieneisen, V. A., Sozzani, R., Zamioudis, C., et al. (2012). A bistable circuit involving SCARECROW-RETINOBLASTOMA integrates cues to inform asymmetric stem cell division. *Cell* 150, 1002–1015. doi: 10.1016/j.cell.2012.07.017

D'Arcy, W. G., and Keating, R. C. (1996). *The Anther: Form, Function, Phylogeny.* Cambridge: Cambridge University Press.

Esau, K. (1965). *Plant Anatomy.* 2nd Edn. New York, NY: John Wiley & Sons, Inc.

Feng, X., and Dickinson, H. (2010). Tapetal cell fate, lineage and proliferation in the Arabidopsis anther. *Development* 137, 2409–2416. doi: 10.1242/dev.049320

Hong, L., Tang, D., Zhu, K., Wang, K., Li, M., and Cheng, Z. (2012). Somatic and reproductive cell development in rice anthers is regulated by a putative glutaredoxin. *Plant Cell* 24, 577–588. doi: 10.1105/tpc.111.093740

Jackson, D. P., Allen, S. M., Johnston, R., Llaca, V., and Yang, F. (2013). *Improving Agronomic Characteristics of Plants Through abph2. WO2013138399.A1, ed USPTO.* Alexandria, VA: Cold Spring Harbor Laboratory; E.I. Du Pont de Nemours and Company.

Johnson, C., Kasprzewska, A., Tennessen, K., Fernandes, J., Nan, G., Walbot, V., et al. (2009). Clusters and superclusters of phased small RNAs in the developing inflorescence of rice. *Genome Res.* 19, 1429–1440. doi: 10.1101/gr.089854.108

Kapoor, M., Arora, R., Lama, T., Nijhawan, A., Khurana, J. P., Tyagi, A. K., et al. (2008). Genome-wide identification, organization and phylogenetic analysis of Dicer-like, Argonaute and RNA-dependent RNA Polymerase gene families and their expression analysis during reproductive development and stress in rice. *BMC Genomics* 9:451. doi: 10.1186/1471-2164-9-451

Kelliher, T., and Walbot, V. (2011). Emergence and patterning of the five cell types of the *Zea mays* anther locule. *Dev. Biol.* 350, 32–49. doi: 10.1016/j.ydbio.2010.11.005

Kelliher, T., and, Walbot, V. (2012). Hypoxia triggers meiotic fate acquisition in maize. *Science* 337, 345–348. doi: 10.1126/science.1220080

Kelliher, T., and Walbot, V. (2014). Germinal cell initials accommodate hypoxia and precociously express meiotic genes. *Plant J.* 77, 639–652. doi: 10.1111/tpj.12414

Komiya, R., Ohyanagi, H., Niihama, M., Watanabe, T., Nakano, M., Kurata, N., et al. (2014). Rice germline-specific Argonaute MEL1 protein binds to phasiR-NAs generated from more than 700 lincRNAs. *Plant J.* 78, 385–397. doi: 10.1111/tpj.12483

Li, J. S., and Torii, K. U. (2012). A tale of two systems: peptide ligand-receptor pairs in plant development. *Cold Spring Harb. Symp. Quant. Biol.* 77, 83–89. doi: 10.1101/sqb.2012.77.014886

Liu, B., Chen, Z., Song, X., Liu, C., Cui, X., Zhao, X., et al. (2007). Oryza sativa *dicer-like4* reveals a key role for small interfering RNA silencing in plant development. *Plant Cell* 19, 2705–2718. doi: 10.1105/tpc.107.052209

Ma, H. (2005). Molecular genetic analyses of microsporogenesis and microgametogenesis in flowering plants. *Annu. Rev. Plant Biol.* 56, 393–434. doi: 10.1146/annurev.arplant.55.031903.141717

Moon, J., Skibbe, D., Timofejeva, L., Wang, C. J., Kelliher, T., Kremling, K., et al. (2013). Regulation of cell divisions and differentiation by MS32 is required for pre-meiotic anther development in *Zea mays. Plant J.* 76, 592–602. doi: 10.1111/tpj.12318

Murmu, J., Bush, M. J., DeLong, C., Li, S., Xu, M., Khan, M., et al. (2010). Arabidopsis basic leucine-zipper transcription factors TGA9 and TGA10 inter- act with floral glutaredoxins ROXY1 and ROXY2 and are redundantly required for anther development. *Plant Physiol.* 154, 1492–1504. doi: 10.1104/pp.110.159111

Nonomura, K.-I., Morohoshi, A., Nakano, M., Eiguchi, M., Miyao, A., Hirochika, H., et al. (2007). A germ cell specific gene of the *ARGONAUTE* family is essential for the progression of premeiotic mitosis and meiosis during sporogenesis in rice. *Plant Cell* 19, 2583–2594. doi: 10.1105/tpc.107.053199

Pekker, I., Alvarez, J. P., and Eshed, Y. (2005). Auxin response factors mediate Arabidopsis organ asymmetry via modulation of KANADI activity. *Plant Cell* 17, 2899–2910. doi: 10.1105/tpc.105.034876

Sanders, P. M., Bui, A. Q., Weterings, K., McIntire, K. N., Hsu, Y. C., Lee, P. Y., et al. (1999). Anther developmental defects in *Arabidopsis thaliana* male-sterile mutants. *Sex. Plant Reprod.* 11, 297–322. doi: 10.1007/s004970050158

Sheridan, W. F., Avalkina, N. A., Shamrov, I. I., Batygina, T. B., and Golubovskaya, I. N. (1996). The *mac1* gene: controlling the commitment to the meiotic pathway in maize. *Genetics* 142, 1009–1020. doi: 10.1534/g3.112.004465

Singh, M., Goel, S., Meeley, R. B., Dantec, C., Parrinello, H., Michaud, C., et al. (2011). Production of viable gametes without meiosis in maize deficient for an ARGONAUTE protein. *Plant Cell* 23, 443–458. doi: 10.1105/tpc.110.079020

Slotkin, R. K., Vaughn, M., Borges, F., Tanurdzić, M., Becker, J. D., Feijó, J. A., et al. (2009). Epigenetic reprogramming and small RNA silencing of transposable elements in pollen. *Cell* 136, 461–472. doi: 10.1016/j.cell.2008.12.038

Smyth, D. R., Bowman, J. L., and Meyerowitz, E. M. (1990). Early flower development in Arabidopsis. *Plant Cell* 2, 755–767. doi: 10.1105/tpc.2.8.755

Song, X., Wang, D., Ma, L., Chen, Z., Li, P., Cui, X., et al. (2012a). Rice RNA-dependent RNA polymerase 6 acts in small RNA biogenesis and spikelet development. *Plant J.* 71, 378–389. doi: 10.1111/j.1365-313X.2012.05001.x

Song, X., Li, P., Zhai, J., Zhou, M., Ma, L., Liu, B., et al. (2012b). Roles of DCL4 and DCL3b in rice phased sRNA biogenesis. *Plant J.* 69, 462–474. doi: 10.1111/j.1365-313X.2011.04805.x

Timofejeva, L., Skibbe, D. S., Lee, S., Golubovskaya, I., Wang, R., Harper, L., et al. (2013). Cytological characterization and allelism testing of pre-meiotic anther developmental mutants identified in a screen of maize male sterile lines. *G3* 3, 231–249. doi: 10.1534/g3.112.004465

Toriba, T., Suzaki, T., Yamaguchi, T., Ohmori, Y., Tsukaya, H., and Hirano, H. Y. (2010). Distinct regulation of adaxial-abaxial polarity in anther patterning in rice. *Plant Cell* 22, 1452–1462. doi: 10.1105/tpc.110.075291

Vernoud, V., Laigle, G., Rozier, F., Meeley, R. B., Perez, P., and Rogowsky, P. M. (2009). The HD-ZIP IV transcription factor OCL4 is necessary for trichome patterning and anther development in maize. *Plant J.* 59, 883–894. doi: 10.1111/j.1365-313X.2009.03916.x

Wang, C.-J. R., Nan, G.-L., Kelliher, T., Timofejeva, L., Vernoud, V., Golubovskaya, I. N., et al. (2012). Maize *multiple archesporial cell 1 (mac1)*, an ortholog of rice *TDL1A*, modulates cell proliferation and identity in early anther development. *Development* 139, 2594–2603. doi: 10.1242/dev.077891

Xing, S., and Zachgo, S. (2008). ROXY1 and ROXY2, two Arabidopsis glutaredoxin genes, are required for anther development. *Plant J.* 53, 790–801. doi: 10.1111/j.1365-313X.2007.03375.x

Yang, W.-C., Ye, D., Xu, J., and Sundaresan, V. (1999). The SPOROCYTELESS gene of Arabidopsis is required for initiation of sporogenesis and encodes a novel nuclear protein. *Genes Dev.* 13, 2108–2117. doi: 10.1101/gad.13.16.2108

Zhai, J., Jeong, D. H., De Paoli, E., Park, S., Rosen, B. D., Li, Y., et al. (2011). MicroRNAs as master regulators of the plant NB-LRR defense gene family via the production of phased, trans-acting siRNAs. *Genes Dev.* 25, 2540–2553. doi: 10.1101/gad.177527.111

Zhai, L., Sun, W., Zhang, K., Jia, H., Liu, L., Liu, Z., et al. (2014). Identification and characterization of Argonaute gene family and meiosis-enriched Argonaute during sporogenesis in maize. *J. Integr. Plant Biol.* doi: 10.1111/jipb.12205. [Epub ahead of print].

Zhang, D., and Li, Y. (2014). Specification of tapetum and microsporocyte cells within the anther. *Curr. Opin. Plant Biol.* 17, 49–55. doi: 10.1016/j.pbi.2013.11.001

Zhang, D., Xue, L., and Zhu, L. (2011). Cytological analysis and genetic control of rice anther development. *J. Genet. Genomics* 38, 379–390. doi: 10.1016/j.jgg.2011.08.001

Zhang, H., Egger, R., Kelliher, T., Morrow, D. J., Fernandes, J., Nan, G.-L., et al. (2014). Transcriptomes and proteomes define gene expression progression in pre-meiotic maize anthers. *G3* 4, 993–1010. doi: 10.1534/g3.113.009738

Zhao, D. Z., Wang, G. F., Speal, B., and Ma, H. (2002). The *EXCESS MICROSPOROCYTES1* gene encodes a putative leucine-rich repeat receptor protein kinase that controls somatic and reproductive cell fates in the Arabidopsis anther. *Genes Dev.* 16, 2021–2031. doi: 10.1101/gad.997902

Zhu, J., Chen, H., Li, H., Gao, J. F., Jiang, H., Wang, C., et al. (2008). *Defective in Tapetal development and function 1* is essential for anther development and tapetal function for microspore maturation in Arabidopsis. *Plant J.* 55, 266–277. doi: 10.1111/j.1365-313X.2008.03500.x

Regulation of inflorescence architecture by cytokinins

Yingying Han[1,2][†], Haibian Yang[1][†] and Yuling Jiao[1]*

[1] State Key Laboratory of Plant Genomics, National Center for Plant Gene Research, Institute of Genetics and Developmental Biology – Chinese Academy of Sciences, Beijing, China
[2] University of Chinese Academy of Sciences, Beijing, China

Edited by:
Dazhong Dave Zhao, University of Wisconsin-Milwaukee, USA

Reviewed by:
Jill Christine Preston, University of Vermont, USA
Beth Thompson, East Carolina University, USA

*Correspondence:
Yuling Jiao, State Key Laboratory of Plant Genomics, National Center for Plant Gene Research, Institute of Genetics and Developmental Biology – Chinese Academy of Sciences, 1 West Beichen Road, Beijing, China
e-mail: yljiao@genetics.ac.cn

[†]*Yingying Han and Haibian Yang have contributed equally to this work.*

In flowering plants, the arrangement of flowers on a stem becomes an inflorescence, and a huge variety of inflorescence architecture occurs in nature. Inflorescence architecture also affects crop yield. In simple inflorescences, flowers form on a main stem; by contrast, in compound inflorescences, flowers form on branched stems and the branching pattern defines the architecture of the inflorescence. In this review, we highlight recent findings on the regulation of inflorescence architecture by cytokinin plant hormones. Results in rice (*Oryza sativa*) and *Arabidopsis thaliana* show that although these two species have distinct inflorescence architectures, cytokinins have a common effect on inflorescence branching. Based on these studies, we discuss how cytokinins regulate distinct types of inflorescence architecture through their effect on meristem activities.

Keywords: branching, cytokinin, floral meristem, inflorescence, shoot apical meristem

INTRODUCTION

Plants have an enormous, striking diversity of forms, with varying numbers and arrangements of organs in different sizes and shapes; this diversity derives from regulation of meristem activity. The aerial organs of a plant come from the shoot apical meristem (SAM) which gives rise to leaves, stem, and axillary meristems during the vegetative stage and transforms into the inflorescence meristem (IM) after the floral transition. The various developmental patterns of the IM in different species produce diverse inflorescence architectures, which not only attract artists and plant scientists, but also draw the attention of plant breeders, because inflorescence traits directly affect crop yields. Branching hierarchy and complexity depend on the species, but are also affected by environmental factors, including nutrition, light, and temperature (Tanaka et al., 2013; Kyozuka et al., 2014; Teo et al., 2014).

The enormous diversity of inflorescence architecture also leads to difficulties in defining consensus criteria to classify these structures. Following Weberling's (1989) suggestions, inflorescence architectures can be broadly grouped into inflorescences without branching (simple) and inflorescences with branching (compound). Another key parameter is whether the IM ends in a terminal flower (determinate) or continues to produce structures, including branches and flowers (indeterminate). Following these key distinctions, at least three typical groups of inflorescence architectures are commonly seen, namely the raceme (simple, indeterminate, as in *Arabidopsis*), the cyme (complex, determinate, as in tomato), and the panicle [complex, determinate, as in wheat (*Triticum aestivum*); or complex, indeterminate, as in maize (*Zea mays*), especially tassel; **Figure 1**; Prusinkiewicz et al., 2007; Kellogg et al., 2013]. These distinct inflorescence architectures

result from different developmental programs that are elaborated below.

Development of the IM conditions the branching of the inflorescence. In *Arabidopsis*, the IM directly initiates floral meristems (FMs, which are determinate meristems) on its flanks; this forms a simple raceme (**Figure 1A**; Benlloch et al., 2007; Tanaka et al., 2013; Teo et al., 2014). The grasses have more diverse inflorescence architectures (Kellogg et al., 2013). In a generalized grass inflorescence, the IM gives rise to several branch meristems (BMs, which are usually indeterminate meristems). These BMs may initiate secondary BMs to form lateral branches and spikelet meristems (SMs) that then initiate FMs (**Figure 1E**). In maize and other Andropogoneae species, determinate spikelet-pair meristems (SPMs) are produced from the IM or BMs, and each SPM makes two SMs. The SM initiates one or more FMs (Kellogg et al., 2013; Kyozuka et al., 2014). These intermediate BMs cause secondary or higher-order branches, which form a compound inflorescence termed the panicle (Benlloch et al., 2007). Therefore, the branch structure determines the final inflorescence pattern, which contributes to the enormous diversity of inflorescence architectures. Specific genetic regulatory networks control every stage and transition of meristem activity, as described in several recent reviews (Tanaka et al., 2013; Kyozuka et al., 2014).

Meristem activity, especially determinacy, fundamentally affects inflorescence architecture. For example, in the raceme-type inflorescence of *Arabidopsis*, the IM continues to initiate FMs; by contrast, in the cyme-type inflorescence of tomato, the IM forms a terminal flower immediately after developing a new IM below it, which reiterates this pattern (**Figures 1A,B**). The panicle-type inflorescence is initially indeterminate and initiates BMs and FMs before it finally terminates in a FM in some species.

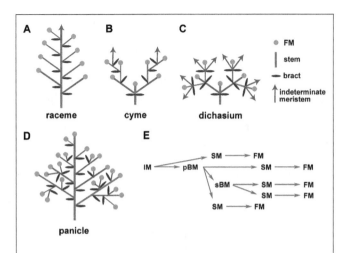

FIGURE 1 | Schematic representation of common types of inflorescences. (A) Simple raceme, which is indeterminate and unbranched, (B) cyme, (C) dichasium, which are determinate and branched, (D) panicle, which is determinate and branched, and (E) Transition of the reproductive meristems in rice panicle.

At least two groups of genes, relatives of *Arabidopsis LEAFY* (*LFY*) and *TERMINAL FLOWER1* (*TFL1*), play a central role in meristem determinacy. *LFY* promotes determinate FM identity and termination of IMs, and *TFL1* maintains the indeterminacy of IMs to prevent termination (Prusinkiewicz et al., 2007).

Recent work has identified cytokinins as key regulators of inflorescence architecture in plants with different inflorescence types, through regulation of meristem activity, which is often also associated with meristem identity. Cytokinins have profound effects on plant development and growth, including meristem activity (Kyozuka, 2007; Werner and Schmulling, 2009; Perilli et al., 2010). Accumulating data point to a role for cytokinins in influencing inflorescence complexity by fine-tuning IM and BM determinacy. Also, recent work reveals that cytokinins can regulate the initiation of meristems from floral organ axils (the junction where the floral organ meets the stem), and thus convert a determinate flower into an inflorescence (Han et al., 2014). Here we review these two mechanisms through which cytokinins regulate inflorescence architecture.

CYTOKININS PROMOTE IM ACTIVITY
Increasing cytokinin concentrations and signaling activity increase meristem size and activity. Reduced meristem activity often leads to conversion of an IM or a BM into a terminal flower, which subsequently affects inflorescence architecture.

Work in rice and in *Arabidopsis* showed that cytokinin levels affects meristem activity and inflorescence complexity. The ATP/ADP isopentenyl transferases (IPTs) catalyze the first step of cytokinin biosynthesis (Miyawaki et al., 2006). *Arabidopsis atipt3 5 7* triple mutants and *atipt1 3 5 7* quadruple mutants have lower levels of cytokinins, which leads to reduced IM size, formation of a terminal flower, and conversion of an indeterminate inflorescence to a determinate inflorescence (Miyawaki et al., 2006). Rice *LONELY GUY* (*LOG*) encodes a cytokinin-activating enzyme catalyzing the final step of cytokinin biosynthesis and *LOG* is strongly

expressed in BMs and FMs of developing panicles. The absence of *LOG* results in early termination of IM and BMs, which reduces branching complexity (Kurakawa et al., 2007). *Arabidopsis* has nine *LOG* homologs and the triple *log3 log4 log7* and septuple *log1 log2 log3 log4 log5 log7 log8* mutants produce fewer FMs, suggesting reduced IM activity (Kuroha et al., 2009; Tokunaga et al., 2012).

In addition to cytokinin homeostasis, defects in cytokinin signaling also leads to simplified inflorescence architecture. Cytokinins are perceived by transmembrane histidine kinase receptors, such as *Arabidopsis* HISTIDINE KINASE 2 (AHK2), AHK3, and AHK4. The *ahk* triple mutants have a smaller IM that terminates early, resulting in a simplified inflorescence with only a few flowers (Nishimura et al., 2004).

Conversely, elevated cytokinin homeostasis results in increased inflorescence complexity. Cytokinin oxidase/dehydrogenase (CKX) plays a major role in the degradation of bioactive cytokinins (Mok and Mok, 2001). *Arabidopsis* plants overexpressing *CKX1* or *CKX3* have dramatically reduced cytokinins contents and IMs that produce very few flowers (Werner et al., 2003). *CKX* overexpression in tobacco plants also leads to fewer flowers and conversion of IMs from indeterminate to determinate (Werner et al., 2001). Similarly, rice varieties with lower *OsCKX2* expression have more elaborated and larger panicles with more primary and secondary branches and higher yield, and rice varieties with higher *OsCKX2* activity have the opposite phenotype, with fewer branches and lower yield (Ashikari et al., 2005; Li et al., 2013).

Cytokinins promote IM activity and affect inflorescence architecture by promoting expression of the meristematic gene *WUSCHEL* (*WUS*) and suppressing the meristem inhibitors *CLAVATA1* (*CLV1*) and *CLV3*. Plants ectopically treated with cytokinins show a *clv*-like phenotype with larger IMs and more floral organs (Venglat and Sawhney, 1996; Lindsay et al., 2006). Cytokinins suppress the expression of *CLV1*; this suppression results in upregulation of *WUS* expression (Brand et al., 2000; Schoof et al., 2000; Lindsay et al., 2006; Gordon et al., 2009). In addition, cytokinins directly induce *WUS* expression, independent of *CLV1*, and *WUS* enhances cytokinin signaling, forming a positive feedback loop (Leibfried et al., 2005; Gordon et al., 2009). Computational modeling shows that a combination of the negative feedback between *WUS* and *CLV*, and the positive feedback of *WUS* and cytokinin signaling determines the fine-scale positioning of the *WUS*-expressing stem cell niche domain (Gordon et al., 2009; Chickarmane et al., 2012).

CYTOKININS PROMOTE LATERAL INDETERMINACY IN DETERMINATE FMs
In indeterminate inflorescences, the periphery of the meristem produces BMs (and also SPMs and SMs for grasses) or FMs. In many determinate inflorescences, such as in wheat spikes, BM, SM, and FM can also initiate from the IM before its termination in a FM. In contrast to this initiation pattern, FM and BM can also initiate laterally from a terminal flower, either from the axil of a leaf-like organ (such as petals) or can initiate without subtending lateral organs. These types of inflorescence are termed dichasium and pleiochasium (**Figure 1C**), depending on the number of lateral branches, and can be considered a specialized cyme,

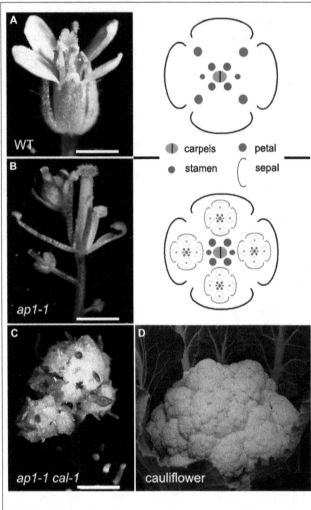

FIGURE 2 | Conversion of a terminate flower into a pleiochasium-like inflorescence. Phenotype of *Arabidopsis* **(A)** Lansberg *erecta* wild-type, **(B)** *ap1-1* **(C)** *ap1-1 cal-1* flowers, and **(D)** a cauliflower with a highly elaborated inflorescence, similar to that seen in the *Arabidopsis ap1-1 cal-1* mutant. Bars = 1 mm.

floral whorl specification of sepals and petals, *AP1* inhibits sepal axil meristem activity (Irish and Sussex, 1990; Mandel et al., 1992). In *ap1* mutants, secondary flowers initiate in the axils of sepals, and tertiary flowers can initiate in the sepal axils of secondary flowers, and so on (**Figure 2B**; Irish and Sussex, 1990; Mandel et al., 1992). This forms a dichasium or pleiochasium-like inflorescence (**Figure 1C**). The inflorescence phenotype in *ap1* is enhanced by *cauliflower* and *fruitful* mutants to form a cauliflower-like, highly elaborated pleiochasium inflorescence (**Figure 2C**; Ferrandiz et al., 2000). Indeed, cauliflower has lost a homolog of *AP1* (**Figure 2D**; Kempin et al., 1995), suggesting that *AP1* function is required to inhibit conversion of a simple raceme to a pleiochasium.

A recent study has shown that *AP1* inhibits lateral inflorescence branching by reducing cytokinin levels. During vegetative stages, leaf axil axillary meristem formation requires cytokinin signaling (Wang et al., 2014) and during reproductive stages, lateral FM formation similarly requires cytokinin signaling (Han et al., 2014). The *ap1* flowers have enhanced cytokinin signaling, as shown by examination of cytokinin-responsive reporter genes, and these flowers also have elevated levels of certain types of cytokinins. In addition, cytokinin treatment or ectopic expression of the cytokinin biosynthesis enzyme *IPT8* in the *AP1*-expressing domain phenocopies the sepal axil secondary flower phenotype (Venglat and Sawhney, 1996; Han et al., 2014). This secondary flower phenotype can be rescued by mutations of cytokinin receptors. Further molecular dissection showed that AP1 suppresses the cytokinin biosynthetic gene *LOG1* and activates the cytokinin degradation gene *CKX3*, through direct binding to the target gene promoters, thus reducing cytokinins levels in the outer whorls of developing flowers. Restoring the expression levels of either *LOG1* or *CKX3* can partially rescue the *ap1* secondary flower phenotype. In addition to affecting cytokinin homeostasis, AP1 also directly downregulates a group of flowering time-related MADS-box genes, including *SHORT VEGETATIVE PHASE* (*SVP*), *AGAMOUS-LIKE 24* (*AGL24*), and *SUPPRESSOR OF OVEREXPRESSION OF CO1* (*SOC1*), to suppress secondary FM formation. Similar to *IPT8* overexpression, overexpression of *SVP*, *AGL24*, or *SOC1* leads to sepal axil secondary FM formation (Liu et al., 2007). There appears to be crosstalk between cytokinin signaling and these flowering time-related MADS-box genes in the regulation of sepal axil secondary FM formation.

Taken together, the results described above show that cytokinins promote inflorescence complexity in different ways, by promoting meristem activity of IMs and BMs in inflorescences that branch iteratively, and by promoting indeterminate lateral meristem formation in inflorescences that branch laterally. Manipulating cytokinin levels directly or indirectly in crops is expected to change inflorescence complexity to increase yields (Kempin et al., 1995; Ashikari et al., 2005; Kurakawa et al., 2007; Zhang et al., 2012; Li et al., 2013).

Common examples include cauliflower and broccoli, which have a phenotype similar to that of the *Arabidopsis apetala1-1 cauliflower-1* (*ap1-1 cal-1*) double mutants (**Figure 2**). In the *ap1-1* single mutant (**Figure 2B**), secondary flowers laterally initiate from sepal axils and from the pedicel. The *ap1-1 cal-1* double mutants have the same but more complicated inflorescence branching pattern. This lateral inflorescence branching mechanism has many similarities to vegetative stage lateral shoot branching. In contrast to vegetative shoot branching, inflorescences like the raceme and panicle develop iteratively, similar to frond development in ferns (Sanders et al., 2011). Despite these differences, cytokinins also regulate this type of lateral inflorescence branching.

Lateral inflorescence branching is controlled by *AP1* and related MADS-box transcription factor genes in *Arabidopsis* and other Brassicaceae species. In the simple indeterminate inflorescence of *Arabidopsis*, the IM gives rise to FMs and each FM differentiates into four whorls of floral organs that occupy precise positions (**Figure 2A**). In addition to promoting FM formation and outer

ACKNOWLEDGMENTS

We apologize to those authors whose work could not be cited due to space limitations. The authors' laboratory is supported by the National Natural Science Foundation of China Grants 31222033 and 31430010, the Strategic Priority Research Program of CAS

Grant XDA08020105, and funds from the State Key Laboratory of Plant Genomics (through Grant SKLPG2011B0103).

REFERENCES

Ashikari, M., Sakakibara, H., Lin, S., Yamamoto, T., Takashi, T., Nishimura, A., et al. (2005). Cytokinin oxidase regulates rice grain production. *Science* 309, 741–745. doi: 10.1126/science.1113373

Benlloch, R., Berbel, A., Serrano-Mislata, A., and Madueno, F. (2007). Floral initiation and inflorescence architecture: a comparative view. *Ann. Bot.* 100, 659–676. doi: 10.1093/aob/mcm146

Brand, U., Fletcher, J. C., Hobe, M., Meyerowitz, E. M., and Simon, R. (2000). Dependence of stem cell fate in *Arabidopsis* on a feedback loop regulated by CLV3 activity. *Science* 289, 617–619. doi: 10.1126/science.289.5479.617

Chickarmane, V. S., Gordon, S. P., Tarr, P. T., Heisler, M. G., and Meyerowitz, E. M. (2012). Cytokinin signaling as a positional cue for patterning the apical-basal axis of the growing *Arabidopsis* shoot meristem. *Proc. Natl. Acad. Sci. U.S.A.* 109, 4002–4007. doi: 10.1073/pnas.1200636109

Ferrandiz, C., Gu, Q., Martiensson, R., and Yanofsky, M. F. (2000). Redundant regulation of meristem identity and plant architecture by FRUITFULL, APETALA1 and CAULIFLOWER. *Development* 127, 725–734.

Gordon, S. P., Chickarmane, V. S., Ohno, C., and Meyerowitz, E. M. (2009). Multiple feedback loops through cytokinin signaling control stem cell number within the *Arabidopsis* shoot meristem. *Proc. Natl. Acad. Sci. U.S.A.* 106, 16529–16534. doi: 10.1073/pnas.0908122106

Han, Y., Zhang, C., Yang, H., and Jiao, Y. (2014). Cytokinin pathway mediates APETALA1 function in the establishment of determinate floral meristems in *Arabidopsis. Proc. Natl. Acad. Sci. U.S.A.* 111, 6840–6845. doi: 10.1073/pnas.1318532111

Irish, V. F., and Sussex, I. M. (1990). Function of the apetala-1 gene during *Arabidopsis* floral development. *Plant Cell* 2, 741–753. doi: 10.1105/tpc.2.8.741

Kellogg, E. A., Camara, P. E., Rudall, P. J., Ladd, P., Malcomber, S. T., Whipple, C. J., et al. (2013). Early inflorescence development in the grasses (Poaceae). *Front. Plant Sci.* 4:250. doi: 10.3389/fpls.2013.00250

Kempin, S. A., Savidge, B., and Yanofsky, M. F. (1995). Molecular basis of the cauliflower phenotype in *Arabidopsis. Science* 267, 522–525. doi: 10.1126/science.7824951

Kurakawa, T., Ueda, N., Maekawa, M., Kobayashi, K., Kojima, M., Nagato, Y., et al. (2007). Direct control of shoot meristem activity by a cytokinin-activating enzyme. *Nature* 445, 652–655. doi: 10.1038/nature05504

Kuroha, T., Tokunaga, H., Kojima, M., Ueda, N., Ishida, T., Nagawa, S., et al. (2009). Functional analyses of LONELY GUY cytokinin-activating enzymes reveal the importance of the direct activation pathway in *Arabidopsis. Plant Cell* 21, 3152–3169. doi: 10.1105/tpc.109.068676

Kyozuka, J. (2007). Control of shoot and root meristem function by cytokinin. *Curr. Opin. Plant Biol.* 10, 442–446. doi: 10.1016/j.pbi.2007.08.010

Kyozuka, J., Tokunaga, H., and Yoshida, A. (2014). Control of grass inflorescence form by the fine-tuning of meristem phase change. *Curr. Opin. Plant Biol.* 17, 110–115. doi: 10.1016/j.pbi.2013.11.010

Leibfried, A., To, J. P., Busch, W., Stehling, S., Kehle, A., Demar, M., et al. (2005). WUSCHEL controls meristem function by direct regulation of cytokinin-inducible response regulators. *Nature* 438, 1172–1175. doi: 10.1038/nature04270

Li, S., Zhao, B., Yuan, D., Duan, M., Qian, Q., Tang, L., et al. (2013). Rice zinc finger protein DST enhances grain production through controlling Gn1a/OsCKX2 expression. *Proc. Natl. Acad. Sci. U.S.A.* 110, 3167–3172. doi: 10.1073/pnas.1300359110

Lindsay, D. L., Sawhney, V. K., and Bonham-Smith, P. C. (2006). Cytokinin-induced changes in CLAVATA1 and WUSCHEL expression temporally coincide with altered floral development in *Arabidopsis. Plant Sci.* 170, 1111–1117. doi: 10.1016/j.plantsci.2006.01.015

Liu, C., Zhou, J., Bracha-Drori, K., Yalovsky, S., Ito, T., and Yu, H. (2007). Specification of *Arabidopsis* floral meristem identity by repression of flowering time genes. *Development* 134, 1901–1910. doi: 10.1242/dev.003103

Mandel, M. A., Gustafson-Brown, C., Savidge, B., and Yanofsky, M. F. (1992). Molecular characterization of the *Arabidopsis* floral homeotic gene APETALA1. *Nature* 360, 273–277. doi: 10.1038/360273a0

Miyawaki, K., Tarkowski, P., Matsumoto-Kitano, M., Kato, T., Sato, S., Tarkowska, D., et al. (2006). Roles of *Arabidopsis* ATP/ADP isopentenyltransferases and tRNA isopentenyltransferases in cytokinin biosynthesis. *Proc. Natl. Acad. Sci. U.S.A.* 103, 16598–16603. doi: 10.1073/pnas.0603522103

Mok, D. W., and Mok, M. C. (2001). *Annu. Rev. Plant Physiol. Plant Mol. Biol.* 52, 89–118. doi: 10.1146/annurev.arplant.52.1.89

Nishimura, C., Ohashi, Y., Sato, S., Kato, T., Tabata, S., and Ueguchi, C. (2004). Histidine kinase homologs that act as cytokinin receptors possess overlapping functions in the regulation of shoot and root growth in *Arabidopsis. Plant Cell* 16, 1365–1377. doi: 10.1105/tpc.021477

Perilli, S., Moubayidin, L., and Sabatini, S. (2010). The molecular basis of cytokinin function. *Curr. Opin. Plant Biol.* 13, 21–26. doi: 10.1016/j.pbi.2009.09.018

Prusinkiewicz, P., Erasmus, Y., Lane, B., Harder, L. D., and Coen, E. (2007). Evolution and development of inflorescence architectures. *Science* 316, 1452–1456. doi: 10.1126/science.1140429

Sanders, H. L., Darrah, P. R., and Langdale, J. A. (2011). Sector analysis and predictive modelling reveal iterative shoot-like development in fern fronds. *Development* 138, 2925–2934. doi: 10.1242/dev.065888

Schoof, H., Lenhard, M., Haecker, A., Mayer, K. F., Jurgens, G., and Laux, T. (2000). The stem cell population of *Arabidopsis* shoot meristems in maintained by a regulatory loop between the CLAVATA and WUSCHEL genes. *Cell* 100, 635–644. doi: 10.1016/S0092-8674(00)80700-X

Tanaka, W., Pautler, M., Jackson, D., and Hirano, H. Y. (2013). Grass meristems II: inflorescence architecture, flower development and meristem fate. *Plant Cell Physiol.* 54, 313–324. doi: 10.1093/pcp/pct016

Teo, Z. W., Song, S., Wang, Y. Q., Liu, J., and Yu, H. (2014). New insights into the regulation of inflorescence architecture. *Trends Plant Sci.* 19, 158–165. doi: 10.1016/j.tplants.2013.11.001

Tokunaga, H., Kojima, M., Kuroha, T., Ishida, T., Sugimoto, K., Kiba, T., et al. (2012). *Arabidopsis* lonely guy (LOG) multiple mutants reveal a central role of the LOG-dependent pathway in cytokinin activation. *Plant J.* 69, 355–365. doi: 10.1111/j.1365-313X.2011.04795.x

Venglat, S. P., and Sawhney, V. K. (1996). Benzylaminopurine induces phenocopies of floral meristem and organ identity mutants in wild-type *Arabidopsis* plants. *Planta* 198, 480–487. doi: 10.1007/BF00620066

Wang, Y., Wang, J., Shi, B., Yu, T., Qi, J., Meyerowitz, E. M., et al. (2014). The stem cell niche in leaf axils is established by auxin and cytokinin in *Arabidopsis. Plant Cell* 26, 2055–2067. doi: 10.1105/tpc.114.123083

Weberling, F. (1989). *Morphology of Flowers and Inflorescences.* Cambridge: Cambridge University Press.

Werner, T., Motyka, V., Laucou, V., Smets, R., Van Onckelen, H., and Schmulling, T. (2003). Cytokinin-deficient transgenic *Arabidopsis* plants show multiple developmental alterations indicating opposite functions of cytokinins in the regulation of shoot and root meristem activity. *Plant Cell* 15, 2532–2550. doi: 10.1105/tpc.014928

Werner, T., Motyka, V., Strnad, M., and Schmulling, T. (2001). Regulation of plant growth by cytokinin. *Proc. Natl. Acad. Sci. U.S.A.* 98, 10487–10492. doi: 10.1073/pnas.171304098

Werner, T., and Schmulling, T. (2009). Cytokinin action in plant development. *Curr. Opin. Plant Biol.* 12, 527–538. doi: 10.1016/j.pbi.2009.07.002

Zhang, L., Zhao, Y. L., Gao, L. F., Zhao, G. Y., Zhou, R. H., Zhang, B. S., et al. (2012). TaCKX6-D1, the ortholog of rice OsCKX2, is associated with grain weight in hexaploid wheat. *New Phytol.* 195, 574–584. doi: 10.1111/j.1469-8137.2012.04194.x

3

Analysis of *Arabidopsis* floral transcriptome: detection of new florally expressed genes and expansion of Brassicaceae-specific gene families

Liangsheng Zhang[1,2], Lei Wang[3,4], Yulin Yang[1], Jie Cui[3], Fang Chang[3], Yingxiang Wang[3]* and Hong Ma[3,4]*

[1] Department of Pharmacy, Shanghai Tenth People's Hospital, School of Life Sciences and Technology, Tongji University, Shanghai, China
[2] Advanced Institute of Translational Medicine, Tongji University, Shanghai, China
[3] State Key Laboratory of Genetic Engineering and Collaborative Innovation Center for Genetics and Development, Ministry of Education Key Laboratory of Biodiversity Science and Ecological Engineering and Institute of Biodiversity Sciences, Institute of Plants Biology, Center for Evolutionary Biology, School of Life Sciences, Fudan University, Shanghai, China
[4] Institutes of Biomedical Sciences, Fudan University, Shanghai, China

Edited by:
Dazhong Dave Zhao, University of Wisconsin-Milwaukee, USA

Reviewed by:
Xuelin Wu, Harvey Mudd College, USA
Yuling Jiao, Chinese Academy of Sciences, China

*Correspondence:
Yingxiang Wang and Hong Ma, School of Life Sciences, Fudan University, 2005 Songhu Road, Shanghai 200433, China
e-mail: yx_wang@fudan.edu.cn; hongma@fudan.edu.cn

The flower is essential for sexual reproduction of flowering plants and has been extensively studied. However, it is still not clear how many genes are expressed in the flower. Here, we performed RNA-seq analysis as a highly sensitive approach to investigate the *Arabidopsis* floral transcriptome at three developmental stages. We provide evidence that at least 23, 961 genes are active in the *Arabidopsis* flower, including 8512 genes that have not been reported as florally expressed previously. We compared gene expression at different stages and found that many genes encoding transcription factors are preferentially expressed in early flower development. Other genes with expression at distinct developmental stages included *DUF577* in meiotic cells and *DUF220, DUF1216,* and *Oleosin* in stage 12 flowers. *DUF1216* and *DUF577* are Brassicaceae specific, and together with other families experienced expansion within the Brassicaceae lineage, suggesting novel/greater roles in Brassicaceae floral development than other plants. The large dataset from this study can serve as a resource for expression analysis of genes involved in flower development in *Arabidopsis* and for comparison with other species. Together, this work provides clues regarding molecular networks underlying flower development.

Keywords: *Arabidopsis thaliana*, RNA-Seq, differentially expressed genes, floral development, gene families

INTRODUCTION

Flower is one of the most complex structures of the angiosperms (flowering plants), and is thought to make great contribution to sexual reproduction in either developmental or evolutionary aspects (Alvarez-Buylla et al., 2010). The basic floral architecture is highly conserved among the core eudicots, including *Arabidopsis thaliana,* which is an important model plant for studying flower development. Over the past three decades, extensive molecular genetic analyses have identified a large number of key floral regulators controlling flower development (O'Maoileidigh et al., 2014), making it one of the best-understood aspects of plant development. However, the present knowledge in understanding gene regulatory network in flower development is incomplete, such as information on genes with low expression levels.

Genome-wide approaches have become valuable tools in characterizing gene expression and in elucidating the genetic networks of flower development at a global level. In the past, large-scale analyses of transcript enrichment among *Arabidopsis* floral organs largely depends on hybridization, such as cDNA and oligonucleotide arrays (Hennig et al., 2004; Wellmer et al., 2004, 2006;

Zhang et al., 2005; Alves-Ferreira et al., 2007; Benedito et al., 2008) and represents a major step in the spatial characterization of floral transcriptome, resulting in identifying many genes important for flower development (Alvarez-Buylla et al., 2010; Irish, 2010). However, array analyses and other hybridization-based approaches have several limitations, including knowledge of genes for probe design, non-specific hybridization, and difficulty in detecting low level expression (Marioni et al., 2008). On the other hand, more recently developed RNA sequencing (RNA-seq) technologies can overcome such limitations of hybridization-based approaches and other conventional large-scale gene expression analysis methods (Marioni et al., 2008; Xiong et al., 2010). It also has great sensitivity, allowing the detection of transcripts with lower expression levels, such as those of many transcription factors (Marioni et al., 2008; Chen et al., 2010). In the last few years, RNA-seq has been extensively applied in the characterization of transcriptome regarding developmental stage, organ, even specific cell types or single cell level, from yeast to human, including several plant species (Jiao et al., 2009; Zhang et al., 2010; Yang et al., 2011). To date, RNA-seq has been used for cell-specific analysis of actively translated mRNAs associated with polyribosomes in developing flowers, providing insights and resources to further study flower development (Jiao and Meyerowitz, 2010).

To further explore the *Arabidopsis* flower transcriptome, we employed RNA-seq for three developmental periods. We detected 8512 additional genes that are not present on previously used microarray experiments, and provide evidence that at least 23,961 genes are truly expressed in the *Arabidopsis* flower. We also identified differentially and specifically expressed genes and gene families during flower development.

MATERIALS AND METHODS
SEQUENCING DATASETS
The inflorescent meristem (IM), stage 1–9 flowers (F1–9) and stage 12 flowers (F12) samples for RNA-seq were collected in our lab, and the three samples were subjected to 50 bp single-end sequencing on a SOLiD 3 platform; details for the methods were recently described in a study for alternative splicing (Wang et al., 2014a). All sequenced short reads were submitted to NCBI Short Read Archive under accession number SRP035230. The datasets for seeding and stage 4 flowers were from previously studies, which generated 36-bp and 42-bp long reads, respectively, using the Illumina genome analyzer (Filichkin et al., 2010; Jiao and Meyerowitz, 2010). The meiocyte datasets were from our previous study, which included two runs (36 and 50 bp) using Life Technologies' SOLiD sequencing platform (Yang et al., 2011).

ALIGNMENT OF SEQUENCING READS
Sequence reads from the three sample plus the three floral samples were mapped using PerM (Chen et al., 2009) to the *Arabidopsis* genome (release 9) from the *Arabidopsis* Information Resource (TAIR) database (TAIR9; www.arabidopsis.org) allowing 5, 4, and 3 mismatches per 50, 42, and 36-bp read, respectively.

DIGITAL GENE EXPRESSION AND EXPRESSION ARRAYS
For the RNA-seq experiments, we used at least 10 reads mapped to a gene as the threshold for being expressed. The raw digital gene expression counts were normalized using the reads per kilo-base of mRNA length per million of mapped reads (RPKM) method. The equation was used:

$$RPKM = \frac{10^9 * C}{N * L}$$

Where C is the uniquely mapped counts determined from mapping results, L is the length of the cDNA for the longest splice variant for a particular gene model and N is the total reads that were mapped to the genome. Log2-transformation of this normalized value was performed as in other analyses.

To test differential expression with mapping data DEGseq was used (Wang et al., 2010). Fisher's Exact Test ($P < 0.01$) method was selected. Microarray results were obtained from a previous study (Zhang et al., 2005). The Microarray experiments have a background value, which was 5 (log value of base 2) as previously described (Zhang et al., 2005) for the evaluation of "expressed" or "unexpressed" genes. Identification of differentially expressed genes according to the microarray data also used the Fisher's Exact Test method.

Z-SCORE
Calculation of the Z-score was based on the log2-transformed RPKM-normalized transcript levels as follows:

$$Z = (X - \mu)/\sigma$$

X is the RPKM of a gene for a specific tissue/developmental stage. μ is the mean RPKM of a gene across all tissues/developmental stages and σ is the RPKM standard deviation of a gene across all tissues/developmental stages. All calculations and plotting were performed by Perl and excel, respectively.

GENE FAMILY AND FUNCTIONAL ANNOTATION
The protein domain annotations were obtained from the Pfam database (http://pfam.sanger.ac.uk) (Punta et al., 2011). *Arabidopsis* protein sequences were then searched against protein family models in the Pfam-A database, resulting in 21102 *Arabidopsis* proteins identified as having at least one Pfam domain. Transcription factor family annotations were from The Database of *Arabidopsis* Transcription Factors (http://datf.cbi. pku.edu.cn/) (Guo et al., 2005), which contains 1922 transcription factors in *Arabidopsis*. Gene ontology (GO) enrichment analysis was performed with the agriGO browser (http://bioinfo. cau.edu.cn/agriGO/) (Du et al., 2010) using Singular Enrichment Analysis.

Multiple sequence alignment was performed in MUSCLE (http://www.drive5.com/muscle/) using the default parameters. Maximum likelihood (ML) trees were constructed by FastTree (www.microbesonline.org/fasttree) with the approximate likelihood ratio test method.

RESULTS AND DISCUSSION
GLOBAL GENE EXPRESSION OF FLOWER TRANSCRIPTOMES IN *ARABIDOPSIS*
To obtain more insights about the overall transcriptome landscape during flower development, we analyzed RNA-seq datasets of the *Arabidopsis* flower at three developmental stages recently generated in our laboratory; these datasets were analyzed for alternative splicing in a separate study (Wang et al., 2014a): inflorescent meristem (IM), stage 1–9 flowers (F1–9) and stage 12 flowers (F12), detecting 21,181 (IM), 22,137 (F1–9), and 22,827 (F12) reliably expressed genes (**Table S1**, **Figure S1**). A recent report summarized that a total of 126 *Arabidopsis* genes have been demonstrated genetically to have a role during flower development (Alvarez-Buylla et al., 2010), 122 of which were also detected as expressed in our dataset (**Table S2**), indicating that our data were very reliable, and can be used for further analysis. To compare gene expression during flower development, besides the three datasets described above, we also included data for *Arabidopsis* male meiocytes that we had generated previously (Yang et al., 2011), and two other public datasets of *Arabidopsis* seedlings and stage 4 flowers (Filichkin et al., 2010; Jiao and Meyerowitz, 2010), the latter of which were from isolated polysomic.

To further explore how many genes are truly expressed in *Arabidopsis* flower, we searched The *Arabidopsis* Information Resource (TAIR) database and obtained a total of 24,570 genes, which are supported by at least one EST. Then, we searched the present tiling array database to find available probes for 30,228 genes. 4734 genes were found to be tiling array-specific compared with ESTs and RNA-seq data. Among them, 2634 and 276 are transposons and pseudogenes, respectively. The other 1824 genes seem to be expressed at very low levels. The average value of the 4734 genes is 5.4, which is regarded as a threshold in this study for the evaluation of "expressed" or "unexpressed" genes. Based on this criterion, we believe that tiling array can detect at least 27, 617 genes. As described previously, RNA-seq detected 24,769 genes in flowers. Comparison of the detected genes among EST,

tiling array and RNA-seq found that 22,440 genes were detected by three data sets and 1521 genes were detected by RNA-seq and either ESTs/tiling array (**Figure 1A**). The results suggest that at least 23,961 genes are reliably detected as expressed in the *Arabidopsis* flower. In addition, 621 genes were only detected by RNA-seq; most of these are low abundance genes that are nearly undetectable by arrays and the others are likely to be stage-specific genes.

Characterization of stage or cell-specific genes provides a foundation for unraveling their molecular mechanisms. Previous studies in multiple plants demonstrated that each stage or tissue

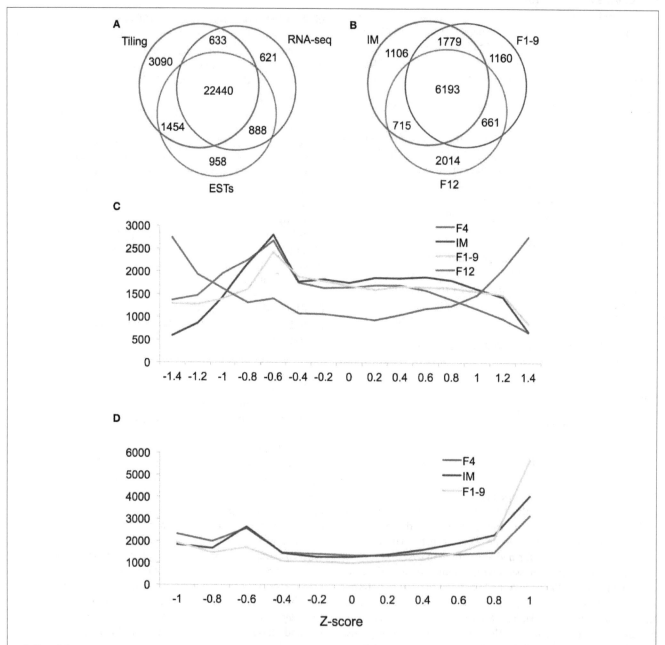

FIGURE 1 | Global gene expression during flower development. (A) A Venn diagram showing the overlap in detected genes between three technologies: the tilling array, RNA-seq and ESTs. **(B)** A Venn diagram showing the overlap between IM, F1–9, and F12 for genes called as DEG by RNA-seq. Histograms of relative expression levels (measured by Z-scores) in four **(C)** and three **(D)** organs. For easy visualization, we plotted Z-score on the x-axis and gene numbers on the y-axis.

Analysis of Arabidopsis floral transcriptome: detection of new florally expressed genes and expansion genes...

17

has specific transcripts (Jiao et al., 2009; Jiao and Meyerowitz, 2010; Yang et al., 2011; Liu et al., 2014). To better establish the genome-wide gene expression pattern of flower development, we conducted a Z-score analysis to assess the extent of differential gene expression for florally expressed genes. Results showed that the Z-score distribution of gene expression in F12 was dramatically different from that for early flower development (F1–9, F4, and IM) (**Figure 1C**), suggesting that nearly mature flowers requires many more specifically or differentially expressed genes than early flowers. In contrast, Z-score distributions were very similar between F1–9, F4, and IM (**Figure 1D**), further supporting the idea that the developmental programs of these stage/organ are similar.

DETECTION OF EXPRESSION OF 8512 GENES IN THE *ARABIDOPSIS* FLOWER NOT REPORTED FROM MICROARRAY ANALYSIS

We first compared the F1–9 and F12 RNA-seq data with the Affymetrix ATH1 array data at similar stages (Zhang et al., 2005). We compared the number of sequencing reads mapped to each gene with the corresponding (normalized) absolute intensities from the array (**Figures 2A,B**), and found that the correlations between the two platforms were high, with Spearman correlation coefficients of 0.82 (F1–9; **Figure 2A**) and 0.80 (F12; **Figure 2B**). Thus, comparison of RNA-seq and microarray identified 15,180 overlapping genes with relatively high expression levels (**Figure 2C**), covering 96% of genes detected using microarray. In addition, our RNA-seq identified additional 8512 genes that were undetected by microarray (Zhang et al., 2005), whose expression levels were obviously lower; the average expression level is 56.12 (F1–9) and 57.82 (F12) RPKM (**Figure S2b**), compared to the average expression level about 250 for the 15,180 genes. Also, the curvature of the comparison toward the microarray axis suggested that microarray possibly underestimated the expression level of genes relative to RNA-seq (**Figures 2A,B**). Together, these results suggest that RNA-seq has a great advantage over microarray in detecting low-abundance transcripts, consistent with previous reports (Marioni et al., 2008; Yang et al., 2011).

To investigate further regarding the 8512 genes, we analyzed the enrichment of protein family (PFAM) domains (gene families) among these genes, and identified several enriched gene families that were not reported previously as enriched, including *F-box*, *NB-ARC*, *C1_3*, *PPR*, *LRR_1*, *Myb*, *bHLH*, and *AP2* gene families (**Table 1**). Previously, many *F-box* genes were reported as unexpressed or undetectable by microarray analysis (Schmid

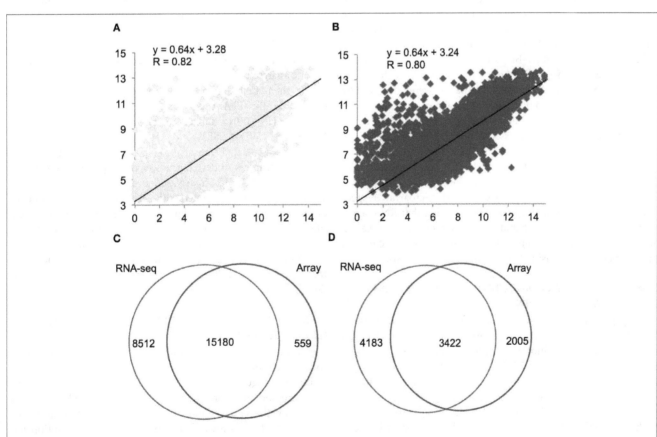

FIGURE 2 | A scatter plot of relative expression values obtained by RNA-seq and microarray for F1–9 and F12. (A) Comparison of expression levels between RNA-seq and microarray for F1–9, RNA-seq and microarray, gene expression levels transformed with log2 were plotted. **(B)** Comparison of expression between RNA-seq and microarray for F12. **(C)** A Venn diagram illustrating genes detected by RNA-seq (left) and microarray (right) analyses. **(D)** A Venn diagram presenting the overlap of differentially expressed genes between F1–9 and F12 from RNA-seq (left) and from microarray analysis (right).

Table 1 | Enriched protein family (Pfam) among 8512 genes that are undetected by microarray analysis.

Family	Total	RNA-seq	Percent	Family	Total	RNA-seq	Percent
NB-ARC	167	108	0.65	bHLH	134	59	0.44
FBD	115	71	0.62	AP2	145	62	0.43
TIR	132	80	0.61	UDPGT	115	49	0.43
SCRL	25	14	0.56	peroxidase	82	33	0.40
U-box	61	31	0.51	Myb	256	103	0.40
MATH	60	30	0.50	Malectin_like	78	31	0.40
DUF26	97	47	0.48	LRR_1	391	154	0.39
Auxin	79	38	0.48	PMEI	122	46	0.38
C1_3	146	70	0.48	SLR1-BP	41	15	0.37
DUF295	78	36	0.46	p450	249	91	0.37
PPR_1	301	138	0.46	zf-rbx1	174	63	0.36
PPR	465	211	0.45	ABC_tran	117	42	0.36
NAC	113	51	0.45	Kelch_1	110	38	0.35
F-box	522	233	0.45	Ank_2	106	36	0.34
FBA_1	176	78	0.44	zf-C3HC4	306	96	0.31

Table 2 | Differentially expressed genes (DEGs) for each floral sample compared with seedling.

	Seedling	IM	F4	F1–9	F12	Meiocytes
Seeding		9793	8703	9583	9340	4966
IM	6627		3747	6866	6943	3900
F4	5812	5354		6016	6460	3449
F1–9	6866	7228	4834		7697	3773
F12	5987	8960	6724	8109		3820
Meiocytes	5110	7167	6154	6628	6401	

et al., 2005), further suggesting that microarray is not as sensitive as RNA-seq for detecting low-abundance transcripts. In addition, we also detected some enriched gene families that belongs to the highly expressed genes; for instance, Plant self-incompatibility response (SCRL) and S locus-related glycoprotein 1 binding pollen coat protein (SLR1-BP) are specifically enriched in F12 (**Table 1** and **Table S3**), suggesting a potential role at this stage. In contrast, 559 genes detected by microarray were not found by RNA-seq, possibly due to difference in growth conditions.

We further employed a widely used Fisher's Exact Test method to identify differentially expressed genes (DEGs) between F1–9 and F12 in RNA-seq and microarray data. Altogether, 7605 and 5327 DEGs were identified in each dataset. Among them, 3422 genes were detected by both platforms (**Figure 2D**), 1272 and 84 DEGs were only detected by RNA-seq and microarray, respectively (**Figure 2D**), and consistent with the fact that RNA-seq is more sensitive for detection and comparison of gene expression. Taken together, these results indicate that deep sequencing can greatly increase the sensitivity of transcriptome analysis.

IDENTIFICATION OF STAGE-DIFFERENTIALLY EXPRESSED GENES DURING FLOWER DEVELOPMENT

Floral organ identity and cell fate determination are highly regulated by the temporal and spatial gene expression, with each organ or cell type having distinct transcriptomes (Jiao et al., 2009; Yang et al., 2011; Wang et al., 2014b). To investigate DEGs between one of the floral stages with seedlings, we compared the flower transcriptomes of IM, F1–9, or F12 with that of seedlings, and identified IM with 9793 DEGs, F1–9 with 9583 DEGs, and F12 with 9340 DEGs (**Table 2**). Furthermore, the intersection between these three sets contained 6193 genes (**Figure 1B**), indicating these three samples are quite similar regarding differentially gene expression compared with seedlings. GO annotation showed that these genes were enriched for categories such as "histone modification" and "methylation" ($p = 3.2E-11$ and $4.7E-9$), suggesting

these genes are involved in the establishment of transcription regulation during flower development. Likewise, 2014 genes were specifically expressed in F12 and showed significant enrichment for genes in reproduction ($p = 2.2E-47$), flower development ($p = 2.9E-22$) and post-embryonic development ($p = 1.3E-194$), which might suggest that genes expressed during gametophyte development can function in later stages.

To further examine the combined set (13,628 genes) of the above floral DEGs, we compared these genes with 4505 genes identified as potential targets of the SEP3 and/or AP1 proteins by ChIP-seq (Immink et al., 2009; Kaufmann et al., 2010). The results showed that 2506 genes overlapped between the floral DEGs and the SEP3/AP1 targets with significance (Fisher's test, $p = 7.03e-08$). It is likely that some of these genes are involved in the regulation of flower development, but the role of these genes in flower development needs to be determined using molecular genetic analyses.

We then analyzed the enrichment of protein domains as defined in the PFAM database, and found several enriched domains ($P \leq 0.01$; **Table 3**), including ATPase, Helicase_C, DEAD box, WD40, SET, and PHD domains, suggesting that chromatin associated transcriptional regulation might be one of the major features underlying flower development. In addition, proteins with "UCH," "hydrolase," and "IQ" domains were also significantly over-represented, although their functions in flower development are largely unknown. Interestingly, we also identified "PPR," "Mito_carr" and "Miro" domains as significantly enriched; members of these genes are involved in gene expression and other functions in mitochondria and plastid, suggesting that such organellar functions might be important for flower development.

DISTINCT ENRICHMENT OF TRANSCRIPTION FACTORS IN EARLY FLOWER DEVELOPMENT

Identification of transcription factors (TFs) expressed in a specificstage provides a foundation for understanding the transcriptional regulatory networks underlying the development, structure and function of thestage. To investigatethe expressed TFs during flower development, we examined the TFs among IM, F1–9, and F12 and identified a total of 1667 transcription factors, 927 of which showed differential expression compared with the seedling (**Figure 3A**). Among the 927 TFs, 70% showed highest expression in IM (designated as D1), whereas 14 and 16% showed highest expression in F1–9 (designated as D2) and F12 (designated as D3), respectively (**Figure 3B**).

Table 3 | Significance of enriched Pfam domains in differentially expressed genes during flower development.

Pfam domain	Total	Num.	Percent	P-vaule	Pfam domain	Total	Num.	Percent	P-vaule
Helicase_C	149	133	0.89	4.00E-07	Proteasome	24	23	0.96	0.02
WD40	234	184	0.79	1.00E-06	Cyclin_C	30	28	0.93	0.02
DEAD	114	97	0.85	5.00E-05	HATPase_c	35	30	0.86	0.02
RRM_1	245	171	0.70	0.0003	AAA_5	45	37	0.82	0.02
Kinesin	61	55	0.90	0.001	Mito_carr	59	46	0.78	0.02
PHD	52	47	0.90	0.002	Galactosyl_T	21	20	0.95	0.03
SNF2_N	45	40	0.89	0.005	HA2	21	20	0.95	0.03
PPR_1	301	188	0.62	0.006	Cyclin_N	52	40	0.77	0.03
IQ	56	46	0.82	0.008	Hydrolase	60	45	0.75	0.03
PPR	465	275	0.59	0.008	Miro	114	76	0.67	0.03
AAA	144	98	0.68	0.009	OB_NTP_bind	20	19	0.95	0.04
LSM	26	26	1.00	0.01	Cpn60_TCP1	23	21	0.91	0.04
UCH	45	39	0.87	0.01	KH_1	25	22	0.88	0.04
ResIII	48	40	0.83	0.01	SET	46	35	0.76	0.04
RuvB_N	46	38	0.83	0.01	Histone	67	48	0.72	0.04

D1 mainly contained members of the homeobox domain (HB), MADS, MYB, AP2, and NAC families, suggesting that floral meristem development largely requires those transcription factor (**Figure 3B**). For example, *homeobox* genes encode transcription factors that contain a classic DNA binding domain with about 60 amino acids and regulate gene expression via Polycomb-dependent modulation of chromatin structure, thereby controlling development in animals, fungi and plants (Zhong and Holland, 2011). Several known members of *HB* (**Figure 3C**) identified in IM support that early floral development requires active *HB* genes, consistent with the finding that epigenetic reprogramming of gene expression is important for the establishment of initial floral identity (Mukherjee et al., 2009). The co-expressed pattern between *HB* genes and chromatin factors in IM is in agreement with previous studies that a number of floral genes with similar expression patterns and/or associated with each other regulate the expression of downstream genes to ensure proper flower development (Kaufmann et al., 2009, 2010; Deng et al., 2011).

D2 included *MADS-box, MYB, AS2, C2H2, bZIP, ABI3,* and *bHLH* families (**Figure 3C**). *MADS-box* genes encode not only key repressors or activators for flowering transition, but also master regulators of reproductive organ identities (Alvarez-Buylla et al., 2010). Our data detected expression of most MADS-box genes known to be involved in flower development (**Figure 3C**), such as *FLC, SHP1/2, AP3, AG, AGL11/15/77, TT16, SEP1-3, STK, AT5G49420; AG, AP3,* and *SEP1-3* are genes for the ABCE model, consistent with their known function in floral organ identity (Smaczniak et al., 2012). In addition, genes coding for transcription factors important for microsporogenesis were also uncovered, such as *AMS, MS1, MYB35,* and *MYB99* (Chang et al., 2011), as well as *MMD1* required for meiosis (Yang et al., 2003).

D3 was enriched in *MADS, MYB, AP2, C2H2, C2C2-CO-like, NAC, AUX-IAA-ARF,* and *bHLH* families (**Figure 3C**). Previous studies showed that auxin-dependent transcriptional regulation requires the auxin/indole-3-acetic acid (Aux/IAA) and auxin response factor (ARF) families of TFs and formation of

Aux/IAA-ARFs heterodimers represses auxin signaling (Reed, 2001), which has been demonstrated to participate in pollen development, pollination and fertilization (Sundberg and Ostergaard, 2009), as well as female gametophyte specification (Pagnussat et al., 2009). Indeed, our data identified several known and unknown ARFs and IAAs factors in G3, suggesting that the Aux/IAA-ARF regulatory pathway is vital for late reproductive development. However, the function of other enriched TFs in flowers is still largely unknown. Together, these results demonstrate that flower development at different stages requires common and distinct transcription factor families.

IDENTIFICATION OF SPECIFIC GENE FAMILIES AT DISTINCT STAGES OF FLOWER DEVELOPMENT

We sought to identify stage-specific genes, which were defined as those genes that were differentially expressed (>4-fold change) at one stage over all other stages studied here using DEG seq. The largest numbers of stage specific genes were identified in the seedling, F12 and meiocytes (1083, 552, and 652 genes, respectively; **Table 4**). Given the lack of correlation in overall gene expression between the floral transcriptome (F12) and the other stages sampled (**Figure 2C**), it was not surprising to identify this stage as having the largest number of organ-specific genes. These genes are strong candidates for determining the specific functional components of the nearly mature flower.

Interestingly, the F1–9 flower-specific genes with 8-fold changes had 26 genes, including 9 transposons and 5 snoRNAs (**Table S4**), consistent with the previous finding that transposons and small RNAs were enriched among genes expressed in male meiocytes (Chen et al., 2010; Yang et al., 2011). There are also 12 coding genes, one of which (*AT5G09780*) codes for a transcription factor of the B3 family and two (*AT1G48700* and *AT4G03050*) are for iron binding proteins.

For meiocyte-specific genes, 424 genes were found with ~8-fold changes and showed enrichment for genes in an insertion of mitochondrial origin on chromosome II, as supported by similar preferential expression in meiocytes reported previously

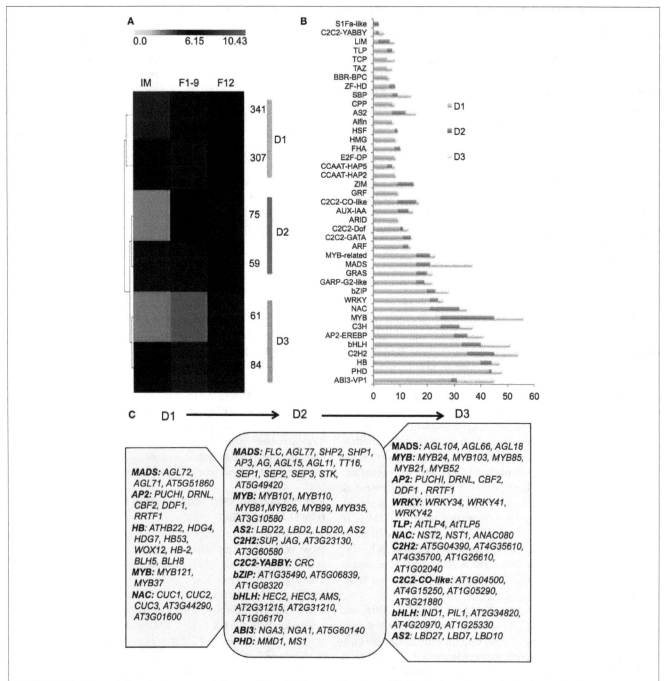

FIGURE 3 | The transcription factor accumulation profiles. (A) A heatmap for the transcription factors. 937 significantly differentially expressed transcription factors from IM, F1–9, and F12 were clustered into three groups (D1–D3) using the Self Organization Tree Algorithm. **(B)** Distribution of transcription factor families among D1–D3. **(C)** Representative functions and genes showing different transcription factor families for developmental stage-dependent groups D1 (IM), D2 (F1–9), and D3 (F12).

Table 4 | The specifically expressed genes in one sample compared with others.

Fold change	Seedling	IM	F4	F1–9	F12	Meiocyte
1	2656	1280	686	1740	1632	1636
2	1871	157	118	110	817	1200
4	1083	27	33	26	552	652
8	695	7	16	13	418	424

(Chen et al., 2010). The enriched genes also included 45 mitochondrial and 28 chloroplast genes, respectively. Moreover, in addition to previously reported gene families (Yang et al., 2011), we also detected several other enriched gene families, such as *Oxidored_q1*, *Oxidored_q2*, *Oxidored_q3*, *Oxidored_q4*, and *NADHdh*. In particular, most members of the *DUF577* family showed specific expression in meiocytes (**Table S5**). To further investigate this gene family, we performed phylogenetic analyses

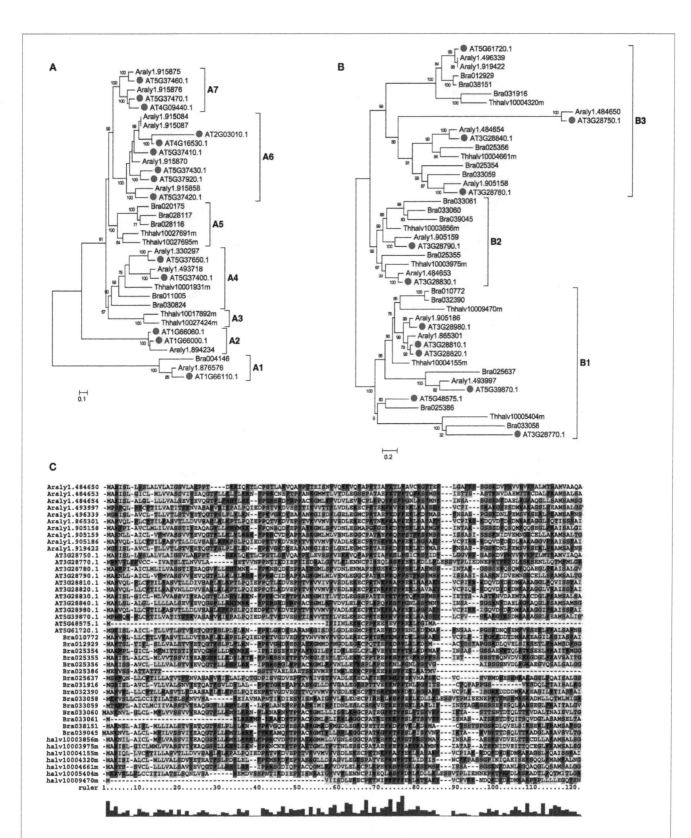

FIGURE 4 | The phylogenetic tree of *DUF577* and DUF1216 gene families in plant. The *DUF577* and *DUF1216* gene families were found only in plants, but not in other eukaryotic groups. **(A)** A Maximum likelihood (ML) tree of the *DUF577* gene family using representative species in eudicots and it can be divided into 8 clades, A1~A7. **(B)** An ML tree of *DUF1216* gene family using representative species in eudicots and it can be divided into 3 clades, B1~B3. **(C)** An alignment of the N-terminal regions of DUF1216 proteins. Species names are abbreviated as below: AT-*Arabidopsis thaliana*, Araly-*Arabidopsis lyrata*, Bra-*Brassica rapa*, Thhalv-*Eutrema salougineuma*

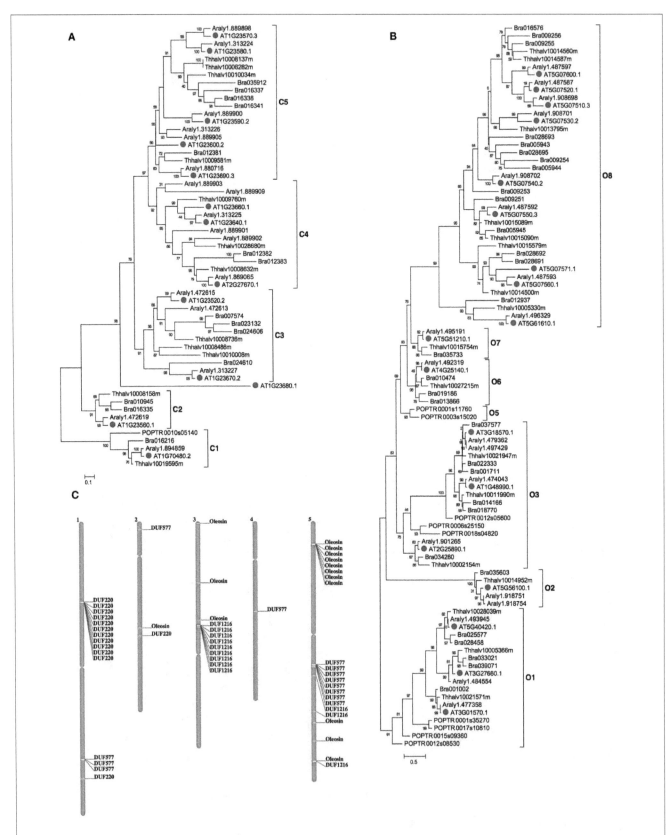

FIGURE 5 | The phylogenetic tree of *DUF220* and *Oleosin* gene families in plant. (A) An ML tree of *DUF220* gene family can be divided into 5 clades, C1~C5. **(B)** An ML tree of *Oleosin* gene family can be divided into 8 clades, O1~O8. Species names are abbreviated as below: AT-*Arabidopsis thaliana*, Araly-*Arabidopsis lyrata*, Bra-*Brassica rapa*, Thhalv-*Eutrema salsugineum*; POPTR-*Populus trichocarpa*. **(C)** The chromosomal positions of the *Arabidopsis DUF577, DUF1216, DUF220*, and *Oleosin* genes. The names of genes refer to locus ID as listed in **Table S5**.

of this gene family with members from several representative plant species, including *Arabidopsis lyrata*, *Eutrema salsugineum*, *Brassica rapa*. As shown in **Figure 4A**, this family can be divided into seven subfamilies, designated as A1-A7. The tree supported that this gene family have experienced expansion and origin in Brassicaceae (**Figure 4A**). The A6 and A7 subfamilies only included *Arabidopsis lyrata* and *Arabidopsis thaliana*, suggesting an expansion that occurred since the divergence of *Arabidopsis* and other Brassicaceae species. Besides, functions of the *DUF577* and other enriched family genes in meiosis need to be tested.

Similarly for F12, the enriched families included *DUF1216*, *Oleosin* and *DUF220*. Most members of the three gene families had specific expression in F12; these gene families contain lineages that originated and expanded within Brassicaceae (**Table S5** and **Figures 4, 5**). Phylogenetic analyses of the *DUF1216* family suggested that this gene family is specific to Brassicaceae, without homologs in other plants, and experienced gene duplication during Brassicaceae history (**Figure 4B**). Interestingly, the N-terminal region of DUF1216 proteins had putative signal peptides with similar sequences, according to the SignalP prediction (http://www.cbs.dtu.dk/services/SignalP/). The predicted signal peptide contains a large number of hydrophobic amino acids, a conserved basic amino acid and a conserved cysteine at the ninth position (**Figure 4C**). *At5g07750* of the *Oleosin* family was reported to have experienced positive selection (Schein et al., 2004). However, expression of each of three tandem duplicated genes in the Oleosin family (*AT5G07510*: 10081.76, *AT5G07550*: 29536.45, *AT5G07560*: 10942.38) had extraordinarily high levels of more than RPKM of 10, 000, suggesting that such high expression levels are important for F12 for later functions. Analysis of the *Arabidopsis* genome indicates that tandem duplication contributed to the expansion of *DUF1216* in Brassicaceae, as well as the expansion of the *Oleosin*, *DUF220* and *DUF577* families (**Figure 5C**). This pattern is different to those of *SET*, *JmjC*, and *Rhomboid* gene families, which are more likely to be retained after whole genome duplication events (Zhou and Ma, 2008; Zhang and Ma, 2012; Li et al., 2014).

CONCLUSIONS

The analysis of *Arabidopsis* floral transcriptome datasets presented here provides a valuable resource of candidate genes for further studies to understand the flower development program. We provided evidence for at least 23,961 genes that are expressed in the *Arabidopsis* flower. Compared with seedling, over 10,000 DEGs were identified, revealing novel and different molecular characteristics in the developing flower such as regulatory genes, genes for high-energy production, and transposable elements. These results showed that flower development at different stages requires common and distinct transcription factor families. The gene expression in F12 was dramatically different from that for early flower development (F1–9, F4, and IM).

In addition to identifying floral developmental gene candidates, we found many genes or gene families specifically expressed at one stage. Many transposable element genes, at least 45 mitochondrial and 28 chloroplast genes showed specific expression in meiocytes. The *SCRL*, *SLR1-BP*, *DUF1216*, *Oleosin*, and *DUF220* gene families showed specific expression in F12 and *DUF577* genes were detected to have specific expression meiosis. These specifically expressed genes have functions that are closely related to reproductive development, showed that mature flowers require many more specifically or differentially expressed genes than early flowers. These gene families expanded dramatically within the Brassicaceae lineage, suggesting novel functions that are possibly important for the origin and evolution of Brassicaceae. This dataset can be useful for discovering functional genes at different stages of the flower development and provide clues for the molecular and regulatory relationships between different stages.

ACKNOWLEDGMENTS

We would like to thank Qi Li for comments on the manuscript and helpful discussions. This work was supported by the National Natural Science Foundation of China (91131007) and Chinese Ministry of Science and Technology (2011CB944600). Liangsheng Zhang was supported by funds from Tongji University (2013KJ052).

REFERENCES

Alvarez-Buylla, E. R., Benitez, M., Corvera-Poire, A., Chaos Cador, A., De Folter, S., Gamboa De Buen, A., et al. (2010). Flower development. *Arabidopsis Book* 8:e0127. doi: 10.1199/tab.0127

Alves-Ferreira, M., Wellmer, F., Banhara, A., Kumar, V., Riechmann, J. L., and Meyerowitz, E. M. (2007). Global expression profiling applied to the analysis of *Arabidopsis* stamen development. *Plant Physiol.* 145, 747–762. doi: 10.1104/pp.107.104422

Benedito, V. A., Torres-Jerez, I., Murray, J. D., Andriankaja, A., Allen, S., Kakar, K., et al. (2008). A gene expression atlas of the model legume Medicago truncatula. *Plant J.* 55, 504–513. doi: 10.1111/j.1365-313X.2008.03519.x

Chang, F., Wang, Y., Wang, S., and Ma, H. (2011). Molecular control of microsporogenesis in *Arabidopsis*. *Curr. Opin. Plant Biol.* 14, 66–73. doi: 10.1016/j.pbi.2010.11.001

Chen, C., Farmer, A. D., Langley, R. J., Mudge, J., Crow, J. A., May, G. D., et al. (2010). Meiosis-specific gene discovery in plants: RNA-Seq applied to isolated *Arabidopsis* male meiocytes. *BMC Plant Biol.* 10:280. doi: 10.1186/1471-2229-10-280

Chen, Y., Souaiaia, T., and Chen, T. (2009). PerM: efficient mapping of short sequencing reads with periodic full sensitive spaced seeds. *Bioinformatics* 25, 2514–2521. doi: 10.1093/bioinformatics/btp486

Deng, W., Ying, H., Helliwell, C. A., Taylor, J. M., Peacock, W. J., and Dennis, E. S. (2011). FLOWERING LOCUS C (FLC) regulates development pathways throughout the life cycle of *Arabidopsis*. *Proc. Natl. Acad. Sci. U.S.A.* 108, 6680–6685. doi: 10.1073/pnas.1103175108

Du, Z., Zhou, X., Ling, Y., Zhang, Z., and Su, Z. (2010). agriGO: a GO analysis toolkit for the agricultural community. *Nucleic Acids Res.* 38, W64–W70. doi: 10.1093/nar/gkq310

Filichkin, S. A., Priest, H. D., Givan, S. A., Shen, R., Bryant, D. W., Fox, S. E., et al. (2010). Genome-wide mapping of alternative splicing in *Arabidopsis thaliana*. *Genome Res.* 20, 45–58. doi: 10.1101/gr.093302.109

Guo, A., He, K., Liu, D., Bai, S., Gu, X., Wei, L., et al. (2005). DATF: a database of *Arabidopsis* transcription factors. *Bioinformatics* 21, 2568–2569. doi: 10.1093/bioinformatics/bti334

Hennig, L., Gruissem, W., Grossniklaus, U., and Kohler, C. (2004). Transcriptional programs of early reproductive stages in *Arabidopsis*. *Plant Physiol.* 135, 1765–1775. doi: 10.1104/pp.104.043182

Immink, R. G., Tonaco, I. A., De Folter, S., Shchennikova, A., Van Dijk, A. D., Busscher-Lange, J., et al. (2009). SEPALLATA3: the "glue" for MADS box transcription factor complex formation. *Genome Biol.* 10:R24. doi: 10.1186/gb-2009-10-2-r24

Irish, V. F. (2010). The flowering of *Arabidopsis* flower development. *Plant J.* 61, 1014 1028. doi: 10.1111/j.1365-313X.2009.04065.x

Jiao, Y., and Meyerowitz, E. M. (2010). Cell-type specific analysis of translating RNAs in developing flowers reveals new levels of control. *Mol. Syst. Biol.* 6, 419. doi: 10.1038/msb.2010.76

Jiao, Y., Tausta, S. L., Gandotra, N., Sun, N., Liu, T., Clay, N. K., et al. (2009). A transcriptome atlas of rice cell types uncovers cellular, functional and developmental hierarchies. *Nat. Genet.* 41, 258–263. doi: 10.1038/ng.282

Kaufmann, K., Muino, J. M., Jauregui, R., Airoldi, C. A., Smaczniak, C., Krajewski, P., et al. (2009). Target genes of the MADS transcription factor SEPALLATA3: integration of developmental and hormonal pathways in the *Arabidopsis* flower. *PLoS Biol.* 7:e1000090. doi: 10.1371/journal.pbio.1000090

Kaufmann, K., Wellmer, F., Muino, J. M., Ferrier, T., Wuest, S. E., Kumar, V., et al. (2010). Orchestration of floral initiation by APETALA1. *Science* 328, 85–89. doi: 10.1126/science.1185244

Li, Q., Zhang, N., Zhang, L., and Ma, H. (2014). Differential evolution of members of the rhomboid gene family with conservative and divergent patterns. *New Phytol.* doi: 10.1111/nph.13174. [Epub ahead of print].

Liu, D., Sui, S., Ma, J., Li, Z., Guo, Y., Luo, D., et al. (2014). Transcriptomic analysis of flower development in wintersweet (*Chimonanthus praecox*). *PLoS ONE* 9:e86976. doi: 10.1371/journal.pone.0086976

Marioni, J. C., Mason, C. E., Mane, S. M., Stephens, M., and Gilad, Y. (2008). RNA-seq: an assessment of technical reproducibility and comparison with gene expression arrays. *Genome Res.* 18, 1509–1517. doi: 10.1101/gr.079558.108

Mukherjee, K., Brocchieri, L., and Burglin, T. R. (2009). A comprehensive classification and evolutionary analysis of plant homeobox genes. *Mol. Biol. Evol.* 26, 2775–2794. doi: 10.1093/molbev/msp201

O'Maoileidigh, D. S., Graciet, E., and Wellmer, F. (2014). Gene networks controlling *Arabidopsis* thaliana flower development. *New Phytol.* 201, 16–30. doi: 10.1111/nph.12444

Pagnussat, G. C., Alandete-Saez, M., Bowman, J. L., and Sundaresan, V. (2009). Auxin-Dependent Patterning and Gamete Specification in the *Arabidopsis* Female Gametophyte. *Science* 324, 1684–1689. doi: 10.1126/science.1167324

Punta, M., Coggill, P. C., Eberhardt, R. Y., Mistry, J., Tate, J., Boursnell, C., et al. (2011). The Pfam protein families database. *Nucleic Acids Res.* 40, D290–D301. doi: 10.1093/nar/gkr1065

Reed, J. W. (2001). Roles and activities of Aux/IAA proteins in *Arabidopsis*. *Trends Plant Sci.* 6, 420–425. doi: 10.1016/S1360-1385(01)02042-8

Schein, M., Yang, Z., Mitchell-Olds, T., and Schmid, K. J. (2004). Rapid evolution of a pollen-specific oleosin-like gene family from *Arabidopsis thaliana* and closely related species. *Mol. Biol. Evol.* 21, 659–669. doi: 10.1093/molbev/msh059

Schmid, M., Davison, T. S., Henz, S. R., Pape, U. J., Demar, M., Vingron, M., et al. (2005). A gene expression map of *Arabidopsis* thaliana development. *Nat. Genet.* 37, 501–506. doi: 10.1038/ng1543

Smaczniak, C., Immink, R. G., Muino, J. M., Blanvillain, R., Busscher, M., Busscher-Lange, J., et al. (2012). Characterization of MADS-domain transcription factor complexes in *Arabidopsis* flower development. *Proc. Natl. Acad. Sci. U.S.A.* 109, 1560–1565. doi: 10.1073/pnas.1112871109

Sundberg, E., and Ostergaard, L. (2009). Distinct and dynamic auxin activities during reproductive development. *Cold Spring Harb. Perspect. Biol.* 1:a001628. doi: 10.1101/cshperspect.a001628

Wang, H., You, C., Chang, F., Wang, Y., Wang, L., Qi, J., et al. (2014a). Alternative splicing during *Arabidopsis* flower development results in constitutive and stage-regulated isoforms. *Front. Genet.* 5:25. doi: 10.3389/fgene.2014.00025

Wang, L., Cao, C., Ma, Q., Zeng, Q., Wang, H., Cheng, Z., et al. (2014b). RNA-seq analyses of multiple meristems of soybean: novel and alternative transcripts, evolutionary and functional implications. *BMC Plant Biol.* 14:169. doi: 10.1186/1471-2229-14-169

Wang, L., Feng, Z., Wang, X., and Zhang, X. (2010). DEGseq: an R package for identifying differentially expressed genes from RNA-seq data. *Bioinformatics* 26, 136–138. doi: 10.1093/bioinformatics/btp612

Wellmer, F., Alves-Ferreira, M., Dubois, A., Riechmann, J. L., and Meyerowitz, E. M. (2006). Genome-wide analysis of gene expression during early *Arabidopsis* flower development. *PLoS Genet.* 2:e117. doi: 10.1371/journal.pgen.0020117

Wellmer, F., Riechmann, J. L., Alves-Ferreira, M., and Meyerowitz, E. M. (2004). Genome-wide analysis of spatial gene expression in *Arabidopsis* flowers. *Plant Cell* 16, 1314–1326. doi: 10.1105/tpc.021741

Xiong, Y., Chen, X., Chen, Z., Wang, X., Shi, S., Zhang, J., et al. (2010). RNA sequencing shows no dosage compensation of the active X-chromosome. *Nat. Genet.* 42, 1043–1047. doi: 10.1038/ng.711

Yang, H., Lu, P., Wang, Y., and Ma, H. (2011). The transcriptome landscape of *Arabidopsis* male meiocytes from high-throughput sequencing: the complexity and evolution of the meiotic process. *Plant J.* 65, 503–516. doi: 10.1111/j.1365-313X.2010.04439.x

Yang, X. H., Makaroff, C. A., and Ma, H. (2003). The *Arabidopsis MALE MEIOCYTE DEATH1* gene encodes a PHD-finger protein that is required for male meiosis. *Plant Cell* 15, 1281–1295. doi: 10.1105/tpc.010447

Zhang, G. J., Guo, G. W., Hu, X. D., Zhang, Y., Li, Q. Y., Li, R. Q., et al. (2010). Deep RNA sequencing at single base-pair resolution reveals high complexity of the rice transcriptome. *Genome Res.* 20, 646–654. doi: 10.1101/gr.100677.109

Zhang, L., and Ma, H. (2012). Complex evolutionary history and diverse domain organization of SET proteins suggest divergent regulatory interactions. *New Phytol.* 195, 248–263. doi: 10.1111/j.1469-8137.2012.04143.x

Zhang, X., Feng, B., Zhang, Q., Zhang, D., Altman, N., and Ma, H. (2005). Genome-wide expression profiling and identification of gene activities during early flower development in *Arabidopsis*. *Plant Mol. Biol.* 58, 401–419. doi: 10.1007/s11103-005-5434-6

Zhong, Y. F., and Holland, P. W. (2011). HomeoDB2: functional expansion of a comparative homeobox gene database for evolutionary developmental biology. *Evol. Dev.* 13, 567–568. doi: 10.1111/j.1525-142X.2011.00513.x

Zhou, X., and Ma, H. (2008). Evolutionary history of histone demethylase families: distinct evolutionary patterns suggest functional divergence. *BMC Evol. Biol.* 8, 294. doi: 10.1186/1471-2148-8-294

4

Establishment of embryonic shoot–root axis is involved in auxin and cytokinin response during *Arabidopsis* somatic embryogenesis

Ying Hua Su‡, Yu Bo Liu‡, Bo Bai† and Xian Sheng Zhang*

State Key Laboratory of Crop Biology, College of Life Sciences, Shandong Agricultural University, Taian, China

Edited by:
Kang Chong, Institute of Botany, The Chinese Academy of Sciences, China

Reviewed by:
Venugopala Reddy Gonehal, University of California, Riverside, USA
Marcelo Carnier Dornelas, Universidade Estadual de Campinas, Brazil

*Correspondence:
Xian Sheng Zhang, State Key Laboratory of Crop Biology, College of Life Sciences, Shandong Agricultural University, Taian, Shandong 271018, China
e-mail: zhangxs@sdau.edu.cn

†Present address:
Bo Bai, Shandong Rice Research Institute, Shandong Academy of Agricultural Science, Jinan, Shandong, China

‡These authors have contributed equally to this work.

Auxin and cytokinin signaling participates in regulating a large spectrum of developmental and physiological processes in plants. The shoots and roots of plants have specific and sometimes even contrary responses to these hormones. Recent studies have clearly shown that establishing the spatiotemporal distribution of auxin and cytokinin response signals is central for the control of shoot apical meristem (SAM) induction in cultured tissues. However, little is known about the role of these hormones in root apical meristem (RAM) initiation. Here, we found that the expression patterns of several regulatory genes critical for RAM formation were correlated with the establishment of the embryonic root meristem during somatic embryogenesis in *Arabidopsis*. Interestingly, the early expression of the *WUS-RELATED HOMEOBOX 5* (*WOX5*) and *WUSCHEL* genes was induced and was nearly overlapped within the embryonic callus when somatic embryos (SEs) could not be identified morphologically. Their correct expression was essential for RAM and SAM initiation and embryonic shoot–root axis establishment. Furthermore, we analyzed the auxin and cytokinin response during SE initiation. Notably, cytokinin response signals were detected in specific regions that were correlated with induced *WOX5* expression and subsequent SE formation. Overexpression of the *ARABIDOPSIS RESPONSE REGULATOR* genes *ARR7* and *ARR15* (feedback repressors of cytokinin signaling), disturbed RAM initiation and SE induction. These results provide new information on auxin and cytokinin-regulated apical–basal polarity formation of shoot–root axis during somatic embryogenesis.

Keywords: shoot–root axis, root apical meristem, cytokinin response, auxin response, somatic embryogenesis, Arabidopsis

INTRODUCTION

The most critical event during embryogenesis appears to be the formation of the shoot apical meristem (SAM) and root apical meristem (RAM), from which almost the entire plant is post-embryonically established (Meinke, 1991; Scheres, 2007). In the SAM of *Arabidopsis*, *WUSCHEL* (*WUS*) is a critical regulator, and encodes a homeodomain protein that is required for stem cell formation and maintenance (Laux et al., 1996). *WUS* is switched on in the four inner cells of the pro-embryo at the 16-cell globular stage, and is an early molecular marker for SAM initiation in the embryo (Weigel and Jürgens, 2002).

Root growth and development are sustained by the RAM, which is formed during embryogenesis (Sabatini et al., 2003; Petricka et al., 2012). Embryonic RAM formation is initiated at the globular stage, when the uppermost cell of the suspensor—the hypophysis—is recruited in the embryo proper. After asymmetrical division of the hypophysis, the small descendant cell gives rise to the quiescent center (QC), which maintains stem cell identity in the surrounding cells of the RAM to produce a set of differentiated tissues (Möller and Weijers, 2009; Peris et al., 2010). Mutants that fail to form the hypophysis often produce rootless seedlings (Möller and Weijers, 2009). An element required in the QC to

maintain columella stem cells is WUS-RELATED HOMEOBOX 5 (WOX5), a putative homeodomain transcription factor (Haecker et al., 2004). In the QC, WOX5 acts in a similar way to WUS in the organizing center (OC) of the SAM, highlighting molecular and developmental similarities between the stem cell niches of both root and shoot meristems (Perilli et al., 2012). In addition, several other putative transcription factors have been shown to contribute to embryonic RAM formation. *PLETHORA* (*PLT*) genes, which belong to the AP2-type transcription factor family, play a key role in the specification and maintenance of root stem cells from early embryogenesis onward (Aida et al., 2004; Galinha et al., 2007). Ectopic *PLT* expression in the embryo induces transformation of apical domain cells into root stem cells (Aida et al., 2004). The SCARECROW (SCR)/SHORTROOT (SHR) transcription factors are required to maintain stem cell activity within the RAM (Di Laurenzio et al., 1996; Helariutta et al., 2000; Sabatini et al., 2003).

Auxin and cytokinin are required for cell differentiation and specification during embryogenesis (Müller and Sheen, 2008; Möller and Weijers, 2009). Asymmetric distribution of auxin mediated by auxin polar transport establishes the apical–basal axis of the embryo, showing that auxin is required for pattern formation of the embryo (Friml et al., 2003; Möller and Weijers, 2009). Cytokinin signaling components function in the hypophysis at the early globular stage of the embryo (Müller and Sheen, 2008). After the first division of the hypophysis, the apical daughter cell maintains the phosphorelay activity of cytokinin signaling, whereas cytokinin signaling is repressed in the basal daughter cell. In early embryogenesis, auxin antagonizes cytokinin signaling through direct transcriptional activation of *ARABIDOPSIS RESPONSE REGULATOR (ARR)7* and *ARR15*, feedback repressors of cytokinin signaling in the basal cell (Hwang and Sheen, 2001; Müller and Sheen, 2008; Buechel et al., 2010).

Somatic embryogenesis is generally believed to be mediated by a signaling cascade triggered by exogenous auxin (Skoog and Miller, 1957; Sugiyama, 1999, 2000). Indeed, our previous work has shown that the establishment of auxin gradients is correlated with induced *WUS* expression and subsequent embryonic SAM formation during somatic embryogenesis (Su et al., 2009). It has also been suggested that the establishment of the RAM in somatic embryos requires an appropriate auxin gradient (Bassuner et al., 2007). However, the mechanism by which root stem cell specification occurs during early somatic embryogenesis is far from understood. Here, we analyzed the expression patterns of a few critical marker genes involved in RAM formation and development during somatic embryogenesis. Besides the auxin gradients that are established in specific regions of embryonic callus, we found that the spatiotemporal cytokinin response was correlated with RAM formation. Such cytokinin response patterns were critical for spatial induction of RAM-specific genes, such as *WOX5* and *PLT*, and subsequent RAM establishment in the embryonic callus. Our results reveal the distinct functions of cytokinin and auxin signaling required for RAM and SAM induction and shoot–root axis establishment during early somatic embryogenesis.

MATERIALS AND METHODS
PLANT MATERIALS
All *Arabidopsis* mutants and transgenic lines used in this study were Columbia ecotypes. The *pWOX5::GFP*, *pPLT2::RFP*, *pSCR::GFP* reporter lines and *plt2-1* mutants were kindly provided by Dr. C. Li (Institute of Genetics and Developmental Biology, Chinese Academy of Sciences). *pWUS::DsRED-N7* and *DR5rev:3XVENUS-N7* seeds were kindly provided by Dr E. M. Meyerowitz (Division of Biology, California Institute of Technology, Pasadena, CA, USA). The *ahk2 ahk4* and *ahk3 ahk4* mutants were obtained from Dr. C. Ueguchi (Bioscience and Biotechnology Center, Nagoya University, Nagoya, Japan). The *DR5rev::GFP* lines were provided by Dr J. Friml (Zentrum für Molekularbiologie der Pflanzen, Universität Tübingen, Germany). *pARR7::GFP* and *pARR15::GFP* seeds were provided by Dr J. Sheen (Harvard Medical School, USA). Double reporter lines were generated as follows: *pWOX5::GFP* lines were crossed with *pWUS::DsRED-N7* lines; *DR5rev:3XVENUS-N7* lines were crossed with *pWOX5::GFP* lines, and *pPLT2::RFP* lines were crossed with *DR5rev::GFP* lines; *pARR7::GFP* lines were crossed with *pWUS::DsRED-N7* lines.

ANTISENSE *WOX5* cDNA PLASMID CONSTRUCTION
To determine the role of *WOX5* during somatic embryo induction, a 754 bp cDNA fragment of the *WOX5* coding region was amplified using the primers 5′-ATACTAGTAAACAGTTGAGGACTTTACA TC-3′ (forward) and 5′-ATCTCGAGTACGCATTCCATAACATAG ATT-3′ (reverse). The *WOX5* antisense cDNA was then cloned into the estradiol inducible XVE binary vector (Zuo et al., 2000), and transformed into *Arabidopsis* plants.

GROWTH CONDITIONS, SE INDUCTION AND CHEMICAL TREATMENTS
The growth conditions for *Arabidopsis* plants and somatic embryo (SE) induction followed Su et al. (2009). To induce the transcription of inserted *WOX5* antisense cDNAs, the primary somatic embryos (PSEs) were cultured in embryonic callus-inducing medium (ECIM) with 10 μM estradiol (prepared in DMSO as 10 mM stock; Sigma) for 14 days. Then, the cultured tissues were transferred to somatic embryo-inducing medium (SEIM) with 10 μM estradiol for another 8 days. Estradiol was added every 2 days. The embryonic calli in SEIM were collected for phenotype observation.

IN SITU HYBRIDIZATION
Embryonic calli were fixed in FAA (10% formaldehyde: 5% acetic acid: 50% alcohol) overnight at 4°C. After dehydration, the fixed callus tissues were embedded in paraffin (Sigma) and sectioned at 8 μm. Antisense and sense *WOX5* probes were used for hybridization as previously described by Zhao et al. (2006). The primers used to amplify the 440-bp *WOX5* probes were 5′-CATCATCATCAACCATCAACT-3′ (forward) and 5′-CCATAACATAGATTCTTATATC-3′ (reverse).

IMAGING CONDITIONS
Somatic embryo morphology was photographed using an Olympus JM dissecting microscope. To detect the fluorescence signals of the marker lines, a Zeiss 510 Meta laser scanning confocal microscope with a 20 × air objective and a 40 × oil-immersion lens was used. Specific sets of filters were selected as described previously by Heisler et al. (2005). The Zeiss LSM software was used to analyze the confocal images. At least 80 samples of each marker line were imaged to confirm the expression patterns at each stage.

RESULTS
GENES FOR RAM SPECIFICATION ARE INDUCED DURING EARLY SOMATIC EMBRYOGENESIS
Previously, we described a highly reproducible somatic embryogenesis system in *Arabidopsis* in detail (Su et al., 2009). Green PSEs can be generated from explants (immature zygotic embryos), and then disk-like embryonic calli are produced from PSEs in ECIM containing 2,4-dichlorophenoxyacetic acid (2,4-D). After the calli are transferred to 2,4-D-free SEIM, secondary somatic embryos (SSEs) are induced.

To analyze the spatiotemporally regulated formation of the root stem cell niche at the early stages of somatic embryogenesis, we examined the expression patterns of genes that play critical roles in root stem cell specification. Weak *pWOX5::GFP* signals were detected in a few internal regions but not in the edge regions of embryonic callus grown in ECIM for 14 days (data not shown). In contrast, after the embryonic calli were transferred to SEIM, stronger GFP signals started to be detected in some small edge regions at around 24 h (**Figure 1A**). At this time, the pro-embryos

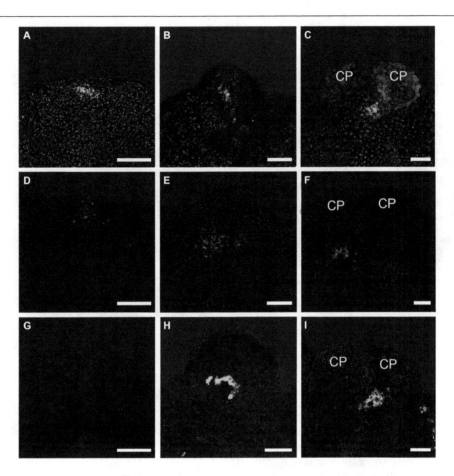

FIGURE 1 | Expression patterns of *WOX5*, *PLT2*, and *SCR* genes in embryonic calli during somatic embryogenesis. (A–C) Expression patterns of *WOX5* indicated by *pWOX5::GFP* in embryonic calli induced in SEIM for 24 h (**A**; 83.72%, *n* = 86), 2 days (**B**; 84.27%, *n* = 89) and 3 days (**C**; 80.21%, *n* = 96). **(D–F)** Expression patterns of *PLT2* indicated by *pPLT2::RFP* in embryonic calli induced in somatic embryo-inducing medium (SEIM) for 24 h (**D**; 87.65%, *n* = 81), 2 days (**E**; 84.95%, *n* = 93) and 3 days (**F**; 90.53%, *n* = 95). **(G–I)** Expression patterns of *SCR* indicated by *pSCR::GFP* in embryonic calli induced in SEIM for 24 h (**G**; 87.36%, *n* = 87), 2 days (**H**; 89.58%, *n* = 96) and 3 days (**I**; 86.90%, *n* = 84). CP, cotyledon primordia. Red signals in **A–C** represent chlorophyll autofluorescence. Scale bars = 80 μm.

were not identifiable morphologically, but the callus cells with *WOX5* activity might have been the initial QC. Later, GFP signals were observed in the basal regions of the globular pro-embryos and then pro-embryos with cotyledon primordia (CP; **Figures 1B,C**). *PLT2* exhibited similar spatial expression patterns to *WOX5* at these stages (**Figures 1D–F**). Different from *WOX5* expression, *PLT2* expression was observed in a relatively large group of cells within the callus, which represented the root stem cell niche of the SE. We also determined the expression patterns of *SCR*, whose expression defines the position of the QC (**Figures 1G–I**). The GFP signals of *SCR* were first detected at 36–48 h after induction in SEIM, which was later than *WOX5* and *PLT2* expression. The expression patterns of genes for the QC and stem cell formation indicated that the RAM is established during SE induction.

THE EMBRYONIC SHOOT–ROOT AXIS OF THE SE IS ESTABLISHED AT EARLY SOMATIC EMBRYOGENESIS

To determine the relative expression domains of *WUS* and *WOX5*, we analyzed their co-localization using a *pWUS::DsRED-N7 pWOX5::GFP* marker line. *WUS* and *WOX5* transcription

signals were first detected nearly overlapped at the edge regions of callus grown in SEIM for around 24 h (**Figure 2A**). At 36 h, the *WOX5* expression domain was just below and adjacent to that of *WUS* (**Figure 2B**). Subsequently, *WUS* transcripts were localized at the top regions between the CP of the pro-embryo, whereas *WOX5* transcripts were localized at the basal regions (**Figures 2C,D**). Thus, the SAM and RAM were initiated early and nearly overlapped in the edge regions of the callus, indicating that the apical–basal polarity of SE is determined and an embryonic shoot–root axis is established at the early stages of somatic embryogenesis.

RAM-SPECIFIC *WOX5* AND *PLT* EXPRESSION IS REQUIRED FOR EMBRYONIC ROOT FORMATION AND SE INDUCTION

To determine the roles of WOX5 during somatic embryogenesis, we constructed a vector carrying antisense *WOX5* driven by an estradiol receptor-based transactivator, XVE (Zuo et al., 2000), and transferred it into plants. To monitor estradiol-induced production of cDNA-encoded transcripts, quantitative real-time PCR (qRT-PCR) was performed to detect expression levels of *WOX5*

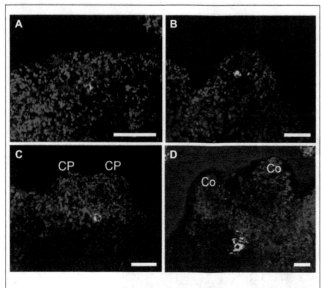

FIGURE 2 | Relative expression domains of *WOX5* and *WUS* genes.
pWOX5::GFP (green) and *pWUS::DsRed-N7* (red) signals in embryonic calli
induced in SEIM for 24 h (**A**; 86.81%, *n* = 91), 2 days (**B**; 85.39%, *n* = 89),
3 days (**C**; 89.66%, *n* = 87) and 4 days (**D**; 87.37%, *n* = 95); CP, cotyledon
primordia; Co, cotyledons. Blue signals represent chlorophyll
autofluorescence. Scale bars = 80 μm.

in 15 days shoots of wild type (WT) and *WOX5* antisense plants
(**Figure 3G**). Estradiol was added every 2 days in medium. Immature zygotic embryos of the transgenic plants were used as explants.
Green PSEs were induced on the shoot meristems of 84.47% of
the explants after 10 days of culture on B5 agar medium containing 4.5 μM 2,4-D in light, without estradiol in the medium
(**Figure 3A**; **Table 1**). After the PSEs were transferred from ECIM
to SEIM, 62.26% of untreated calli produced SSEs and each embryonic callus generated 52.6 ± 7.6 normal SSEs (**Figure 3B**; **Table 1**).
However, most explants carrying the antisense *WOX5* construct
produced abnormal PSEs with deficient hypocotyl elongation and
embryonic root formation in the presence of estradiol (**Figure 3C**;
Table 1). Only 8.61% of embryonic calli carrying the antisense
WOX5 construct produced SSEs, and each embryonic callus generated only 3.3 ± 2.2 normal SSEs (**Figure 3D**; **Table 1**). We
also examined PLT2's function during SE induction. The *plt2-1*
mutants generated PSEs with abnormal hypocotyls and embryonic
roots (**Figure 3E**), as described for *plt1-1 plt2-1* double mutants
(Su and Zhang, 2014). Embryonic calli of *plt2-1* mutants produced
severely abnormal SSEs, without cotyledons or SAMs (**Figure 3F**;
Table 1). Likewise, no hypocotyl- or root-like structures were
observed on these SSEs. These results suggested that both WOX5
and PLT2 are required for SE formation.

SPATIOTEMPORAL DISTRIBUTION OF AUXIN RESPONSES IN EARLY SE INDUCTION

Previously, we reported that auxin response gradients were established in specific regions of the embryonic callus, and were
responsible for SE formation (Su et al., 2009). Furthermore,
we showed that the spatiotemporal distribution of the auxin
response was correlated with the induced *WUS* expression at

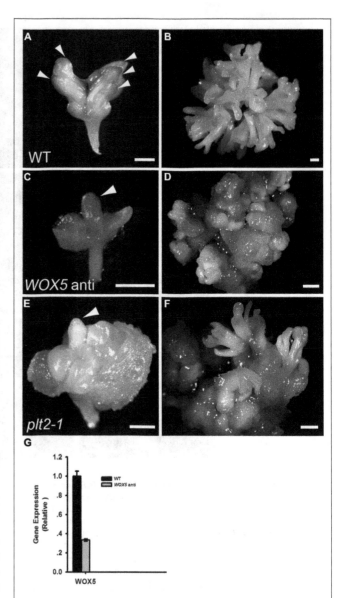

FIGURE 3 | Functional analysis of both *WOX5* and *PLT2* during somatic embryogenesis. (A,C,E) Phenotypes of primary somatic embryos (PSE)
induction from wild type (WT; **A**), *WOX5* antisense (**C**) and *plt2-1* mutant (**E**)
explants. Arrowheads indicate the PSE. **(B,D,F)** Phenotypes of SSE
induction from WT (**B**), *WOX5* antisense (**D**) and *plt2-1* mutant (**F**) calli
grown on SEIM for 8 days. **(G)** Expression levels of *WOX5* in
estradiol-induced 15 days shoots of WT and *WOX5* antisense plants. Scale
bars = 0.5 mm **(A,C,E)** and 1.2 mm **(B,D,F)**.

early somatic embryogenesis. Interestingly, we observed hardly any
auxin response signals in the basal part of the somatic pro-embryo.
We analyzed the auxin response signals and *WOX5* expression
within the callus by double labeling with *DR5rev:3XVENUS-
N7* and *pWOX5::GFP*. After 16 h incubation in SEIM, auxin
response signals were detectable at the edge regions of the callus, but no *WOX5* signal could be detected (**Figure 4A**). At
24 h after induction, *WOX5* signals were detected in the region
just beneath the outermost cell layers, where auxin response
signals were identified (**Figure 4B**). After 48 h incubation,

Establishment of embryonic shoot–root axis is involved in auxin and cytokinin response during...

29

Table 1 | Somatic embryo (SE) regeneration frequencies of different mutants and transgenetic lines.

Mutant [a]	Wild type	WOX5 anti	plt2-1	35S::ARR7	35S::ARR15	ahk2 ahk4	ahk3 ahk4
Ratio[b]	84.47%	18.79%	25.56%	34.14%	26.67%	32.23%	30.54%
Ratio[c]	62.26%	8.61%	23.50%	21.05%	17.70%	15.95%	15.15%
Number[d]	52.6 ± 7.6	3.3 ± 2.2	17.2 ± 7.3	15.3 ± 5.6	13.5 ± 7.1	15.5 ± 7.6	12.2 ± 4.3

[a] Mutant name; [b] Proportion of explants that produced normal PSEs; [c] Proportion of embryonic calli that produced normal SEs following culture in SEIM for 8 days;
[d] Number of normal SEs produced per embryonic callus following culture in SEIM for 8 days (mean ± SD, n ≥ 90).

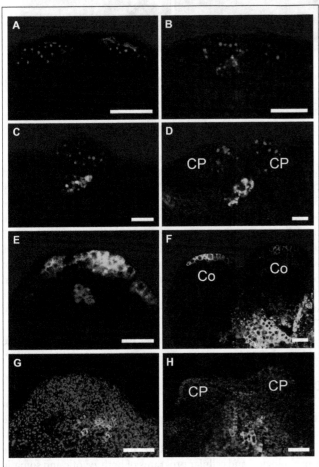

FIGURE 4 | Auxin and cytokinin responses in early somatic embryogenesis. (A–D) Auxin response represented by *DR5rev:3XVENUS-N7* correlated with *WOX5* induction represented by *pWOX5::GFP* in embryonic calli induced in SEIM for 16 h (**A**; 88.24%, n = 85), 24 h (**B**; 83.33%, n = 96), 2 days (**C**; 87.50%, n = 88) and 3 days (**D**; 89.66%, n = 87). *pWOX5::GFP* signals are in green, *DR5rev:3XVENUS-N7* fluorescence signals are in red. (**E,F**) Auxin response represented by *DR5rev::GFP* correlated with *PLT2* induction represented by *pPLT2::RFP* in embryonic calli induced in SEIM for 3 days (**E**; 86.73%, n = 98) and 4 days (**F**; 90.91%, n = 77). *DR5rev::GFP* fluorescence signals are in green, *pPLT2::RFP* signals are in red. (**G,H**) Cytokinin response represented by *pARR7::GFP* correlated with *WUS* induction represented by *pWUS::DsRed-N7* in embryonic calli induced in SEIM for 2 days (**G**; 90.43%, n = 94) and 3 days (**H**; 85.37%, n = 82). *pARR7::GFP* fluorescence signals are in green, *pWUS::DsRed-N7* signals are in red, and chlorophyll autofluorescence is shown in blue. CP, cotyledon primordia; Co, cotyledons. Scale bars = 80 μm

strong auxin response signals were detected at the upper part of the pro-embryo, but *WOX5* signals were localized at the basal part (**Figure 4C**). Later, auxin signals were redistributed to the top regions of the CP, and *WOX5* was continuously expressed in the basal part of the pro-embryo (**Figure 4D**). We also examined the auxin response in relation to *PLT2* expression through double labeling with *DR5rev:GFP* and the *PLT2* reporter *pPLT2:RFP*. Until 4 days after induction in SEIM, there were auxin signals distributed at the basal region where *PLT2* was expressed (**Figures 4E,F**). These results suggested that auxin response gradients were established in the SAM but not in the RAM of the pro-embryo during the early stages of somatic embryogenesis.

CYTOKININ RESPONSES ARE SPATIOTEMPORALLY CORRELATED WITH RAM FORMATION

It has been reported that both auxin and cytokinin responses are critical for specifying the root stem cell niche in embryos (Müller and Sheen, 2008). To determine how the cytokinin response occurs in callus when SEs are induced, we analyzed the spatiotemporal expression patterns of *ARR7* and *ARR15*, which are primary responsive genes in cytokinin signaling and can be rapidly induced by cytokinin (To and Kieber, 2008; Werner and Schmülling, 2009). Signals of *pARR7::GFP* were first detected at some small regions of the calli near the edge at 24 and 36 h after induction in SEIM (**Figures 5A,B**), which was similar to the auxin response at these stages. Interestingly, after 2 days induction in SEIM, the signals were restricted to the basal part of the pro-embryo rather than the top (**Figures 5C,D**). We also used the *pARR15::GFP* reporter to examine the expression patterns of *ARR15*, and found similar distribution patterns of GFP signals to those of *ARR7* (**Figures 5E–H**). These results showed that the cytokinin response occurs in the regions of SE initiation, but the cytokinin response patterns are different from those of the auxin response. To examine whether the cytokinin response is correlated with the establishment of the embryonic SAM, we visualized the cytokinin response using a *pARR7::GFP pWUS::DsRED-N7* marker line. The distribution regions of cytokinin signaling were quite different from those of *WUS* expression, which was localized in the opposite pole of the pro-embryos (**Figures 4G,H**). We further examined the cytokinin response in relation to *WOX5* expression through double labeling with the *pARR7::GFP* and *PWOX5::RFP* reporters (Su and Zhang, 2014). Strong cytokinin responses were induced in the restrictive regions substantially overlapping with the *WOX5* signals. Thus, the results suggest that establishment of the cytokinin response is

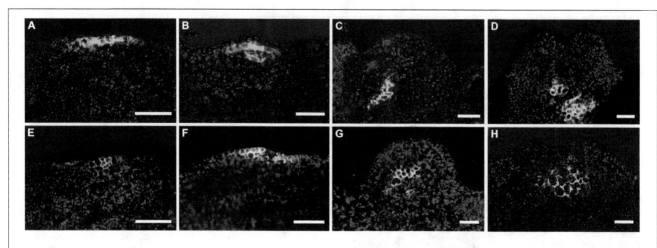

FIGURE 5 | Expression patterns of *ARR7* and *ARR15* during early somatic embryogenesis. (A–D) Expression patterns of *ARR7* indicated by *pARR7::GFP* in embryonic calli induced in SEIM for 16 h (**A**; 93.02%, n = 86), 24 h (**B**; 90.53%, n = 95), 2 days (**C**; 85.88%, n = 85) and 3 days (**D**; 87.63%, n = 97). **(E–H)** Expression patterns of *ARR15* indicated by *pARR15::GFP* in embryonic calli induced in SEIM for 16 h (**E**; 88.24%, n = 85), 24 h (**F**; 82.29%, n = 96), 2 days (**G**; 94.05%, n = 84) and 3 days (**H**; 86.96%, n = 92). Red signals represent chlorophyll autofluorescence. Scale bars = 80 μm.

correlated not with *WUS* but with *WOX5* induction within the callus, implying that cytokinin is required for embryonic RAM initiation.

CYTOKININ SIGNALING IS REQUIRED FOR EMBRYONIC RAM REGENERATION AND SE INDUCTION

The spatial distribution of cytokinin responses detected through *ARR7* and *ARR15* transcriptional signal profiles prompted us to confirm whether a functional cytokinin signaling mechanism is necessary for normal somatic embryogenesis. *ARR7* and *ARR15* act as negative regulators of cytokinin signaling by repressing type-B *ARRs* via unknown mechanisms (To and Kieber, 2008; Werner and Schmülling, 2009). Overexpression of *ARR7* and *ARR15* can attenuate cytokinin signaling to a sufficiently low level, resulting in reduced sensitivity to cytokinin in root elongation and shoot formation or an early flowering phenotype (Werner and Schmülling, 2009). To facilitate functional analysis of cytokinin signaling, we generated transgenic plants overexpressing *ARR7* or *ARR15* under the control of the CaMV 35S promoter. The *ARR7*-overexpressing explants showed a severely defective PSE phenotype without normal elongated hypocotyls or obvious embryonic roots (Su and Zhang, 2014). Subsequently, 78.95% of PSEs generated severely abnormal SSEs after induction in SEIM (**Figure 6A; Table 1**). Similar to *ARR7*-overexpressing plants, *ARR15*-overexpressing plants also generated abnormal PSEs with defective hypocotyls and embryonic roots, and subsequently, abnormal SSEs (Su and Zhang, 2014; **Figure 6B**). In *Arabidopsis*, three histidine kinases (AHKs), AHK2, AHK3, and AHK4, positively regulate cytokinin-signaling as direct receptors of cytokinin (To and Kieber, 2008). Thus, we further analyzed the developmental characteristics of *ahk2 ahk4* and *ahk3 ahk4* double mutant calli during SE regeneration. The phenotypes of both double mutants were consistent with previous descriptions of *ARR7*- and *ARR15*-overexpressing plants (**Figures 6C,D; Table 1**). Interestingly, we found that SE regeneration was impaired with defective cytokinin signaling, which was similar to *WOX5*-antisense plants and *plt2-1* mutants.

Because the expression patterns of *WOX5* were quite similar to the cytokinin response distribution, we hypothesized that cytokinin signaling might regulate *WOX5* expression for early SE induction. We performed *in situ* hybridization to analyze *WOX5* expression in the *ahk3 ahk4* double mutant. Indeed, the *WOX5* expression pattern was greatly disrupted in the *ahk3 ahk4* double mutant compared with the WT after the callus was transferred into SEIM (**Figures 6E,F**). *WOX5* signals were restricted to the site of the future embryonic root meristem in the WT (**Figure 6E**), whereas the localization of *WOX5* signals was stronger and more dispersed in the *ahk3 ahk4* mutant (**Figure 6F**). These results indicated that cytokinin signaling negatively regulates *WOX5* expression in the proper pattern for initiation of the embryonic RAM.

DISCUSSION

During somatic embryogenesis, the developmental process from the globular stage to the torpedo stage shares considerable similarity with that of zygotic embryogenesis (Meinke, 1991; Zimmerman, 1993). Although there are many similarities in the morphological and cellular programs of both zygotic and somatic embryogenesis, the mechanisms determining the initiation of these two processes might be different. The specific characteristics of somatic embryogenesis could be due to the origination of the somatic embryo from embryonic callus, not a zygote as in zygotic embryogenesis.

THE ORIGINS OF EMBRYONIC SAM AND RAM WERE QUITE DIFFERENT BETWEEN SOMATIC EMBRYOS AND ZYGOTIC EMBRYOS

In *Arabidopsis*, the mature embryo displays a main shoot–root axis of polarity, with the correct relative positioning of the embryonic SAM at the top and RAM at the opposite pole, separated by the hypocotyl (embryonic stem; Jürgens, 2001; Friml et al., 2003). The origin of this apical–basal pattern of shoot–root axis has been traced back to early embryogenesis, when zygotic division generates a smaller apical and a larger basal cell. After the apical domain

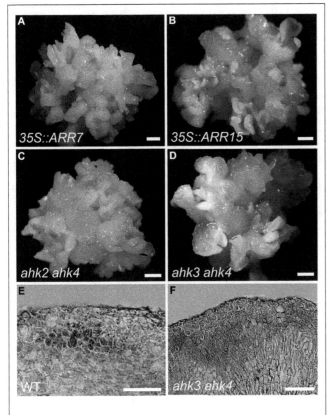

FIGURE 6 | Functions of cytokinin signaling on SE induction. (A–D)
Phenotypes of SSE induction from 35S::ARR7 **(A)**, 35S::ARR15 **(B)**, *ahk2 ahk4* mutant **(C)**, and *ahk3 ahk4* mutant **(D)** calli grown on SEIM for 8 days.
(E) *WOX5* transcript signals in WT embryonic callus cultured in SEIM for 24 h; 86.15%, *n* = 35. **(F)** Dispersed *WOX5* transcript signals in *ahk3 ahk4* double mutant callus following culture in SEIM for 24 h; 87.30%, *n* = 33. Scale bars = 1.2 mm **(A–D)** and 80 μm **(E,F)**.

niches of the RAM and the SAM share developmental correlations during SE initiation.

CYTOKININ RESPONSE WAS INVOLVED IN INDUCING CORRECT *WOX5* EXPRESSION AND RAM FORMATION

The patterns of embryonic SAM and RAM establishment in SE initiation suggest the presence of inductive hormonal signals to position them within the embryonic callus. Given the positive effects of auxin on *WUS* expression and SE induction (Su et al., 2009; **Figure 4**), it is likely a candidate factor that is required for embryonic RAM formation. Here, we found a cytokinin response distribution established in the regions where *WOX5* and *PLT2* were initiated (**Figures 1** and **5**). RAM formation and SE regeneration were severely inhibited in transgenic plants overexpressing *ARR7* or *ARR15* and in the *ahk* mutants, in which cytokinin signaling was inhibited (**Figure 6**). Moreover, in cultured tissues of the *ahk* mutants, *WOX5* expression patterns were seriously disturbed compared with control tissues (**Figure 6**). Thus, we hypothesize that removal of exogenous auxin may be a stress factor that causes cytokinin polar distribution and responses in specific regions, which induce correct *WOX5* expression and subsequent SE initiation. Induced *WOX5* transcripts were continuously detectable in areas of high cytokinin response (Su and Zhang, 2014), suggesting that cytokinin functions in the initiation and maintenance of the embryonic RAM during somatic embryogenesis. The positive action of cytokinin in SAM regeneration has been reported in several studies (Pernisová et al., 2009; Buechel et al., 2010). Treatment with high levels of exogenous cytokinin induces cell proliferation and stimulates shoot regeneration (Skoog and Miller, 1957). Cytokinin induces *WUS* expression during *in vitro* establishment of the SAM from cultured root explants (Gordon et al., 2009). A cytokinin response occurs in the center of the regenerated SAM, overlapping with *WUS* expression regions (Cheng et al., 2013). In contrast, an opposite effect of cytokinin in root regeneration has been observed. Cytokinin influences auxin-induced RAM regeneration via regulation of PIN-mediated auxin polar transport (Pernisová et al., 2009). Therefore, the functions of cytokinin in RAM establishment during SE initiation differ from those in shoot and root regeneration.

SPATIOTEMPORAL DISTRIBUTION OF AUXIN AND CYTOKININ RESPONSE IN EMBRYONIC CALLUS DETERMINES ESTABLISHMENT OF SE SHOOT–ROOT AXIS

Cytokinin and auxin appear to be the most important hormones in the regulation of organ regeneration (Moubayidin et al., 2009; Su et al., 2011). A high exogenous auxin/cytokinin ratio induces root regeneration, whereas a low ratio promotes shoot induction (Skoog and Miller, 1957). Recent studies have suggested that exogenous hormones treatment is the critical factor triggering biosynthesis and response of endogenous hormones in early developmental events of *in vitro* regeneration. Specialized endogenous hormonal signaling is required for specific cell differentiation that determines the developmental fate of callus cells (Gordon et al., 2007; Su et al., 2009, 2011). During early somatic embryogenesis, removal of exogenous auxin triggers the regional distribution of endogenous auxin response

of the pro-embryo has been specified, the embryonic SAM is initiated by the onset of *WUS* expression in the four subepidermal apical cells of the 16-cell embryo (Mayer et al., 1998; Laux et al., 2004). Subsequently, the QC of the RAM is established at approximately late globular embryo stage and marked by the expression of *WOX5* (Wysocka-Diller et al., 2000; Haecker et al., 2004).

During somatic embryogenesis, *WUS* and *WOX5* were simultaneously activated in nearly overlapped callus cells, when somatic pro-embryos could not be identified morphologically (**Figure 2**; Su et al., 2009). The nearly overlapped spatial relationship between regenerated embryonic SAM and RAM during SE initiation represents the different origins of apical–basal pattern between somatic embryos and zygotic embryos. It is likely that SEs initiate from specific embryonic callus cells which acquire features similar to meristematic cells. These specific embryonic cells of callus are reprogrammed and determined to form cells of both OC and QC for embryonic SAM and RAM formation. In addition, early defects in RAM initiation with inhibited *WOX5* expression also affected the initiation of the SAM, probably by disrupting the apical–basal pattern of early somatic embryogenesis. These results suggest that QC signaling not only maintains stem cell identity in the RAM but also is crucial for OC cells initiation, implying that the stem cell in callus surrounding areas of *WUS* expression initiation (Su et al., 2009). Following *WUS* induction, distribution of auxin response was re-established in the SAM region. In contrast, the distribution of cytokinin-response signal in callus overlapped with the areas of *WOX5* expression (Su and Zhang, 2014). These results imply that establishment of auxin and cytokinin response patterns within callus plays an important role in *WUS* and *WOX5* regional expression and shoot–root axis formation. Furthermore, auxin response signals accumulated at the basal region of pro-embryos following prolonged incubation in SEIM (**Figures 4E,F**). The redistributed auxin response corresponded to *PLT2* expression at the later stages of SE development. Therefore, our results suggest that cytokinin and auxin are key players in axial patterning of the SE, especially in shoot and root meristem initiation. The mechanisms of hormonal regulation in SE initiation are quite different from those in shoot or root regeneration individually, which remains a major challenge for the future.

AUTHOR CONTRIBUTIONS

Ying Hua Su and Xian Sheng Zhang designed the research. Ying Hua Su and Yu Bo Liu performed the research. Bo Bai analyzed the data. Ying Hua Su and Xian Sheng Zhang wrote the paper.

ACKNOWLEDGMENT

We are grateful to all members for their assistance in our laboratory. This research was supported by grants from the National Natural Science Foundation of China (90917015, 91217308, and 31170272).

REFERENCES

Aida, M., Beis, D., Heidstra, R., Willemsen, V., Blilou, I., Galinha, C., et al. (2004). The PLETHORA genes mediate patterning of the *Arabidopsis* root stem cell niche. *Cell* 119, 109–120. doi: 10.1016/j.cell.2004.09.018

Bassuner, B. M., Lam, R., Lukowitz, W., and Yeung, E. C. (2007). Auxin and root initiation in somatic embryos of *Arabidopsis*. *Plant Cell Rep.* 26, 1–11. doi: 10.1007/s00299-006-0207-5

Buechel, S., Leibfried, A., To, J. P., Zhao, Z., Andersen, S. U., Kieber, J. J., et al. (2010). Role of A-type *ARABIDOPSIS* RESPONSE REGULATORS in meristem maintenance and regeneration. *Eur. J. Cell Biol.* 89, 279–284. doi: 10.1016/j.ejcb.2009.11.016

Cheng, Z. J., Wang, L., Sun, W., Zhang, Y., Zhou, C., Su, Y. H., et al. (2013). Pattern of auxin and cytokinin responses for shoot meristem induction results from the regulation of cytokinin biosynthesis by AUXIN RESPONSE FACTOR3. *Plant Physiol.* 161, 240–251. doi: 10.1104/pp.112.203166

Di Laurenzio, L., Wysocka-Diller, J., Malamy, J. E., Pysh, L., Helariutta, Y., Freshour, G., et al. (1996). The SCARECROW gene regulates an asymmetric cell division that is essential for generating the radial organization of the *Arabidopsis* root. *Cell* 86, 423–433. doi: 10.1016/S0092-8674(00)80115-4

Friml, J., Vieten, A., Sauer, M., Weijers, D., Schwarz, H., Hamann, T., et al. (2003). Efflux-dependent auxin gradients establish the apical-basal axis of *Arabidopsis*. *Nature* 426, 147–153. doi: 10.1038/nature02085

Galinha, C., Hofhuis, H., Luijten, M., Willemsen, V., Blilou, I., Heidstra, R., et al. (2007). PLETHORA proteins as dose-dependent master regulators of *Arabidopsis* root development. *Nature* 449, 1053–1057. doi: 10.1038/nature06206

Gordon, S. P., Chickarmane, V. S., Ohno, C., and Meyerowitz, E. M. (2009). Multiple feedback loops through cytokinin signaling control stem cell number within the *Arabidopsis* shoot meristem. *Proc. Natl. Acad. Sci. U.S.A.* 106, 16529–16534. doi: 10.1073/pnas.0908122106

Gordon, S. P., Heisler, M. G., Reddy, G. V., Ohno, C., Das, P., and Meyerowitz, E. M. (2007). Pattern formation during de novo assembly of the *Arabidopsis* shoot meristem. *Development* 134, 3539–3548. doi: 10.1242/dev.010298

Haecker, A., Gross-Hardt, R., Geiges, B., Sarkar, A., Breuninger, H., Herrmann, M., et al. (2004). Expression dynamics of WOX genes mark cell fate decisions during early embryonic patterning in *Arabidopsis thaliana*. *Development* 131, 657–668. doi: 10.1242/dev.00963

Heisler, M. G., Ohno, C., Das, P., Sieber, P., Reddy, G. V., Long, J. A., et al. (2005). Patterns of auxin transport and gene expression during primordium development revealed by live imaging of the *Arabidopsis* inflorescence meristem. *Curr. Biol.* 15, 1899–1911. doi: 10.1016/j.cub.2005.09.052.

Helariutta, Y., Fukaki, H., Wysocka-Diller, J., Nakajima, K., Jung, J., Sena, G., et al. (2000). The SHORT-ROOT gene controls radial patterning of the *Arabidopsis* root through radial signaling. *Cell* 101, 555–567. doi: 10.1016/S0092-8674(00)80865-X

Hwang, I., and Sheen, J. (2001). Two-component circuitry in *Arabidopsis* cytokinin signal transduction. *Nature* 413, 383–389. doi: 10.1038/35096500

Jürgens, G. (2001). Apical-basal pattern formation in *Arabidopsis* embryogenesis. *EMBO J.* 20, 3609–3616. doi: 10.1093/emboj/20.14.3609

Laux, T., Mayer, K. F., Berger, J., and Jurgens, G. (1996). The WUSCHEL gene is required for shoot and floral meristem integrity in *Arabidopsis*. *Development* 122, 87–96.

Laux, T., Wurschum, T., and Breuninger, H. (2004). Genetic regulation of embryonic pattern formation. *Plant Cell* 16, S190–S202. doi: 10.1105/tpc.016014

Mayer, K. F., Schoof, H., Haecker, A., Lenhard, M., Jurgens, G., and Laux, T. (1998). Role of WUSCHEL in regulating stem cell fate in the *Arabidopsis* shoot meristem. *Cell* 95, 805–815. doi: 10.1016/S0092-8674(00)81703-1

Meinke, D. W. (1991). Perspectives on genetic analysis of plant embryogenesis. *Plant Cell* 3, 857–866. doi: 10.1105/tpc.3.9.857

Möller, B., and Weijers, D. (2009). Auxin control of embryo patterning. *Cold Spring Harb. Perspect. Biol.* 1:a001545. doi: 10.1101/cshperspect.a001545

Moubayidin, L., Di Mambro, R., and Sabatini, S. (2009). Cytokinin-auxin crosstalk. *Trends Plant Sci.* 14, 557–562. doi: 10.1016/j.tplants.2009.06.010

Müller, B., and Sheen, J. (2008). Cytokinin and auxin interaction in root stem-cell specification during early embryogenesis. *Nature* 453, 1094–1097. doi: 10.1038/nature06943

Perilli, S., Di Mambro, R., and Sabatini, S. (2012). Growth and development of the root apical meristem. *Curr. Opin. Plant Biol.* 15, 17–23. doi: 10.1016/j.pbi.2011.10.006

Peris, C. I., Rademacher, E. H., and Weijers, D. (2010). Green beginnings – pattern formation in the early plant embryo. *Curr. Top. Dev. Biol.* 91, 1–27. doi: 10.1016/S0070-2153(10)91001-6

Pernisová, M., Klima, P., Horak, J., Valkova, M., Malbeck, J., Soucek, P., et al. (2009). Cytokinins modulate auxin-induced organogenesis in plants via regulation of the auxin efflux. *Proc. Natl. Acad. Sci. U.S.A.* 106, 3609–3614. doi: 10.1073/pnas.0811539106

Petricka, J. J., Winter, C. M., and Benfey, P. N. (2012). Control of *Arabidopsis* root development. *Annu. Rev. Plant Biol.* 63, 563–590. doi: 10.1146/annurev-arplant-042811-105501

Sabatini, S., Heidstra, R., Wildwater, M., and Scheres, B. (2003). SCARECROW is involved in positioning the stem cell niche in the *Arabidopsis* root meristem. *Genes Dev.* 17, 354–358. doi: 10.1101/gad.252503.

Scheres, B. (2007). Stem-cell niches: nursery rhymes across kingdoms. *Nat. Rev. Mol. Cell Biol.* 8, 345–354. doi: 10.1038/nrm2164

Skoog, F., and Miller, C. O. (1957). Chemical regulation of growth and organ formation in plant tissues cultured in vitro. *Symp. Soc. Exp. Biol.* 11, 118–130.

Su, Y. H., Liu, Y. B., and Zhang, X. S. (2011). Auxin-cytokinin interaction regulates meristem development. *Mol. Plant.* 4, 616–625. doi: 10.1093/mp/ssr007

Su, Y. H., and Zhang, X. S. (2014). The hormonal control of regeneration in plants. *Curr. Top. Dev. Biol.* 108, 35–69. doi: 10.1016/B978-0-12-391498-9.00010-3

Su, Y. H., Zhao, X. Y., Liu, Y. B., Zhang, C. L., O'neill, S. D., and Zhang, X. S. (2009). Auxin-induced WUS expression is essential for embryonic stem cell renewal during somatic embryogenesis in *Arabidopsis*. *Plant J.* 59, 448–460. doi: 10.1111/j.1365-313X.2009.03880.x

Sugiyama, M. (1999). Organogenesis in vitro. *Curr. Opin. Plant Biol.* 2, 61–64. doi: 10.1016/S1369-5266(99)80012-0

Sugiyama, M. (2000). Genetic analysis of plant morphogenesis in vitro. *Int. Rev. Cytol.* 196, 67–84. doi: 10.1016/S0074-7696(00)96002-9

To, J. P. C., and Kieber, J. J. (2008). Cytokinin signaling: two-components and more. *Trends Plant Sci.* 13, 85–92. doi: 10.1016/j.tplants.2007.11.005

Weigel, D., and Jürgens, G. (2002). Stem cells that make stems. *Nature* 415, 751–754. doi: 10.1038/415751a

Werner, T., and Schmülling, T. (2009). Cytokinin action in plant development. *Curr. Opin. Plant Biol.* 12, 527–538. doi: 10.1016/j.pbi.2009.07.002

Wysocka-Diller, J. W., Helariutta, Y., Fukaki, H., Malamy, J. E., and Benfey, P. N. (2000). Molecular analysis of SCARECROW function reveals a radial patterning mechanism common to root and shoot. *Development* 127, 595–603.

Zhao, X. Y., Cheng, Z. J., and Zhang, X. S. (2006). Overexpression of TaMADS1, a SEPALLATA-like gene in wheat, causes early flowering and the abnormal development of floral organs in *Arabidopsis*. *Planta* 223, 698–707. doi: 10.1007/s00425-005-0123-x

Zimmerman, J. L. (1993). Somatic embryogenesis: a model for early development in higher plants. *Plant Cell* 5, 1411–1423. doi: 10.1105/tpc.5.10.1411

Zuo, J., Niu, Q. W., and Chua, N. H. (2000). Technical advance: an estrogen receptor-based transactivator XVE mediates highly inducible gene expression in transgenic plants. *Plant J.* 24, 265–273. doi: 10.1046/j.1365-313x.2000.00868.x

5

The concept of the sexual reproduction cycle and its evolutionary significance

Shu-Nong Bai*

State Key Laboratory of Protein & Plant Gene Research, Quantitative Biology Center, College of Life Science, Peking University, Beijing, China

Edited by:
Dazhong Dave Zhao, University of Wisconsin–Milwaukee, USA

Reviewed by:
Kang Chong, Institue of Botany – The Chinese Academy of Sciences, China
Marcelo Carnier Dornelas, Universidade Estadual de Campinas, Brazil
Manyuan Long, The University of Chicago, USA

*Correspondence:
Shu-Nong Bai, State Key Laboratory of Protein & Plant Gene Research, Quantitative Biology Center, College of Life Science, Peking University, Beijing 100871, China
e-mail: shunongb@pku.edu.cn

The concept of a "sexual reproduction cycle (SRC)" was first proposed by Bai and Xu (2013) to describe the integration of meiosis, sex differentiation, and fertilization. This review discusses the evolutionary and scientific implications of considering these three events as part of a single process. Viewed in this way, the SRC is revealed to be a mechanism for efficiently increasing genetic variation, facilitating adaptation to environmental challenges. It also becomes clear that, in terms of cell proliferation, it is appropriate to contrast mitosis with the entire SRC, rather than with meiosis alone. Evolutionarily, it appears that the SRC was first established in unicellular eukaryotes and that all multicellular organisms evolved within that framework. This concept provides a new perspective into how sexual reproduction evolved, how generations should be defined, and how developmental processes of various multicellular organisms should properly be compared.

Keywords: sexual reproduction cycle, meiosis, heterogametogenesis, fertilization, generation

Sex is always a hot topic in human society and is an intensively investigated phenomenon in biology. Studies on sex determination in plants have focused on regulation of unisexual flower development (Ainsworth, 1999; Bai and Xu, 2013) since plant sex was defined based on unisexual flowers by Robbins and Pearson (1933), i.e., "a flower or plant is male if it bears only stamen and that is female if it bears only pistils." However, plants with unisexual flowers account for only a small percentage of angiosperms, and angiosperms themselves are only some of the species in the plant kingdom. This raises the question of what sex is in angiosperms with perfect flowers (with both pistils and stamens) or in plants without flowers. Even further, do sex and sexual differentiation share any features in common in the plant and animal kingdoms?

To understand the regulatory mechanism of sex differentiation, following the concept originated from Robbins and Pearson (1933), we have investigated the regulation of unisexual flower development using cucumber for more than a decade. Cucumber is monoecious, and naturally bears both male and female flowers and, rarely, even hermaphrodite flowers, on the same plant. The ratio of male and female flowers can be affected by application of phytohormones such as ethylene (increasing the proportion of female flowers) or GA (increasing the proportion of male flowers). As hormones play key roles in mammalian sex expression, this phytohormonal regulation of the ratio of male and female flowers led to the use of cucumber a model system for research into plant sex determination starting in the 1960s (see Bai and Xu, 2013 and references there in). However, following the discovery that both male and female flowers contain initiated stamen and carpel primordia, we found that flowers become female because stamen development is inhibited early at stage 6, and that ethylene is involved in this inhibition (for detailed review, see Bai and Xu,

2013). These findings indicated that analysis of regulatory mechanisms in unisexual flowers will allow us to understand only how the inappropriate organs are inhibited, not how the appropriate organs are differentiated. We therefore referred to this situation as a "bird-nest puzzle," meaning that it is not adequate to understand how a bird lays and hatches eggs through investigating how the nest was ruined (Bai and Xu, 2012).

Unisexual cucumber flowers are not the only example of this type of puzzle. All unisexual flowers in monoecious plants and many in dioecious plants for which the developmental mechanisms are known result from inhibition of one type of sexual organs (Bai and Xu, 2013; Akagi et al., 2014). To solve the "bird-nest puzzle," we proposed a new way to define sex as a "heterogamete-centered dimorphic phenomenon," sex differentiation as "the key divergence point(s) leading to the heterogamete differentiation" (Bai and Xu, 2013), which is in line with a definition previously suggested (Juraze and Banks, 1998), and generally applicable not only to all species in plant kingdom, but also in other kingdoms. We also hypothesized that before multicellular organisms emerged, a process called the sexual reproduction cycle (SRC) evolved based on the existing mitotic cell cycle in unicellular eukaryotes. This SRC starts from one diploid zygote, goes through meiosis, gametogenesis, and fertilization, and ends with two diploid zygotes. This review addresses the question of whether adopting the SRC concept can facilitate our understanding of sex.

PREMISES FOR PROPOSAL OF THE SRC

We begin with the basic facts based on which the SRC was hypothesized.

Firstly, although DNA transmission in prokaryotes is sometimes referred to as "recombination" (Cavalier-Smith, 2002 and references therein), it is widely accepted that sex and sex differentiation are phenomena occurring in eukaryotes, which contain chromosomes, nuclei, cell skeletons, and the mitotic cell cycle, regardless of the obscurity of their evolutionary origins (Cavalier-Smith, 2002; Kirschner and Gerhart, 2005; Knoll, 2014). We would restrict our discussion in eukaryotes.

Secondly, meiosis is highly conserved in almost all known eukaryotes, including animals, plants, fungi, and protists (Logsdon, 2007; Schurko and Logsdon, 2008). It is also widely accepted that meiosis may originate from mitosis, probably via occasional mistakes in cohesin binding and/or digestion on chromosomes (Cavalier-Smith, 2002; Marston and Amon, 2004).

Thirdly, despite the extreme diversity of morphology and recognition mechanisms of gametes, cell fusion of two haploid gametes into a new diploid zygote is conserved in all eukaryotes.

Fourthly, while dimorphism of gametes is common in the organisms with which people are familiar, the smaller gametes being referred to as sperm and the bigger ones as eggs, there are also heterogametes that can be distinguished only at the molecular level, with different mating types such as $MATa/\alpha$ in yeast and MTA/MTD in *Chlamydomonas* (Goodenough et al., 2007). Furthermore, heterogamy is not restricted to dimorphism, but rather can include multiple mating types, e.g., three in *Dictyostelium* and seven in *Tetrahymena* (Nanney and Caughey, 1953; Bloomfield et al., 2010; Cervantes et al., 2013). It is worth noting that such multiple mating types are mainly found in protists and fungi, but not in plant and animal kingdoms.

Fifthly, regardless of the presence or absence of germlines, e.g., in animals or plants, respectively, (Evens and Walbot, 2003), new generations in all multicellular organisms are generated through sexual reproduction consisting of the three key events: meiosis, sex differentiation, and fertilization. With this view, sexual reproduction is predicted to be more ancient than multicellular structures as sexual reproduction already existed in unicellular eukaryotic organisms.

There has been much debate about how sexual reproduction evolved (e.g., http://en.wikipedia.org/wiki/Sex; http://www.britannica.com/EBchecked/topic/536936/sex). It seems nearly impossible to obtain direct evidence regarding what happened during the period when sexual reproduction emerged, unless someday people can artificially "recapitulate" the evolutionary process. Nonetheless, we can try to explore the events involved in such evolution by analyzing the benefits for which the ancient evolutionary innovations could have been selected.

MEIOSIS: A LUCKY MISTAKE?

The first indispensable event in sexual reproduction is considered to be meiosis. It is currently agreed that the most important benefit of meiosis is increasing genetic variation through recombination. However, by definition, meiosis is characterized by a reduction of chromosome numbers from diploid to haploid. How, then, did meiosis emerge and become selected?

It is known that haploid cells, like diploid cells, can undergo mitosis, such as in budding yeast. Considering the complexity of chromosome organization, it would be reasonable to speculate that the earliest eukaryotic cells were haploid. If that were the case, meiosis would be predicted to have evolved not only after

the emergence of mitosis, but also after the emergence of diploid cells, which may have arisen from cell fusion or chromosome duplication in haploid cells.

While many organisms in the protist and fungus kingdoms live mainly in a haploid state (Campbell and Reece, 2005), almost all multicellular organisms in the animal and plant kingdoms use diploid cells as their building blocks. The prevalence of diploidy in the latter kingdoms suggests that diploidy must confer some advantages. If we naively believe that diploidy can doubly secure the genome stability of eukaryotic cells, then it follows that haploidy provides little leeway for mistakes. From this perspective, reduction of chromosome number would not be a good reason for meiosis to be selected. Instead, meiosis must occur and be selected for other reasons.

Based on Marston and Amon's (2004) comparison of mitosis and meiosis, cohesins play important roles in both processes. In mitosis, cohesins like Scc3, Smc1, and Smc3 facilitate the cohesion of the two sister chromosomes, whereas in meiosis, the cohesins can hold together non-sister chromosomes from two different chromosomes (**Figure 1**). This might be analogous to playing ringtoss: the ring is thrown to capture a target, but sometimes the ring mistakenly captures something else together with the target. If cohesion is required for mitosis, mistaken association of non-sister but homologous chromosomes by cohesins may possibly occur like an off-target ring toss, and this may result in meiosis, facilitated by an ultimately meiosis-specific cohesin Rec8 and a kinetochore-associated protein MEI-S332/Sgo1 (Marston and Amon, 2004) and with abnormal degradation of cohesins afterward (Cavalier-Smith, 2002). Recently Ross et al. (2013) reported an evolutionary analysis on how a neogene acquired an essential function for chromosome segregation in *Drosophila melanogaster*, opening up a new perspective for investigating how the molecular mechanism of fundamental events like meiosis was evolved.

Why would a mistakenly occurring, "unnecessary" cell division be selected evolutionarily? Probably because of meiotic recombination. Although DNA transmission from cell to cell already existed in prokaryotes, meiotic recombination is considered to be the first efficient mechanism evolved for autonomously increasing genetic variation. This begs the question of why genetic variation would be so important for a cell that meiosis conferred an advantage during evolution. Regardless of how the first cell arose from an RNA world or pre-cellular biosystem, afterward the cells were relatively isolated from the environment from which they emerged. Although the advent of the cell granted the biosystem tremendous independence and the ability to proliferate itself through cell division, it created a problem of adapting to the unpredictable changes in its environment. Spontaneous DNA mutation is the original way to adapt, but with low efficiency. By contrast, meiotic recombination can generate numerous genetic variations more efficiently. Among the variations randomly generated during meiosis could be those that are adaptive to the prevailing environmental conditions and enable cell survival in a changed environment. Therefore, increasing genetic variation for adaptation might be a primary reason for meiosis to be selected. The stress-induced meiosis observed in protists is consistent with this speculation.

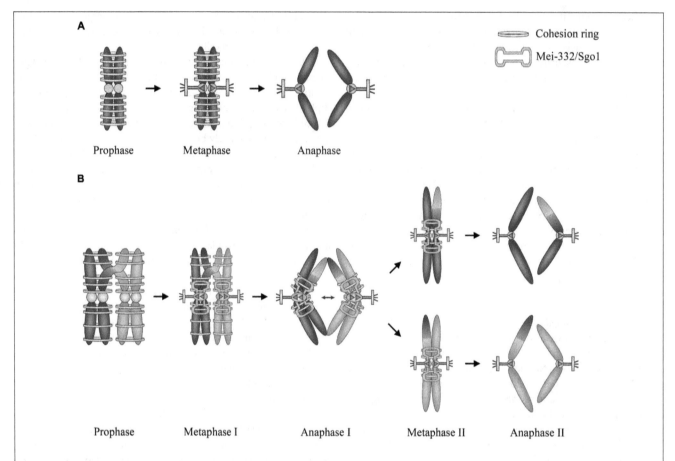

FIGURE 1 | Illustration of the role of cohesins in the origin of meiosis. (A) Cohesins hold sister chromosome together during mitosis. **(B)** Cohesins mistakenly hold non-sister but homologous chromosomes together, enhanced by kinetochore-associated protein Mei-332/Sgo1 during meiosis. Modified from Marston and Amon (2004), by permission of Nature Publishing Group.

DID FERTILIZATION EMERGE PRIOR TO MEIOSIS?

It is conventionally understood that the sexual reproduction starts from meiosis and ends at fertilization. However, as we speculated previously, it is likely that eukaryotes first emerged as haploid. If that is the case, cell fusion, rather than cell division, should be an ancient event in the emergence of diploid cells. Regardless of whether the first eukaryotic cell emerged when one prokaryotic organism engulfed another as Margulis and Sagan (1991) summarized and of how chromosome duplication originated, without diploid cells there would be no meiosis. Since fertilization is essentially a cell fusion process, we can predict that it is derived from the ancient cell fusion mechanism.

If indeed cell fusion arose prior to meiosis, once meiosis emerged in the resulting diploid cells, fertilization could be readily used as a mechanism to restore genome diploidy. Furthermore, if we think about the genetic variation randomly increased by meiosis, fertilization actually retains the variation derived from both cells that are the products of meiosis. In addition, considering the interaction between the meiotically produced cells and their environment, fertilization between the surviving haploid cells would actually execute a selective function in maintaining variations adaptive to that environment, as cells carrying non-adaptive variations would have not survived to participate in fertilization.

Therefore, the advantages of fertilization include not only restoration of genome diploidy, but more significantly, double retention of the selected genetic variations generated through meiosis.

However, each meiotic cell would generate four types of resulting cells after random recombination. If those four cells randomly paired and fused, too much variation would rapidly diversify the characteristic genome structure of the species along with the increase of round from one zygote to zygotes of next generations. This does not even take into account the genetic complexity from a population perspective, in which the meiotically produced cells could pair and fuse with those arising from other meiotic cells. Is there any way to solve that problem?

A BENEFIT OF HETEROGAMY: LABELING MEIOTICALLY PRODUCED CELLS TO HARNESS VARIATION WHILE ENHANCING HETEROGENEITY

Heterogametes in animals and plants generally display morphological differences, i.e., small sperm and large egg cells in comparison with their progenitor meiotic cells. However, as mentioned above, heterogamy in unicellular eukaryotic organisms is frequently determined at the molecular level by a single genetic locus, and can include more than two mating types. If we use the assumption that unicellular eukaryotes evolved prior to multicellular organisms, we may infer that the morphological and/or physiological differentiation of heterogametes in multicellular organisms should be considered elaborations of the very simple differences, such as at a single genetic locus, observed in gametes of unicellular eukaryotes. This inference is consistent with the recent finding that in *Volvox*, expansion of a mating locus causes heterogametes to change from being equal in size to being dimorphic (Ferris et al., 2010).

What is the advantage of heterogamy that enables the mating loci to be selected among the enormous genetic variation? If we remember the problems mentioned above regarding random pairing and fusion of meiotically produced cells in fertilization, we can speculate that heterogamy would significantly restrict diversity. With heterogamy, meiotically produced cells are classified into different groups that prevent pairing and fusion of those among the same group. In other words, heterogamy is essentially a mechanism of labeling meiotically produced cells to prevent "self-mating," therefore "harnessing," while enhancing, heterogeneity generated from meiosis and fertilization. As it allows pairing and fusion only between haploid cells from different groups, this harnessing mechanism creates a relatively stable interval during which the adaptive cells can be selected.

If the essential function of heterogamy is the labeling of meiotically produced cells and thereby harnessing variation while enhancing heterogeneity, there should be a multitude of ways to achieve this. Genetic loci for mating types probably represent the most ancient and simple way, but there could be many other modifications to enhance the differentiation for higher efficiency.

In majority of animals familiar to human experience, heterogametes are differentiated from germlines that migrate into and complete the differentiation in dimorphic gonads during embryogenesis. Heterogamy is determined mainly in gonad differentiation prior to germ cells undergoing meiosis. One may therefore believe that sex determination or differentiation is a precondition of the occurrence of meiosis. The situation appears similar in plants if we examine only angiosperms. However, if we take ferns and mosses in consideration, it is easily found that dimorphism of multicellular structures is not necessarily required for meiosis and that heterogamy in multicellular organisms can be achieved after meiosis (**Figure 2**), similar to what takes place in *Chlamydomonas* and *Volvox* (Goodenough et al., 2007; Ferris et al., 2010). If we compare the divergence points in green algae and the four groups of land plants, we see a trend in which the divergence point(s) that leads to the heterogamete differentiation shifted from gametophytes after meiosis to sporophytes before meiosis in green algae and angiosperms, respectively. Little is known regarding how this shift evolved. However, efficiency in gamete distribution and meeting might contribute to the shift: in *Chlamydomonas*, the two types of gametes differentiate in water and their pairing and fusion occurs randomly. In mosses and ferns, sperm cells are shed into water as well and swim to archegonia and eggs with water as the medium. In gymnosperms and angiosperms, in which the divergence points are shifted to sporophytes, the delivery of sperm is no longer restricted to water. This may allow these two groups of plants to increase their spatial distribution.

Similarly, if we examine mechanisms of animal sex differentiation with a broader view, there are also diversifications worth noting: although gametogenesis is carried out in the germlines, sex differentiation mainly occurs at the gonads. While mammalian gonad differentiation from bipotential to unisexual is triggered by sex-determining genes, a similar gonad differentiation is induced by environmental temperature in some reptiles (Ramsey and Crews, 2009). This implies that over the course of evolution there might be a trend in which determination of heterogametogenesis shifted from germ cells in *cis* to somatic gonads in *trans*, and further that the trigger(s) for gonad differentiation shifted from environmental signals to genetic factors encoded in chromosomes, and even further that the chromosomes bearing genetic factors determining sex evolved into sex chromosomes, as suggested by Charlesworth et al. (2005).

If the above speculation is accurate, sex differentiation indeed can be considered essentially a labeling mechanism for heterogamy, regardless of how diversified in form and complicated in regulation, in a wide spectrum of organisms from unicellular eukaryotes to multicellular animals, plants, and fungi.

SEQUENTIAL DIFFERENTIATION OR INTEGRATION OF INDEPENDENTLY EVOLVED EVENTS?

In animals, meiosis and gametogenesis occur sequentially in germlines and sex differentiation occurs in somatic gonads into which the germline migrated. In plants, meiosis and gametogenesis occur separately from somatic cells of the sporophyte and gametophyte, while sex differentiation could occur either in sporophyte or gametophyte. How were meiosis, gametogenesis, and sex differentiation originally integrated? Considering that all three processes exist in protists, one possible scenario is that each evolved independently, and they were integrated together as a coordinated process by chance and thereafter genetically fixed as a program in protists. This scenario is possible because all three processes, meiosis, heterogametogenesis (including sex differentiation and gametogenesis), and fertilization, occur at the cellular level. Protists are unicellular eukaryotes and live in a population. These two characteristics provide the required conditions for SRC emergence: on one hand each cell can behave independently for emergence of meiosis and gametogenesis, and on the other, all cells live together closely enough to make both cell fusion and cell–cell recognition possible. Integration of the three events would have brought all of their selective advantages together and such integration, now referred as SRC, would be therefore selected during evolution.

MODIFIED CELL DIVISION: ORIGIN OF GENERATIONS

In nearly all biology textbooks, meiosis is introduced in comparison with mitosis, whereas fertilization and sex determination or differentiation are introduced elsewhere. However, if we view meiosis and fertilization together, we find that one cell becomes four (except in some particular cases, such as angiosperms and mammalian, only one female meiotically produced cell remaining alive to differentiate into female gamete) through meiosis and two cells become one through fertilization. Thus, the net result of the entire SRC is that one cell becomes two, just like one round of mitosis. The fundamental difference between the SRC and the mitotic cell cycle is that the genetic compositions of the two cells

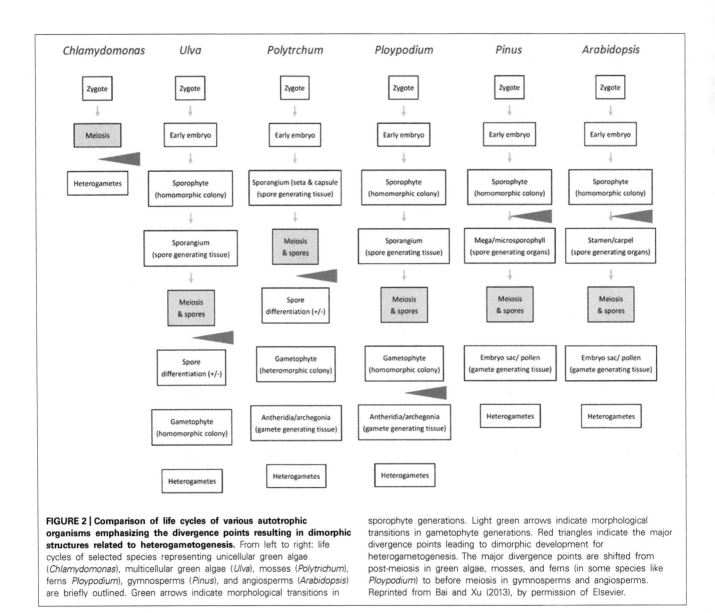

FIGURE 2 | Comparison of life cycles of various autotrophic organisms emphasizing the divergence points resulting in dimorphic structures related to heterogametogenesis. From left to right: life cycles of selected species representing unicellular green algae (*Chlamydomonas*), multicellular green algae (*Ulva*), mosses (*Polytrichum*), ferns *Ploypodium*), gymnosperms (*Pinus*), and angiosperms (*Arabidopsis*) are briefly outlined. Green arrows indicate morphological transitions in sporophyte generations. Light green arrows indicate morphological transitions in gametophyte generations. Red triangles indicate the major divergence points leading to dimorphic development for heterogametogenesis. The major divergence points are shifted from post-meiosis in green algae, mosses, and ferns (in some species like *Ploypodium*) to before meiosis in gymnosperms and angiosperms. Reprinted from Bai and Xu (2013), by permission of Elsevier.

resulting from the SRC are different from that of the progenitor while the products of the mitotic cell cycle remain similar to that of their progenitor (**Figure 3**). It needs to be emphasized that what can appropriately be compared with the mitotic cell cycle is not meiosis alone, but rather the entire process of SRC, including meiosis, heterogametogenesis, and fertilization.

Cell division occurred well before eukaryotes evolved. Despite the difference in complexity between mitotic cell division in eukaryotes and cell fission in prokaryotes, the two processes are similar in that the two resulting cells retain the same genome structure as the starting cell. By contrast, the genome structures of the two cells resulting from the SRC are no longer the same as that of the original cell, as described above. They are genetically a new "generation." From this perspective, although the terms "mother cells" and "daughter cells" are often used in describing the beginning and resulting cells in mitosis, these cells do not truly represent two generations. They are actually clones, the same as in the cell division observed in *Escherichia coli* leading to a proliferation of

the same generation. Only through the SRC is a new generation created.

According to Chen et al. (2013), genetic variations are generated in several ways, such as mutation and new gene origination, in addition to meiotic recombination. However, only variations retained through the SRC can be maintained from one generation to the next, rather than being diluted and ultimately disappearing through continuous cell divisions. In that sense, mainly because it was integrated as part of the SRC, meiotic recombination took a prominent position among the various ways of creating genetic variations.

THE SRC: AN APPENDIX OF A MULTICELLULAR ORGANISM OR A FRAMEWORK ONTO WHICH MULTICELLULAR STRUCTURES ARE INTERPOLATED?

When animal development is discussed, an organism and embryogenesis takes center stage. Germline initiation is an appendant event during embryogenesis, while meiosis and gametogenesis are

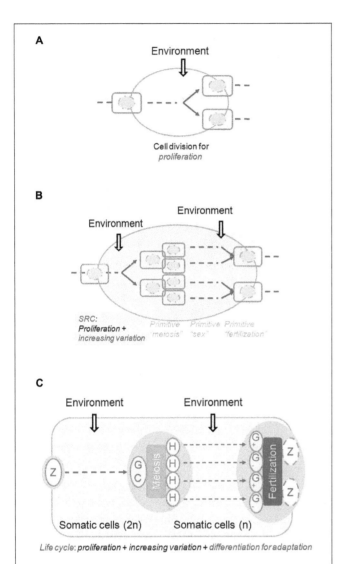

FIGURE 3 | Diagram of the sexual reproduction cycle (SRC). (A) A regular cell cycle for proliferation, through which one cell becomes two, and environmental conditions trigger or affect the cycle at various points in the process. **(B)** A "SRC." Three hypothetically random and independent events, "primitive meiosis," "primitive fertilization," and "primitive sex," occasionally integrated and were selected for advantages in adaptation. The net result of the SRC is that one cell becomes two, just as in the regular cell cycle, regardless of how these events evolved and were integrated (of which little is known). **(C)** Core processes of the life cycles of multicellular organisms. From the perspective of the SRC, it is clear that all multicellular structures arose in the interval phases of the SRC through the regular cell cycle and cellular differentiation, whether diploid (in almost all organisms) or haploid (mainly in plants and fungi). Modified from Bai and Xu (2013), by permission of Elsevier.

only two events of germline differentiation. A similar situation occurs in our understanding of plant development. The focus is mainly on the morphogenesis of multicellular structures. However, if we take the standpoint that the SRC emerged in protists, an obvious inference is that multicellular structures all emerged within the framework of the SRC. Is this possible?

If we accept the argument that cell growth and division are coupled with optimal cell volume (Buchanan et al., 2000), and

the inference that meiosis is a stress-induced specific cell division for adapting to unpredicted environmental challenges, we can imagine that under optimal environmental conditions, a cell would keep dividing in order to maintain optimal cell volume. That means that between a zygote and a meiotic cell, and between meiotically produced cells and gametes, there would be two intervals during the process completing the SRC (**Figure 3C**). In unicellular protists, cells in these two intervals would be mainly freely living in a population. However, under certain conditions, for example nutritional shortages, the free-living cells might be aggregated or organized, such as in *Dictyostelium* (Dormann et al., 2002) and *Volvox* (Kirk, 2005). What would result if the conditional aggregations or organizations were genetically fixed? Multicellular organisms! If this speculation is valid, what is the relationship between the core cells involved in the SRC, such as the zygote, meiotic cells and gametes, and those involved in multicellular structure morphogenesis? A reasonable answer is that the SRC serves as a framework or backbone, within or on which multicellular structures consisting of "somatic cells" could interpolated into the two intervals when the environment was optimal. Different from simply maintaining cell volume through cell division, organized multicellular structures ultimately facilitate energy acquisition and environmental adaptation. All elaborated sex differentiation and mating behaviors can be viewed as nothing more than modifications of multicellular structures evolved afterward to facilitate meeting, recognition, and proper fusion of the two gametes.

From this perspective, we can also see a relatively novel scenario in which to compare the core developmental processes of animals, plants, and fungi using the SRC as a common reference framework (**Figure 4**). From this comparison, we observe three different strategies of morphogenesis. In animals, the multicellular structures (soma) are interpolated at the first interval between zygote and meiotic cells as rest of embryos in addition to germlines. In fungi, the multicellular structures are interpolated at the second interval between meiotically produced cells and gametogenic cells, with unknown mechanisms of cell aggregation (Meskauskas et al., 2004; Lee et al., 2010). In plants, multicellular structures are interpolated at both intervals, with an additive strategy (Bai and Xu, 2013).

CONCLUSION: THE SRC AS A KEY EVOLUTIONARY INNOVATION

Taking the above analysis together, several speculations can be made: first, meiosis is the most effective mechanism evolved to autonomously increasing genetic variation; second, the SRC is the first known genetic program of cell differentiation and coordination, integrated with the three independently evolved events, i.e., meiosis, sex differentiation, and fertilization, among cell populations; third, with the SRC, eukaryotes were first equipped with an internal mechanism to adapt effectively to unpredictable environmental challenges. From this perspective, the SRC deserves to be recognized as a key evolutionary innovation. This innovation first arose in unicellular eukaryotes, and then, because of its advantages in adaptation to environmental challenges, was adopted by subsequent multicellular organisms as a conserved program. Only based on the SRC can generations be properly defined, thereby

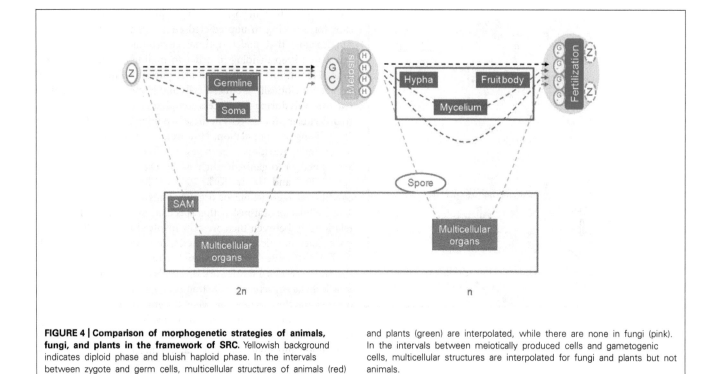

FIGURE 4 | Comparison of morphogenetic strategies of animals, fungi, and plants in the framework of SRC. Yellowish background indicates diploid phase and bluish haploid phase. In the intervals between zygote and germ cells, multicellular structures of animals (red) and plants (green) are interpolated, while there are none in fungi (pink). In the intervals between meiotically produced cells and gametogenic cells, multicellular structures are interpolated for fungi and plants but not animals.

allowing genetic analysis of biological processes. In addition, SRC establishes a reasonable framework in which to understand how multicellular organisms evolved in various kingdoms, including animals, plants and fungi.

I used to be highly curious about why the developmental processes of multicellular organisms are unidirectional, i.e., from zygotes to gametes. After I became aware of the existence of the SRC, this puzzle seemed easy enough to be solved because the SRC is a program that originally emerged and was selected as a response to environmental stresses. As environmental stresses occur ultimately independent of organisms and irreversible, the SRC process is irreversible and therefore unidirectional. Although the three core events in the SRC of many multicellular organisms have been genetically encoded during evolution and may not directly respond to environmental stresses, the unidirectionality was inherited from their unicellular eukaryotic ancestors.

Another issue confusing for a long time is how to understand the role of "asexual reproduction" in biological processes, such as speciation (Coney and Orr, 2004). With the concept of the SRC, it is clear that only the organism multiplication that is coupled with the SRC should be regarded as "reproduction," as new generations are produced. Multiplication of organisms not coupled with SRC should not be regarded as "reproduction," as no new generations are produced. Instead, the products of the latter are properly referred to as clones, equivalent to mitotic cell division, regardless of how complicated and similar the process is in comparison to the process starting from a zygote. In that sense, the past comparisons of "sexual reproduction" vs. "asexual reproduction" should more correctly be of "reproduction" vs. "proliferation." The former refers to a multiplication of organisms based on the SRC, from one generation to the next in a lineage, while the latter refers to a multiplication of organisms remaining within the same generation.

One more issue worth noting here is that the SRC is essentially a mechanism allowing diploid cells to adapt to unpredictable environmental challenges. If an organism adapts to its environment in the haploid form, there would be no selective pressure for it to adopt SRC. This can explain why there are numerous examples of meiosis and fertilization occurring only rarely in many haploid-dominant protists (Logsdon, 2007), and no stable SRC identified in many haploid-dominant fungi (Heitman et al., 2013).

In his book "Primitive Land Plants" first published in 1935, Bower (1935) had advised that in understanding plant morphology, we should adopt "upward comparative analysis," i.e., use forms of organisms that emerged early as a reference framework, and treat the new traits in organisms that emerged later as modifications upon the earlier forms. Similar principles can also be applied to understanding sex and sexual reproduction.

Lastly, I would like to emphasize that without the investigation of unisexual flowers, which brought about the "bird-nest puzzle," and subsequent historical retrospection of the definition of sex in plants, I would not have this chance to think about the essential role of sex in living systems. In addition, without analysis of highly diversified sex differentiation, i.e., divergence points leading to heterogametes in various plant groups, it would be impossible to disconnect the sex or sex differentiation from meiosis, as in all animal systems, sex differentiation always occurs prior to meiosis, and therefore meiosis appears undeniably linked to sex differentiation. In many aspects, our views of biological processes in plants have been borrowed from those developed in animal research, into which vastly more resources have been invested since we, as humans, belong to the animal kingdom. Sometimes, however, plants provide an invaluable alternative due to their different morphogenetic strategies. Ultimately, how we look at biological events depends on what questions we address. If we want to understand fundamental principles of living systems, we need to properly compare animals and plants at least, as each can be used as a control for the other. Alternatively, if one only wants to know the mechanism of animal disease or crop yield for instance, that comparison might be not necessary. For questions about sex *per se*, proper comparisons among animals and plants, plus fungi, and more importantly protists, seem indispensable!

ACKNOWLEDGMENTS

I cordially thank Prof. Manyuan Long of University of Chicago. After he kindly helped me in editing my comprehensive review on unisexual flowers, he advised me that the review was too long and few people would have time to read it through and get the main idea. It is at his suggestion that I write this short version focusing on my opinions about sex and sex differentiation. I also cordially thank Prof. Dave Zhao of University of Wisconsin at Milwaukee, who kindly invited me to contribute a review or opinion article to the Frontier as he is a co-editor for a special issue on the topic "Molecular and Cellular Plant Reproduction." Without their encouragement and invitation, there would be no such article. I also thank Prof. Qi-Chang Fan and Ming-Xiao Ding of Peking University, Prof. Jody Banks of Purdue University for their critical reading and valuable comments and suggestions, and Prof. Qun He of China Agriculture University and Prof. Feng-Yan Bai of the Institute of Microbiology, CAS for their helpful discussion on sex differentiation in fungi. Finally, I have to thank Dr. Nancy Hofmann for her editing of the manuscript and very valuable comments and suggestions.

REFERENCES

Ainsworth, C. C. (1999). *Sex Determination in Plants*. Oxford: BIOS Scientific Publishers.

Akagi, T., Henry, I. M., Tao, R., and Comai, L. (2014). A Y-chromosome-encoded small RNA acts as a sex determinant in persimmons. *Science* 346, 646–650. doi: 10.1126/science.1257225

Bai, S. N., and Xu, Z. H. (2012). Bird-nest puzzle: can the study of unisexual flowers such as cucumber solve the problem of plant sex determination? *Protoplasma* 249(Suppl. 2), S119–S123. doi: 10.1007/s00709-012-0396-4

Bai, S. N., and Xu, Z. H. (2013). "Unisexual cucumber flowers, sex and sex differentiation," in *International Review of Cell and Molecular Biology*, Vol. 304, ed. K. Jeon (London: Academic Press), 1–56.

Bloomfield, G., Skelton, J., Ivens, A., Tanaka, Y., and Kay, R. R. (2010). Sex determination in the social amoeba *Dictyostelium discoideum. Science* 330, 1533–1536. doi: 10.1126/science.1197423

Bower, F. O. (1935). *Primitive Land Plants: Also Known as the Archegoniatae*. New York: Macmillan.

Buchanan, B. B., Gruissem, W., and Jones, R. L. (2000). *Biochemistry and Molecular Biology of Plants*. Rockville, MA: American Society of Plant Physiology.

Campbell, N. A., and Reece, J. B. (2005). *Biology*, 7th Edn. San Fransisco: Benjamin Cunnings.

Cavalier-Smith, T. (2002). Origins of the machinery of recombination and sex. *Heredity (Edinb.)* 88, 125–141. doi: 10.1038/sj.hdy.6800034

Cervantes, M. D., Hamilton, E. P., Xiong, J., Lawson, M. J., Yuan, D. X., Hadjithomas, M., et al. (2013). Selecting one of several mating types through gene segment joining and deletion in *Tetrahymena thermophila. PLoS Biol.* 11:e1001518. doi: 10.1371/journal.pbio.1001518

Charlesworth, D., Charlesworth, B., and Marais, G. (2005). Steps in the evolution of heteromorphic sex chromosomes. *Heredity (Edinb.)* 95, 118–128. doi: 10.1038/sj.hdy.6800697

Chen, S. D., Krinsky, B. H., and Long, M. Y. (2013). New genes as drivers of phenotypic evolution. *Nat. Rev. Genet.* 14, 645–660. doi: 10.1038/nrg3521

Coney, J. A., and Orr, H. A. (2004). *Speciation*. Sunderland, MA: Sinauer Associates, Inc.

Dormann, D., Vasiev, B., and Weijer, C. J. (2002). Becoming multicellular by aggregation; the morphogenesis of the social amoebae *Dictyostelium discoideum. J. Biol. Phys.* 28, 765–780. doi: 10.1023/A:1021259326918

Evens, M. M., and Walbot, V. (2003). Unique feature of the plant life cycle and their consequences. *Nat. Rev. Genet.* 4, 369–379. doi: 10.1038/nrg1064

Ferris, P., Olson, B. J. S. C., De Hoff, P. L., Douglass, S., Casero, D., Prochnik, S., et al. (2010). Evolution of an expanded sex-determing locus in *Volvox. Science* 328, 351–353. doi: 10.1126/science.1186222

Goodenough, U., Lin, H., and Lee, J. H. (2007). Sex determination in *Chlamydomonas. Semin. Cell Dev. Biol.* 18, 350–361. doi: 10.1016/j.semcdb.2007.02.006

Heitman, J., Sun, S., and James, T. Y. (2013). Evolution of fungal sexual reproduction. *Mycologia* 105, 1–27. doi: 10.3852/12-253

Juraze, C., and Banks, J. A. (1998). Sex determination in plants. *Curr. Opin. Plant Biol.* 1, 68–72. doi: 10.1016/S1369-5266(98)80130-1

Kirk, D. L. (2005). A twelve-step program for evolving multicellularity and a division of labor. *Bioessay* 27, 290–310. doi: 10.1002/bies.20197

Kirschner, M. W., and Gerhart, J. C. (2005). *The Plausibility of Life*. New Haven: Yale University Press.

Knoll, A. H. (2014). Paleobiological perspectives on early eukaryotic evolution. *Cold Spring Harb. Perspect. Biol.* 6, a016121. doi: 10.1101/cshperspect.a016121

Lee, S. C., Ni, M., Li, W. J., Shertz, C., and Heitman, J. (2010). The evolution of sex: a perspective from the fungal kingdom. *Microbiol. Mol. Biol. Rev.* 74, 298–340. doi: 10.1128/MMBR.00005-10

Logsdon, J. M. Jr. (2007). Evolutionary genetics: sex happens in *Giardia. Curr. Biol.* 18, R66–R68. doi: 10.1016/j.cub.2007.11.019

Margulis, L., and Sagan, D. (1991). *Mystery Dance: On the Evolution of Human Sexuality*. New York: Simon & Schuster.

Marston, A. L., and Amon, A. (2004). Meiosis: cell cycle controls shuffle and deal. *Nat. Rev. Mol. Cell Biol.* 5, 983–997. doi: 10.1038/nrm1526

Meskauskas, A., McNulty, L. J., and Moore, D. (2004). Concerted regulation of all hyphal tips generates fungal fruit body structures: experiments with computer visualizations produced by a new mathematical model of hyphal growth. *Mycol. Res.* 108, 341–353. doi: 10.1017/S0953756204009670

Nanney, D. L., and Caughey, P. A. (1953). Mating types determination in *Tetrahymena pyriformis. Genetics* 39, 1057–1063.

Ramsey, M., and Crews, D. (2009). Steroid signaling and temperature-dependent sex determination – Reviewing the evidence for early action of estrogen during ovarian determination in turtles. *Semin. Cell Dev. Biol.* 20, 283–292. doi: 10.1016/j.semcdb.2008.10.004

Robbins, W. W., and Pearson, H. M. (1933). *Sex in Plant World*. New York: D. Appleton-Century Company.

Ross, B. D., Rosin, L., Thomae, A. W., Hiatt, M. A., Vermaak, D., de la Cruz, A. F. A., et al. (2013). Stepwise evolution of essential centromere function in a *Drosophila neogene. Science* 340, 1211–1214. doi: 10.1126/science.1234393

Schurko, A. M., and Logsdon, J. M. Jr. (2008). Using a meiosis detection toolkit to investigate ancient asexual "scandals" and the evolution of sex. *Bioessay* 30, 579–589. doi: 10.1002/bies.20764

6

Regulation of floral stem cell termination in *Arabidopsis*

Bo Sun[1] and Toshiro Ito[1,2]*

[1] Temasek Life Sciences Laboratory, 1 Research Link, National University of Singapore, Singapore
[2] Department of Biological Sciences, National University of Singapore, Singapore

Edited by:
Dazhong Dave Zhao, University of
Wisconsin-Milwaukee, USA

Reviewed by:
Robert G. Franks, North Carolina
State University, USA
Xiaoyu Zhang, UGA, USA

*Correspondence:
Toshiro Ito, Temasek Life Sciences
Laboratory, 1 Research Link,
National University of Singapore,
Singapore 117604, Republic of
Singapore;
Department of Biological Sciences,
National University of Singapore,
Singapore 117543, Republic of
Singapore
e-mail: itot@tll.org.sg

In *Arabidopsis*, floral stem cells are maintained only at the initial stages of flower development, and they are terminated at a specific time to ensure proper development of the reproductive organs. Floral stem cell termination is a dynamic and multi-step process involving many transcription factors, chromatin remodeling factors and signaling pathways. In this review, we discuss the mechanisms involved in floral stem cell maintenance and termination, highlighting the interplay between transcriptional regulation and epigenetic machinery in the control of specific floral developmental genes. In addition, we discuss additional factors involved in floral stem cell regulation, with the goal of untangling the complexity of the floral stem cell regulatory network.

Keywords: ***Arabidopsis*, floral meristem, stem cell, determinacy, flower development**

INTRODUCTION

The flower is an elegant structure produced by angiosperms for effective reproduction. In *Arabidopsis*, floral organs are built in four whorls of concentric circles. From outermost to innermost, they consist of four sepals, four petals, six stamens and two fused carpels. The molecular mechanism specifying the identity of each whorl of floral organs is explained by the genetic ABCE model (Krizek and Fletcher, 2005). All four whorls of floral organs are derived from a self-sustaining stem cell pool named the floral meristem (FM), which arises from the peripheral regions of the shoot apical meristem (SAM). Much like the stem cells in the SAM, the stem cells in the FM are maintained by a signaling pathway involving the homeodomain protein WUSCHEL (WUS) and the CLAVATA (CLV) ligand-receptor system (Fletcher et al., 1999; Brand et al., 2000; Schoof et al., 2000). *WUS* is expressed in the organizing center, and it specifies and maintains the stem cell identity of the overlying cells. Expansion of *WUS* expression is prevented by the CLV signaling pathway, in which the CLV3 peptide is transcriptionally induced by WUS in the stem cells (Yadav et al., 2011; Daum et al., 2014). Due to the negative feedback regulatory loop of *CLV3* and *WUS*, the stem cell pool remains constant in the initial floral developmental stages (stage 1~2) (Smyth et al., 1990; Schoof et al., 2000).

In the stage 3 floral bud, the C class gene *AGAMOUS (AG)* is induced by LEAFY (LFY) together with WUS in whorls 3 and 4 (Lenhard et al., 2001; Lohmann et al., 2001). AG has two major roles. It specifies reproductive organs, and it also regulates floral stem cell activity (Lenhard et al., 2001; Lohmann et al., 2001). In stage 6, floral stem cells are terminated in an AG-dependent manner to ensure proper development of the carpels. With respect to floral stem cell regulation, the major two pathways, the *AG-WUS* pathway and the *CLV-WUS* pathway, seem to function independently. The double mutant *ag clv1* shows an additive phenotype of *ag* and *clv1*, and it expresses *WUS* in a broader domain than the *ag* mutant flower (Lohmann et al., 2001). In fact, the *CLV-WUS* pathway regulates floral stem cells spatially to restrict and maintain the stem cell pool in the early floral stages (stage 1–6), whereas the *AG-WUS* pathway provides temporal regulation to shut off stem cell activity at floral stage 6 (**Figure 1**). The precise timing of *WUS* repression is a key factor that determines the number of cells produced for reproductive organ development.

DIRECT AND INDIRECT ROLES OF AG IN *WUS* REPRESSION

AG is reported to directly bind to the *WUS* locus to repress *WUS* expression (Liu et al., 2011). Based on an ethyl methane-sulfonate mutagenesis screening of enhancer mutants of a weak allele, *ag-10*, which has only a moderate effect on floral meristem determinacy, one *CURLY LEAF (CLF)* mutant allele, *clf-47*, was identified (Liu et al., 2011). This suggests that *CLF* is required for floral meristem determinacy. CLF is a core component of poly-comb repressive complex 2 (PRC2), which suggests that *WUS* repression is associated with the deposition of the repressive mark H3 lysine 27 tri-methylation (H3K27me3), a mark that is mediated by the polycomb group proteins (PcG). Consistent with this, one mutant allele of *TERMINAL FLOWER 2 (TFL2)*, a PRC1 factor in *Arabidopsis*, can enhance the *ag-10* indeterminate phenotype (Liu et al., 2011). The *ag-10 tfl2-2* double mutant flowers show enlarged carpels bearing ectopic internal organs, as observed in *ag-10 clf-47*. These results indicate that *WUS* is a target of PcG during flower development. AG binds to the two CArG boxes in the *WUS* 3′ non-coding region, and TFL2 occupancy at *WUS* is largely compromised in the *ag-1* null mutant background. These results suggest that AG has a role in the recruitment of PcG

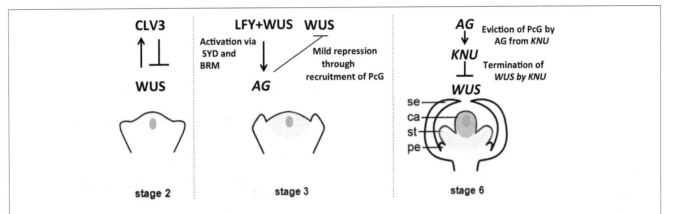

FIGURE 1 | Regulation of the timing of floral stem cell termination. Signaling cascades, transcriptional regulation and epigenetic regulation of the key proteins involved in floral meristem regulation are illustrated in a stage-specific manner. Orange indicates the domain of *WUS* expression, and pink, blue and green indicate the expression domains of *CLV3*, *AG*, and *KNU*, respectively.

to repress *WUS*. However, whether AG recruits PcG directly is still an open question.

35S::AG transgenic plants do not show any obvious floral meristem defects (Mizukami and Ma, 1997), and *WUS* is only mildly repressed after stage 3 directly by AG. For the termination of *WUS* at floral stage 6, a C2H2 zinc finger repressor protein, KNUCKLES (KNU), plays a pivotal role (Payne et al., 2004; Sun et al., 2009). *KNU* expression starts in stage 5–6, and mutation of *KNU* leads to enlarged carpels and repeated ectopic growth of stamens and carpels. This indeterminate floral phenotype is caused by the prolonged activity of *WUS*, showing that KNU is necessary for floral stem cell termination. *KNU* is directly induced by AG, and mutations in three CArG box sequences on the *KNU* promoter can abolish *KNU* induction (Sun et al., 2009). Timed induction of *KNU* by AG in stage 6 of flower development ensures floral meristem termination and proper development of the female reproductive organs. The timing of *KNU* expression is important for balancing floral stem cell proliferation and differentiation. Delayed *KNU* expression leads to indeterminate flowers with more stamens, and ectopic *KNU* activity can terminate floral meristem precociously and produce flowers without carpels. *KNU* is also regulated by PcG-mediated H3K27me3, and the removal of the repressive marks of H3K27me3 is AG-dependent. It takes approximately 2 days for AG to induce *KNU* in stage 6. During these 2 days, the H3K27me3 level on the *KNU* locus is progressively reduced, revealing a potential link between the transcriptional activation of *KNU* by AG and AG-dependent removal of H3K27me3 from the *KNU* chromatin (Sun et al., 2009).

EPIGENETIC REGULATION OF TERMINATION TIMING IN FLORAL STEM CELLS

In floral meristems, cell division take 1–2 days on average (Reddy et al., 2004). Therefore, the 2-day of delay in *KNU* induction corresponds to 1–2 rounds of cell division. Through cell division, the pre-existing H3K27me3 on the *KNU* locus may be passively diluted by incorporation of unmodified histone H3, enabling *KNU* expression (Sun et al., 2014). The core components of PcG, FIE and EMF2 are associated with specific promoter regions of

KNU, which include the binding sites of AG. Indeed, this region contains a 153 bp fragment that is the minimal sequence of a functional polycomb response element (PRE). This sequence is both necessary and sufficient for PcG-mediated silencing of a ubiquitous promoter. This raises the possibility that AG plays a role in removing PcG to activate *KNU*. By simulating AG's physical blocking of the site with an artificially-designed TAL protein (a effector-based synthetic DNA binding protein designed to recognize the sequences around the first AG binding site), we showed that a YFP reporter could be activated in a cell cycle-dependent manner, even though it had been silenced by the minimal PRE sequence.

PRE was first identified in the fruit fly *Drosophila*, and it is targeted by the Pho-repressive complex (PhoRC) (Muller and Kassis, 2006). In *Arabidopsis*, homologs of PhoRC have not been identified, but in a genome-wide analysis of FIE binding sites, GA-repeat motifs appeared frequently, much like the *Drosophila* PRE (Deng et al., 2013). The *KNU* PRE is located near the 1kb upstream promoter region of the *KNU* transcriptional start site (Sun et al., 2014). Although the entire *KNU* locus is found to be bound by FIE and EMF2, only the transcribed region is covered by the repressive mark H3K27me3, and the PRE is not covered by the repressive mark. The indispensable role of the *KNU* PRE in recruiting PRC2 and establishing the FIE and EMF2 binding pattern on *KNU* indicates that PcG is first recruited to the *KNU* PRE and may later act on the *KNU* transcribed region to establish the H3K27me3 marks by sliding or by DNA looping. When the AG protein binds to the CArG box sequences that overlap the *KNU* PRE, the occupancy of AG triggers the displacement of PRC2, which leads to the loss of the H3K27me3 marks on *KNU*. Through cell division, H3K27me3 is diluted due to the lack of PcG activity, and *KNU* become de-repressed. Delayed reporter induction has been reported following artificial removal of a PRE by a cre-lox system in *Drosophila*, supporting this model of *KNU* de-repression (Beuchle et al., 2001; Muller et al., 2002).

Alternatively, the H3K27me3 mark can be erased by the JmjC-domain-containing histone demethylases REF6, EFL6, JMJ30 and JMJ32 (Lu et al., 2011; Crevillen et al., 2014; Gan et al., 2014). It has been reported that AG, REF6 and some other MADS-domain proteins may form a large protein complex whose function has not been characterized (Smaczniak et al., 2012). Therefore, it is also possible that, in parallel with H3K27me3 passive dilution, AG may recruit REF6 to the *KNU* promoter to actively remove H3K27me3. However, this hypothesis does not explain why cell cycle progression is required for AG to induce *KNU*. Also, the known mutants for these demethylases show no meristematic defects. Hence, we propose that REF6 might be involved in the regulation of some other direct downstream targets that are induced by AG.

To remove H3K27me3 marks and activate gene expression, other transcription factors or chromatin remodeling factors may perform functions similar to those that AG does. One such example is LFY in the control of the *AG* locus. For *AG* expression, the repressive mark H3K27me3 is removed by LEAFY (LFY), which recruits the SWI/SNF chromatin remodeling factors SPLAYED (SYD) and BRAHMA (BRM) on the *AG* second intron (Wu et al., 2012). Notably, GA-repeat motifs located near PREs are enriched at LFY targets (Wu et al., 2012; Zhang, 2014).

WUS, which is required in the organizing center to stimulate the maintenance of stem cell properties in the overlying cells (Yadav et al., 2011; Daum et al., 2014), is negatively regulated by PcG-mediated H3K27me3 (Zhang et al., 2007). In the SAM, the *WUS-CLV* signaling pathway works to maintain an appropriately sized stem cell. The signaling pathway remains active in floral stem cells and works to maintain their identity. It is interesting that *WUS* is re-activated and the signaling pathway is re-established in the stage 1 floral primordia (Mayer et al., 1998) and that the SWI/SNF chromatin remodeling factor SYD plays an important role in *WUS* activation (Wagner and Meyerowitz, 2002; Kwon et al., 2005). In floral stage 6, *WUS* is terminated by KNU and later silenced by PcG-mediated H3K27me3 marks (Sun et al., 2009; Liu et al., 2011). Because transcriptional repression of *WUS* and epigenetic silencing of *WUS* both occur at floral stage 6, we suggest that the transcriptional repressor KNU may integrate

the two processes. During reproductive development, *WUS* is activated in developing stamens at stages 7-8, and later it is activated in developing ovules (Gross-Hardt et al., 2002; Deyhle et al., 2007). How the repressive mark H3K27me3 is removed from the *WUS* locus in those specific tissues and cell types is another open question that will require further investigation.

OTHER FACTORS INVOLVED IN FLORAL MERISTEM REGULATION

In addition to the known *CLV-WUS* signaling pathway that is responsible for the spatial maintenance of the floral stem cell niche, and in addition to the *AG-KNU-WUS* pathway for the timed termination of floral stem cells, other factors are known to be required for fine-tuning floral stem cell activities (**Figure 2**).

ULTRAPETALA1 (ULT1), a SAND domain containing protein (Carles and Fletcher, 2009), functions to induce *AG* in floral stem cells in an LFY-independent manner (Engelhorn et al., 2014). ULT1 may negatively regulate floral stem cell proliferation. The *ult1* mutant flowers have bigger floral meristems and prolonged *WUS* activity, resulting in five petals instead of the usual four (Fletcher, 2001). Thus, both genetic and molecular studies indicate that ULT1 negatively regulates the *WUS* expressing domain in floral buds, potentially through the *AG-WUS* regulatory pathway (Carles et al., 2004). ULT1 is reported be a trithorax group (trxG) protein that can physically interact with another trxG protein, ATX1, a H3K4me3 methyltransferase (Alvarez-Venegas et al., 2003; Carles and Fletcher, 2009). By binding directly to *AG* regulatory sequences, ULT1 may recruit ATX1 to actively modulate the methylation status of nucleosomes at the *AG* locus.

Two other factors, REBELOTE (RBL) and SQUINT (SQN), can redundantly regulate floral stem cells in addition to ULT1 (Prunet et al., 2008). Reiterative reproductive floral organs are observed in flowers of the double mutants *rbl sqn*, *rbl ult1*, and *sqn ult1*. In the double mutant flowers, *WUS* activity is prolonged. Presumably, *RBL* and *SQN* both regulate the floral meristem by

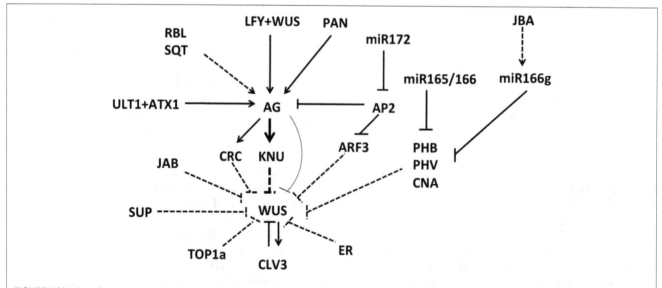

FIGURE 2 | Various factors involved in floral meristem control. Many regulators control the expression of the stem cell identity gene *WUS* in direct and indirect ways. Solid lines indicate direct activation or repression, and dashed lines indicate a proposed type of regulation that has yet to be confirmed.

reinforcing *AG* expression. As a cyclophilin protein, SQN was recently found to bind the protein chaperone Hsp90 and promote microRNA activity via AGO1 (Earley and Poethig, 2011). The *sqn* single mutant displays increased carpel number relative to wild-type, and it has abnormal phyllotaxy of the flowers. This phenotype increased expression of *SPL* family transcription factors, which are targeted by the microRNA miR156 (Smith et al., 2009). *PERIANTHIA* (*PAN*), a bZIP transcription factor, also affects floral stem cell activity through direct activation of *AG* (Running and Meyerowitz, 1996; Chuang et al., 1999; Das et al., 2009; Maier et al., 2009). In *pan* mutant flowers, *AG* mRNA levels are reduced in short-day conditions, resulting in flowers with an increased number of floral organs. In addition, increased floral meristem indeterminacy is observed in *lfy pan* and *seuss* (*seu*) *pan* double mutant flowers. Ectopic floral organs continue to grow inside the fourth whorl floral organs of *lfy pan* and *seu pan* plants, suggesting a potential effect of the floral identity gene *LFY* and the adaptor-like transcriptional repressor *SEU* in floral meristem regulation (Das et al., 2009; Wynn et al., 2014).

SUPERMAN (*SUP*), which encodes a C2H2 zinc finger protein with a C-terminal EAR-like repression motif, is thought to function as a transcriptional repressor during flower development (Hiratsu et al., 2002). Loss-of-function mutants of *SUP* produce supernumerary stamens at the expense of carpels, indicating that *SUP* has a role in maintaining the boundary between the 3rd and 4th whorl floral organs (Sakai et al., 1995). Compared to the *ag-1* mutant flowers, flowers of the double mutant *ag-1 sup* produce greatly enlarged floral meristems, generating reiterating whorls of petals, indicating the role of *SUP* in floral stem cell regulation in parallel with *AG* (Bowman et al., 1992).

CRABS CLAW (*CRC*), which is a direct downstream target of AG, is reported to be involved in floral meristem control. Null mutants for *crc-1* do not show floral meristem defects; instead, the apical part of the mutant carpel is unfused. However, in combination with certain other mutants, supernumerary whorls of floral organs are observed; this occurs in *crc-1 spatula-2*, *crc-1 ag-1/+*, *crc-1 rbl-1*, *crc-1 sqn-4*, *crc-1 ult1-4*, *crc-1 pan-3* and *crc-1 jaiba* double mutant flowers (Prunet et al., 2008; Zuniga-Mayo et al., 2012). *CRC* encodes a YABBY family transcription factor, and its expression begins in floral stage 5-6 on the abaxial side of the carpel primordia. CRC may regulate *WUS* activity in a non-cell autonomous manner (Bowman and Smyth, 1999; Lee et al., 2005).

Various microRNAs are reported to be involved in floral meristem determinacy control. For instance, *miR172* promotes termination of floral stem cells by reducing the expression of its target, *AP2* (Chen, 2004). Over-expression of a *miR172*-resistant version of *AP2* (*35S::AP2m1/3*) leads to indeterminate stamens and petals (Chen, 2004; Zhao et al., 2007). The class III HD-ZIP genes, including *PHABULOSA* (*PHB*) and *PHAVOLUTA* (*PHV*), are targeted by *miR165/166*. Over-expression of *miR165/166* in an *ag-10* background, a weak allele of *ag*, or alleles of *PHB* and *PHV* that are resistant to *miR165/166* can lead to indeterminate growth of floral organs (Ji et al., 2011). A proper balance of *PHB/PHV* and *mir165/166* is important for floral meristem determinacy control. Consistent with this, in the triple mutant of *phb phv cna*, floral carpel number is increased (Prigge et al., 2005). Similarly,

enlarged shoot meristems caused by increased *WUS* expression are observed in the *jabba1-D* mutant, a dominant allele of *JABBA* (*JBA*) that produces an increased amount of *miR166g* to regulate *PHB*, *PHV* and *CORONA* (*CNA*) expression (Williams et al., 2005).

The ERECTA (ER) receptor kinase-mediated regulation of *WUS* expression was recently reported to be mediated by a pathway parallel to the *WUS-CLV* pathway in both SAM and FM (Mandel et al., 2014). As a secondary signaling factor, ER works together with the nuclear protein JBA to repress *WUS*. In a *jba-1D/+ er-20* double mutant background, the SAM and floral meristem are greatly enlarged, and the spiral vegetative phyllotaxy switches to whorled patterns. In the *jba-1D/+ er-20* background, *AG* is ectopically expressed at a level that produces ectopic fused carpels from the inflorescence meristem, indicating an indirect role of ER in floral meristem identity control.

Recently, a mutation in the DNA topoisomerase gene *TOPOISOMERASE1a* (*TOP1a*) was shown to increase floral meristem indeterminacy in an *ag-10* background, as the *ag-10 top1a-2* double mutant exhibits an indeterminate floral meristem (Liu et al., 2014a). In floral stem cell regulation, TOP1a may function to reduce nucleosome density, thus facilitating PcG-mediated H3K27me3 deposition on *WUS*. Mutations in another gene *AUXIN RESPONSE FACTOR 3* (*ARF3*), have also been reported to enhance the *ag-10* indeterminate phenotype (Liu et al., 2014b). Double mutant *ag-10 arf3-29* flowers produce additional floral organs that grow inside of the unfused sepaloid carpels, suggesting that *ARF3* may reinforce floral meristem determinacy through *WUS* repression. The *ARF3* locus is directly bound by AP2, indicating that AP2's role in floral stem cell regulation is also partially mediated by ARF3.

CONCLUSION

The complex regulatory network controlling floral meristem development produces elegant flowers with defined numbers and whorls of floral organs, thus ensuring that plant reproduction can occur (**Figure 2**). With knowledge of the spatial and temporal control of floral stem cells, as well as knowledge of the many factors responsible for fine-tuning floral stem cell activity, steady progress will be made in unraveling the mysteries of floral meristem regulation. Recently developed techniques, including ChIP-seq, RNA-seq, TALENs, CRISPR/Cas9, confocal live imaging and mathematical modeling, will help to provide further insights into the intriguing nature of flower development.

ACKNOWLEDGMENTS

The authors apologize for references not cited because of space limitations. This work was supported by research grants to Toshiro Ito from Temasek Life Sciences Laboratory (TLL).

REFERENCES

Alvarez-Venegas, R., Pien, S., Sadder, M., Witmer, X., Grossniklaus, U., and Avramova, Z. (2003). ATX-1, an *Arabidopsis* homolog of trithorax, activates flower homeotic genes. *Curr. Biol.* 13, 627–637. doi: 10.1016/S0960-9822(03)00243-4

Beuchle, D., Struhl, G., and Muller, J. (2001). Polycomb group proteins and heritable silencing of *Drosophila* Hox genes. *Development* 128, 993–1004.

Bowman, J. L., Sakai, H., Jack, T., Weigel, D., Mayer, U., and Meyerowitz, E. M. (1992). SUPERMAN, a regulator of floral homeotic genes in *Arabidopsis*. *Development* 114, 599–615.

Bowman, J. L., and Smyth, D. R. (1999). CRABS CLAW, a gene that regulates carpel and nectary development in *Arabidopsis*, encodes a novel protein with zinc finger and helix-loop-helix domains. *Development* 126, 2387–2396.

Brand, U., Fletcher, J. C., Hobe, M., Meyerowitz, E. M., and Simon, R. (2000). Dependence of stem cell fate in *Arabidopsis* on a feedback loop regulated by CLV3 activity. *Science* 289, 617–619. doi: 10.1126/science.289.5479.617

Carles, C. C., and Fletcher, J. C. (2009). The SAND domain protein ULTRAPETALA1 acts as a trithorax group factor to regulate cell fate in plants. *Genes Dev.* 23, 2723–2728. doi: 10.1101/gad.1812609

Carles, C. C., Lertpiriyapong, K., Reville, K., and Fletcher, J. C. (2004). The ULTRAPETALA1 gene functions early in *Arabidopsis* development to restrict shoot apical meristem activity and acts through WUSCHEL to regulate floral meristem determinacy. *Genetics* 167, 1893–1903. doi: 10.1534/genetics.104.028787

Chen, X. (2004). A microRNA as a translational repressor of APETALA2 in *Arabidopsis* flower development. *Science* 303, 2022–2025. doi: 10.1126/science.1088060

Chuang, C. F., Running, M. P., Williams, R. W., and Meyerowitz, E. M. (1999). The PERIANTHIA gene encodes a bZIP protein involved in the determination of floral organ number in *Arabidopsis thaliana*. *Genes Dev.* 13, 334–344. doi: 10.1101/gad.13.3.334

Crevillen, P., Yang, H., Cui, X., Greeff, C., Trick, M., Qiu, Q., et al. (2014). Epigenetic reprogramming that prevents transgenerational inheritance of the vernalized state. *Nature* 515, 587–590. doi: 10.1038/nature13722

Das, P., Ito, T., Wellmer, F., Vernoux, T., Dedieu, A., Traas, J., et al. (2009). Floral stem cell termination involves the direct regulation of AGAMOUS by PERIANTHIA. *Development* 136, 1605–1611. doi: 10.1242/dev.035436

Daum, G., Medzihradszky, A., Suzaki, T., and Lohmann, J. U. (2014). A mechanistic framework for noncell autonomous stem cell induction in *Arabidopsis*. *Proc. Natl. Acad. Sci. U.S.A.* 111, 14619–14624. doi: 10.1073/pnas.1406446111

Deng, W., Buzas, D. M., Ying, H., Robertson, M., Taylor, J., Peacock, W. J., et al. (2013). *Arabidopsis* Polycomb Repressive Complex 2 binding sites contain putative GAGA factor binding motifs within coding regions of genes. *BMC Genomics* 14:593. doi: 10.1186/1471-2164-14-593

Deyhle, F., Sarkar, A. K., Tucker, E. J., and Laux, T. (2007). WUSCHEL regulates cell differentiation during anther development. *Dev. Biol.* 302, 154–159. doi: 10.1016/j.ydbio.2006.09.013

Earley, K. W., and Poethig, R. S. (2011). Binding of the cyclophilin 40 ortholog SQUINT to Hsp90 protein is required for SQUINT function in *Arabidopsis*. *J. Biol. Chem.* 286, 38184–38189. doi: 10.1074/jbc.M111.290130

Engelhorn, J., Moreau, F., Fletcher, J. C., and Carles, C. C. (2014). ULTRAPETALA1 and LEAFY pathways function independently in specifying identity and determinacy at the Arabidopsis floral meristem. *Ann. Bot.* 114, 1497–1505. doi: 10.1093/aob/mcu185

Fletcher, J. C. (2001). The ULTRAPETALA gene controls shoot and floral meristem size in *Arabidopsis*. *Development* 128, 1323–1333.

Fletcher, J. C., Brand, U., Running, M. P., Simon, R., and Meyerowitz, E. M. (1999). Signaling of cell fate decisions by CLAVATA3 in *Arabidopsis* shoot meristems. *Science* 283, 1911–1914. doi: 10.1126/science.283.5409.1911

Gan, E. S., Xu, Y., Wong, J. Y., Geraldine Goh, J., Sun, B., Wee, W. Y., et al. (2014). Jumonji demethylases moderate precocious flowering at elevated temperature via regulation of FLC in *Arabidopsis*. *Nat. Commun.* 5, 5098. doi: 10.1038/ncomms6098

Gross-Hardt, R., Lenhard, M., and Laux, T. (2002). WUSCHEL signaling functions in interregional communication during *Arabidopsis* ovule development. *Genes Dev.* 16, 1129–1138. doi: 10.1101/gad.225202

Hiratsu, K., Ohta, M., Matsui, K., and Ohme-Takagi, M. (2002). The SUPERMAN protein is an active repressor whose carboxy-terminal repression domain is required for the development of normal flowers. *FEBS Lett.* 514, 351–354. doi: 10.1016/S0014-5793(02)02435-3

Ji, L., Liu, X., Yan, J., Wang, W., Yumul, R. E., Kim, Y. J., et al. (2011). ARGONAUTE10 and ARGONAUTE1 regulate the termination of floral stem cells through two microRNAs in *Arabidopsis*. *PLoS Genet.* 7:e1001358. doi: 10.1371/journal.pgen.1001358

Krizek, B. A., and Fletcher, J. C. (2005). Molecular mechanisms of flower development: an armchair guide. *Nat. Rev. Genet.* 6, 688–698. doi: 10.1038/nrg1675

Kwon, C. S., Chen, C., and Wagner, D. (2005). WUSCHEL is a primary target for transcriptional regulation by SPLAYED in dynamic control of stem cell fate in *Arabidopsis*. *Genes Dev.* 19, 992–1003. doi: 10.1101/gad.1276305

Lee, J. Y., Baum, S. F., Alvarez, J., Patel, A., Chitwood, D. H., and Bowman, J. L. (2005). Activation of CRABS CLAW in the nectaries and carpels of *Arabidopsis*. *Plant Cell* 17, 25–36. doi: 10.1105/tpc.104.026666

Lenhard, M., Bohnert, A., Jurgens, G., and Laux, T. (2001). Termination of stem cell maintenance in *Arabidopsis* floral meristems by interactions between WUSCHEL and AGAMOUS. *Cell* 105, 805–814. doi: 10.1016/S0092-8674(01)00390-7

Liu, X., Dinh, T. T., Li, D., Shi, B., Li, Y., Cao, X., et al. (2014b). AUXIN RESPONSE FACTOR 3 integrates the functions of AGAMOUS and APETALA2 in floral meristem determinacy. *Plant J.* 80, 629–641. doi: 10.1111/tpj.12658

Liu, X., Gao, L., Dinh, T. T., Shi, T., Li, D., Wang, R., et al. (2014a). DNA topoisomerase I affects polycomb group protein-mediated epigenetic regulation and plant development by altering nucleosome distribution in *Arabidopsis*. *Plant Cell* 26, 2803–2817. doi: 10.1105/tpc.114.124941

Liu, X., Kim, Y. J., Muller, R., Yumul, R. E., Liu, C., Pan, Y., et al. (2011). AGAMOUS terminates floral stem cell maintenance in *Arabidopsis* by directly repressing WUSCHEL through recruitment of Polycomb Group proteins. *Plant Cell* 23, 3654–3670. doi: 10.1105/tpc.111.091538

Lohmann, J. U., Hong, R. L., Hobe, M., Busch, M. A., Parcy, F., Simon, R., et al. (2001). A molecular link between stem cell regulation and floral patterning in *Arabidopsis*. *Cell* 105, 793–803. doi: 10.1016/S0092-8674(01)00384-1

Lu, F., Cui, X., Zhang, S., Jenuwein, T., and Cao, X. (2011). *Arabidopsis* REF6 is a histone H3 lysine 27 demethylase. *Nat. Genet.* 43, 715–719. doi: 10.1038/ng.854

Maier, A. T., Stehling-Sun, S., Wollmann, H., Demar, M., Hong, R. L., Haubeiss, S., et al. (2009). Dual roles of the bZIP transcription factor PERIANTHIA in the control of floral architecture and homeotic gene expression. *Development* 136, 1613–1620. doi: 10.1242/dev.033647

Mandel, T., Moreau, F., Kutsher, Y., Fletcher, J. C., Carles, C. C., and Eshed Williams, L. (2014). The ERECTA receptor kinase regulates *Arabidopsis* shoot apical meristem size, phyllotaxy and floral meristem identity. *Development* 141, 830–841. doi: 10.1242/dev.104687

Mayer, K. F., Schoof, H., Haecker, A., Lenhard, M., Jurgens, G., and Laux, T. (1998). Role of WUSCHEL in regulating stem cell fate in the *Arabidopsis* shoot meristem. *Cell* 95, 805–815. doi: 10.1016/S0092-8674(00)81703-1

Mizukami, Y., and Ma, H. (1997). Determination of *Arabidopsis* floral meristem identity by AGAMOUS. *Plant Cell* 9, 393–408. doi: 10.1105/tpc.9.3.393

Muller, J., Hart, C. M., Francis, N. J., Vargas, M. L., Sengupta, A., Wild, B., et al. (2002). Histone methyltransferase activity of a *Drosophila* Polycomb group repressor complex. *Cell* 111, 197–208. doi: 10.1016/S0092-8674(02)00976-5

Muller, J., and Kassis, J. A. (2006). Polycomb response elements and targeting of Polycomb group proteins in *Drosophila*. *Curr. Opin. Genet. Dev.* 16, 476–484. doi: 10.1016/j.gde.2006.08.005

Payne, T., Johnson, S. D., and Koltunow, A. M. (2004). KNUCKLES (KNU) encodes a C2H2 zinc-finger protein that regulates development of basal pattern elements of the *Arabidopsis* gynoecium. *Development* 131, 3737–3749. doi: 10.1242/dev.01216

Prigge, M. J., Otsuga, D., Alonso, J. M., Ecker, J. R., Drews, G. N., and Clark, S. E. (2005). Class III homeodomain-leucine zipper gene family members have overlapping, antagonistic, and distinct roles in *Arabidopsis* development. *Plant Cell* 17, 61–76. doi: 10.1105/tpc.104.026161

Prunet, N., Morel, P., Thierry, A. M., Eshed, Y., Bowman, J. L., Negrutiu, I., et al. (2008). REBELOTE, SQUINT, and ULTRAPETALA1 function redundantly in the temporal regulation of floral meristem termination in *Arabidopsis thaliana*. *Plant Cell* 20, 901–919. doi: 10.1105/tpc.107.053306

Reddy, G. V., Heisler, M. G., Ehrhardt, D. W., and Meyerowitz, E. M. (2004). Real-time lineage analysis reveals oriented cell divisions associated with morphogenesis at the shoot apex of *Arabidopsis thaliana*. *Development* 131, 4225–4237. doi: 10.1242/dev.01261

Running, M. P., and Meyerowitz, E. M. (1996). Mutations in the PERIANTHIA gene of *Arabidopsis* specifically alter floral organ number and initiation pattern. *Development* 122, 1261–1269.

Sakai, H., Medrano, L. J., and Meyerowitz, E. M. (1995). Role of SUPERMAN in maintaining *Arabidopsis* floral whorl boundaries. *Nature* 378, 199–203. doi: 10.1038/378199a0

Schoof, H., Lenhard, M., Haecker, A., Mayer, K. F., Jurgens, G., and Laux, T. (2000). The stem cell population of *Arabidopsis* shoot meristems in maintained by a regulatory loop between the CLAVATA and WUSCHEL genes. *Cell* 100, 635–644. doi: 10.1016/S0092-8674(00)80700-X

Smaczniak, C., Immink, R. G., Muino, J. M., Blanvillain, R., Busscher, M., Busscher-Lange, J., et al. (2012). Characterization of MADS-domain transcription factor complexes in *Arabidopsis* flower development. *Proc. Natl. Acad. Sci. U.S.A.* 109, 1560–1565. doi: 10.1073/pnas.1112871109

Smith, M. R., Willmann, M. R., Wu, G., Berardini, T. Z., Moller, B., Weijers, D., et al. (2009). Cyclophilin 40 is required for microRNA activity in *Arabidopsis*. *Proc. Natl. Acad. Sci. U.S.A.* 106, 5424–5429. doi: 10.1073/pnas.0812729106

Smyth, D. R., Bowman, J. L., and Meyerowitz, E. M. (1990). Early flower development in *Arabidopsis*. *Plant Cell* 2, 755–767. doi: 10.1105/tpc.2.8.755

Sun, B., Looi, L. S., Guo, S., He, Z., Gan, E. S., Huang, J., et al. (2014). Timing mechanism dependent on cell division is invoked by Polycomb eviction in plant stem cells. *Science* 343, 1248559. doi: 10.1126/science.1248559

Sun, B., Xu, Y., Ng, K. H., and Ito, T. (2009). A timing mechanism for stem cell maintenance and differentiation in the *Arabidopsis* floral meristem. *Genes Dev.* 23, 1791–1804. doi: 10.1101/gad.1800409

Wagner, D., and Meyerowitz, E. M. (2002). SPLAYED, a novel SWI/SNF ATPase homolog, controls reproductive development in *Arabidopsis*. *Curr. Biol.* 12, 85–94. doi: 10.1016/S0960-9822(01)00651-0

Williams, L., Grigg, S. P., Xie, M., Christensen, S., and Fletcher, J. C. (2005). Regulation of *Arabidopsis* shoot apical meristem and lateral organ formation by microRNA miR166g and its AtHD-ZIP target genes. *Development* 132, 3657–3668. doi: 10.1242/dev.01942

Wu, M. F., Sang, Y., Bezhani, S., Yamaguchi, N., Han, S. K., Li, Z., et al. (2012). SWI2/SNF2 chromatin remodeling ATPases overcome polycomb repression and control floral organ identity with the LEAFY and SEPALLATA3 transcription factors. *Proc. Natl. Acad. Sci. U.S.A.* 109, 3576–3581. doi: 10.1073/pnas.1113409109

Wynn, A. N., Seaman, A. A., Jones, A. L., and Franks, R. G. (2014). Novel functional roles for PERIANTHIA and SEUSS during floral organ identity specification, floral meristem termination, and gynoecial development. *Front. Plant Sci.* 5:130. doi: 10.3389/fpls.2014.00130

Yadav, R. K., Perales, M., Gruel, J., Girke, T., Jonsson, H., and Reddy, G. V. (2011). WUSCHEL protein movement mediates stem cell homeostasis in the *Arabidopsis* shoot apex. *Genes Dev.* 25, 2025–2030. doi: 10.1101/gad.17258511

Zhang, X. (2014). Plant science. Delayed gratification—waiting to terminate stem cell identity. *Science* 343, 498–499. doi: 10.1126/science.1249343

Zhang, X., Clarenz, O., Cokus, S., Bernatavichute, Y. V., Pellegrini, M., Goodrich, J., et al. (2007). Whole-genome analysis of histone H3 lysine 27 trimethylation in *Arabidopsis*. *PLoS Biol.* 5:e129. doi: 10.1371/journal.pbio.0050129

Zhao, L., Kim, Y., Dinh, T. T., and Chen, X. (2007). miR172 regulates stem cell fate and defines the inner boundary of APETALA3 and PISTILLATA expression domain in *Arabidopsis* floral meristems. *Plant J.* 51, 840–849. doi: 10.1111/j.1365-313X.2007.03181.x

Zuniga-Mayo, V. M., Marsch-Martinez, N., and de Folter, S. (2012). JAIBA, a class-II HD-ZIP transcription factor involved in the regulation of meristematic activity, and important for correct gynoecium and fruit development in *Arabidopsis*. *Plant J.* 71, 314–326. doi: 10.1111/j.1365-313X.2012.04990.x

7

OsSDS is essential for DSB formation in rice meiosis

Zhigang Wu[1†], Jianhui Ji[2,3†], Ding Tang[2], Hongjun Wang[2], Yi Shen[2], Wenqing Shi[2], Yafei Li[2], Xuelin Tan[1], Zhukuan Cheng[2]* and Qiong Luo[1]*

[1] Ministry of Education Key Laboratory of Agriculture Biodiversity for Plant Disease Management, Yunnan Agricultural University, Kunming, China
[2] State Key Laboratory of Plant Genomics and Center for Plant Gene Research, Institute of Genetics and Developmental Biology, Chinese Academy of Sciences, Beijing, China
[3] School of Life Sciences, Huaiyin Normal University, Huaian, China

Edited by:
Kang Chong, Chinese Academy of Sciences, China

Reviewed by:
Pingli Lu, Fudan Universtiy, China
Weiwei Jin, China Agricultural University, China

*Correspondence:
Zhukuan Cheng, State Key Laboratory of Plant Genomics and Center for Plant Gene Research, Institute of Genetics and Developmental Biology, Chinese Academy of Sciences, No.1 West Beichen Road, Chaoyang District, Beijing 100101, China
e-mail: zkcheng@genetics.ac.cn;
Qiong Luo, Ministry of Education Key Laboratory of Agriculture Biodiversity for Plant Disease Management, Yunnan Agricultural University, Heilongtan, Guandu District, Kunming 650201, China
e-mail: qiongbf@aliyun.com

[†] These authors have contributed equally to this work.

SDS is a meiosis specific cyclin-like protein and required for DMC1 mediated double-strand break (DSB) repairing in *Arabidopsis*. Here, we found its rice homolog, OsSDS, is essential for meiotic DSB formation. The *Ossds* mutant is normal in vegetative growth but both male and female gametes are inviable. The *Ossds* meiocytes exhibit severe defects in homologous pairing and synapsis. No γH2AX immunosignals in *Ossds* meiocytes together with the suppression of chromosome fragmentation in *Ossds-1 Osrad51c*, both provide strong evidences that OsSDS is essential for meiotic DSB formation. Immunostaining investigations revealed that meiotic chromosome axes are normally formed but both SC installation and localization of recombination elements are failed in *Ossds*. We suspected that this cyclin protein has been differentiated pretty much between monocots and dicots on its function in meiosis.

Keywords: rice, OsSDS, meiosis, DSB formation

INTRODUCTION

Meiosis is one of the key processes in sexual reproduction for all sexually propagating eukaryotic organisms. Meiosis includes one single round of DNA replication but followed by two successive rounds of nuclear segregations (meiosis I and II), and finally produces four haploid gametes with halved chromosomes. The meiotic prophase I is a complicated and prolonged stage, which can be divided into five substages, like leptotene, zygotene, pachytene, diplotene, and diakinesis, based on chromosome characterizations (Ashley and Plug, 1998; Dawe, 1998). During meiotic prophase I, pairing, synapsis and recombination of homologous chromosomes are coordinately accomplished. These events make sure the precise segregation of homologs, and generate both genetic conservation and diverse individuals in the future generations (Zickler and Kleckner, 1999; Page and Hawley, 2003).

In meiosis, DSBs are purposely produced to initiate homologous recombination. The formation of DSBs in meiosis is catalyzed by a type-II topoisomerase-like enzyme Spo11 (Bergerat et al., 1997; Keeney et al., 1997). Meanwhile, a series of cofactors are also required for this process. In budding yeast, the formation of meiotic DSBs requires at least nine other proteins

(Rec102, Rec104, Rec114, Mei4, Mer2, Rad50, Mre11, Xrs2, and Ski8) for the cleavage mediated by SPO11 and further broken end resection (Paques and Haber, 1999; Keeney, 2001, 2008). The budding yeast Mre11-Rad50-Xrs2 (MRX) complex is homologous with the mammalian Mre11-Rad50-Nbs1 (MRN), which is required to incise the 5′ end of the break and then form 3′ single-strand tails (Symington, 2002; Mimitou and Symington, 2009). After that, one 3′ free single-strand DNA end recruits two RecA homologs, Rad51 and Dmc1, to mediate the single-end invasion (SEI) with its homologous duplex DNA (Bishop, 1994; Hunter and Kleckner, 2001), and the other 3′ free single-strand DNA end on the other side of the nick is captured simultaneously to form the Double Holliday Junction (DHJ). And then, the DHJ is exclusively processed into crossovers (COs), which represents the accomplishment of homologous recombination (Allers and Lichten, 2001; Bishop and Zickler, 2004; Borner et al., 2004). Consequently, meiotic DSBs are finally repaired during this process.

The function of SPO11 initiating meiotic recombination seems to be widely conserved within eukaryotes, as more and more homologs of SPO11 were identified in a wide range of organisms covering yeasts, flies, mice, humans, and plants

(Dernburg et al., 1998; McKim and Hayashi-Hagihara, 1998; Celerin et al., 2000; Hartung and Puchta, 2000; Romanienko and Camerini-Otero, 2000; Grelon et al., 2001; Yu et al., 2010). Unlike animals and fungi where a single SPO11 is sufficient for meiotic DSBs formation, higher plants always possess multiple SPO11 homologs (Keeney et al., 1997; Grelon et al., 2001; Hartung et al., 2007; Shingu et al., 2012; Sprink and Hartung, 2014). But not every SPO11 homolog has the function to cleavage double-strand DNA and generate DSBs in plants. *Arabidopsis* owns three SPO11 homologs and they appear to function in two distinct processes, AtSPO11-1 and AtSPO11-2 in DSB formation, while AtSPO11-3 in DNA replication (Stacey et al., 2006; Shingu et al., 2012). While in monocot rice, there are five SPO11 homologs have been identified (Jain et al., 2006, 2008). Among them, only OsSPO11-4 has been proved to be with double-strand DNA cleavage activity (An et al., 2011). OsSPO11-1 is essential for homologous pairing, recombination and SC installation (Yu et al., 2010; Luo et al., 2014). So, it seems that the formation of meiotic DSBs is more complicated in plants.

Besides SPO11, several other DSB formation proteins have been identified recently in multicellular eukaryotes. Mei1 and Mei4 were shown to be required for DSB formation in mice (Libby et al., 2003; Kumar et al., 2010). Using a high throughput genetic screen, AtPRD1, AtPRD2, and AtPRD3 were identified to be essential for DSB formation in *Arabidopsis* (De Muyt et al., 2009). Nevertheless, AtDFO was also found to be necessary for DSB formation in *Arabidopsis* (Zhang et al., 2012). Studies in rice revealed that CRC1 works together with PAIR1 as a complex to regulate meiotic DSB formation (Miao et al., 2013).

Studies in budding yeast demonstrated that cyclin-dependent kinase Cdc7 and Cdc28 can directly regulate the meiotic DSB formation via the phosphorylation of Mer2 (Henderson et al., 2006; Sasanuma et al., 2008; Wan et al., 2008). However, in plants, only a few cyclins have been found involved in meiosis (Bulankova et al., 2013). The meiosis specific cyclin SDS was first found in *A. thaliana*, which play a specific role in regulating synapsis in prophase I (Azumi et al., 2002; Wang et al., 2004). In a recent study, SDS was found to be required for DMC1-mediated DSB repair (De Muyt et al., 2009). Although the rice *SDS*-RNAi plants showed the similar meiotic defects with those in *Arabidopsis* (Chang et al., 2009), the molecular mechanism of SDS in rice meiosis remains to be clear. Here, we identified the SDS homolog in rice by map-based cloning. Surprisingly, we found OsSDS is essential for DSB formation during rice meiosis, which is much different from that in *Arabidopsis*. We suspected this cyclin protein had been differentiated pretty much between monocots and dicots on its function in meiosis.

MATERIALS AND METHODS

PLANT MATERIALS

The rice (*Oryza sativa* L.) spontaneous mutant *Ossds-1* was isolated from an *indica* rice, Zhongxian 3037. The F2 and F3 mapping populations were generated by crossing the *Ossds-1*$^{\pm}$ heterozygous plants with a *japonica* cultivar, Zhonghua 11. The other two mutant alleles, *Ossds-2* and *Ossds-3*, both were spontaneous mutants arose in tissue culture of Nipponbare. The meiotic

mutant *Osrad51c* has been reported previously (Tang et al., 2014). The *Ossds-1 Osrad51c* double mutant was generated by crossing the two heterozygous *Ossds-1*$^{\pm}$ and *Osrad51c*$^{\pm}$, and further identified from the F2 progeny. All plant materials were grown in paddy fields in the summer in Beijing or in the winter in Hainan.

MOLECULAR CLONING OF *OsSDS*

Total 861 sterile plants segregated from the F2 and F3 mapping populations were used for isolation the target gene. Sequence-tagged site (STS) markers were developed according to sequence differences between the japonica variety Nipponbare and the indica variety 9311, using the data published on the NCBI website (http://www.ncbi.nlm.nih.gov). All primers are listed in Supplemental Table 1.

RNAi ANALYSIS

In the first exon, a 336-bp fragment of *OsSDS* cDNA sequence was chosen and amplified with the primers SDS-RNAi-F (adding a *Bam*HI site) and SDS-RNAi-R (adding a *Sal*I site) (Supplemental Table 1). RNAi vector construction and transformation were performed as described (Wang et al., 2009).

COMPLEMENTATION TEST

The complementary plasmid was constructed by cloning the 10.8 kb OSJNBa0081P02 genomic DNA fragment containing the entire *OsSDS* coding region into the pCAMBIA-1300 vector. A control plasmid, containing 7.8 kb of the truncated *OsSDS* gene was also constructed. Both of these plasmids were transformed into EHA105 and then into embryonic calli of *OsSDS*$^{\pm}$ plants. The genotypes of the transgenic plants were further identify using the Primers SDS-JD (Supplemental Table 1).

REAL-TIME PCR FOR TRANSCRIPT EXPRESSION ASSAY

Total RNA was extracted from the root, internode, leaf and panicle of Zhongxian 3037. Real-time PCR analysis was performed using the Bio-Rad CFX96 real-time PCR instrument (Bio-Rad, http://www.bio-rad.com/) and EvaGreen (Biotium, http://www.biotium.com/). The RT-PCR was carried out using the gene-specific primer pairs SDS-RT-F and SDS-RT-R for *OsSDS*. The primers Ubi-RT-F and Ubi-RT-R for ubiquitin were used as an internal control for the normalization of RNA sample. The results were analyzed using OPTICON MONITOR 3.1 (Bio-Rad). Each experiment had three replicates.

CLONING THE FULL-LENGTH *OsSDS* cDNA

Total RNA was extracted from the panicle of Zhongxian 3037. The 3′ RACE and 5′ RACE were performed according to the protocol of the kit (3′-Full RACE Core Set and 5′-Full RACE Core Set; Takara, http://www.takara-bio.com/). 3′ RACE was carried out using primers 3R-1F, 3R-2F, 3R-3F, and adaptor primer (P-ada). During 5′RACE, the RNA was reverse transcribed with 5′ (P)-labeled primer (SDS-4Rb); the first and second PCRs were performed using two sets of *OsSDS* specific primers (5R-1 and 5R-2). The 3′ RACE-PCR and 5′ RACE-PCR products were cloned and sequenced. OsSDS amino acid sequence translation and alignment were completed with the Vector NTI 11.5

(Invitrogen). *OsSDS* gene structure diagram was generated from GSDS (http://gsds.cbi.pku.edu.cn/index.php).

MEIOTIC CHROMOSOME PREPARATION

Young panicles with appropriate size of both *Ossds* mutants and wild type were collected, fixed in Carnoy's solution (ethanol:glacial acetic, 3:1) and stored at −20°C. Microsporocytes at the appropriate meiotic stage were squashed and stained with acetocarmine. After washing the chromosome preparations with 45% acetic acid and freezing them in liquid nitrogen, the coverslips were quickly removed with a razor blade and the slides harbored with samples were dehydrated through an ethanol series (70, 90, and 100%) for 5 min each and finally air-dried. Chromosomes on slides were counterstained with 4, 6-diamidinophenylindole (DAPI) in an anti-fade solution (Vector Laboratories, Burlingame, CA, USA). Finally, images were captured under the ZEISS A2 fluorescence microscope with a micro CCD camera (Zeiss, http://www.zeiss.de/en).

FLUORESCENCE IMMUNOLOCALIZATION

Fresh young panicles (40–60 mm) were fixed with 4% (w/v) paraformaldehyde for 30 min at room temperature. Anthers with appropriate stage were squashed into one drop of 1 × PBS solution added on a slide. Then, covering the slide with a coverslip and pressing it with appropriate strength, the slide together with the coverslip was frozen thoroughly in liquid nitrogen. After quickly prizing up the coverslip, the slide was dehydrated through an ethanol series (70, 90, and 100%). The following immunolocalization procedure was performed as described (Tang et al., 2014).

The polyclonal antibodies against γ-H2AX, OsMSH5, OsREC8, PAIR2, PAIR3, ZEP1, OsMER3, and OsZIP4 used in this study have been described previously (Wang et al., 2009, 2010; Shao et al., 2011; Shen et al., 2012; Zhang et al., 2012; Luo et al., 2013; Miao et al., 2013).

RESULTS
CHARACTERIZATION OF A STERILE MUTANT

We identified a spontaneous mutant exhibiting complete sterility in a rice field of Zhongxian 3037. From the heterozygous plant progeny related to this mutation, the normal fertile plants and the sterile plants were segregated in a 3:1 ratio, indicating that it was a single recessive mutation ($\chi^2 = 0.57$; $P > 0.05$). The mutant plant was normal during vegetative growth and could not be distinguished from the wild type based on plant morphology (Supplemental Figure 1A). However, when come into reproductive stage, its spikelets exhibited complete sterility (Supplemental Figure 1B). So we further examined the mature pollen viability of the mutant by staining with 1% iodine potassium iodide solution (I$_2$-KI) (Supplemental Figures 1C,D). Only empty and shrunken pollen grains were observed in the mutant plant, indicating that microspores of the mutant are all abnormal and inviable. Moreover, when pollinated with wild-type pollens, the mutant spikelets still did not set any seeds, indicating that its female gametes were also affected.

MAP-BASED CLONING OF *OsSDS*

To isolate the mutated gene, we constructed a population by crossing heterozygous plants with a japonica cultivar Zhonghua 11. A total of 861 sterile plants segregated from the F2 and F3 populations were used for mapping the target gene. Linkage analysis initially mapped the target gene onto the long arm of chromosome 3, which subsequently further delimited to a 130 kb region. Within this region, we found one candidate gene (*Os03g12414*) annotated as a putative cyclin with high similarity with *SOLO DANCERS* (*SDS*) gene in *Arabidopsis*. Thus, this candidate gene was chosen to be amplified and sequenced. Sequencing analysis revealed that there was a transversion happened from base C to T in the first exon, which produces a premature termination codon (TGA) and causes early termination (**Figure 1A**). We named the mutant here *Ossds-1* and suspected the mutation of *Os03g12414* leading to the sterile phenotype. We also obtained two other alleles arose from tissue culture of Nipponbare, named *Ossds-2* and *Ossds-3*, respectively. They all showed the same defects as *Ossds-1*. Sequencing analysis showed that there were a transversion from base G to C at the third exon causing corresponding amino acid A replaced by P in *Ossds-2* and two bases (GC) deletion at the fifth exon resulting in frame-shift mutation causing a premature termination codon (TGA) in *Ossds-3* (**Figure 1A**). As earlier termination close to the start codon in *Ossds-1* compared with the other mutants, it was selected for the subsequent experiments described blow. To further confirm the mutant phenotype was resulted by the mutation of *OsSDS* gene, RNA interference (RNAi) approach was carried out to down-regulate *SDS* in rice variety Yandao 8. We got 27 transgenic plants with 19 plants exhibited complete sterility. Additionally, the transformation of plasmid pCAMBIA-1300 containing the whole *OsSDS* gene was successful in rescuing the sterility of the mutant plants, just as respected. These results strongly confirmed that the mutation of *OsSDS* gene led to the sterile phenotype as described above.

Expression of *OsSDS* was also analyzed by real-time PCR (RT-PCR). As shown in Supplemental Figure 2, transcripts were all detected at root, internode, leaf and panicle, indicating that *OsSDS* is not a meiosis-specific gene.

FULL-LENGTH cDNA CLONING AND DEDUCED PROTEIN SEQUENCE OF *OsSDS*

The full-length cDNA of *OsSDS* gene was obtained by performing RT-PCR combined with 5′ and 3′ rapid amplification of cDNA ends PCR (RACE-PCR) using specific primers. We found that the *OsSDS* cDNA is comprised of 2786 bp containing an open reading frame (ORF) of 1410 bp and 1376 bp 5′ and 3′ untranslated regions (UTRs). The *OsSDS* cDNA sequence obtained is consistent with one mRNA (GenBank: AK065907.1) from the public network database (http://www.ncbi.nlm.nih.gov/). As shown in **Figure 1A**, the *OsSDS* gene contains seven exons and six introns. The predicted 469 amino acid protein of OsSDS shares as low as 30.6% identity with SDS in *Arabidopsis*, but with high similarity at the C-terminal (**Figure 1B**). Compared with the dicots *Arabidopsis*, OsSDS shares more than 60% identity with those in monocots, such as *Zea mays*, *Sorghum bicolor*, and *Brachypodium*. Conserved domain searching in NCBI revealed there are two conserved domains in SDS, namely, cyclin box fold domain

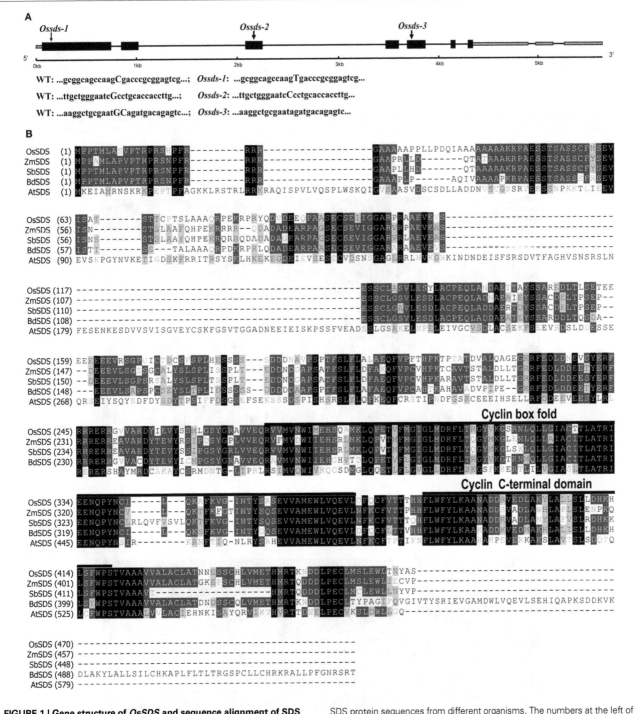

FIGURE 1 | Gene structure of *OsSDS* and sequence alignment of SDS proteins. (A) Schematic of representation of gene structure and three mutation sites in *OsSDS*. Exons are represented by black boxes, intones are represented by black lines, 5′UTR and 3′UTR are represented by gray boxes. The capitalized bases in sequences of wild type and mutants represent accurate mutation sites of the three allelic mutants. **(B)** Multiple alignment of SDS protein sequences from different organisms. The numbers at the left of the sequences are the amino acid numbers. The black boxes represent identical sequences; the dark gray boxes represent conservative sequences; the light gray boxes represent weakly similar sequences. A predicted cyclin box fold domain (276–366 amino acids) and a predicted cyclin C-terminal domain (378–419 amino acids) are underlined.

(residues 276–366) and cyclin C-terminal domain (residues 378–419) (**Figure 1B**). Blast searching in NCBI revealed that SDS is a plant specific protein and owns one single copy in the plant kingdom.

MEIOTIC DEFECTS IN *Ossds*

To clarify whether the sterile phenotype of *Ossds* is caused by meiosis defects, chromosome behaviors at different stages of pollen mother cells (PMCs) from both wild type and *Ossds-1* mutants were investigated. In wild type, chromosomes began to condense and became visible as thin thread-like structures at leptotene. After that, homologous chromosomes started to pair at zygotene. Fully synapsis between homologs formed at pachytene (**Figure 2A**). After further chromosome condensation, 12 bivalents were clearly observed at diakinesis (**Figure 2B**). Accompanying with the spindle installation, these bivalents aligned on the equatorial plate at metaphase I (**Figure 2C**). Then, homologous chromosomes separated and moved to the two opposite poles from anaphase I to telophase I (**Figures 2D,E**). During meiosis II, sister chromatids of each chromosome

separated from each other and finally tetrads formed (**Figure 2F**).

In *Ossds-1* PMCs, chromosome behaviors were similar to the wild type from leptotene to zygotene. However, obvious abnormities were first observed at pachytene stage, where the *Ossds-1* mutant shows severely defects in homologous chromosome pairing and synapsis, and fully synapsed homologs were never observed (**Figure 3A**). Due to the lack of chromosome pairing, only univalents were observed at diakinesis (**Figure 3B**). During metaphase I, the univalents were unable to align on the equator plate (**Figure 3C**). From anaphase I to

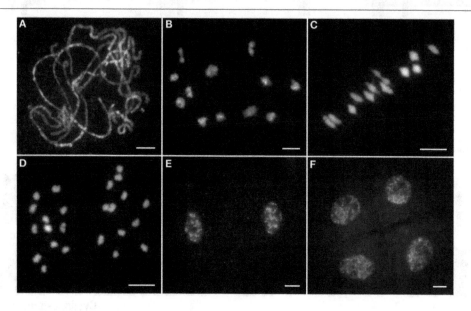

FIGURE 2 | Male meiosis of the wild type. (A) Pachytene; **(B)** Diakinesis; **(C)** Metaphase I; **(D)** Anaphase I; **(E)** Dyad; **(F)** Tetrad. Chromosomes stained with 4,6-diamidino-2-phenylindole (DAPI). Bars = 5 μm.

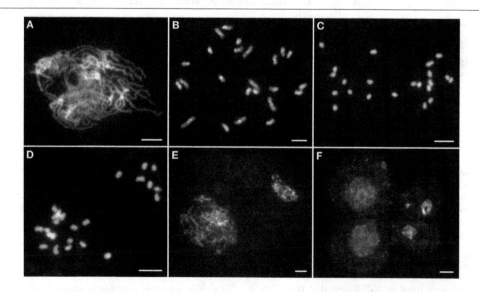

FIGURE 3 | Male meiosis of the *Ossds-1* mutant. (A) Pachytene; **(B)** Diakinesis; **(C)** Metaphase I; **(D)** Anaphase I; **(E)** Dyad; **(F)** Tetrad. Chromosomes stained with DAPI. Bars = 5 μm.

telophase I, they randomly segregated into two daughter cells (**Figure 3D**). In meiosis II, dyads and tetrads always exhibited different sizes caused by uneven chromosome segregation (**Figures 3E,F**). Cytological observation of meiocytes from the other two alleles, *Ossds-2* and *Ossds-3*, as well as *OsSDS* RNAi plants showed the same meiotic defects as described in *Ossds-1* (Supplemental Figure 3). Therefore, we proposed that the sterility of *Ossds* was caused by uneven segregation of homologous chromosomes without pairing and recombination during propose I.

OsSDS IS ESSENTIAL FOR MEIOTIC DSB FORMATION

The above cytological observation indicated *Ossds* showed similar defects with the loss of function of CRC1, a meiotic DSB formation protein in rice (Miao et al., 2013). We wondered whether OsSDS is also required for meiotic DSB formation. Phosphorylation of histone H2AX occurs within a few minutes after a DSB initiated in mitosis (Banath et al., 2010), and the same kind phosphorylation takes place during the meiotic DSB formation (Dickey et al., 2009). To verify this hypothesis, a dual-immunostaining experiment was carried out utilizing antibodies specific for γH2AX and OsREC8 raised from rabbit and mouse, respectively, in the meiocytes both from wild type and *Ossds-1*. OsREC8, one of the key cohesion proteins in rice meiosis, is localized onto meiotic chromosomes from leptotene to metaphase I (Wang et al., 2009). Here, we used it as a biomarker to indicate the rice meiotic chromosomes in the prophase I specifically. Results showed that numerous dot and patchy immunosignals of γH2AX were detected at zygotene in wild type (**Figure 4A**). However, no γH2AX immunosignals was detected in *Ossds-1* meiocytes at the corresponding stage (**Figure 4B**), showing that OsSDS is essential for meiotic DSB formation.

During meiosis, all DSBs will be repaired by the repair system for maintaining genome stability. The loss function of repair proteins always results in chromosome fragmentation. OsRAD51C

has been proved to be involved in meiotic DSB repair in rice (Tang et al., 2014). To verify this speculation, the *Ossds-1 Osrad51c* double mutant was generated by crossing the two heterozygous *Ossds-1*$^{\pm}$ and *Osrad51c*$^{\pm}$, and further identified from their F2 progeny. In the *Osrad51c* mutant meiocytes, cytological observation shows 24 irregularly univalents appeared at diakinesis (**Figure 5A**). These univalents did not align well along the equatorial plate, and several chromosome fragments started to be shown at metaphase I (**Figure 5B**). Numerous chromosome fragments detained at the equatorial plate, while those massive chromosome bodies with centromeres moved into the two opposite poles at anaphase I (**Figure 5C**). While in the *Ossds-1 Osrad51c* meiocytes, chromosome behaviors were very much similar to *Ossds-1* at the corresponding stages (**Figures 5D–F**). Together with the above γH2AX immunostaining data, we demonstrated very strong evidence that OsSDS is essential for DSB formation during rice meiosis.

MEIOTIC CHROMOSOME AXES NORMALLY FORMED BUT SC INSTALLATION FAILED IN *Ossds*

In rice, there are three axis-associated proteins OsREC8, PAIR2 and PAIR3 have been reported playing important roles in SC assembly (Nonomura et al., 2006; Shao et al., 2011; Wang et al., 2011). Except for those axial elements, ZEP1, the central element of SC, has also been identified (Wang et al., 2010). To investigate what kind of SC installation defects happened due to the loss function of OsSDS, we conducted immunodetection experiments using antibodies against PAIR2, PAIR3, and ZEP1 in *Ossds-1* microsporocytes. The results showed that OsREC8 was normally localized onto chromosomes in *Ossds-1* meiocytes. Moreover, both PAIR2 and PAIR3 were overlapped very well with OsREC8 at zygotene, indicating their normal localization along the chromosome axis (**Figures 6A,B**). However, ZEP1 signals always appeared as short dots or discontinuous lines at early prophase I (**Figure 6C**), showing the deficient central element

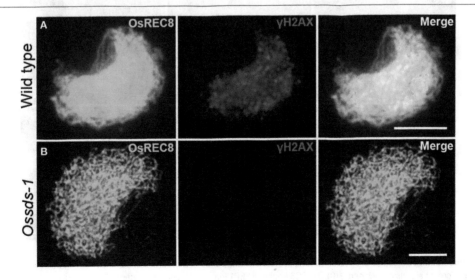

FIGURE 4 | Immunostaining of γ-H2AX at zygotene in the wild type and *Ossds-1* mutant. OsREC8 signals were used to indicate the chromosome axes. Bars = 5 μm.

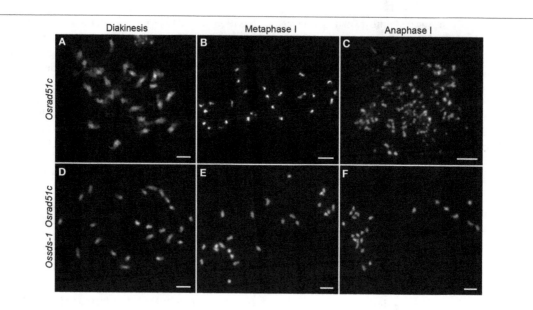

FIGURE 5 | Comparison of chromosome behaviors between *Osrad51c* and the *Ossds-1 Osrad51c* double mutant. Chromosomes were stained with DAPI. Bars = 5 μm.

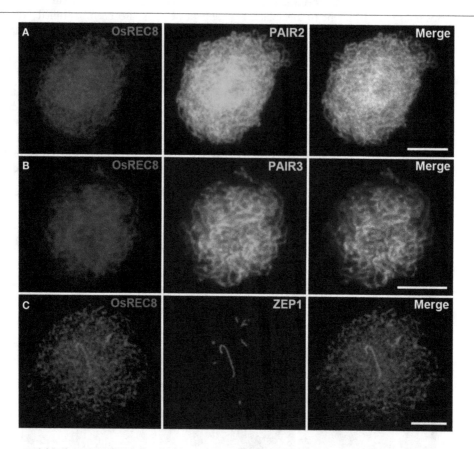

FIGURE 6 | Dual immunostaining detection of several meiotic proteins in the *Ossds-1*. (A) OsREC8 (red) and PAIR2 (green) signals at late zygotene; **(B)** OsREC8 (red) and PAIR3 (green) signals at pachytene; **(C)** OsREC8 (red) and ZEP1 (green) signals at pachytene. Bars = 5 μm.

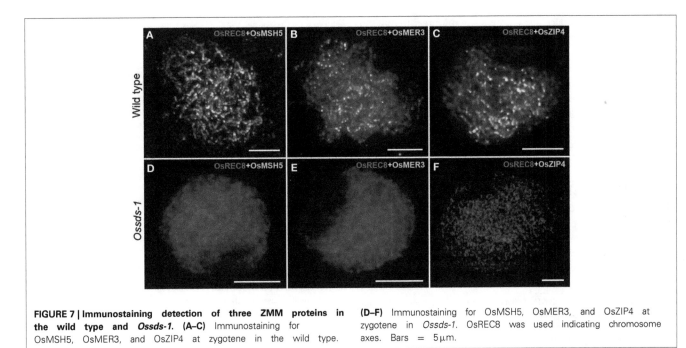

FIGURE 7 | Immunostaining detection of three ZMM proteins in the wild type and *Ossds-1*. (A–C) Immunostaining for OsMSH5, OsMER3, and OsZIP4 at zygotene in the wild type. **(D–F)** Immunostaining for OsMSH5, OsMER3, and OsZIP4 at zygotene in *Ossds-1*. OsREC8 was used indicating chromosome axes. Bars = 5 μm.

installation of SC in the mutant. Thus, the meiotic chromosome axes normally formed but SC installation failed in *Ossds*.

OsSDS IS CRITICAL FOR FAITHFUL LOCALIZATION OF RECOMBINATION ELEMENTS ONTO MEIOTIC CHROMOSOMES

Meiotic homologous recombination finally falls into two recombination pathways by forming two type crossovers, interference-sensitive COs (class I) and interference-insensitive COs (class II). ZMM proteins are closely associated with class I crossovers formation (Chen et al., 2008; Shinohara et al., 2008). In rice, OsMSH5, OsMER3 and OsZEP4 are three ZMM proteins participating in the class I COs formation (Wang et al., 2009; Shen et al., 2012; Luo et al., 2013). To determine whether the defective OsSDS also affects the localization of OsMSH5, OsMER3, and OsZEP4, dual immunolocalization were carried out using antibodies against OsMSH5, OsMER3, and OsZEP4. In the wild-type microsporocytes, immunostaining experiments showed that OsMSH5, OsMER3, and OsZEP4 were observed as punctate foci on chromosomes at zygotene (**Figures 7A–C**), and these foci persisted until late pachytene stage (Wang et al., 2009; Shen et al., 2012; Luo et al., 2013). However, no obvious signal foci of OsMSH5, OsMER3, and OsZEP4 were observed on chromosomes in *Ossds-1* microsporocytes at the corresponding stage (**Figures 7D–F**), indicating that OsSDS is critical for the localization of OsMSH5, OsMER3 and OsZEP4.

DISCUSSION

During meiosis, Spo11-catalyzed DSB formation is a highly conserved biological process among eukaryotes (Li and Ma, 2006). As increased data on plant meiosis research, several DSB formation proteins have been identified in higher plants, such as PRD1, PRD2, PRD3, and AtDFO in *Arabidopsis* (De Muyt et al., 2009; Zhang et al., 2012). And in rice, CRC1 was also reported to be involved in meiotic DSB formation besides those OsSPO11

homologs (Miao et al., 2013; Luo et al., 2014). Deficiency of these proteins always cause severe defects in homologous chromosome pairing, synapsis and nondisjunction.

MRE11 is known as an important DSB processing protein. The *Atmre11 sds* double mutant exhibited similar phenotype with the *Atmre11* single mutant (De Muyt et al., 2009), proposing that SDS may be not required for meiotic DSB formation in *Arabidopsis*. Moreover, in *sds* mutant, the localization of DMC1 was abnormal while RAD51 was normally loaded. Thus, the role of SDS in *Arabidopsis* was thought to be necessary for DMC1-mediated DSB repair utilizing the homologous chromosome (De Muyt et al., 2009). While in rice, we provided evidences that OsSDS is essential for meiotic DSB formation. SDS is a plant specific cyclin protein. Amino acid sequences alignment revealed that SDS orthologs between dicots and monocots showed very low identities, *Arabidopsis* SDS sharing 30.6% identity with OsSDS, and as low as 26.1% with BdSDS. While among monocots, they share high identities. For example, ZmSDS shares 69.1% with OsSDS, and 83.1% with SbSDS. The diverged sequences between dicots and monocots suggest that SDS function in meiosis has been differentiated a lot during its evolution.

To date, the precise mechanisms of DSB formation are still unclear. It has been reported that the NBS1 protein is recruited to the end of the DSB soon after a DSB formed, which then initiates the formation of the NBS1/MRE11/RAD50 complex (Kobayashi, 2004). After that, the ATM protein phosphorylates the serine 139 residue of H2AX through its auto-phosphorlation and leading to γH2AX formation (Kinner et al., 2008). Studies revealed that γH2AX plays dual role in the DSB triggered signaling pathway: one is recruiting more MRN complex to the DSB site thus enhancing the signalization by a positive feedback loop; the other one is binding the DNA damage repair proteins (Paull et al., 2000, Minter Dykhouse et al., 2008). The process of H2AX phosphorlation takes place just within a few minutes after a DSB occurrence (Banath et al., 2010). We did not detect any γH2AX immunosignals in *Ossds*, indicating OsSDS is required for DSB formation during rice meiosis.

It has been reported that cyclin-dependent kinase Cdc7 and Cdc28 can directly regulate the meiotic DSB formation via the phosphorylation of Mer2 in budding yeast (Henderson et al., 2006; Sasanuma et al., 2008; Wan et al., 2008). However, we still do not know the molecular mechanism of how rice cyclin OsSDS being involved in meiotic DSB formation.

ACKNOWLEDGMENTS

This work was supported by grants from the National Natural Science Foundation of China (U1302261, 31360260, and 31400169), and the Natural Science Foundation of Jiangsu Province (BK20140454).

REFERENCES

Allers, T., and Lichten, M. (2001). Differential timing and control of non-crossover and crossover recombination during meiosis. *Cell* 106, 47–57. doi: 10.1016/S0092-8674(01)00416-0

An, X. J., Deng, Z. Y., and Wang, T. (2011). OsSpo11-4, a rice homologue of the archaeal TopVIA protein, mediates double-strand DNA cleavage and interacts with OsTopVIB. *PLoS ONE* 6:e20327. doi: 10.1371/journal.pone.0020327

Ashley, T., and Plug, A. (1998). Caught in the act: deducing meiotic function from protein immunolocalization. *Curr. Top. Dev. Biol.* 37, 201–239. doi: 10.1016/S0070-2153(08)60175-1

Azumi, Y., Liu, D., Zhao, D., Li, W., Wang, G., Hu, Y., et al. (2002). Homolog interaction during meiotic prophase I in *Arabidopsis* requires the *SOLO DANCERS* gene encoding a novel cyclin-like protein. *EMBO J.* 21, 3081–3095. doi: 10.1093/emboj/cdf285

Banath, J. P., Klokov, D., Macphail, S. H., Banuelos, C. A., and Olive, P. L. (2010). Residual gamma H2AX foci as an indication of lethal DNA lesions. *BMC Cancer* 10:4. doi: 10.1186/1471-2407-10-4

Bergerat, A., De Massy, B., Gadelle, D., Varoutas, P. C., Nicolas, A., and Forterre, P. (1997). An atypical topoisomerase II from Archaea with implications for meiotic recombination. *Nature* 386, 414–417. doi: 10.1038/386414a0

Bishop, D. K. (1994). RecA homologs Dmc1 and Rad51 interact to form multiple nuclear-complexes prior to meiotic chromosome synapsis. *Cell* 79, 1081–1092. doi: 10.1016/0092-8674(94)90038-8

Bishop, D. K., and Zickler, D. (2004). Early decision: meiotic crossover interference prior to stable strand exchange and synapsis. *Cell* 117, 9–15. doi: 10.1016/S0092-8674(04)00297-1

Borner, G. V., Kleckner, N., and Hunter, N. (2004). Crossover/noncrossover differentiation, synaptonemal complex formation, and regulatory surveillance at the leptotene/zygotene transition of meiosis. *Cell* 117, 29–45. doi: 10.1016/S0092-8674(04)00292-2

Bulankova, P., Akimcheva, S., Fellner, N., and Riha, K. (2013). Identification of *Arabidopsis* meiotic cyclins reveals functional diversification among plant cyclin genes. *PLoS Genet.* 9:e1003508. doi: 10.1371/journal.pgen.1003508

Celerin, M., Merino, S. T., Stone, J. E., Menzie, A. M., and Zolan, M. E. (2000). Multiple roles of Spo11 in meiotic chromosome behavior. *EMBO J.* 19, 2739–2750. doi: 10.1093/emboj/19.11.2739

Chang, L., Ma, H., and Xue, H. W. (2009). Functional conservation of the meiotic genes SDS and RCK in male meiosis in the monocot rice. *Cell Res.* 19, 768–782. doi: 10.1038/cr.2009.52

Chen, S. Y., Tsubouchi, T., Rockmill, B., Sandler, J. S., Richards, D. R., Vader, G., et al. (2008). Global analysis of the meiotic crossover landscape. *Dev. Cell* 15, 401–415. doi: 10.1016/j.devcel.2008.07.006

Dawe, R. K. (1998). Meiotic chromosome organization and segregation in plants. *Annu. Rev. Plant Physiol. Plant Mol. Biol.* 49, 371–395. doi: 10.1146/annurev.arplant.49.1.371

De Muyt, A., Pereira, L., Vezon, D., Chelysheva, L., Gendrot, G., Chambon, A., et al. (2009). A high throughput genetic screen identifies new early meiotic recombination functions in *Arabidopsis thaliana*. *PLoS Genet.* 5:e1000654. doi: 10.1371/journal.pgen.1000654

Dernburg, A. F., McDonald, K., Moulder, G., Barstead, R., Dresser, M., and Villeneuve, A. M. (1998). Meiotic recombination in *C-elegans* initiates by a conserved mechanism and is dispensable for homologous chromosome synapsis. *Cell* 94, 387–398. doi: 10.1016/S0092-8674(00)81481-6

Dickey, J. S., Redon, C. E., Nakamura, A. J., Baird, B. J., Sedelnikova, O. A., and Bonner, W. M. (2009). H2AX: functional roles and potential applications. *Chromosoma* 118, 683–692. doi: 10.1007/s00412-009-0234-4

Grelon, M., Vezon, D., Gendrot, G., and Pelletier, G. (2001). AtSPO11-1 is necessary for efficient meiotic recombination in plants. *EMBO J.* 20, 589–600. doi: 10.1093/emboj/20.3.589

Hartung, F., and Puchta, H. (2000). Molecular characterisation of two paralogous *SPO11* homologues in *Arabidopsis thaliana*. *Nucleic Acids Res.* 28, 1548–1554. doi: 10.1093/nar/28.7.1548

Hartung, F., Wurz-Wildersinn, R., Fuchs, J., Schubert, I., Suer, S., and Puchta, H. (2007). The catalytically active tyrosine residues of both SPO11-1 and SPO11-2 are required for meiotic double-strand break induction in *Arabidopsis*. *Plant Cell* 19, 3090–3099. doi: 10.1105/tpc.107.054817

Henderson, K. A., Kee, K., Maleki, S., Santini, P. A., and Keeney, S. (2006). Cyclin-dependent kinase directly regulates initiation of meiotic recombination. *Cell* 125, 1321–1332. doi: 10.1016/j.cell.2006.04.039

Hunter, N., and Kleckner, N. (2001). The single-end invasion: an asymmetric intermediate at the double-strand break to double-holliday junction transition of meiotic recombination. *Cell* 106, 59–70. doi: 10.1016/S0092-8674(01)00430-5

Jain, M., Tyagi, A. K., and Khurana, J. P. (2006). Overexpression of putative *topoisomerase 6* genes from rice confers stress tolerance in transgenic *Arabidopsis* plants. *FEBS J.* 273, 5245–5260. doi: 10.1111/j.1742-4658.2006.05518.x

Jain, M., Tyagi, A. K., and Khurana, J. P. (2008). Constitutive expression of a meiotic recombination protein gene homolog, *OsTOP6A1*, from rice confers abiotic stress tolerance in transgenic *Arabidopsis* plants. *Plant Cell Rep.* 27, 767–778. doi: 10.1007/s00299-007-0491-8

Keeney, S. (2001). Mechanism and control of meiotic recombination initiation. *Curr. Top. Dev. Biol.* 52, 1–53. doi: 10.1016/S0070-2153(01)52008-6

Keeney, S. (2008). Spo11 and the formation of DNA double-strand breaks in Meiosis. *Genome Dyn. Stab.* 2, 81–123. doi: 10.1007/7050_2007_026

Keeney, S., Giroux, C. N., and Kleckner, N. (1997). Meiosis-specific DNA double-strand breaks are catalyzed by Spo11, a member of a widely conserved protein family. *Cell* 88, 375–384. doi: 10.1016/S0092-8674(00)81876-0

Kinner, A., Wu, W. Q., Staudt, C., and Iliakis, G. (2008). gamma-H2AX in recognition and signaling of DNA double-strand breaks in the context of chromatin. *Nucleic Acids Res.* 36, 5678–5694. doi: 10.1093/nar/gkn550

Kobayashi, J. (2004). Molecular mechanism of the recruitment of NBS1/hMRE11/hRAD50 complex to DNA double-strand breaks: NBS1 binds to gamma-H2AX through FHA/BRCT domain. *J. Radiat. Res.* 45, 473–478. doi: 10.1269/jrr.45.473

Kumar, R., Bourbon, H. M., and De Massy, B. (2010). Functional conservation of Mei4 for meiotic DNA double-strand break formation from yeasts to mice. *Genes Dev.* 24, 1266–1280. doi: 10.1101/gad.571710

Li, W. X., and Ma, H. (2006). Double-stranded DNA breaks and gene functions in recombination and meiosis. *Cell Res.* 16, 402–412. doi: 10.1038/sj.cr.7310052

Libby, B. J., Reinholdt, L. G., and Schimenti, J. C. (2003). Positional cloning and characterization of *Mei1*, a vertebrate-specific gene required for normal meiotic chromosome synapsis in mice. *Proc. Natl. Acad. Sci. U.S.A.* 100, 15706–15711. doi: 10.1073/pnas.2432067100

Luo, Q., Li, Y., Shen, Y., and Cheng, Z. (2014). Ten years of gene discovery for meiotic event control in rice. *J. Genet. Genomics* 41, 125–137. doi: 10.1016/j.jgg.2014.02.002

Luo, Q., Tang, D., Wang, M., Luo, W. X., Zhang, L., Qin, B. X., et al. (2013). The role of OsMSH5 in crossover formation during rice meiosis. *Mol. Plant* 6, 729–742. doi: 10.1093/mp/sss145

McKim, K. S., and Hayashi-Hagihara, A. (1998). *mei-W68* in *Drosophila melanogaster* encodes a Spo11 homolog: evidence that the mechanism for initiating meiotic recombination is conserved. *Genes Dev.* 12, 2932–2942. doi: 10.1101/gad.12.18.2932

Miao, C. B., Tang, D., Zhang, H. G., Wang, M., Li, Y. F., Tang, S. Z., et al. (2013). CENTRAL REGION COMPONENT1, a novel synaptonemal complex component, is essential for meiotic recombination initiation in rice. *Plant Cell* 25, 2998–3009. doi: 10.1105/tpc.113.113175

Mimitou, E. P., and Symington, L. S. (2009). DNA end resection: many nucleases make light work. *DNA Repair* 8, 983–995. doi: 10.1016/j.dnarep.2009.04.017

Minter-Dykhouse, K., Ward, I., Huen, M. S. Y., Chen, J., and Lou, Z. K. (2008). Distinct versus overlapping functions of MDC1 and 53BP1 in DNA damage response and tumorigenesis. *J. Cell Biol.* 181, 727–735. doi: 10.1083/jcb.200801083

Nonomura, K. I., Nakano, M., Eiguchi, M., Suzuki, T., and Kurata, N. (2006). PAIR2 is essential for homologous chromosome synapsis in rice meiosis I. *J. Cell Sci.* 119, 217–225. doi: 10.1242/jcs.02736

Page, S. L., and Hawley, R. S. (2003). Chromosome choreography: the meiotic ballet. *Science* 301, 785–789. doi: 10.1126/science.1086605

Paques, F., and Haber, J. E. (1999). Multiple pathways of recombination induced by double-strand breaks in *Saccharomyces cerevisiae*. *Microbiol. Mol. Biol. Rev.* 63, 349–404.

Paull, T. T., Rogakou, E. P., Yamazaki, V., Kirchgessner, C. U., Gellert, M., and Bonner, W. M. (2000). A critical role for histone H2AX in recruitment of repair factors to nuclear foci after DNA damage. *Curr. Biol.* 10, 886–895. doi: 10.1016/S0960-9822(00)00610-2

Romanienko, P. J., and Camerini-Otero, R. D. (2000). The mouse *Spo11* gene is required for meiotic chromosome synapsis. *Mol. Cell* 6, 975–987. doi: 10.1016/S1097-2765(00)00097-6

Sasanuma, H., Hirota, K., Fukuda, T., Kakusho, N., Kugou, K., Kawasaki, Y., et al. (2008). Cdc7-dependent phosphorylation of Mer2 facilitates initiation of yeast meiotic recombination. *Genes Dev.* 22, 398–410. doi: 10.1101/gad.1626608

Shao, T., Tang, D., Wang, K. J., Wang, M., Che, L. X., Qin, B. X., et al. (2011). OsREC8 is essential for chromatid cohesion and metaphase I monopolar orientation in rice meiosis. *Plant Physiol.* 156, 1386–1396. doi: 10.1104/pp.111.177428

Shen, Y., Tang, D., Wang, K. J., Wang, M., Huang, J., Luo, W. X., et al. (2012). ZIP4 in homologous chromosome synapsis and crossover formation in rice meiosis. *J. Cell Sci.* 125, 2581–2591. doi: 10.1242/jcs.090993

Shingu, Y., Tokai, T., Agawa, Y., Toyota, K., Ahamed, S., Kawagishi-Kobayashi, M., et al. (2012). The double-stranded break-forming activity of plant SPO11s and a novel rice SPO11 revealed by a *Drosophila* bioassay. *BMC Mol. Biol.* 13:1. doi: 10.1186/1471-2199-13-1

Shinohara, M., Oh, S. D., Hunter, N., and Shinohara, A. (2008). Crossover assurance and crossover interference are distinctly regulated by the ZMM proteins during yeast meiosis. *Nat. Genet.* 40, 299–309. doi: 10.1038/ng.83

Sprink, T., and Hartung, F. (2014). The splicing fate of plant *SPO11* genes. *Front. Plant Sci.* 5:214. doi: 10.3389/fpls.2014.00214

Stacey, N. J., Kuromori, T., Azumi, Y., Roberts, G., Breuer, C., Wada, T., et al. (2006). *Arabidopsis* SPO11-2 functions with SPO11-1 in meiotic recombination. *Plant J.* 48, 206–216. doi: 10.1111/j.1365-313X.2006.02867.x

Symington, L. S. (2002). Role of *RAD52* epistasis group genes in homologous recombination and double-strand break repair. *Microbiol. Mol. Biol. Rev.* 66, 630–670. doi: 10.1128/MMBR.66.4.630-670.2002

Tang, D., Miao, C. B., Li, Y. F., Wang, H. J., Liu, X. F., Yu, H. X., et al. (2014). OsRAD51C is essential for double-strand break repair in rice meiosis. *Front. Plant Sci.* 5:167. doi: 10.3389/fpls.2014.00167

Wan, L., Niu, H., Futcher, B., Zhang, C., Shokat, K. M., Boulton, S. J., et al. (2008). Cdc28-Clb5 (CDK-S) and Cdc7-Dbf4 (DDK) collaborate to initiate meiotic recombination in yeast. *Genes Dev.* 22, 386–397. doi: 10.1101/gad.1626408

Wang, G. F., Kong, H. Z., Sun, Y. J., Zhang, X. H., Zhang, W., Altman, N., et al. (2004). Genome-wide analysis of the cyclin family in *Arabidopsis* and comparative phylogenetic analysis of plant cyclin-like proteins. *Plant Physiol.* 135, 1084–1099. doi: 10.1104/pp.104.040436

Wang, K. J., Tang, D., Wang, M., Lu, J. F., Yu, H. X., Liu, J. F., et al. (2009). MER3 is required for normal meiotic crossover formation, but not for presynaptic alignment in rice. *J. Cell Sci.* 122, 2055–2063. doi: 10.1242/jcs.049080

Wang, K. J., Wang, M., Tang, D., Shen, Y., Qin, B. X., Li, M., et al. (2011). PAIR3, an axis-associated protein, is essential for the recruitment of recombination elements onto meiotic chromosomes in rice. *Mol. Biol. Cell* 22, 12–19. doi: 10.1091/mbc.E10-08-0667

Wang, M., Wang, K. J., Tang, D., Wei, C. X., Li, M., Shen, Y., et al. (2010). The central element protein ZEP1 of the synaptonemal complex regulates the number of crossovers during meiosis in rice. *Plant Cell* 22, 417–430. doi: 10.1105/tpc.109.070789

Yu, H. X., Wang, M., Tang, D., Wang, K. J., Chen, F. L., Gong, Z. Y., et al. (2010). OsSPO11-1 is essential for both homologous chromosome pairing and crossover formation in rice. *Chromosoma* 119, 625–636. doi: 10.1007/s00412-010-0284-7

Zhang, C., Song, Y., Cheng, Z. H., Wang, Y. X., Zhu, J., Ma, H., et al. (2012). The *Arabidopsis thaliana* DSB formation (*AtDFO*) gene is required for meiotic double-strand break formation. *Plant J.* 72, 271–281. doi: 10.1111/j.1365-313X.2012.05075.x

Zickler, D., and Kleckner, N. (1999). Meiotic chromosomes: integrating structure and function. *Annu. Rev. Genet.* 33, 603–754. doi: 10.1146/annurev.genet.33.1.603

8

A little bit of sex matters for genome evolution in asexual plants

Diego Hojsgaard* and Elvira Hörandl*

Department of Systematics, Biodiversity and Evolution of Plants, Albrecht-von-Haller Institute for Plant Sciences, Georg-August University of Göttingen, Göttingen, Germany

Edited by:
Ravishankar Palanivelu, University of Arizona, USA

Reviewed by:
John E. Fowler, Oregon State University, USA
Richard D. Noyes, University of Central Arkansas, USA

*Correspondence:
Diego Hojsgaard and Elvira Hörandl, Department of Systematics, Biodiversity and Evolution of Plants, Albrecht-von-Haller Institute for Plant Sciences, Georg-August University of Göttingen, Untere Karspüle 2, D-37073 Göttingen, Germany
e-mail: diego.hojsgaard@ biologie.uni-goettingen.de; elvira.hoerandl@ biologie.uni-goettingen.de

Genome evolution in asexual organisms is theoretically expected to be shaped by various factors: first, hybrid origin, and polyploidy confer a genomic constitution of highly heterozygous genotypes with multiple copies of genes; second, asexuality confers a lack of recombination and variation in populations, which reduces the efficiency of selection against deleterious mutations; hence, the accumulation of mutations and a gradual increase in mutational load (Muller's ratchet) would lead to rapid extinction of asexual lineages; third, allelic sequence divergence is expected to result in rapid divergence of lineages (Meselson effect). Recent transcriptome studies on the asexual polyploid complex *Ranunculus auricomus* using single-nucleotide polymorphisms confirmed neutral allelic sequence divergence within a short time frame, but rejected a hypothesis of a genome-wide accumulation of mutations in asexuals compared to sexuals, except for a few genes related to reproductive development. We discuss a general model that the observed incidence of facultative sexuality in plants may unmask deleterious mutations with partial dominance and expose them efficiently to purging selection. A little bit of sex may help to avoid genomic decay and extinction.

Keywords: apomixis, Muller's ratchet, Meselson effect, polyploidy, heterozygosity

INTRODUCTION

Currently, the understanding of evolution patterns of genomes on different phylogenetic groups is a hot topic in evolutionary biology. The arrival of next generation sequencing (NGS) technologies and the generation of huge amounts of genomic data is allowing researchers to dig in the past and better resolve organisms' natural history as well as evolutionary enigmas. One of such enigmas is the predominance of sex in nature (Otto, 2009). One of the most prominent theories explaining the benefits of sex (for broad analyses see, e.g., Bell, 1982; Birdsell and Wills, 2003) proposes that sexuality protects the genome from the accumulation of deleterious mutations (Muller, 1964; Kondrashov, 1988; Hörandl, 2009; **Figure 1**). Here we will discuss theoretical assumptions and empirical possibilities of presence/absence of meiosis for asexual plant genome evolution in the light of unexpected recent findings on sexual/asexual taxa of *Ranunculus*.

Sexuality is a crucial factor molding the genomic features of eukaryotes. In plants, the formation of a new individual through sexuality involves an alternation between the sporophytic (2n) and the gametophytic (n) generations via meiosis and gamete fusion, the two mechanisms that create new genetic combinations. Additionally, outcrossing further potentiates genetic variation in populations. Thus, with few exceptions, every single sexual organism has a distinctive genotype that differentiates it from parents and siblings. Therefore, meiosis is the main source of genetic recombination and mixis, and segregates genetic factors in the offspring creating genetic variation. By doing so, meiosis and

sexuality allows natural selection purging a lineage from harmful mutations. Because plant meiosis produces spores (mega- and microspores) and these spores develop into female and male gametophytes, in which considerable percentages of genes are being expressed (Joseph and Kirkpatrick, 2004), selection in a sexual plant will act at two developmental stages: during gametophyte development (haploid gametophytic selection) and after the formation of the zygote (sporophytic selection) Hörandl (2013).

In contrast, by circumventing or suppressing meiosis and syngamy, asexual organisms skip the alternation of generation cycle and hence elude the ploidy-phase change step. In angiosperms, asexually-derived individuals can be formed either as consequence of vegetative propagation, or of asexual seed formation (apomixis), a trait that is taxonomically widespread in plants (Hojsgaard et al., 2014a). While the first involves extra vegetative growth and fragmentation without undergoing the single-cell stage and embryogenesis, the latter comprises the development of a new organism out of an unreduced, unfertilized egg cell, embryogenesis and seed formation (Mogie, 1992). In apomictic plants, a combination of complex developmental features avoid recombination and reductional steps present in the normal sexual reproductive process, thus developing a seed carrying a clonal embryo (Asker and Jerling, 1992).

A central fact for genome evolution, however, is that apomixis in angiosperms is rarely obligate. Apomictic plants produce asexual and sexual progeny within the same offspring generation, i.e., from different ovules and seeds in the same mother plant,

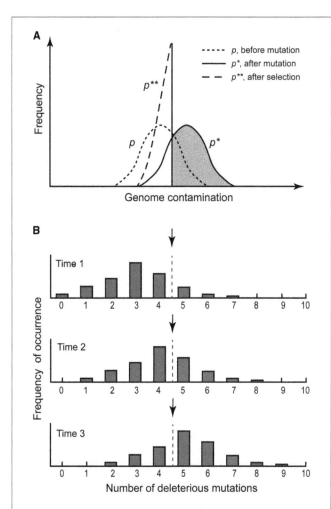

FIGURE 1 | Principles of Muller's ratchet. (A) Scheme of distributions (p) of mutations in a sexual population. Before mutation, distribution in the population is (p), after mutation, distribution shifts upward to p*. After recombination and selection against mutants, individuals in the gray part remain sterile and die, and the distribution goes backward to p**. At equilibrium the means of p and p** are equal (redrawn after Kondrashov, 1988). (B) Scheme of mutational load distributions in an asexual population. Initially, genotypes with zero mutations exist in the population, but are lost over time by drift. Without recombination, the class with zero or few mutations cannot be restored, and consequently mutations accumulate until a threshold level of extinction (arrow) is reached (redrawn after Maynard Smith, 1988).

and therefore asexuality is facultative. Consequently, a proportion of the offspring represents recombinants, but frequencies of sexuality vary a lot among genera, species and different modes of apomixis (e.g., Aliyu et al., 2010; Sartor et al., 2011; Šarhanová et al., 2012; Noyes and Givens, 2013; Hojsgaard et al., 2013, 2014a). The role that facultative sexuality and genetically highly diverse apomictic populations play in the evolution of angiosperms is still unclear.

THEORETICAL SIGNIFICANCE OF APOMIXIS FOR THE EVOLUTION OF THE PLANT GENOME

Sexuality has the effect that deleterious mutations appear in various genotypic configurations in the offspring. Thus, harmful mutations expressed under different states (e.g., homozygous, dominant heterozygous, etc.) will negatively affect the genotype's fitness and natural selection will remove such genotypes and purge the lineage from an increase in the mutational load. Evolutionary benefits of purging a lineage from mutation accumulation have long been seen as a major advantage of sexuality (Muller, 1964; Kondrashov, 1988; **Figure 1A**). Apomictic plants, by circumventing genetic reshuffling mechanisms, inherit the same genomic features of their female parental genome. Without recombination, once a genotype acquire a spontaneous mutation cannot reconstitute a non-mutated genotypic state (**Figure 1B and 2**). Thus, a deleterious mutation in any asexual individual will be transmitted to all the offspring and loaded onto the gene pool of that clonal lineage. In any segment of a genome with absence of recombination, the number of random independent mutations is expected to increase because a mutational load smaller than the least-loaded lineage can never be generated (the ratchet mechanism, Muller, 1964; the hatchet mechanism, Kondrashov, 1988). Over time, drift will ultimately lead to a loss of genotypes with a lower mutational load and, once a threshold on mutational load is reached, to extinction of the asexual lineage (**Figure 1B**; Kimura et al., 1963; Muller, 1964; Felsenstein, 1974). Moreover, genetic interference effects between loci (Hill-Robertson effect, Hill and Robertson, 1966) may increase effects of mutations (Kondrashov, 1988). The genetic load of clonal lineages will reduce their fitness and obstruct further adaptation, driving those lineages to an early extinction (Maynard Smith, 1978; Bell, 1982).

Apomixis in plants is nearly exclusive associated polyploids, and often it is a result of hybridity (e.g., Koch et al., 2003; Paun et al., 2006a; Pellino et al., 2013). Polyploidy can accelerate mutation accumulation since additional gene copies represent additional mutational sites. Effects of deleterious mutations will reduce the mean fitness of any individual (and ultimately of the population) with a rate c U, where c stand for the ploidy level and U for the mutation rate per haploid genome (Gerstein and Otto, 2009). However, the effects of single deleterious recessive mutations in heterozygous states can be masked by a functional gene copy of the wild or dominant allele in a diploid organism (Crow and Kimura, 1970; Kondrashov and Crow, 1991; Otto and Whitton, 2000). After a prolonged diploid stage, the return to haploidy leads to the expression of accumulated, but previously masked deleterious recessive alleles, and selection against mutations (Crow and Kimura, 1970; Hörandl, 2009). However, in a polyploid plant with more than two allele copies per locus, accumulated recessive mutations in heterozygous states may still be masked after a return to haploidy. Thus, masking effects will be stronger and recessive mutations may not be effective unless they show certain level of dominance (i.e., partial dominance). So far few genomic studies are available to understand the forces underlying mutation dynamics in polyploid plants. In theory, absence of haploid gametophytic selection plus masking of recessive alleles in polyploid condition should increase the mutational load compared to sexuals.

Allelic sequence divergence, the so-called Meselson effect, is another consequence of long-term asexuality (Mark Welch and Meselson, 2000). Asexual seed reproduction, by suppressing meiosis promotes the divergence between allelic sequences by neutral mutations (Kimura and Crow, 1964). Due to the loss of sexuality, alleles within lineages will gain neutral differences at a much higher rate than the normal substitution rates observed between alleles in sexual populations (Birky, 1996). Until now, the presence of genome-wide allele sequence divergence significantly larger than those in inter-lineages have been hard to prove. Processes such as gene conversion, mitotic recombination, efficient DNA repair, meiotic parthenogenesis in animals (automixis), occasional sex, ploidy reduction, and hybridization can moderate or remove sequence divergence (e.g., Schön and Martens, 1998; Schaefer et al., 2006; Liu et al., 2007; Mark Welch et al., 2008; Flot et al., 2013). Effects of facultative sexuality on allelic sequence divergence in plants remain unknown.

IS A LITTLE BIT OF SEX SUFFICIENT TO AVOID MUTATION ACCUMULATION?

A recent transcriptome study of the *Ranunculus auricomus* complex, a system of diploid sexual species and hexaploid apomictic hybrids, in fact showed Meselson-like sequence divergence effects, but data did not support the idea of mutation accumulation (Pellino et al., 2013). Analyses of Muller's ratchet on high-quality single nucleotide polymorphisms (SNPs) and indels obtained from RNA-seq revealed for 1231 annotated genes that ratios of non-synonymous vs. synonymous substitutions (dN/dS) were mostly clearly below one, and did not differ significantly between apomictic-apomictic, apomictic-sexual and sexual-sexual comparisons (Pellino et al., 2013). A number of all annotated genes showed high dN/dS ratios (outlier values; see Figure 3 in Pellino et al., 2013), and hence appeared to be under divergent selection. A gene ontology analysis of outliers showed that a small proportion of those genes ($n = 62$; 6.7%; Table 5 in Pellino et al., 2013) were associated with processes involved in meiosis and gametogenesis. Strikingly, most such outlier genes ($n = 41$; 66.1%) were found in the apomictic-sexual comparison and thus indicated a significant enrichment of genes associated to reproductive shifts during ovule development when compared to those outliers in sexual (or apomictic) genomes. Whether these mutations have positive or negative effects, needs further investigations. However, most mutations that are under selection do have strongly deleterious effects (e.g., Loewe and Hill, 2010). Since we analyzed only RNA sequences we assume that the observed non-synonymous substitutions mostly have negative effects (dominant or partial dominant). If deleterious mutations would have accumulated genome-wide in the apomicts–a situation expected following the nature of the genetic code and transition/transversion rates (e.g., Yang and Bielawski, 2000)—they would drive the dN/dS ratios over 1. One interpretation of such results is that apomicts in *Ranunculus* are evolutionarily too young (c. 70,000 years) to have accumulated significant mutations and hence dN/dS ratios are still similar to those in sexual putative parentals.

The alternative explanation, however, assumes that apomicts do not accumulate genome-wide deleterious mutations because facultative sexuality purges deleterious mutations (**Figure 2**). Detailed developmental studies revealed that apomictic hexaploid

Ranunculus hybrids show varying proportions of sexually formed seed in all genotypes, with a grand mean of 29.1% (Hojsgaard et al., 2014b). Reduced seed set and lower pollen quality of apomicts compared to sexuals (Hörandl, 2008; Hojsgaard et al., 2014b) indicate negative effects of apomixis on fitness parameters. Population genetic studies data indicate considerable genetic diversity within and among populations (Paun et al., 2006a). Considering the expected turn-over of recombinant individuals in natural populations, regular sexuality in facultative apomicts can purge mutations via two mechanisms: First, ploidy reduction in a diploid plant can already unmask recessive deleterious mutations in the gametophyte and expose it to purging selection (Hörandl, 2009), while in a polyploid mutations showing partial dominance would be exposed to selection during haploidy (see **Figure 2**, stages 1 and 2); since the gametophytes represent few-celled mini-organisms, a high proportion of the genome is expressed and exposed to selection at this stage. In fact, proportions of sexual development decrease during haploid gametophyte development (Hojsgaard et al., 2013, 2014a). Second, via recombination, mutations will segregate and offspring with variable mutational load will be formed. Additionally, self-fertilization will generate zygotes with higher doses (e.g., hemizygous) of partially dominant mutated alleles, and consequently mutations will become "unmasked" and fully exposed to purging selection in the offspring (see **Figure 2**, stage 3). Thus, those genotypes where deleterious mutations are being expressed will be eliminated upon dosage level and only individuals carrying mutated alleles at low dosages will persist in the population (see **Figure 2**, stage 4). The lineage will consequently be regularly purged by eliminating genotypes carrying these mutations. Hence, novel mutations in a facultative will add up slower than in obligate apomicts in which each mutation is added to the mutational load (**Figure 2**, stages 5 and 6).

The ultimate efficacy will certainly depend on the level of functional meiosis and sexuality occurring in the population. Besides this, purging is expected to be even more efficient under diverse conditions. For example, depending on the level of penetrance and dominance of the mutation, purging would be faster as phenotypic effects would become exposed to natural selection at different dosages. A higher purging efficacy is expected with inbreeding (e.g., Agrawal and Chasnov, 2001). In fact, in a clonal population established from an apomictic mother, all neighboring individuals will carry the same deleterious mutation in their gametes. Occasional facultative sex among individuals will occur among the same genotypes, which is possible because of self-compatibility of apomictic plants (including hexaploid *Ranunculus*; Hörandl, 2008, 2010). Hence, plants effectively conduct self-fertilization, even if cross-pollination takes place among individuals (clone-mates); thus, alleles carrying deleterious mutations may rapidly increase in their dosage, and consequently their effects will be exposed to selection (**Figure 2**).

Another mechanism to increase efficacy of selection can be assumed from epistasis. If additional (recessive or non-recessive) deleterious mutations lead to a larger decrease of fitness because of negative interactions of these genes, then even truncating selec-

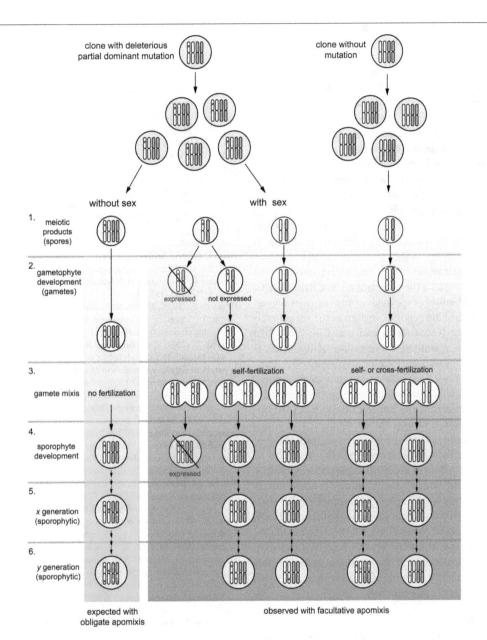

FIGURE 2 | Model of purging mutations in a tetraploid, facultative apomictic plant lineage (blue column) compared to an obligate apomict (without meiosis; green column). For simplicity, the model is presented for a new self-fertile allotetraploid lineage with regularly reduced male gametes; and partial dominant mutations are considered to be deleterious and expressed a 50% penetrance. Moreover, preferred homolog pairing is assumed during meiosis I (e.g., Comai, 2005) and considers only the perspective of a mutated deleterious allele, all other alleles pondered to be functionally equivalent. The effects of absence or presence of meiosis on mutation accumulation are illustrated after one generation following the occurrence of mutation (stages 1–4), and after several generations of obligate (without sex) or facultative apomixis (with residual levels of sex; stages 5–6). 1. Once a deleterious mutation (red star) with a 50% penetrance is loaded onto the clonal offspring, without sex only unreduced female gametes rise (clonal) progeny. With sex recombinant spores are formed. 2. Expression of mutated alleles and deleterious effects would appear only in those gametophytes with a ploidy-phase change; thus, 50% of haploid gametes would be eliminated, biasing expected progeny proportions (but not progeny types). 3. During gamete mixis, parthenogenetic embryo development avoids

egg-cell fertilization in apomictic female gametophytes while meiotic ones can produce an array of progeny types upon self- or cross-fertilization syndromes. 4. A dosage increase (to duplex condition) and full expression of deleterious effects is expected in some recombinant offspring during sporophyte development, but not in non-recombinant ones. Only individuals carrying a low allele dosage (simplex condition) will remain in the population together with those without the mutation. 5–6. After a number of generations, mutations will gradually appear and added up to the genetic load in the obligate apomictic lineage. In the facultative apomictic lineage, occasional sex will segregate mutated alleles and purging selection will eliminate gametophytes and sporophytes with certain allelic dosages (as in stages 1–4). On the long run, an obligate apomictic genotype (left) will become sooner extinct compared to a facultative apomictic lineage which is continuously purged. The model does not yet consider possible purging effects via conversion during meiosis, and does not quantify facultative sexuality and actual frequencies of spore formation. The model fits to higher ploidy levels if the same penetrance level is assumed in mutated alleles. Assorted colored mutations represent independent events arisen randomly in the genome at different times.

tion can act and rapidly eliminate this genotype (Kondrashov, 1988). Hence, despite nearly-obligate apomictic clones within a population could accumulate mutations for some generations, fitter recombinant genotypes with a lower mutational load will continuously replace them (see **Figure 2**). Population genetic data on *Ranunculus* strongly support this hypothesis of clonal turnover (Paun et al., 2006b). Consequently, Muller's ratchet is halted or at least slowed down for the lineage as whole. A fourth possibility to avoid mutation accumulation is a large effective population size (Kondrashov, 1988) and/or high migration rates amid populations (Whitton et al., 2008). Hence, within populations genetic variability will be kept at high levels and selection will act on different genotypes. Large geographical distribution and clonal diversity of the hexaploid hybrids (see Paun et al., 2006a,b) suggest that also this factor contribute to genome evolution in *Ranunculus*.

The evolutionary consequences of facultative apomixis in plants have so far received little attention. Mendelian genetic studies on control mechanism of apomixis in angiosperms suggest that the apomixis-controlling genomic regions occur -in general- in a heterozygous state (Ozias-Akins and van Dijk, 2007). In *Ranunculus auricomus*, quantitative expression of apomixis is dosage-dependent on the apospory factor (A), which is a dominant Mendelian factor with variable penetrance in the sporophyte, but with lethal effects in haploid or homozygous states (Nogler, 1984). Consequently, (A) appears always heterozygous with the wild type allele a in various allelic configurations, which means that apomixis cannot become completely obligate (Nogler, 1984). The long term effects of facultative sexuality remain to be studied. Overall, data (Pellino et al., 2013) suggest that apomictic polyploid lineages on the one hand accumulate Meselson-effect-like neutral substitutions in divergent gene copies, and on the other hand, mask partially dominant deleterious alleles in clones, which may become exposed to purifying selection via facultative sexuality. Comprehensive empirical studies will be needed to further test theoretical models and answer the question of how much and to what extend "a little bit of sex" protects apomictic plants from genomic decay and extinction.

ACKNOWLEDGMENTS

We thank TF Sharbel and J.-F Flot for insightful discussions. For financial support we thank the Austrian Research Foundation (FWF) under project I 310-B16 (to EH).

REFERENCES

Agrawal, A. F., and Chasnov, J. R. (2001). Recessive mutations and maintenance of sex in structured populations. *Genetics* 158, 913–917.

Aliyu, O. M., Schranz, M. E., and Sharbel, T. (2010). Quantitative variation for apomictic reproduction in the genus *Boechera* (Brassicaceae). *Am. J. Bot.* 97, 1719–1731. doi: 10.3732/ajb.1000188

Asker, S., and Jerling, L. (1992). *Apomixis in Plants*. Boca Raton: CRC press.

Bell, G. (1982). *The Masterpiece of Nature: The Evolution and Genetics of Sexuality*. Berkeley: University of California Press.

Birdsell, J. A., and Wills, C. (2003). The evolutionary origin and maintenance of sexual recombination: a review of contemporary models. *Evol. Biol.* 33, 27–138. doi: 10.1007/978-1-4757-5190-1_2

Birky, C. W. (1996). Heterozygosity, heteromorphy, and phylogenetic trees in asexual eukaryotes. *Genetics* 144, 427–437.

Comai, L. (2005). The advantages and disadvantages of being polyploid. *Nat. Rev. Genet.* 6, 836–846.

Crow, J. F., and Kimura. M. (1970). *An Introduction to Population Genetics Theory*. New York: Harper and Row.

Felsenstein, J. (1974). The evolutionary advantage of recombination. *Genetics* 78, 737–756.

Flot, J.-F., Hespeels, B., Li, X., Noel, B., Arkhipova, I., Danchin, E. G. J., et al. (2013). Genomic evidence for ameiotic evolution in the bdelloid rotifer *Adineta vaga*. *Nature* 500, 453–457. doi: 10.1038/nature12326

Gerstein, A. C., and Otto, S. P. (2009). Ploidy and the causes of genomic evolution. *J. Hered* 100, 571–581. doi: 10.1093/jhered/esp057

Hill, W. G., and Robertson, A. (1966). The effect of linkage on limits to artificial selection. *Genetic Res.* 8, 269–294. doi: 10.1017/S0016672300010156

Hojsgaard, D. H., Martinez, E. J., and Quarin, C. L. (2013). Competition between meiotic and apomictic pathways during ovule and seed development results in clonality. *New Phytol.* 197, 336–347. doi: 10.1111/j.1469-8137.2012.04381.x

Hojsgaard, D. H., Klatt, S., Baier, R., Carman, J. G., and Hörandl, E. (2014a). Taxonomy and biogeography of apomixis in angiosperms and associated biodiversity characteristics. *Crit. Rev. Plant Sci.* 33, 1–14. doi: 10.1080/07352689.2014.898488

Hojsgaard, D. H., Greilhuber, J., Pellino, M., Paun, O., Sharbel, T. F., and Hörandl, E. (2014b). Emergence of apospory and bypass of meiosis via apomixis after sexual hybridisation and polyploidization. *New Phytol.* 204, 1000–1012. doi: 10.1111/nph.12954

Hörandl, E. (2008). Evolutionary implications of self-compatibility and reproductive fitness in the apomictic *Ranunculus auricomus* polyploid complex (Ranunculaceae). *Int. J. Plant Sci.* 169, 1219–1228. doi: 10.1086/591980

Hörandl, E. (2009). A combinational theory for maintenance of sex. *Heredity* 103, 445–457. doi: 10.1038/hdy.2009.85

Hörandl, E. (2010). The evolution of self-fertility in apomictic plants. *Sex. Plant Reprod.* 23, 73–86. doi: 10.1007/s00497-009-0122-3

Hörandl, E. (2013). "Meiosis and the paradox of sex in nature," in *Meiosis*, eds C. Bernstein and H. Bernstein (Rijeka: InTech—Open Access Publisher), 17–39. doi: 10.5772/56542

Joseph, S. B., and Kirkpatrick, M. (2004). Haploid selection in animals. *Trends Ecol. Evol.* 19, 592–597. doi: 10.1016/j.tree.2004.08.004

Kimura, M., and Crow, J. F. (1964). The number of alleles that can be maintained in a finite population. *Genetics* 49, 725–738.

Kimura, M., Maruyama, T., and Crow, J. F. (1963). The mutation load in small populations. *Genetics* 48, 1303–1312.

Koch, M. A., Dobeš, C., and Mitchell-Olds, T. (2003). Multiple hybrid formation in natural populations: concerted evolution of the Internal Transcribed Spacer of nuclear ribosomal DNA (ITS) in North American Arabis divaricarpa (Brassicaceae). *Mol. Biol. Evol.* 20, 338–350. doi: 10.1093/molbev/msg046

Kondrashov, A. S. (1988). Deleterious mutations and the evolution of sexual reproduction. *Nature* 336, 435–440. doi: 10.1038/336435a0

Kondrashov A. S., and Crow, J. F. (1991). Haploidy or diploidy: which is better? *Nature* 351, 314–315. doi: 10.1038/351314a0

Liu, Q. L., Varghese, P. T., and Williamson, V. M. (2007). Meiotic parthenogenesis in a root-knot nematode results in rapid genomic homozygosity. *Genetics* 176, 1483–1490. doi: 10.1534/genetics.107.071134

Loewe, L., and Hill, W. G. (2010). The population genetics of mutations: good, bad, and indifferent. *Phil. Trans. Roy. Soc. B* 365, 1153–1167. doi: 10.1098/rstb.2009.0317

Mark Welch, D. B., Mark Welch, J. L., and Meselson, M. (2008). Evidence for degenerate tetraploidy in bdelloid rotifers. *Proc. Natl. Acad. Sci. U.S.A.* 105, 5145–5149. doi: 10.1073/pnas.0800972105

Mark Welch, D., and Meselson, M. (2000). Evidence for the evolution of bdelloid rotifers without sexual reproduction or genetic exchange. *Science* 288, 1211–1215. doi: 10.1126/science.288.5469.1211

Maynard Smith, J. (1978). *The Evolution of Sex*. Cambridge: Cambridge University Press.

Maynard Smith, J. (1988). "The evolution of recombination" in *The Evolution of Sex. An Examination of Current Ideas*, eds R.E. Michod and B.R. Levin (Sinauer, Sunderland, MA), 106–125.

Mogie, M. (1992). *The Evolution of Asexual Reproduction in Plants*. London: Chapman and Hall.

Muller, H. J. (1964). The relation of recombination to mutational advance. *Mutat. Res.* 106, 2–9. doi: 10.1016/0027-5107(64)90047-8

Nogler, G. A. (1984). Genetics of apospory in apomictic *Ranunculus auricomus*: 5. Conclusion. *Bot. Helv.* 94, 411–423.

Noyes, R. D., and Givens, A. D. (2013). Quantitative assessment of megasporogenesis for the facultative apomicts *Erigeron annuus* and *Erigeron strigosus* (Asteraceae). *Int. J. Plant Sci.* 174, 1239–1250. doi: 10.1086/673243

Otto, S. P. (2009). The evolutionary enigma of sex. *Am. Nat.* 174, S1–S14. doi: 10.1086/599084

Otto, S. P., and Whitton, J. (2000). Polyploidy incidence and evolution. *Annu. Rev. Genet.* 34, 401–437. doi: 10.1146/annurev.genet.34.1.401

Ozias-Akins, P., and van Dijk, P. J. (2007). Mendelian genetics of apomixis in plants. *Ann. Rev. Genetics* 41, 509–537. doi: 10.1146/annurev.genet.40.110405.090511

Paun, O., Stuessy, T. F., and Hörandl, E. (2006a). The role of hybridization, polyploidization and glaciation in the origin and evolution of the apomictic *Ranunculus cassubicus* complex. *New Phytol.* 171, 223–236. doi: 10.1111/j.1469-8137.2006.01738.x

Paun, O., Greilhuber, J., Temsch, E., and Hörandl, E. (2006b). Patterns, sources and ecological implications of clonal diversity in apomictic Ranunculus carpaticola (*Ranunculus auricomus*) complex, Ranunculaceae. *Mol. Ecol.* 15, 897–930. doi: 10.1111/j.1365-294X.2006.02800.x

Pellino, M., Hojsgaard, D., Schmutzer, T., Scholz, U., Hörandl, E., Vogel, H., et al. (2013). Asexual genome evolution in the apomictic *Ranunculus auricomus* complex: examining the effects of hybridization and mutation accumulation. *Mol. Ecol.* 22, 5908–5921. doi: 10.1111/mec.12533

Šarhanová, P., Vašut, R. J., Dančák, M., Bureš, P., and Trávníček, B. (2012). New insights into the variability of reproduction modes in European populations of *Rubus* subgen. *Rubus*: how sexual are polyploid brambles? *Sex. Plant Reprod.* 25, 319–335. doi: 10.1007/s00497-012-0200-9

Sartor, M. E., Quarin, C. L., Urbani, M. H., and Espinoza, F. (2011). Ploidy levels and reproductive behaviour in natural populations of five *Paspalum* species. *Plant Syst. Evol.* 293, 31–41. doi: 10.1007/s00606-011-0416-4

Schaefer, I., Domes, K., Heethoff, M., Schnider, K., Schön, I., Norton, R. A., et al. (2006). No evidence for the "Meselson effect" in parthenogenetic oribatid mites (Oribatida, Acari). *J. Evol. Biol.* 19, 184–193. doi: 10.1111/j.1420-9101.2005.00975.x

Schön, I. and Martens, K. (1998). Opinion: DNA-repair in ancient asexuals: a new solution to an old problem? *J. Nat. Hist.* 32, 943–948.

Whitton, J., Sears, C. J., Baack, E. J., and Otto, S. P. (2008). The dynamic nature of apomixis in the angiosperms. *Int. J. Plant Sci.* 169, 169–182. doi: 10.1086/523369

Yang, Z., and Bielawski, J. P. (2000). Statistical methods for detecting molecular adaptation. *Trends Ecol. Evol.* 15, 496–503. doi: 10.1016/S0169-5347(00)01994-7

9

Using giant scarlet runner bean embryos to uncover regulatory networks controlling suspensor gene activity

Kelli F. Henry and Robert B. Goldberg*

Department of Molecular, Cell and Developmental Biology, University of California, Los Angeles, Los Angeles, CA, USA

Edited by:
Ravishankar Palanivelu, University of Arizona, USA

Reviewed by:
Stewart Gillmor, Centro de Investigación y de Estudios Avanzados del Instituto Politécnico Nacional, Mexico
Martin Bayer, Max Planck Institute for Developmental Biology, Germany

*Correspondence:
Robert B. Goldberg, Department of Molecular, Cell and Developmental Biology, University of California, Los Angeles, 610 Charles E. Young Drive East, Los Angeles, CA 90095, USA
e-mail: bobg@ucla.edu

One of the major unsolved issues in plant development is understanding the regulatory networks that control the differential gene activity that is required for the specification and development of the two major embryonic regions, the embryo proper and suspensor. Historically, the giant embryo of scarlet runner bean (SRB), *Phaseolus coccineus*, has been used as a model system to investigate the physiological events that occur early in embryogenesis—focusing on the question of what role the suspensor region plays. A major feature distinguishing SRB embryos from those of other plants is a highly enlarged suspensor containing at least 200 cells that synthesize growth regulators required for subsequent embryonic development. Recent studies have exploited the giant size of the SRB embryo to micro-dissect the embryo proper and suspensor regions in order to use genomics-based approaches to identify regulatory genes that may be involved in controlling suspensor and embryo proper differentiation, as well as the cellular processes that may be unique to each embryonic region. Here we review the current genomics resources that make SRB embryos a compelling model system for studying the early events required to program embryo development.

Keywords: *Phaseolus coccineus*, scarlet runner bean, suspensor, gene regulatory network, *cis*-regulatory elements, transcriptome, comparative genomics

WHY STUDY THE SUSPENSOR?

Embryogenesis in most higher plants begins with a double fertilization event, in which one sperm cell fertilizes the egg cell to form the zygote, and the other fertilizes the central cell to form the endosperm (Bleckmann et al., 2014). The zygote undergoes an asymmetric cell division, giving rise to a small, cytoplasm-rich apical cell and a large, vacuolated basal cell (West and Harada, 1993). The apical cell divides to form the embryo proper, which becomes the next generation plant, whereas the basal cell divides to form the suspensor, a terminally differentiated structure that transports nutrients to the embryo proper (Yeung, 1980; Nagl, 1990) and degenerates as the embryo matures (Yeung and Meinke, 1993). The uppermost cell of the suspensor, the hypophysis, contributes to the root meristem of the embryo (Dolan et al., 1993). While much is known about embryo proper development, comparatively little is known about the suspensor (Lau et al., 2012; Wendrich and Weijers, 2013). Genetic studies in *Arabidopsis* have illuminated some processes leading to suspensor differentiation. The molecular pathways involved in elongation of the zygote, the asymmetric division that forms the two-cell embryo, and apical and basal cell fate specification require (1) auxin signaling (Friml et al., 2003), (2) the YDA/MAPK signaling pathway (Bayer et al., 2009), and (3) the transcriptional networks involving RKD4 (Waki et al., 2011), WRKY2, WOX2, WOX8, and WOX9 (Ueda et al., 2011). However, genes in these pathways account for a very small percentage of the ~11,000 diverse mRNAs detected

in the *Arabidopsis* suspensor (Belmonte et al., 2013), and the molecular mechanisms governing suspensor development and function remain largely elusive. In addition, little is known about (1) the regulatory networks controlling suspensor differentiation and development in species with diverse suspensor morphologies, (2) the mechanisms activating different gene sets in the embryo proper and suspensor after fertilization, and (3) the cellular processes governing suspensor degeneration in later embryo development.

WHY USE SRB TO STUDY SUSPENSOR DIFFERENTIATION?

The physical features of the SRB suspensor (**Figure 1A**), including its massive size, enlarged basal cells, and polytene chromosomes (Nagl, 1962) provide a unique system to study the functional significance of this highly specialized suspensor, the cellular processes shared by all suspensors, and suspensor differentiation events. Additionally, SRB seeds are a protein-rich legume crop, closely related to soybean, common bean, and cowpea in the economically important *Phaseoleae* clade of legumes, and thus can serve as a model for legume seed development. Common bean (*Phaseolus vulgaris*), which is a major source of calories in many developing countries[1] and a $1B crop in the United States[2], and SRB are congeneric species that diverged less than eight million years ago (mya; Lavin et al., 2005) and can form

---

[1]http://faostat.fao.org
[2]http://www.nass.usda.gov

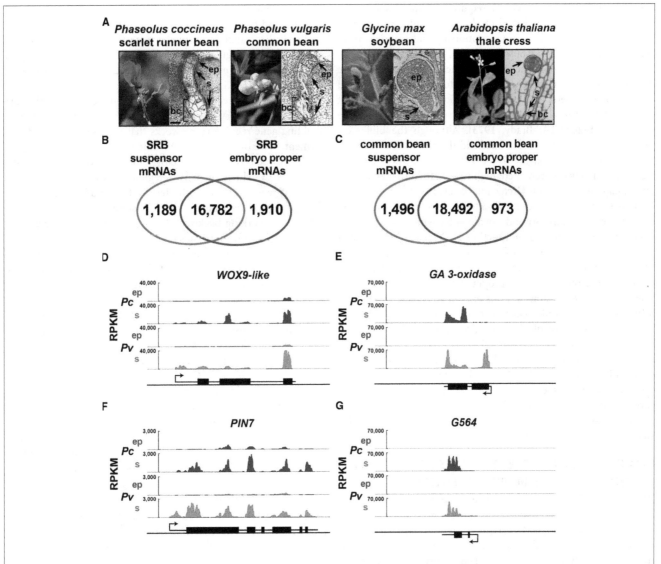

FIGURE 1 | Suspensors with diverse morphologies and bean suspensor-specific gene activity. (A) Scarlet runner bean, common bean, soybean, and *Arabidopsis* plants, plastic sections of globular-stage embryo of SRB, common bean and *Arabidopsis*, and paraffin section of globular-stage embryo of soybean. Common bean flower image was taken from http://www.pbase.com/valterj/dsch9 photographed by Valter Jacinto, and *Arabidopsis* flower image was taken from Kawashima and Goldberg (2010). **(B,C)** Venn diagrams representing the mRNAs detected in SRB **(B)** and common bean **(C)** suspensor and embryo proper. RNA-Seq data for SRB and common bean are from GEO accession GSE57537. **(D–G)** Genome browser views of WOX9-like, GA 3-oxidase, PIN7 and G564 mRNA accumulation levels in SRB and common bean suspensor and embryo proper. Each panel depicts a 5 kb window including the gene structure. Black boxes represent exons. Black lines represent UTRs and introns. Arrows indicate the transcription start site. bc, basal cell(s); ep, embryo proper; RPKM, reads per kilobase per million; Pc, *P. coccineus*; Pv, *P. vulgaris*; s, suspensor. Scale bar: 100 μm.

successful hybrids (Lamprecht, 1941; Thomas, 1964), as was first reported by Mendel in 1865 (cited by Mok et al., 1986). SRB diverged ∼19 mya from soybean (Lavin et al., 2005), the second largest crop in the United States (see text footnote 2). Taken together, SRB is an excellent plant in which to study suspensor development because of (1) its specialized structure, (2) its phylogenetic placement in the legume family, (3) a 40-year history of use as a model for embryo development (Yeung and Meinke, 1993; Kawashima and Goldberg, 2010), and (4) new genomic resources, including (i) the common bean genome sequence (Schmutz et al., 2014) and (ii) gene expression profiles for the SRB suspensor and embryo proper during early embryogenesis

(Le et al., 2007; Kawashima and Goldberg, 2010; GEO accession GSE57536).

WHAT WAS LEARNED FROM USING SRB AS A MODEL FOR SUSPENSOR DEVELOPMENT FOR OVER 40 YEARS?

The first experimental studies of suspensor function were performed by Ian Sussex and collaborators using SRB because the large size of the SRB embryo allows hand-dissection of the suspensor and embryo proper, and facilitates the collection of large amounts of suspensor tissue for use in biochemical studies (Clutter and Sussex, 1968; Walbot et al., 1972; Sussex et al., 1973; Clutter et al., 1974; Lorenzi et al., 1978). Early SRB experiments determined that the suspensor is required for the development of the embryo proper (Cionini et al., 1976; Yeung and Sussex, 1979), and that it is highly transcriptionally and translationally active (Walbot et al., 1972; Sussex et al., 1973; Clutter et al., 1974), in part due to its polytene chromosomes which can increase the DNA content of suspensor cells up to 8,192C (Nagl, 1962; Brady, 1973). There is a progressive increase in the level of polyteny from the chalazal pole of the suspensor to the large basal cells (**Figure 1A**; Brady, 1973). Although the biological function of polytene chromosomes and their puffs and loops in SRB suspensor cells (Nagl, 1970, 1974) is unclear, polyteny is a sign of terminally differentiated, highly specialized tissues such as *Drosophila* salivary glands (Heitz and Bauer, 1933).

Two specialized suspensor functions uncovered from early SRB studies are transport and hormone biosynthesis. The giant basal cells of the suspensor function as "transfer cells," using their enlarged membrane surfaces and prominent ingrowths to absorb solutes from the surrounding seed tissues and transport them to the growing embryo proper (Gunning and Pate, 1969; Nagl, 1974, 1990; Yeung and Sussex, 1979; Yeung, 1980). The SRB suspensor not only acts as a conduit for nutrients, but also synthesizes growth regulators, e.g., gibberellic acid (GA) needed by the embryo proper in early development (Alpi et al., 1975). In fact, biochemical studies showed that SRB suspensors are a rich source of GAs (Alpi et al., 1975) and contain enzymes for synthesizing GAs (Ceccarelli et al., 1979, 1981). Classical approaches carried out 40 years ago revealed that the transport of nutrients and GA biosynthesis are essential processes carried out by the SRB suspensor for embryo development.

HOW HAS GENOMICS BEEN USED TO DISSECT EARLY SRB SUSPENSOR DIFFERENTIATION AND DEVELOPMENT?

Because the SRB embryo is uniquely large, our laboratory was able to hand dissect globular-stage embryo-proper and suspensor regions and use pre-NextGen sequencing approaches—such as differential display, in situ hybridization, EST sequencing, and microarray analysis—to study the gene expression events that occur shortly after fertilization (Weterings et al., 2001; Le et al., 2007; Kawashima and Goldberg, 2010; Le, 2013). These experiments showed that the SRB embryo apical and basal regions transcribe different genes as early as the four-cell stage, suggesting that these regions are specified at the molecular level after division of the zygote (Weterings et al., 2001). At the globular stage there is a large overlap in genes expressed in the embryo proper and suspensor regions that are derived from the apical and basal cells, respectively (Le et al., 2007). Many suspensor-specific SRB genes were identified, however, including (1) all genes in the GA biosynthesis pathway, (2) a *WOX9-like* homeodomain transcription factor gene (*PcWox9-like*), and (3) *PcG564*, a gene of unknown function, among many others (Weterings et al., 2001; Le et al., 2007; Kawashima and Goldberg, 2010; Le, 2013; Henry, 2014). We confirmed these observations by using laser-capture micro-dissection (LCM) technology to collect SRB globular-stage embryo proper and suspensor regions with more precision (Le et al., 2007), RNA-Seq for transcriptome profiling (GEO accession GSE57536), and the common bean (*Phaseolus vulgaris*) as a reference genome (Schmutz et al., 2014; **Figure 1**). The genome

browser view illustrates the up-regulation of *PcGA 3-oxidase*, *PcG564*, and *PcWox9-like* genes in the SRB suspensor, in addition to the *PcPIN7* auxin transporter gene that has been shown by others to be up-regulated in the *Arabidopsis* suspensor and play an essential role in establishing apical-basal polarity (Friml et al., 2003; **Figures 1D–G**).

Knowing the spectrum of transcription factor genes that are active in the embryo proper and suspensor is a first step to building gene regulatory networks that program embryo development. One or more mRNAs unique to each embryo region could encode transcription factors that are directly linked to the processes by which these two regions of the embryo activate different gene sets shortly after fertilization and become specified for different developmental fates (Weterings et al., 2001). Our strategy of working backward from globular-stage gene activity to cell-fate specification is particularly amenable to the suspensor because its differentiation precedes that of the embryo proper, and the suspensor cells are direct clonal descendants of the basal cell of the two-cell embryo (Weterings et al., 2001; Kawashima and Goldberg, 2010). Thus, the factors that activate genes in the suspensor might be directly linked to the basal cell specification mechanism. For example the globular-stage expression pattern of the SRB *PcWOX9-like* gene is remarkably similar to its *Arabidopsis* counterparts *AtWOX8* and *AtWOX9* (Haecker et al., 2004). In *Arabidopsis*, WOX8 mRNA accumulates in the zygote, and is then confined to the basal cell of the two-cell embryo and the globular-stage suspensor (Haecker et al., 2004). *AtWOX8* transcription is regulated, in part, by the WRKY2 transcription factor (Ueda et al., 2011). Thus, the WRKY2-WOX8 pathway functions in establishing zygote polarity by initiating a shift in organelle positions in the zygote enabling asymmetric division to occur (Ueda et al., 2011). Identifying the downstream target genes of *PcWOX9-like*, and other SRB suspensor-specific transcription factors, should facilitate building regulatory networks that program suspensor gene activity and uncovering the cellular events that are responsible for suspensor differentiation (Le et al., 2007).

WHAT HAS BEEN LEARNED FROM USING COMPARATIVE GENOMICS TO IDENTIFY CONSERVED SUSPENSOR FUNCTIONS?

The suspensor is an evolutionarily conserved structure present in most seed-bearing plants and even some mosses, which diverged ~425 mya (Wardlaw, 1955; Kawashima and Goldberg, 2010). To understand more broadly the core functions carried out by all suspensors, the transcriptomes of suspensors from various species can be compared to identify conserved metabolic processes and transcription factors that may regulate conserved suspensor functions. We have previously reported that *PcG564* mRNA is also localized specifically in the basal region and suspensor of a transgenic globular-stage tobacco embryo transformed with an intact *PcG564* gene (Weterings et al., 2001). This shows that the suspensor transcriptional machinery regulating *PcG564* expression is conserved in plants that diverged ~150 mya (Paterson et al., 2004). It remains to be determined what other transcription factors are conserved in the suspensors of divergent species and what their downstream target genes are.

We have laid the foundation for a comparative genomics analysis of the SRB suspensor transcriptome with that of common bean, soybean, and *Arabidopsis*. Our laboratory has used LCM and RNA-Seq to profile the globular-stage suspensor and embryo proper transcriptomes of SRB and common bean (GEO accession GSE57537; **Figure 1**). WOX9-like, GA3-oxidase, PIN7 and G564 mRNAs are up-regulated similarly in both SRB and common bean suspensors (**Figures 1D–G**), demonstrating the conservation of gene activity and cellular functions carried out by giant bean suspensors. In collaboration with John Harada's laboratory at UC Davis, we have profiled the transcriptomes of the suspensor and embryo proper of soybean (*Glycine max*; GEO accession GSE57349) and *Arabidopsis* (Belmonte et al., 2013) embryos. Recently Slane et al. (2014) profiled *Arabidopsis* globular-stage embryo proper and suspensor nuclear transcriptomes using fluorescence-activated nuclear sorting (FANS). These datasets should illuminate several important questions regarding higher plant suspensors. What functions are conserved in the SRB and common bean giant suspensors? What are the functions of conserved transcription factors in legume suspensors? What functions are evolutionarily conserved in all suspensors regardless of size, morphology, or specialized function?

WHAT UNIQUE PROCESSES OCCUR IN GIANT BEAN SUSPENSORS THAT DIFFER FROM LESS SPECIALIZED SUSPENSORS?

Suspensors display a wide range of morphological diversity in higher plants (Kawashima and Goldberg, 2010; **Figure 1A**). For example, closely related legume species, soybean and SRB, have distinct suspensors. The soybean suspensor is small, consisting of a few cells, whereas the SRB suspensor is huge containing ~200 cells (Sussex et al., 1973). The *Arabidopsis* suspensor, which is even smaller than that of soybean, is a single file of 7–10 cells. There may be several biological processes unique to giant bean suspensors and absent in smaller suspensors, such as those of *Arabidopsis* and soybean. One of the first characterized functions of the SRB suspensor, the synthesis of GA, may be unique to giant, highly specialized bean suspensors (Kawashima and Goldberg, 2010). In fact, GA also accumulates in the massive suspensor of the legume *Cytisus laburnum* (Picciarelli et al., 1984). While *GA 3-oxidase* mRNA (encoding the last enzyme in the GA biosynthesis pathway) accumulates to a high level in both SRB and common bean suspensors at the globular stage (**Figure 1E**; Solfanelli et al., 2005), mRNAs representing the *Arabidopsis* homologs of *GA 3-oxidase* do not accumulate in the suspensor; instead, they accumulate in the endosperm of globular-stage seeds (Belmonte et al., 2013). In dicots, the suspensor and the endosperm are both short-lived structures that degenerate once they have accomplished their function of nourishing the developing embryo proper. It has been suggested that in species with massive suspensors, such as SRB and common bean, the suspensor takes over endosperm functions, resulting in delayed endosperm cellularization and a decreased amount of endosperm (Tison, 1919; Schnarf, 1929; Newman, 1934; Lorenzi et al., 1978; Guignard, 1880). Although there are specific examples that do not support this hypothesis in all plants (Lersten, 1983), it may apply in some cases. Thus,

the endosperm GA biosynthesis gene regulatory network in *Arabidopsis* might have been co-opted by the giant bean suspensors, or vice versa. In *Arabidopsis* seeds, only the location of GA hormone synthesis has changed relative to giant bean seeds, not the developmental time at which hormone accumulation occurs. Perhaps the site of GA synthesis within the seed is not important, as long as the hormone is transported to the embryo proper at the globular stage of development.

Comparative studies of the gene regulatory networks controlling the development and differentiation of suspensors of divergent species will help to unlock the changes that occurred in evolution to produce morphologically and functionally distinct suspensors. A change in gene expression between species could be attributed to an alteration in a transcription factor protein, but more commonly it has been shown to result from changes in gene promoters (Pina et al., 2014). Identifying functional *cis*-regulatory elements and transcription factors that program suspensor gene activity, and comparison between different species will help to trace how novelties arose in gene regulatory networks, which may have led to the evolution of morphologically and functionally distinct suspensors across species.

WHAT ARE THE *CIS*-REGULATORY ELEMENTS CONTAINED WITHIN THE GENOME THAT PROGRAM SUSPENSOR-SPECIFIC TRANSCRIPTION?

DNA sequence comparisons between related species have the potential to identify *cis*-regulatory elements that may regulate suspensor-specific gene transcription (Haeussler and Joly, 2011). However, wet-bench studies are required to determine whether predicted suspensor *cis*-regulatory elements are functional. Previously, we identified five *cis*-regulatory elements in the upstream region of the *PcG564* gene (**Figure 2A**) that activate transcription in transgenic tobacco and *Arabidopsis* suspensors (Kawashima et al., 2009; Henry, 2014). It remains unknown what other genes are regulated by *PcG564* suspensor *cis*-regulatory elements. The simplest hypothesis is that SRB suspensor up-regulated genes, such as *PcGA 20-oxidase* and *PcWOX9-like* (Le et al., 2007), are activated by the same suspensor *cis*-regulatory elements. Indeed, the *PcG564* suspensor *cis*-regulatory elements are found in the *PcGA 20-oxidase* and *PcWOX9-like* gene upstream regions (Kawashima et al., 2009; Henry, 2014), suggesting that these genes may comprise a suspensor gene regulatory network.

The common bean genome sequence (Schmutz et al., 2014) allows us to scan the upstream regions of all suspensor-specific genes for the presence of the five known suspensor *cis*-regulatory elements identified in *PcG564*. The common bean genome sequence can be used as a surrogate for the SRB genome because the two species diverged relatively recently (Lavin et al., 2005) and have similar gene expression profiles for the suspensor and embryo proper at the globular stage (**Figures 1B–G**). For example, G564 mRNA is up-regulated in the suspensor of both SRB and common bean relative to the embryo proper (**Figure 1G**), and the *G564* upstream region is highly conserved in these two species (**Figure 2**; Henry, 2014). The *PcG564* and *PvG564* upstream regions contain five tandem repeats of 150-bp, and each repeat contains the five known suspensor *cis*-regulatory elements,

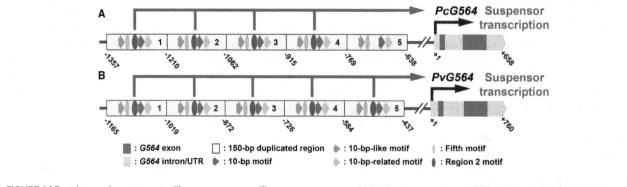

FIGURE 2 | Regulatory elements controlling suspensor-specific gene expression. (A,B) Conceptual representations of the *G564* gene and upstream region in SRB **(A)** and common bean **(B)** taken from Henry (2014). Suspensor *cis*-regulatory sequences were identified by functional *PcG564* promoter/*GUS* gene fusion and mutagenesis experiments in transgenic tobacco (Kawashima et al., 2009; Henry, 2014). Dark blue boxes represent exons. Light blue boxes represent UTRs and introns. Yellow boxes represent 150-bp tandem repeats in the upstream region (Kawashima et al., 2009). Red, orange and green arrows indicate the 10-bp motif, 10-bp-like motif and 10-bp-related motif. Purple ovals indicate the Region two motif. Blue ovals indicate the Fifth motif. Numbers indicate positions relative to the transcription start site (+1). Pc, *P. coccineus*; Pv, *P. vulgaris*.

with the exception of the fifth repeat in *PcG564* (**Figure 2**). The suspensor *cis*-regulatory elements most likely function in *PvG564* because motif sequences and *G564* expression patterns are conserved in both bean species. The identities of the *trans*-acting factors that bind to the bean *G564* suspensor *cis*-regulatory elements remain a mystery. What other genes are regulated by the transcription factors that activate *G564*, what additional regulatory circuits control suspensor gene activity, and how these regulatory circuits are activated after fertilization remain unanswered questions.

FUTURE PERSPECTIVES

The sequence of the common bean genome opens the door to *Phaseolus* suspensor gene regulatory network analysis on a genome-wide scale. Comparison of SRB and common bean suspensor transcriptomes with their embryo proper counterparts can identify suspensor-specific mRNAs that may be involved in processes specific to suspensor differentiation (**Figure 1**). The next major step is to identify suspensor-specific transcription factors, and determine their binding sites across the genome using, for example, ChIP-Seq. The power of the SRB system lies in its giant suspensor and polytene chromosomes, which can facilitate chromatin collection. Functional analysis of binding sites will also have to be carried out through promoter studies, as was done for *PcG564*, because transcription factor occupancy does not necessarily predict enhancer function *in vivo* (Peter and Davidson, 2009; Sanalkumar et al., 2014). The giant bean suspensor system has been resurrected, and should reveal new clues regarding processes that control suspensor differentiation and function in the near future.

ACKNOWLEDGMENTS

We are grateful to the current and former members of our laboratory who have helped to establish SRB as a powerful genomics system to investigate suspensor development, as well as Professors John Harada and Bob Fischer for invaluable insights and discussion on seed development. Our suspensor research has been funded by grants from the National Science Foundation Plant Genome Program, United States Department of Energy, and United States Department of Agriculture. Kelli Henry was supported, in part, by a National Institutes of Health Pre-doctoral Traineeship.

REFERENCES

Alpi, A., Tognoni, F., and Damato, F. (1975). Growth-regulator levels in embryo and suspensor of *Phaseolus coccineus* at 2 stages of development. *Planta* 127, 153–162. doi: 10.1007/BF00388376

Bayer, M., Nawy, T., Giglione, C., Galli, M., Meinnel, T., and Lukowitz, W. (2009). Paternal control of embryonic patterning in *Arabidopsis thaliana*. *Science* 323, 1485–1488. doi: 10.1126/science.1167784

Belmonte, M. F., Kirkbride, R. C., Stone, S. L., Pelletier, J. M., Bui, A. Q., Yeung, E. C., et al. (2013). Comprehensive developmental profiles of gene activity in regions and subregions of the *Arabidopsis* seed. *Proc. Natl. Acad. Sci. U.S.A.* 110, E435–E444. doi: 10.1073/pnas.1222061110

Bleckmann, A., Alter, S., and Dresselhaus, T. (2014). The beginning of a seed: regulatory mechanisms of double fertilization. *Front. Plant Sci.* 5:452. doi: 10.3389/fpls.2014.00452

Brady, T. (1973). Feulgen cytophotometric determination of the DNA content of the embryo proper and suspensor cells of *Phaseolus coccineus*. *Cell Differ.* 2, 65–75. doi: 10.1016/0045-6039(73)90022-5

Ceccarelli, N., Lorenzi, R., and Alpi, A. (1979). Kaurene and kaurenol biosynthesis in cell-free system of *Phaseolus coccineus* suspensor. *Phytochemistry* 18, 1657–1658. doi: 10.1016/0031-9422(79)80178-8

Ceccarelli, N., Lorenzi, R., and Alpi, A. (1981). Gibberellin biosynthesis in *Phaseolus coccineus* suspensor. *Z. Pflanzenphysiol.* 102, 37–44. doi: 10.1016/S0044-328X(81)80215-2

Cionini, P. G., Bennici, A., Alpi, A., and D'amato, F. (1976). Suspensor, gibberellin and in vitro development of *Phaseolus coccineus* embryos. *Planta* 131, 115–117. doi: 10.1007/BF00389979

Clutter, M., Brady, T., Walbot, V., and Sussex, I. (1974). Macromolecular synthesis during plant embryogeny. Cellular rates of RNA synthesis in diploid and polytene cells in bean embryos. *J. Cell Biol.* 63, 1097–1102. doi: 10.1083/jcb.63.3.1097

Clutter, M. E., and Sussex, I. M. (1968). Ultrastructural development of bean embryo cells containing polytene chromosomes. *J. Cell Biol.* 39:26a.

Dolan, L., Janmaat, K., Willemsen, V., Linstead, P., Poethig, S., Roberts, K., et al. (1993). Cellular organisation of the *Arabidopsis thaliana* root. *Development* 119, 71–84.

Friml, J., Vieten, A., Sauer, M., Weijers, D., Schwarz, H., Hamann, T., et al. (2003). Efflux-dependent auxin gradients establish the apical-basal axis of *Arabidopsis*. *Nature* 426, 147–153. doi: 10.1038/nature02085

Guignard, L. (1880). Note sur la structure el les functions du suspensor chez quelques Legumineuses. *Bull. Soc. Bot. Fr.* 27, 253–257. doi: 10.1080/00378941.1880.10825895

Gunning, B. E. S., and Pate, J. S. (1969). "Transfer cells". Plant cells with wall ingrowths, specialized in relation to short distance transport of solutes—their occurrence, structure, and development. *Protoplasma* 68, 107–133. doi: 10.1007/BF01247900

Haecker, A., Gross-Hardt, R., Geiges, B., Sarkar, A., Breuninger, H., Herrmann, M., et al. (2004). Expression dynamics of WOX genes mark cell fate decisions during early embryonic patterning in *Arabidopsis thaliana*. *Development* 131, 657–668. doi: 10.1242/dev.00963

Haeussler, M., and Joly, J. S. (2011). When needles look like hay: how to find tissue-specific enhancers in model organism genomes. *Dev. Biol.* 350, 239–254. doi: 10.1016/j.ydbio.2010.11.026

Heitz, E., and Bauer, H. (1933). Beweise für die chromosomennatur der kernschleifen in den knäuelkernen von bibio hortulanus L. *Z. Zellforsch. Mikrosk. Anat.* 17, 67–82. doi: 10.1007/BF00403356

Henry, K. F. (2014). *Identifying cis-Regulatory Elements and trans-Acting Factors that Activate Transcription in the Suspensor of Plant Embryos.* Ph.D. thesis, University of California, Los Angeles.

Kawashima, T., and Goldberg, R. B. (2010). The suspensor: not just suspending the embryo. *Trends Plant Sci.* 15, 23–30. doi: 10.1016/j.tplants.2009.11.002

Kawashima, T., Wang, X., Henry, K. F., Bi, Y., Weterings, K., and Goldberg, R. B. (2009). Identification of cis-regulatory sequences that activate transcription in the suspensor of plant embryos. *Proc. Natl Acad. Sci. U.S.A.* 106, 3627–3632. doi: 10.1073/pnas.0813276106

Lamprecht, H. (1941). Die Artgrenze zwischen *Phaseolus vulgaris* L. und P. multiflorus Lam. *Hereditas* 27, 51–175. doi: 10.1111/j.1601-5223.1941.tb03251.x

Lau, S., Slane, D., Herud, O., Kong, J., and Jürgens, G. (2012). Early embryogenesis in flowering plants: setting up the basic body pattern. *Annu. Rev. Plant Biol.* 63, 483–506. doi: 10.1146/annurev-arplant-042811-105507

Lavin, M., Herendeen, P. S., and Wojciechowski, M. F. (2005). Evolutionary rates analysis of Leguminosae implicates a rapid diversification of lineages during the tertiary. *Syst. Biol.* 54, 575–594. doi: 10.1080/10635150590947131

Le, B. H. (2013). *Using Genome-wide Approches to Dissect Seed Development.* Ph.D. thesis, University of California, Los Angeles.

Le, B. H., Wagmaister, J. A., Kawashima, T., Bui, A. Q., Harada, J. J., and Goldberg, R. B. (2007). Using genomics to study legume seed development. *Plant Physiol.* 144, 562–574. doi: 10.1104/pp.107.100362

Lersten, N. R. (1983). Suspensors in Leguminosae. *Bot. Rev.* 49, 233–257. doi: 10.1007/BF02861088

Lorenzi, R., Bennici, A., Cionini, P. G., Alpi, A., and D'amato, F. (1978). Embryosuspensor relations in *Phaseolus coccineus*: cytokinins during seed development. *Planta* 143, 59–62. doi: 10.1007/BF00389052

Mok, D. W. S., Mok, M. C., Rabakoarihanta, A., and Shii, C. T. (1986). *Phaseolus*: wide hybridization through embryo culture. *Biotechnol. Agric. For.* 2, 309–318.

Nagl, W. (1962). 4096-Ploidie und "Riesenchromosomen" im Suspensor von *Phaseolus coccineus*. *Naturwissenschaften* 49, 261–262. doi: 10.1007/BF00601428

Nagl, W. (1970). Temperature-dependent functional structures in the polyteene chromosomes of *Phaseolus*, with special reference to the nucleolus organizers. *J. Cell Sci.* 6, 87–107.

Nagl, W. (1974). *Phaseolus* suspensor and its polytene chromosomes. *Z. Pflanzenphysiol.* 73, 1–44. doi: 10.1016/S0044-328X(74)80142-X

Nagl, W. (1990). Translocation of putrescine in the ovule, suspensor and embryo of *Phaseolus coccineus*. *J. Plant Physiol.* 136, 587–591. doi: 10.1016/S0176-1617(11)80218-X

Newman, I. V. (1934). Studies in the Australian acacias. IV. The life history of Acacia baileyana F.V.M. Part 2. Gametophytes, fertilization, seed production and germinationand general conclusion. *Proc. Linn. Soc. N. S. W.* 59, 277–313.

Paterson, A. H., Bowers, J. E., and Chapman, B. A. (2004). Ancient polyploidization predating divergence of the cereals, and its consequences for comparative genomics. *Proc. Natl. Acad. Sci. U.S.A.* 101, 9903–9908. doi: 10.1073/pnas.0307901101

Peter, I. S., and Davidson, E. H. (2009). Modularity and design principles in the sea urchin embryo gene regulatory network. *FEBS Lett.* 583, 3948–3958. doi: 10.1016/j.febslet.2009.11.060

Picciarelli, P., Alpi, A., Pistelli, L., and Scalet, M. (1984). Gibberellin-like activity in suspensors of *Tropaeolum majus* L. and *Cytisus laburnum* L. *Planta* 162, 566–588. doi: 10.1007/BF00399924

Pina, S. D., Souer, E., and Koes, R. (2014). Arguments in the evo-devo debate: say it with flowers! *J. Exp. Bot.* 65, 2231–2242. doi: 10.1093/jxb/eru111

Sanalkumar, R., Johnson, K. D., Gao, X., Boyer, M. E., Chang, Y. I., Hewitt, K. J., et al. (2014). Mechanism governing a stem cell-generating cis-regulatory element. *Proc. Natl. Acad. Sci. U.S.A.* 111, E1091–E1100. doi: 10.1073/pnas.1400065111

Schmutz, J., Mcclean, P. E., Mamidi, S., Wu, G. A., Cannon, S. B., Grimwood, J., et al. (2014). A reference genome for common bean and genome-wide analysis of dual domestications. *Nat. Genet.* 46, 707–713. doi: 10.1038/ng.3008

Schnarf, K. (1929). *Embryologie der Angiospermen.* Berlin: Gebruder Borntraeger.

Slane, D., Kong, J., Berendzen, K. W., Kilian, J., Henschen, A., Kolb, M., et al. (2014). Cell type-specific transcriptome analysis in the early *Arabidopsis thaliana* embryo. *Development* 141, 4831–4840. doi: 10.1242/dev.116459

Solfanelli, C., Ceron, F., Paolicchi, F., Giorgetti, L., Geri, C., Ceccarelli, N., et al. (2005). Expression of two genes encoding gibberellin 2- and 3-oxidases in developing seeds of *Phaseolus coccineus*. *Plant Cell Physiol.* 46, 1116–1124. doi: 10.1093/pcp/pci124

Sussex, I., Clutter, M., Walbot, V., and Brady, T. (1973). Biosynthetic activity of the suspensor of *Phaseolus coccineus*. *Caryologia* 25, 261–272. doi: 10.1080/00087114.1973.10797129

Thomas, H. (1964). Investigations into the inter-relationships of *Phaseolus vulgaris* L. and *P. coccineus* Lam. *Genetica* 35, 59–74. doi: 10.1007/BF01804875

Tison, M. A. (1919). Sur le suspenseur du Trapa natans L. *Rev. Générale Bot.* 31, 219–228.

Ueda, M., Zhang, Z., and Laux, T. (2011). Transcriptional activation of *Arabidopsis* axis patterning genes WOX8/9 links zygote polarity to embryo development. *Dev. Cell* 20, 264–270. doi: 10.1016/j.devcel.2011.01.009

Waki, T., Hiki, T., Watanabe, R., Hashimoto, T., and Nakajima, K. (2011). The *Arabidopsis* RWP-RK protein RKD4 triggers gene expression and pattern formation in early embryogenesis. *Curr. Biol.* 21, 1277–1281. doi: 10.1016/j.cub.2011.07.001

Walbot, V., Brady, T., Clutter, M., and Sussex, I. (1972). Macromolecular synthesis during plant embryogeny: rates of RNA synthesis in *Phaseolus coccineus* embryos and suspensors. *Dev. Biol.* 29, 104–111. doi: 10.1016/0012-1606(72)90047-4

Wardlaw, C. W. (1955). *Embryogenesis in Plants.* London: Methuen.

Wendrich, J. R., and Weijers, D. (2013). The *Arabidopsis* embryo as a miniature morphogenesis model. *New Phytol.* 199, 14–25. doi: 10.1111/nph.12267

West, M., and Harada, J. J. (1993). Embryogenesis in higher plants: an overview. *Plant Cell* 5, 1361–1369. doi: 10.1105/tpc.5.10.1361

Weterings, K., Apuya, N. R., Bi, Y., Fischer, R. L., Harada, J. J., and Goldberg, R. B. (2001). Regional localization of suspensor mRNAs during early embryo development. *Plant Cell* 13, 2409–2425. doi: 10.1105/tpc.13.11.2409

Yeung, E. C. (1980). Embryogeny of *Phaseolus*: the role of the suspensor. *Z. Pflanzenphysiol.* 96, 17–28. doi: 10.1016/S0044-328X(80)80096-1

Yeung, E. C., and Meinke, D. W. (1993). Embryogenesis in angiosperms: development of the suspensor. *Plant Cell* 5, 1371–1381. doi: 10.1105/tpc.5.10.1371

Yeung, E. C., and Sussex, I. M. (1979). Embryogeny of *Phaseolus coccineus*: the suspensor and the growth of the embryo-proper in vitro. *Z. Pflanzenphysiol.* 91, 423–433. doi: 10.1016/S0044-328X(79)80256-1

10

Insight into S-RNase-based self-incompatibility in *Petunia*: recent findings and future directions

Justin S. Williams[1], Lihua Wu[2], Shu Li[2], Penglin Sun[2] and Teh-Hui Kao[1,2]*

[1] *Department of Biochemistry and Molecular Biology, Pennsylvania State University, University Park, PA, USA*
[2] *Intercollege Graduate Degree Program in Plant Biology, Pennsylvania State University, University Park, PA, USA*

Edited by:
Dazhong D. Zhao, University of Wisconsin at Milwaukee, USA

Reviewed by:
Tim P. Robbins, The University of Nottingham, UK
Thomas L. Sims, Northern Illinois University, USA

*Correspondence:
Teh-Hui Kao, Department of Biochemistry and Molecular Biology, Pennsylvania State University, 333 South Frear Laboratory, University Park, PA 16802, USA
email: txk3@psu.edu

S-RNase-based self-incompatibility in *Petunia* is a self/non-self recognition system that allows the pistil to reject self-pollen to prevent inbreeding and to accept non-self pollen for outcrossing. Cloning of *S-RNase* in 1986 marked the beginning of nearly three decades of intensive research into the mechanism of this complex system. S-RNase was shown to be the sole female determinant in 1994, and the first male determinant, S-locus F-box protein1 (SLF1), was identified in 2004. It was discovered in 2010 that additional SLF proteins are involved in pollen specificity, and recently two *S*-haplotypes of *Petunia inflata* were found to possess 17 *SLF* genes based on pollen transcriptome analysis, further increasing the complexity of the system. Here, we first summarize the current understanding of how the interplay between SLF proteins and S-RNase in the pollen tube allows cross-compatible pollination, but results in self-incompatible pollination. We then discuss some of the aspects that are not yet elucidated, including uptake of S-RNase into the pollen tube, nature, and assembly of SLF-containing complexes, the biochemical basis for differential interactions between SLF proteins and S-RNase, and fate of non-self S-RNases in the pollen tube.

Keywords: *Petunia*, SCF[SLF] complex, self-incompatibility, S-locus F-box protein, Solanaceae, S-RNase

INTRODUCTION

Self-incompatibility (SI) is a pre-zygotic reproductive barrier, which prevents inbreeding in many families of angiosperms (Takayama and Isogai, 2005; Franklin-Tong, 2008). *Petunia* possesses the Solanaceae type SI in which this reproductive barrier is regulated by the highly polymorphic *S*-locus. The *S*-locus houses the female determinant gene, *S-RNase* (Lee et al., 1994; Murfett et al., 1994), and multiple male determinant genes, *S-locus F-box* (*SLF*) genes (Sijacic et al., 2004; Kubo et al., 2010). In *Petunia*, 32 *S*-haplotypes have been reported (Sims and Robbins, 2009). A diploid pistil carries two different *S*-haplotypes, each producing an allelic variant of S-RNase. S-RNase is synthesized in the transmitting tissue of the style and secreted into the transmitting tract where pollen tubes grow from the stigma to the ovary. A pollen tube takes up both self S-RNase (product of the same *S*-haplotype as that carried by pollen) and non-self S-RNase (product of a different *S*-haplotype from that carried by pollen; Luu et al., 2000; Goldraij et al., 2006); however, only self S-RNase can inhibit the growth of the pollen tube (in the upper one-third of the style) through its RNase activity (Huang et al., 1994).

The understanding of how a pollen tube escapes the cytotoxic effect of non-self S-RNase has undergone several major developments and revisions in the past decade. SLF was first discovered in *Antirrhinum hispanicum* (Lai et al., 2002; Qiao et al., 2004b), which possesses the same type of SI. Subsequently, the first SLF in *Petunia*, now named SLF1, was confirmed as a male determinant via an *in vivo* functional assay (Sijacic et al.,

2004). The presence of an F-box domain in the N-terminal region of SLF led to the proposal that SLF, like conventional F-box proteins, is a component of a class of E3 ubiquitin ligase, the SCF (Skp1-Cullin1-F-box protein) complex, involved in ubiquitin-mediated protein degradation by the 26S proteasome (Lai et al., 2002; Qiao et al., 2004a; Hua and Kao, 2006). The substrate of an SCF[SLF] complex appears to be non-self S-RNase(s) for the specific allelic variant of SLF in the complex, as an *in vitro* protein pull-down assay showed that non-self interactions between allelic variants of SLF and S-RNase were stronger than self-interactions. This could explain why only self S-RNase can exert a cytotoxic effect on the pollen tube, as it is not ubiquitinated or degraded in the pollen tube. However, given that there are a large number of *S*-haplotypes in *Petunia* and given that allelic variants of S-RNase exhibit a high degree of sequence diversity, it is difficult to envision how an allelic variant of SLF could interact with so many non-self S-RNases, but not with a single self S-RNase. This conundrum was solved when it was discovered that at least two paralogous genes of *SLF1* are also involved in pollen specificity (Kubo et al., 2010). A new model, "collaborative non-self recognition," proposes that multiple SLF proteins produced by pollen of a given *S*-haplotype collaboratively recognize and detoxify all non-self S-RNases (i.e., each SLF is only capable of interacting with a subset of its non-self S-RNases), but none can interact with their self S-RNase (Kubo et al., 2010). To date, S_2-haplotype and S_3-haplotype of *Petunia inflata* have been shown to possess the same 17 *SLF* genes based on pollen

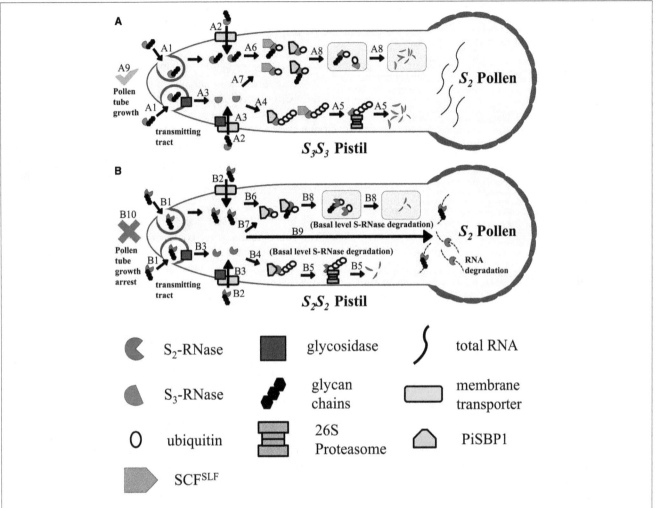

FIGURE 1 | Model for uptake of S-RNase by the pollen tube in the transmitting tract of the pistil, and fates of self and non-self S-RNases after uptake. (A) An S_2 pollen tube, growing in an S_3S_3 pistil, takes up S_3-RNase (a non-self S-RNase). Two possible types of uptake mechanisms are depicted: clathrin-dependent or clathrin-independent endocytosis (A1) and membrane-transporter mediated (A2). During uptake, the N-linked glycan chains of S_3-RNase may be removed by a membrane-associated glycosidase (A3). The deglycosylated S_3-RNase becomes poly-ubiquitinated (A4), mediated largely by the conventional SCF^{SLF} complex and to a much lesser extent by PiSBP1 and the PiSBP1-containing novel SCF^{SLF} complex (not shown). The poly-ubiquitinated S_3-RNase is destined for degradation by the 26S proteasome (A5). S_3-RNase may remain glycosylated and be mono-ubiquitinated (A6), again mediated largely by the conventional SCF^{SLF} complex and to some extent by PiSBP1 and the PiSBP1-containing novel SCF^{SLF} complex (not shown). The deglycosylated S_3-RNase may also be similarly mono-ubiquitinated (A7). The mono-ubiquitinated (deglycosylated) S_3-RNase is then targeted to vacuoles or vacuole-like organelles for degradation (A8). All the steps depicted result in detoxification of the S_3-RNase molecules inside the S_2 pollen tube, allowing it to reach the ovary to effect fertilization (A9). **(B)** An S_2 pollen tube, growing in an S_2S_2 pistil, takes up S_2-RNase (self S-RNase). S_2-RNase is taken up by the same mechanisms (B1 and B2) as those depicted for S_3-RNase in **(A)**, and may also be subjected to deglycoslyation (B3). None of the SLF proteins in the SCF^{SLF} complexes are able to interact with their self S-RNase to mediate its degradation or compartmentalization. However, similar to the scenarios depicted in **(A)**, PiSBP1 may mediate poly-ubiquitination of S_2-RNase (B4) for basal-level degradation by the 26S proteasome (B5), and may mediate mono-ubiquitination of both S_2-RNase (B6) and deglycosylated S_2-RNase (B7). The mono-ubiquitinated (deglycosylated) S_2-RNase is then targeted to vacuoles or vacuole-like organelles for degradation (B8). However, the majority of S_2-RNase molecules remain intact and will degrade RNA (B9) to result in growth arrest of the pollen tube (B10). Note: To make it easier to follow the different fates of self and non-self S-RNase, S_2S_2 and S_3S_3 pistils are used in this figure; however, in nature, pistils of a self-incompatible species are normally heterozygous for the S-locus.

transcriptome analysis (Williams et al., 2014b). Moreover, eight other *Petunia* S-haplotypes have recently been shown to possess 16–20 *SLF* genes (Kubo et al., 2015). So far, eight of them (*SLF1, SLF2, SLF3, SLF4, SLF5, SLF6, SLF8,* and *SLF9*) have been confirmed by an *in vivo* functional assay to be involved in pollen specificity (Sijacic et al., 2004; Kubo et al., 2010, 2015; Williams et al., 2014a).

Despite the impressive progress made in understanding the complex non-self recognition between the female determinant and multiple male determinants, there are still many aspects of S-RNase-based SI that remain unknown. In this article, we discuss the current understanding of some of these aspects, summarize the key features of the discussion in the model shown in **Figure 1**, and list the known and putative proteins involved in **Table 1**.

Table 1 | Known and putative proteins involved in S-RNase-based self-incompatibility.

Component of SI process	Known proteins	Putative proteins
S-RNase uptake	S-RNase	Membrane transporters, endocytosis-related proteins
SCF^{SLF} complex	PiCUL1-P, PiSSK1, SLF proteins, PiRBX1	PiSBP1
Novel SCF^{SLF} complex		SLF1, PiCUL1-G, PiSBP1
S-locus F-box proteins	SLF1–SLF17	
Fate of non-self S-RNase in the pollen tube	SCF^{SLF} complex, 26S proteasome	Glycosidase, PiSBP1, PiCUL1-G, vacuolar proteases

UPTAKE OF S-RNase INTO POLLEN TUBES

For S-RNase to exert its cytotoxic effect it must be taken up by the pollen tube in the transmitting tract; however, the mechanism of this critical step in SI is not yet known. S-RNase could be taken up by endocytosis and/or membrane transporters, both of which are used by the pollen tube to take up extracellular proteins (Moscatelli et al., 2007; Chen et al., 2010).

A delicate equilibrium of exocytosis and endocytosis is required for the rapid polarized growth of the pollen tube (Camacho and Malhó, 2003). Both clathrin-dependent and clathrin-independent endocytosis have been shown to be involved in recycling plasma membrane proteins/lipids and in regulating pollen tube growth (Onelli and Moscatelli, 2013). One notable example is the uptake of SCA (stigma/style cysteine-rich adhesin), a protein that guides the pollen tube toward the ovary; it is internalized in the tip region of a growing pollen tube through clathrin-dependent endocytosis and sorted to the multi-vescular bodies and vacuoles (Kim et al., 2006; Chae et al., 2007).

ATP-binding cassette (ABC) transporters constitute a large protein family in plants (Geisler, 2014), and have been reported to be involved in various processes, e.g., pathogen attack response, deposition of plasma membrane lipids, nutrient accumulation in seeds, and transport of phytohormones (Kang et al., 2010; Tunc-Ozdemir et al., 2013). ABC transporters are active transporters, deriving energy from ATP hydrolysis, and they are either exporters or importers (Geisler, 2014). In plants, ABC transporters have been classified into 13 subfamilies (Rea, 2007). An ABC transporter of apple, MdABCF, has been implicated in the transport of S-RNase into the pollen tube (Meng et al., 2014). Analysis of the S_2-pollen, S_3-pollen, and S_3S_3 leaf transcriptomes of *P. inflata* (Williams et al., 2014b) reveals the presence of 476, 334, and 851 BLAST annotated transporters, and interestingly, ~59% of the potential transporters in each pollen transcriptome are annotated as ABC transporters, whereas only ~25% of the potential transporters in the leaf transcriptome are annotated as ABC transporters (our unpublished data). Given that S-RNase is pistil-specific and is taken up only by the pollen tube, if ABC transporters are involved in the uptake of S-RNases, they would likely be among those that are pollen-specific.

Regardless of how S-RNase is taken up by the pollen tube, the uptake machinery must be able to interact with a large number of highly divergent allelic variants of S-RNase. Despite the sequence diversity, S-RNase contains five conserved regions (C1–C5; Ioerger et al., 1991). Alignment of the amino acid sequences of 20 allelic variants of S-RNase in *Petunia* available in the NCBI non-redundant nucleotide database reveals 21 perfectly conserved amino acid residues. As the crystal structure of *Nicotiana alata* S_{F11}-RNase has already been determined (Ida et al., 2001), those conserved residues that are exposed on the outside surface would be good candidates to use for investigating their role in interaction with a transporter and/or a transmembrane receptor involved in uptake.

THE SCF^{SLF} COMPLEX

The SLF-containing SCF complex, SCF^{SLF}, of *P. inflata* has been shown by co-immunoprecipitation (Co-IP) followed by mass-spectrometry (MS) to contain a conventional Rbx1 (PiRBX1; a RING-finger protein), a pollen-specific Cullin1 (PiCUL1-P), and a pollen-specific Skp1-like protein (PiSSK1; Li et al., 2014). Similar components have been identified in the SCF complex of *Petunia hybrida* (Entani et al., 2014; Liu et al., 2014) and *Pyrus bretschneideri* (Xu et al., 2013). Phylogenetic studies showed that SSK1 and Cullin1 proteins implicated in SI form their own monoclades (Xu et al., 2013; Yuan et al., 2014), and that the 17 SLF proteins of *P. inflata* form a monoclade. Also, SLF proteins were the only F-box proteins that co-immunoprecipitated with PiSSK1 (Li et al., 2014). Interestingly, tomato SpCUL1, sharing 91% sequence identity with PiCUL1-P, is involved in unilateral incompatibility between tomato species (Li and Chetelat, 2010) and is also required for compatible pollination in *Solanum arcanum* (Li and Chetelat, 2013). Thus, three of the four components of the SCF^{SLF} complex (PiSSK1, PiCUL1-P, and SLF) appear to have evolved specifically to function in SI.

A *P. hybrida* RING-finger protein, termed PhSBP1, was found to interact with S-RNase by a yeast two-hybrid assay (Sims and Ordanic, 2001). Subsequently, the PhSBP1 homolog from *P. inflata*, termed PiSBP1, was found to interact with an SLF protein (S_2-SLF1), a different Cullin1 (PiCUL1-G) and S-RNase by a yeast two-hybrid assay (Hua and Kao, 2006; Meng et al., 2011), leading to the suggestion that PiSBP1 plays the roles of both Skp1 and Rbx1, and forms a novel SCF complex with PiCUL1-G and SLF1 (Hua and Kao, 2006). In *Malus × domestica* (apple), both homologs of PiSBP1 and PiSSK1 were found to interact with an SLF (named SFB for S-Locus F-Box) and a Cullin1 (MdCUL1) by an *in vitro* protein binding assay (Minamikawa et al., 2014). MdSSK1 interacted with the SLF protein more strongly than did MdSBP1, and the transcript level of *MdSSK1* was >100 times higher than that of *MdSBP1*. Thus, the conventional SCF^{SLF} complex is thought to play a major role in mediating ubiquitination and degradation of non-self S-RNases. This finding may explain why PiSBP1 was not identified from the Co-IP products using either S_2-SLF1 or PiSSK1 as bait (Li et al., 2014). PiSBP1 may also function as a mono-subunit E3 ubiquitin ligase as it catalyzed ubiquitination of S_3-RNase in the presence of E1, E2, and ubiquitin in an *in vitro* assay (Hua and Kao, 2008). However, the interaction of PiSBP1 with S-RNase does not show allele specificity, so it may mediate basal level S-RNase degradation, perhaps as a safety mechanism to ensure that all non-self S-RNases are cleared from the pollen tube.

The approach of using Co-IP followed by MS has been successful in identifying three (SLF1, SLF4, and SLF13) of the 17 SLF proteins of *P. inflata* as the F-box component of the SCF^{SLF} complexes (Li et al., 2014). Among these 17 SLF proteins, eight (including SLF1 and SLF4) have so far been confirmed to be involved in pollen specificity via an *in vivo* functional assay (Sijacic et al., 2004; Williams et al., 2014a; Kubo et al., 2015). This assay involves raising transgenic plants to examine the effect of expressing a particular allelic variant of an SLF on the SI behavior of the transgenic pollen. In cases when breakdown of SI is observed in the transgenic plants, progeny from crosses with wild-type plants of appropriate S-genotypes have to be examined as well. The Co-IP-MS results suggest that this approach may be a much less time-consuming and labor-intensive alternative to the *in vivo* functional assay for assessing the SI function of SLF proteins. If all SLF proteins are assembled into similar SCF complexes, this would raise the questions of how all these sequence-divergent SLF proteins (e.g., 45.3–87.7% sequence identity between the 17 SLF proteins of S_2-haplotype; Williams et al., 2014b) are capable of being assembled into their respective SCF^{SLF} complexes, and whether the allelic variants of S-RNase taken into a pollen tube may favor the "selection" of particular SLF proteins that can interact with and detoxify the non-self S-RNases, especially when the common components of the complex are limited. It would be interesting to study as well the dynamics of the SLF-containing SCF complexes during growth of compatible pollen tubes in the pistil.

THE S-LOCUS F-BOX PROTEINS

For the SLF proteins of *Petunia* that have been studied so far, each interacts with only one, or a few, of the S-RNases examined (Sijacic et al., 2004; Kubo et al., 2010; Williams et al., 2014a). This pattern of interactions between SLF proteins and S-RNases is consistent with the prediction by the collaborative non-self recognition model (Kubo et al., 2010). How then does one SLF, but not other SLF proteins, interact with a certain S-RNase? How have all these multiple SLF proteins evolved so that pollen of a given S-haplotype has a complete arsenal to counter the cytotoxic effect of all non-self S-RNases, but avoid interacting with their self S-RNase?

Typically, F-box proteins contain two domains: the F-box domain in the N-terminal region and a protein–protein interaction domain located in the C-terminal region (C-terminal domain or CTD; Gagne et al., 2002; Cardozo and Pagano, 2004; Wang et al., 2004). Thus, it is reasonable to examine the CTD of an SLF to identify the amino acids that are involved in its interaction with a particular S-RNase. One approach to identify such amino acids would be to compare the amino acid sequences of SLF proteins that interact with the same S-RNase with the amino acid sequences of SLF proteins that do not. The amino acids in the CTD that are conserved among all the SLF proteins that interact with the same S-RNase, but are divergent among all those that do not, are likely important for the specific interaction with that S-RNase. This approach will benefit from knowing interaction

relationships between as many SLF proteins and S-RNases, as the information can then be used to design strategies to determine the biochemical basis for the differential interactions.

Among the interaction relationships established between SLF proteins and S-RNases of *P. inflata*, S_2-SLF1 interacts with the largest number, four, of the S-RNases examined and all the other SLF proteins interact with none or at most one S-RNase (Sijacic et al., 2004; Kubo et al., 2010; Sun and Kao, 2013; Williams et al., 2014a). The SLF that interacts with more S-RNases than do all other SLF proteins might be the first to have come into existence during the evolution of the SI system. If the first SLF could interact with a number of non-self S-RNases, it would allow pollen to detoxify new non-self S-RNases as more S-haplotypes evolved, without having to generate a new SLF. However, there might be a practical limit as to the number of non-self S-RNases with which each SLF could interact, so that when the maximum capacity is reached, a new SLF would be needed to allow pollen to recognize and detoxify additional non-self S-RNases as more S-haplotypes continued to evolve.

If an SLF has evolved to interact with and detoxify a particular S-RNase, there would be no selective pressure to generate another SLF with the same function. However, from the standpoint of defense against the toxic effect of non-self S-RNases, it would be beneficial to pollen if more than one SLF were capable of detoxifying any particular non-self S-RNase, as this will minimize the deleterious effect caused by mutations that render an SLF incapable of interacting with and detoxifying a non-self S-RNase. In order to maintain an SI system over a long period of time, not only must self-pollen be rejected by the pistil, but also non-self pollen must be accepted by the pistil through the collective effort of all SLF proteins to detoxify all non-self S-RNases. It would be of interest to determine whether, during the evolution of the *SLF* genes, there has been indeed such a redundancy built in for pollen to deal with every non-self S-RNase. The results from studying the effect of silencing the expression of S_2-SLF1 in pollen of S_2S_3 transgenic plants are consistent with the presence of additional SLF proteins for detoxifying S_3-RNase, S_7-RNase, and S_{13}-RNases, as transgenic pollen producing very low levels of, if any, S_2-SLF1, remained compatible with S_3-, S_7-, and S_{13}-carrying pistils (Sun and Kao, 2013). Moreover, Kubo et al. (2010) found that two SLF proteins produced by S_5 pollen of *P. hybrida* interacted with the same S-RNase, S_9-RNase.

THE FATE OF NON-SELF S-RNases IN THE POLLEN TUBE

The current view about what happens to non-self S-RNases, after being taken up by the pollen tube, is that they become ubiquitinated in the cytosol through mediation by appropriate SCF^{SLF} complexes, and subsequently degraded by the 26S proteasome. However, the fate of non-self S-RNases may need a closer examination, considering the following observations in *Petunia*. A cell-free protein degradation system showed that recombinant S_1-, S_2-, and S_3-RNases (expressed in *E. coli* and thus non-glycosylated) were all degraded by the 26S proteasome in pollen tube extracts; however, native glycosylated S_3-RNase purified from styles was not degraded to any significant extent (Hua and Kao, 2006). After purified native S_3-RNase was deglycosylated, the deglycosylated protein was degraded as efficiently as recombinant

S-RNases (Hua and Kao, 2006). Moreover, native S_9-RNase was ubiquitinated to various extents, with most being mono-ubiquitinated, by an $SCF^{S7\text{-}SLF2}$ complex in an *in vitro* ubiquitination assay, but the degradation of the mono-ubiquitinated S_9-RNase in an *in vitro* protein degradation assay was not efficient (Entani et al., 2014).

It is possible that S-RNase is deglycosylated once taken into the pollen tube. Non-glycosylated S-RNase has been shown to function normally in rejecting self-pollen (Karunanandaa et al., 1994), so deglycosylation should not affect the function of S-RNase. Deglycosylated non-self S-RNase then becomes poly-ubiquitinated and degraded in the cytosol. TTS, a tobacco transmitting tissue glycoprotein, has been shown to be incorporated into the pollen tube wall and deglycosylated (Wu et al., 1995), suggesting that deglycosylation of S-RNase by the pollen tube is possible. It is also possible that poly-ubiquitination and degradation by the 26S proteasome is not the only pathway for detoxification of S-RNase. S-RNase (and deglycosylated S-RNase) may also be mono-ubiquitinated and transported to a compartment for further degradation. Mono-ubiquitination has been shown to play an important role in fast labeling of proteins for bulk degradation by autophagy (Kim et al., 2008). Two plant plasma membrane proteins have been shown to be degraded in the vacuole (Cai et al., 2012). This possible fate of S-RNase can explain two contrasting findings about the destination of S-RNase after its uptake into the pollen tube. Luu et al. (2000) observed that S-RNase of *Solanum chacoense* was mainly localized in the cytosol of both compatible and incompatible pollen tubes, whereas Goldraij et al. (2006) observed that S-RNase of *N. alata* was sorted into vacuole-like organelles in both compatible and incompatible pollen tubes. S-RNase may be first taken up into the cytosol of the pollen tubes as observed by Luu et al. (2000), and after the interaction with appropriate SCF^{SLF} complexes (and to a much lesser extent with PiSBP1 and the PiSBP1-containing SLF complex), those non-self S-RNases that are mono-ubiquitinated are sorted to vacuoles or vacuole-like organelles, as observed by Goldraij et al. (2006), for degradation. In the above-mentioned *in vitro* protein degradation assay of Entani et al. (2014), perhaps the compartments might be disrupted and the vacuolar proteases might not be active under the assay conditions, such that no significant degradation was observed for mono-ubiquitinated S_9-RNase. Self S-RNase may follow similar pathways as non-self S-RNase, except that the bulk of its molecules remain stable due to the inability of any of the SLF proteins to interact with and detoxify it.

CONCLUSION

In summary, there are several facets of S-RNase-based SI that require further investigation in order to obtain a comprehensive understanding of this complex self/non-self-recognition system. It is clear that S-RNase functions inside the pollen tube to exert its cytotoxicity, but how S-RNase is taken up into the pollen tube is completely unknown. The initial fates of both self and non-self S-RNase after entering the pollen tube also remain unclear. The model presented in **Figure 1** depicts several possible pathways for uptake and subsequent fates of self and non-self S-RNases. We believe that the difference in the ultimate fate of self and non-

self S-RNase is due in large part to non-self recognition between SLF proteins and S-RNases. A non-self S-RNase interacts with at least one of the SLF proteins and becomes ubiquitinated and suffers degradation, whereas a self S-RNase does not interact with any of the SLF proteins and thus remains stable. The existence of multiple SLF proteins in pollen of any *S*-haplotype and the highly polymorphic nature of the *S*-locus suggest that there are hundreds of SLF proteins (allelic variants of each of the multiple SLF proteins) with a wide range of sequence similarity available for studying interactions with S-RNases. However, so far, the interaction relationships have only been determined between a very small number of SLF proteins and S-RNases. A comprehensive study is necessary to understand the biochemical basis of differential interactions between SLF proteins and S-RNases, and the information obtained will not only further our understanding of S-RNase-based SI, but also lay the foundation for studying the structural basis of the interactions between F-box proteins and their substrates.

REFERENCES

Cai, Y., Zhuang, X., Wang, J., Wang, H., Lam, S. K., Gao, C., et al. (2012). Vacuolar degradation of two integral plasma membrane proteins, AtLRR84A and OsSCAMP1, is cargo ubiquitination-independent and prevacuolar compartment-mediated in plant cells. *Traffic* 13, 1023–1040. doi: 10.1111/j.1600-0854.2012.01360.x

Camacho, L., and Malhó, R. (2003). Endo/exocytosis in the pollen tube apex is differentially regulated by Ca^{2+} and GTPases. *J. Exp. Bot.* 54, 83–92. doi: 10.1093/jxb/erg043

Cardozo, T., and Pagano, M. (2004). The SCF ubiquitin ligase: insights into a molecular machine. *Nat. Rev. Mol. Cell Biol.* 5, 739–751. doi: 10.1038/nrm1471

Chae, K., Zhang, K., Zhang, L., Morikis, D., Kim, S. T., Mollet, J. C., et al. (2007). Two SCA (stigma/style cysteine-rich adhesin) isoforms show structural differences that correlate with their levels of in vitro pollen tube adhesion activity. *J. Biol. Chem.* 282, 33845–33858. doi: 10.1074/jbc.M703997200

Chen, L. Q., Hou, B. H., Lalonde, S., Takanaga, H., Hartung, M. L., Qu, X. Q., et al. (2010). Sugar transporters for intercellular exchange and nutrition of pathogens. *Nature* 468, 527–532. doi: 10.1038/nature09606

Entani, T., Kubo, K. I., Isogai, S., Fukao, Y., Shirakawa, M., Isogai, A., et al. (2014). Ubiquitin-proteasome-mediated degradation of S-RNase in a solanaceous cross-compatibility reaction. *Plant J.* 78, 1014–1021. doi: 10.1111/tpj.12528

Franklin-Tong, V. E. (2008). *Self-Incompatibility in Flowing Plants-Evolution, Diversity, and Mechanism*. Berlin: Springer.

Gagne, J. M., Downes, B. P., Shiu, S.-H., Durski, A. M., and Vierstra, R. D. (2002). The F-box subunit of the SCF E3 complex is encoded by a diverse superfamily of genes in *Arabidopsis. Proc. Natl. Acad. Sci. U.S.A.* 99, 11519–11524. doi: 10.1073/pnas.162339999

Geisler, M. (2014). *Plant ABC Transporters*. Cham: Springer International Publishing.

Goldraij, A., Kondo, K., Lee, C. B., Hancock, C. N., Sivaguru, M., and Vazquez-Santana, S. (2006). Compartmentalization of S-RNase and HT-B degradation in self-incompatible *Nicotiana. Nature* 439, 805–810. doi: 10.1038/nature04491

Hua, Z. H., and Kao, T.-H. (2006). Identification and characterization of components of a putative *Petunia* S-locus F-box-containing E3 ligase complex involved in S-RNase-based self-incompatibility. *Plant Cell* 18, 2531–2553. doi: 10.1105/tpc.106.041061

Hua, Z. H., and Kao, T.-H. (2008). Identification of major lysine residues of S3-RNase of *Petunia inflata* involved in ubiquitin-26S proteasome-mediated degradation in vitro. *Plant J.* 54, 1094–1104. doi: 10.1111/j.1365-313X.2008.03487.x

Huang, S., Lee, H. S., Karunanandaa, B., and Kao, T.-H. (1994). Ribonuclease activity of *Petunia inflata* S proteins is essential for rejection of self-pollen. *Plant Cell* 6, 1021–1028. doi: 10.1105/tpc.6.7.1021

Ida, K., Norioka, S., Yamamoto, M., Kumasaka, T., Yamashita, E., Newbigin, E., et al. (2001). The 1.55 Å resolution structure of *Nicotiana alata* SF11-RNase associated with gametophytic self-incompatibility. *J. Mol. Biol.* 314, 103–112. doi: 10.1006/jmbi.2001.5127

Ioerger, T. R., Gohlke, J. R., Xu, B., and Kao, T.-H. (1991). Primary structural features of the self-incompatibility protein in Solanaceae. *Sex. Plant Reprod.* 4, 81–87. doi: 10.1007/BF00196492

Kang, J., Hwang, J. U., Lee, M., Kim, Y. Y., Assmann, S. M., Martinoia, E., et al. (2010). PDR-type ABC transporter mediates cellular uptake of the phytohormone abscisic acid. *Proc. Natl. Acad. Sci. U.S.A.* 107, 2355–2360. doi: 10.1073/pnas.0909222107

Karunanandaa, B., Huang, S., and Kao, T.-H. (1994). Carbohydrate moiety of the *Petunia inflata* S3 protein is not required for self-incompatibility interactions between pollen and pistil. *Plant Cell* 6, 1933–1940. doi: 10.1105/tpc.6.12.1933

Kim, P. K., Hailey, D. W., Mullen, R. T., and Lippincott-Schwartz, J. (2008). Ubiquitin signals autophagic degradation of cytosolic proteins and peroxisomes. *Proc. Natl. Acad. Sci. U.S.A* 105, 20567–20574. doi: 10.1073/pnas.0810611105

Kim, S. T., Zhang, K., Dong, J., and Lord, E. M. (2006). Exogenous free ubiquitin enhances lily pollen tube adhesion to an in vitro stylar matrix and may facilitate endocytosis of SCA. *Plant Physiol.* 142, 1397–1411. doi: 10.1104/pp.106.086801

Kubo, K., Entani, T., Takara, A., Wang, N., Fields, A. M., Hua, Z. H., et al. (2010). Collaborative non-self recognition system in S-RNase-based self-incompatibility. *Science* 330, 796–799. doi: 10.1126/science.1195243

Kubo, K., Paape, T., Hatakeyama, M., Entani, T., Takara, A., Kajihara, K., et al. (2015). Gene duplication and genetic exchange drive the evolution of S-RNase-based self-incompatibility in Petunia. *Nat. Plants* doi: 10.1038/nplants.2014.5 [Epub ahead of print].

Lai, Z., Ma, W., Han, B., Liang, L., Zhang, Y., Hong, G., et al. (2002). An F-box gene linked to the self-incompatibility (S) locus of Antirrhinum is expressed specifically in pollen and tapetum. *Plant Mol. Biol.* 50, 29–42. doi: 10.1023/A:1016050018779

Lee, H., Huang, S., and Kao, T.-H. (1994). S proteins control rejection of Incompatible pollen in *Petunia inflata*. *Nature* 367, 560–563. doi: 10.1038/367560a0

Li, S., Sun, P., Williams, J. S., and Kao, T.-H. (2014). Identification of the self-incompatibility locus F-box protein-containing complex in *Petunia inflata*. *Plant Reprod.* 27, 31–45. doi: 10.1007/s00497-013-0238-3

Li, W., and Chetelat, R. T. (2010). A pollen factor linking inter- and intraspecific pollen rejection in tomato. *Science* 330, 1827–1830. doi: 10.1126/science.1197908

Li, W., and Chetelat, R. T. (2013). The role of a pollen-expressed Cullin1 protein in gametophytic self-incompatibility in *Solanum*. *Genetics* 196, 439–442. doi: 10.1534/genetics.113.158279

Liu, W., Fan, J., Li, J., Song, Y., Li, Q., Zhang, Y. E., et al. (2014). SCFSLF-mediated cytosolic degradation of S-RNase is required for cross-pollen compatibility in S-RNase-based self-incompatibility in *Petunia hybrida*. *Front. Genet.* 5:228. doi: 10.3389/fgene.2014.00228

Luu, D. T., Qin, X., Morse, D., and Cappadocia, M. (2000). S-RNase uptake by compatible pollen tubes in gametophytic self-incompatibility. *Nature* 407, 649–651. doi: 10.1038/35036623

Meng, D., Gu, Z., Li, W., Wang, A., Yuan, H., Yang, Q., et al. (2014). Apple MdABCF assists in the transportation of S-RNase into pollen tubes. *Plant J.* 78, 990–1002. doi: 10.1111/tpj.12524

Meng, X., Hua, Z., Sun, P., and Kao, T.-H. (2011). The amino terminal F-box domain of *Petunia inflata* S-locus F-box protein is involved in the S-RNase-based self-incompatibility mechanism. *AoB Plants* 2011:plr016. doi: 10.1093/aobpla/plr016

Minamikawa, M. F., Koyano, R., Kikuchi, S., Koba, T., and Sassa, H. (2014). Identification of SFBB-containing canonical and noncanonical SCF complexes in pollen of apple (*Malus × domestica*). *PLoS ONE* 9:e97642. doi: 10.1371/journal.pone.0097642

Moscatelli, A., Ciampolini, F., Rodighiero, S., Onelli, E., Cresti, M., Santo, N., et al. (2007). Distinct endocytic pathways identified in tobacco pollen tubes using charged nanogold. *J. Cell Sci.* 120, 3804–3819. doi: 10.1242/jcs.012138

Murfett, J., Atherton, T. L., Mou, B., Gasser, C. S., and McClure, B. A. (1994). S-RNase expressed in transgenic *Nicotiana* causes S-allele-specific pollen rejection. *Nature* 367, 563–566. doi: 10.1038/367563a0

Onelli, E., and Moscatelli, A. (2013). Endocytic pathways and recycling in growing pollen tubes. *Plants* 2, 211–229. doi: 10.3390/plants2020211

Qiao, H., Wang, H., Zhao, L., Zhou, J., Huang, J., Zhang, Y., et al. (2004a). The F-box protein AhSLF-S2 physically interacts with S-RNases that may be inhibited by the ubiquitin/26S proteasome pathway of protein degradation during compatible pollination in Antirrhinum. *Plant Cell* 16, 582–595. doi: 10.1105/tpc.017673

Qiao, H., Wang, F., Zhao, L., Zhou, J., Lai, Z., Zhang, Y., et al. (2004b). The F-box protein AhSLF-S2 controls the pollen function of S-RNase-based self-incompatibility. *Plant Cell* 16, 2307–2322. doi: 10.1105/tpc.104.024919

Rea, P. A. (2007). Plant ATP-binding cassette transporters. *Annu. Rev. Plant Biol.* 58, 347–375. doi: 10.1146/annurev.arplant.57.032905.105406

Sijacic, P., Wang, X., Skirpan, A., Wang, Y., Dowd, P., McCubbin, A. G., et al. (2004). Identification of the pollen determinant of S-RNase-mediated self-incompatibility. *Nature* 429, 302–305. doi: 10.1038/nature02523

Sims, T. L., and Ordanic, M. (2001). Identification of a S-ribonuclease binding protein in *Petunia hybrida*. *Plant Mol. Biol.* 47, 771–783. doi: 10.1023/A:1013639528858

Sims, T. L., and Robbins, T. P. (2009). "Gametophytic self-incompatibility in Petunia," in *Petunia Evolutionary, Developmental and Physiological Genetics*, eds T. Gerats and J. Strommer (New York, NY: Springer), 85–106.

Sun, P., and Kao, T.-H. (2013). Self-incompatibility in *Petunia inflata*: the relationship between a self-incompatibility locus F-box protein and its non-self S-RNases. *Plant Cell* 25, 470–485. doi: 10.1105/tpc.112.106294

Takayama, S., and Isogai, A. (2005). Self-incompatibility in plants. *Annu. Rev. Plant Biol.* 56, 467–489. doi: 10.1146/annurev.arplant.56.032604.144249

Tunc-Ozdemir, M., Rato, C., Brown, E., Rogers, S., Mooneyham, A., Frietsch, S., et al. (2013). Cyclic nucleotide gated channels 7 and 8 are essential for male reproductive fertility. *PLoS ONE* 8:e55277. doi: 10.1371/journal.pone.0055277

Wang, Y., Tsukamoto, T., Yi, K.-W., Wang, X., Huang, S., McCubbin, A. G., et al. (2004). Chromosome walking in the *Petunia inflata* self-incompatibility (S-) locus and gene identification in an 881-kb contig containing S2-RNase. *Plant Mol. Biol.* 54, 727–742. doi: 10.1023/B:PLAN.0000040901.98982.82

Williams, J. S., Natale, C. A., Wang, N., Li, S., Brubaker, T. R., Sun, P., et al. (2014a). Four previously identified *Petunia inflata* S-locus F-box genes are involved in pollen specificity in self-incompatibility. *Mol. Plant* 7, 567–569. doi: 10.1093/mp/sst155

Williams, J. S., Der, J. P., dePamphilis, C. W., and Kao, T.-H. (2014b). Transcriptome analysis reveals the same 17 S-locus F-box genes in two haplotypes of the self-incompatibility locus of *Petunia inflata*. *Plant Cell* 26, 2873–2888. doi: 10.1105/tpc.114.126920

Wu, H. M., Wang, H., and Cheung, A. Y. (1995). A pollen tube growth stimulatory glycoprotein is deglycosylated by pollen tubes and displays a glycosylation gradient in the flower. *Cell* 82, 395–403. doi: 10.1016/0092-8674(95)90428-X

Xu, C., Li, M., Wu, J., Guo, H., Li, Q., Zhang, Y., et al. (2013). Identification of a canonical SCFSLF complex involved in S-RNase-based self-incompatibility of *Pyrus* (Rosaceae). *Plant Mol. Biol.* 81, 245–257. doi: 10.1007/s11103-012-9995-x

Yuan, H., Meng, D., Gu, Z. Y., Li, W., Wang, A. D., Yang, Q., et al. (2014). A novel gene, MdSSK1, as a component of the SCF complex rather than MdSBP1 can mediate the ubiquitination of S-RNase in apple. *J. Exp. Bot.* 65, 3121–3131. doi: 10.1093/jxb/eru164

Epigenetic regulation of rice flowering and reproduction

Jinlei Shi[1,2], Aiwu Dong[1] and Wen-Hui Shen[1,2] *

[1] State Key Laboratory of Genetic Engineering, Collaborative Innovation Center of Genetics and Development, International Associated Laboratory of CNRS-Fudan-HUNAU on Plant Epigenome Research, Department of Biochemistry, Institute of Plant Biology, School of Life Sciences, Fudan University, Shanghai, China
[2] CNRS, Institut de Biologie Moléculaire des Plantes, Université de Strasbourg, Strasbourg, France

Edited by:
Dazhong Dave Zhao, University of Wisconsin-Milwaukee, USA

Reviewed by:
Yanhai Yin, Iowa State University, USA
Mieke Van Lijsebettens, Flanders Institute for Biotechnology, Belgium

*Correspondence:
*Wen-Hui Shen, CNRS, Institut de Biologie Moléculaire des Plantes, Université de Strasbourg, 12 Rue du Général Zimmer, 67084 Strasbourg Cédex, France
e-mail: wen-hui.shen@ibmp-cnrs.unistra.fr*

Current understanding of the epigenetic regulator roles in plant growth and development has largely derived from studies in the dicotyledonous model plant *Arabidopsis thaliana*. Rice (*Oryza sativa*) is one of the most important food crops in the world and has more recently becoming a monocotyledonous model plant in functional genomics research. During the past few years, an increasing number of studies have reported the impact of DNA methylation, non-coding RNAs and histone modifications on transcription regulation, flowering time control, and reproduction in rice. Here, we review these studies to provide an updated complete view about chromatin modifiers characterized in rice and in particular on their roles in epigenetic regulation of flowering time, reproduction, and seed development.

Keywords: chromatin, epigenetics, flowering time, histone modification, DNA methylation, non-coding RNA, reproduction, *Oryza sativa*

INTRODUCTION

Epigenetics is defined as nucleotide sequence-independent changes in the gene expression that are mitotically and/or meiotically heritable. The fundamental repeating unit of chromatin is nucleosome. The nucleosome contains 145–147 base pairs (bp) of DNA wrapped around an octamer of histone proteins, comprising two copies of each of the four core histones, H2A, H2B, H3, and H4 (McGinty and Tan, 2014). The linker histone H1 associates with DNA inbetween the two nucleosomes and participates in higher order chromatin structure formation and remodeling. The structure of chromatin can be subjected to panoply of epigenetic regulations including DNA methylation, histone covalent modifications, histone variants, and ATP-dependent chromatin remodeling. DNA methylation has been widely considered as a heritable epigenetic mark that regulates expression of genes in both plants and mammals (Law and Jacobsen, 2010; Furner and Matzke, 2011; Wu and Zhang, 2014). Histone modifications including methylation, acetylation, phosphorylation, ubiquitination, and sumoylation, play critical roles in regulating chromatin structure and gene expression, mainly by altering nucleosome stability and positioning that affect DNA accessibility for regulatory proteins or protein complexes involved in transcription, DNA replication and repair (Pikaard and Scheid, 2013; To and Kim, 2014; Van Lijsebettens and Grasser, 2014). ATP-dependent chromatin remodeling factors control relocation or dissociation of nucleosomes, and histone chaperones bind histones and play crucial roles in nucleosome assembly/disassembly in diverse chromatin metabolism and epigenetic regulation (Zhu et al., 2012; Gentry and Hennig, 2014).

Rice (*Oryza sativa*) is a worldwide crop and represents a valuable model plant for monocots, to which many of our food crops belong. Compared to the extensively studied dicot model plant *Arabidopsis thaliana*, rice has only been more recently studied in epigenetic modifications (reviewed in Chen and Zhou,

2013). Genome-wide analyses of DNA methylations have revealed conservation as well as distinct differences between rice and *Arabidopsis*, and that a much higher level of DNA methylation is observed in association with more numerous transposable elements present in the rice genome (Yan et al., 2010; Zemach et al., 2010; Chodavarapu et al., 2012; Li et al., 2012). Genome-wide analyses by chromatin immunoprecipitation combined with high-throughput sequencing (ChIP-Seq) have shown that several types of histone modifications, e.g., histone H3 lysine 9 acetylation (H3K9ac) and H4K12ac, H3K4 di-/tri-methylation (H3K4me2/3), H3K27me3, and H3K36me3, are broadly distributed with distinct patterns within the rice genome (He et al., 2010; Malone et al., 2011; Du et al., 2013). In this review, we summarize and discuss regulators involved in different types of chromatin modifications and their roles in rice plant flowering time control and reproduction.

REGULATION OF DIFFERENT TYPES OF CHROMATIN MODIFICATIONS IN RICE

Different types of chromatin modifications are regulated by specific factors that are generally conserved in rice and other plant species (www.chromdb.org). So far, only some of the rice chromatin modifiers are functionally characterized by analysis of loss-of-function mutants and RNAi or overexpression transgenic plants (**Table 1**).

DNA METHYLATION

In plants, DNA methylation occurs at cytosine residues in symmetric, CG and CHG, as well as asymmetric, CHH, contexts (where H = A, T or C; Law and Jacobsen, 2010). In *Arabidopsis*, CG methylation is maintained by METHYLTRANSFERASE 1 (MET1; Saze et al., 2003), whereas CHG methylation is mediated by CHROMOMETHYLASE 3 (CMT3; Lindroth et al., 2001). The maintenance of CHH methylation is carried

Table 1 | Chromatin modifiers functionally characterized in rice.

	Name	Gene locus	Molecular function	Biological role	Reference
DNA methylation	OsMET1b/OsMET1-2	LOC_Os07g08500	DNA methyltransferase	Seed development	Hu et al. (2014), Yamauchi et al. (2014)
	OsDRM2	LOC_Os03g02010	*De novo* DNA methyltransferase	Pleiotrpic effects on development	Moritoh et al. (2012), Pang et al. (2013)
	OsDDM1	LOC_Os09g27060	DNA methylation maintenance	Transposon repression, growth inhibition	Higo et al. (2012)
DNA demethylation	OsROS1a	LOC_Os01g11900	DNA demethylase	Plant reproduction	Zemach et al. (2010), Ono et al. (2012)
	OsROS1c	LOC_Os05g37350	DNA demethylase	Transposon activation	La et al. (2011)
Histone methylation	SDG714	LOC_Os01g70220	H3K9 methyltransferase	Transposon repression, trichome devopment	Ding et al. (2007b)
	SDG728	LOC_Os05g41170	H3K9 methyltransferase	Transposon repression, seed development	Qin et al. (2010)
	SDG725	LOC_Os02g34850	H3K36 methyltransferase	Hormone regulatory gene activation, flowering	Sui et al. (2012, 2013)
	SDG724	LOC_Os09gl3740	H3K36 methyltransferase	Flowering	Sun et al. (2012)
	SDG723/OsTrx1	LOC_Os09g04890	H3K4 methyltransferase	Flowering	Choi et al. (2014)
Histone demethylation	JMJ706	LOC_Os10g42690	H3K9 demethylase	Floral organ development	Sun and Zhou (2008)
	JMJ705	LOC_Os01g67970	H3K27 demethylase	Biotic stress response, plant reproduction	Li et al. (2013)
	JMJ703	LOC_Os05g10770	H3K4 demethylase	Stem elongation, transposon repression	Chen et al. (2013), Cui et al. (2013)
	JMJ701	LOC_Os03g05680	H3K4 demethylase	Flowering	Yokoo et al. (2014)
Polycomb silencing	OsiEZ1/SDG718	LOC_Os03g19480	H3K27 methyltransferase	Flowering	Liu et al. (2014)
	OsCLF/SDG711	LOC_Os06gl6390	H3K27 methyltransferase	Flowering	Liu et al. (2014)
	OsFIE1	LOC_Os08g04290	*Drosophila* ESC homolog	Pleiotrpic effects on development	Zhang et al. (2012b), Nallamilli et al. (2013), Folsom et al. (2014)
	OsFIE2	LOC_Os08g04270	*Drosophila* ESC homolog	Organ generation, reproduction	Luo et al. (2009), Li et al. (2014)
	OsEMF2b	LOC_Os09g13630	*Drosophila* Su(z)12 homolog	Floral organ development	Yang et al. (2013), Conrad et al. (2014)
Histone deacetylation	OsHDT1/HDT701	LOC_Os05g51840	H4 deacetylase	Biotic stress response, heterosis	Li et al. (2011a), Ding et al. (2012a)
	OsSRT1	LOC_Os04g20270	H3K9 deacetylase	Cell death, transposon repression	Huang et al. (2007), Zhong et al. (2013)
Others	CHD3/CHR729	LOC_Os07g31450	Chromodomain and PHD-domain protein	Pleiotrpic effects on development	Hu et al. (2012)
	MEL1	LOC_Os03g58600	AGO-family protein	Meiosis progression	Nonomura et al. (2007), Komiya et al. (2014)
	SHO1	LOC_Os04g43050	Homolog of DICER-LIKE 4	Pleiotrpic effects on development	Abe et al. (2010)
	SHL2	LOC_Os01g34350	RDR6 homolog	Floral organ development	Toriba et al. (2010)
	WAF1	LOC_Os07g06970	HEN1 homolog	Pleiotrpic effects on development	Abe et al. (2010)
	BRK1	LOC_Os07g32480	H2A phosphorylation	Meiosis progression	Wang et al. (2012)

out by CMT2 and DOMAINS REARRANGED METHYLTRANS-FERASE 2 (DRM2), an ortholog of mammalian Dnmt3 (Law and Jacobsen, 2010; Stroud et al., 2014). DRM2 is required for *de novo* cytosine methylation in both symmetric and asymmetric sequence contexts, which is guided to the target region by RNA-directed DNA methylation (RdDM) pathway (Cao and Jacobsen, 2002; Law and Jacobsen, 2010; Stroud et al., 2014). While *Arabidopsis* contains only one *MET1* gene, rice has two *MET1* genes, *MET1a* (also named *OsMET1-1*) and *MET1b/OsMET1-2* (Teerawanichpan et al., 2004; Yamauchi et al., 2008). The transcripts of *MET1b* accumulate more abundantly than those of *MET1a* in all of the examined rice tissues, indicating that *MET1b* may play a more important role in maintaining DNA methylation (Yamauchi et al., 2008). Consistently, more recent studies demonstrate that *MET1b* is an essential gene and its loss causes genome-wide reduction of CG methylation in rice seedlings (Hu et al., 2014; Yamauchi et al., 2014). Rice contains also one *DRM2* gene, *OsDRM2*, and the recombinant OsDRM2 protein expressed in *Escherichia coli* or *Saccharomyces cerevisiae* exhibits stochastic *de novo* DNA methyltransferase activity *in vitro* at CG, CHG, and CHH (Sharma et al., 2009; Pang et al., 2013). Interestingly, OsDRM2 was found to interact with the ATP-dependent RNA helicase, OseIF4A, in both *in vitro* and *in vivo* assays (Dangwal et al., 2013). The interaction specifically depends on the ubiquitin-associated domain of OsDRM2, pointing to a mechanism in which OsDRM2 is recruited to specific chromatin sites by eIF4A together with other cellular proteins for catalyzing DNA methylation (Dangwal et al., 2013). Similar to the *Arabidopsis DECREASE IN DNA METHYLATION 1* (*DDM1*), which encodes a nucleosome remodeling ATPase, *OsDDM1* is also necessary for maintenance of DNA methylation in transposons and repetitive sequences (Higo et al., 2012). The rice genome contains three putative CMT3 homologs (Sharma et al., 2009), yet their functions remain to be characterized.

DNA methylation can be removed passively through dilution during replication as well as actively through catalysis by demethylation enzymes (La et al., 2011; Ono et al., 2012). In *Arabidopsis*, active demethylation is catalyzed by REPRESSOR OF SILENCING 1 (ROS1; Gong et al., 2002; Agius et al., 2006), DEMETER (DME; Choi et al., 2002; Gehring et al., 2006), and DEMETER-LIKE 2 (DML2) and DML3 (Choi et al., 2002; Ortega-Galisteo et al., 2008). Phylogenetic analysis showed that the rice genome encodes six putative bi-functional DNA glycosylases that catalyze cytosine DNA demethylation: four ROS1 orthologs (ROS1a to ROS1d) and two DML3 orthologs (DML3a and DML3b), but no DME orthologs (Zemach et al., 2010). *ROS1c* has been shown to be involved in DNA demethylation and control of the retrotransposon *Tos17* activity (La et al., 2011). Quantitative RT-PCR analysis revealed that *ROS1a*, *ROS1d,* and *DML3a* are expressed in different examined plant tissues, including anthers and pistils, whereas *ROS1b* and *DML3b* are scarcely expressed in these tissues (Ono et al., 2012). Future studies are necessary to investigate the role of these different genes in rice genome DNA methylation.

HISTONE METHYLATION

Histone methylation marks are established on lysine (K) and arginine (R) residues by distinct enzymes, namely histone lysine methyltransferases (HKMTs) and protein arginine methyltransferases (PRMTs), respectively (Liu et al., 2010; Yao and Shen, 2011). In general, H3K9, H3K27, and H4K20 methylations are associated with transcriptional repression, whereas methylation on H3K4 and H3K36 correlates with gene activation. Furthermore, each K residue can be mono-, di-, or tri-methylated, and different methylation status may have different functional implications (Yu et al., 2009).

All known plant HKMTs contain an evolutionarily conserved SET domain (reviewed in Berr et al., 2011). The rice genome encodes at least 37 SET domain proteins, grouped into distinct families (Ng et al., 2007; Huang et al., 2011; Thorstensen et al., 2011). To date, several members belonging to different families are characterized (**Table 1**). Analyses of SET DOMAIN GROUP 714 (SDG714) and its close homologs (e.g., SDG728) showed that these rice SDG proteins have either specific or redundant functions in regulating histone H3K9 methylation and retrotransposon repression (Ding et al., 2007a,b, 2010; Qin et al., 2010). Knockdown of *SDG714* leads to decreased H3K9 methylation levels accompanied by a reduction of CG and CHG methylation, suggesting that H3K9 methylation and DNA methylation act closely together to stably repress the transposition of transposons to maintain genome stability (Ding et al., 2007b). Ectopic expression of *SDG714* in *Arabidopsis* can cause a global elevation of H3K9me2 (Ding et al., 2010). Knockdown of *SDG725* impairs deposition of H3K36me2/3 at several examined gene loci (Sui et al., 2012, 2013). SDG724 is also involved in H3K36me2/3 deposition (Sun et al., 2012). SDG723/OsTrx1 is a close homolog of the *Arabidopsis* H3K4-methyltransferase ATX1 and can methylate *in vitro* H3 within oligonucleosomes (Choi et al., 2014). The rice genome contains two genes encoding putative H3K27 methyltransferases, OsiEZ1/SDG718 (also named OsSET1) and OsCLF/SDG711, which likely work in protein complexes in Polycomb silencing pathway (see Section below).

Histone lysine methylation can be removed by histone demethylases, which consist of two classes: Lysine Specific Demethylase 1 (LSD1) and Jumonji C (jmjC) domain-containing proteins (Tsukada et al., 2006; Mosammaparast and Shi, 2010). LSD1, a flavin-dependent amine oxidase, has been the first histone demethylase reported (Shi et al., 2004) and *Arabidopsis* contains three LSD1 homologs, which are involved in flowering time regulation (Jiang et al., 2007; Liu et al., 2007; Shafiq et al., 2014). Three rice genes (*Os02g0755200*, *Os04g0560300*, and *Os08g0143400*) encode LSD1 homologs, but their functions remain uncharacterized. There are at least 20 jmjC domain-containing proteins in rice, and the first characterized JMJ706 specifically demethylates H3K9me2/me3 (Sun and Zhou, 2008). More recently, several other rice jmjC-encoding genes have been characterized. *JMJ705* encodes a histone lysine demethylase that specifically removes H3K27me2/3, and the expression of *JMJ705* is induced by stress signals and during pathogen infection (Li et al., 2013). For active histone marks, JMJ703 is involved in the removal of H3K4me1/me2/me3 (Chen et al., 2013; Cui et al., 2013), and JMJ701 in removal of H3K4me3 (Yokoo et al., 2014). So far, however, histone demethylase(s) involved in removal of H3K36 methylation is(are) unknown.

POLYCOMB SILENCING

Polycomb Group (PcG) proteins were first identified as master regulators and suppressors of homeotic genes in *Drosophila melanogaster*. Polycomb Repressive Complex 2 (PRC2) has four core components: ENHANCEROF ZESTE (E[z]), SUPPRESSOR OF ZESTE 12 (Su[z]12), EXTRA SEX COMBS (ESC), and the 55 kDa WD40-repeat protein N55 (Schuettengruber and Cavalli, 2009). PRC2 mediates H3K27me3 deposition *via* the catalytic subunit E[z], a SET-domain containing protein (Czermin et al., 2002). The four core subunits of the PRC2 complex are well conserved in animals as well as in plants (Chen and Rasmuson-Lestander, 2009; He et al., 2013). While in *Drosophila* all but one subunit is encoded by a single gene, most of the plant PRC2 core subunits are encoded by small gene families. In *Arabidopsis*, MEDEA (MEA)/FERTILIZATION INDEPENDENT SEED 1 (FIS1), CURLY LEAF (CLF), and SWINGER (SWN) are the three homologs of E[z]; FIS2, VERNALIZATION 2 (VRN2), and EMBRYONIC FLOWER 2 (EMF2) are the three homologs of Su[z]12; MULTICOPY SUPPRESSOR OF IRA1 (MSI1) to MSI5 are the five homologs of N55; and FERTILIZATION INDEPENDENT ENDOSPERM (FIE) is the only homolog of ESC. Remarkably, MEA/FIS1 and FIS2, which are important for endosperm and seed development in *Arabidopsis*, are absent from rice, and rice has two E[z] homologs: OsiEZ1/SDG718 and OsCLF/SDG711, two Su[z]12 homologs: OsEMF2a and OsEMF2b, but also two FIE homologs: OsFIE1 and OsFIE2 (Luo et al., 2009). Functional roles of some of these rice PcG proteins have been characterized (**Table 1**). The expression of *OsiEZ1/SDG718* and *OsCLF/SDG711* is induced by and represses flowering genes in long day and short day, respectively (Liu et al., 2014). While *OsFIE2* is expressed broadly in all examined rice tissues, *OsFIE1* is expressed specifically in the rice endosperm and its expression in vegetative tissues is likely to be silenced by promoter DNA methylation (Zhang et al., 2012b; Nallamilli et al., 2013). Furthermore, *OsFIE1* is imprinted and only the maternal allele is expressed in endosperm (Luo et al., 2009). More recently, it was reported that *OsFIE1* is responsive to temperature changes and its expression negatively correlates with the duration of the syncytial seed developmental stage during heat stress (Folsom et al., 2014). DNA methylation, H3K9me2 and/or H3K27me3 are likely involved in regulation of varied repressive status of *OsFIE1* (Zhang et al., 2012b; Nallamilli et al., 2013; Folsom et al., 2014). Functional characterization of *OsEMF2b* revealed that PRC2 plays a major role in modulation of the expression of E-function MADS-box transcription factor genes required for floral organ specification and floral meristem determinacy (Luo et al., 2009; Yang et al., 2013; Conrad et al., 2014). Very importantly, OsFIE2 interacts with OsiEZ1/SDG718 and the OsFIE2-associated complex purified from transgenic rice suspension cells (containing OsEMF2b, OsCLF, OsiEZ1/SDG718) can methylate H3K27 in *in vitro* histone methyltransferase assay (Nallamilli et al., 2013).

HISTONE ACETYLATION

Histone lysine acetylation is generally associated with transcription activation and is dynamically regulated by the antagonistic activities between histone acetyltransferases (HATs) and histone deacetylases (HDACs; Chen and Tian, 2007). All four core histones can be acetylated and a nucleosome contains 26 putative acetylation sites (Lusser et al., 2001). Global analysis of lysine acetylation demonstrates the involvement of protein acetylation in diverse biological processes in rice (Nallamilli et al., 2014). The rice genome contains eight HATs and 19 HDACs (Hu et al., 2009; Liu et al., 2012). The eight HATs can be divided into four groups, namely the CREB-Binding Protein (CBP) group, the TAFII-associated factor (TAFII250) group, the GCN5-related *N*-terminal acetyltransferase (GNAT) group, and the MYST (named for the founding members MOZ, Ybf2/Sas3, Sas2, and Tip60) group (Liu et al., 2012). The 19 HDACs are grouped into three distinct families, namely the Reduced Potassium Deficiency 3 (RPD3) family, the Silent Information Regulator 2 (SIR2) family, and the type-II HDAC (HD2) family which is plant specific (Ma et al., 2013). Reversible and dynamic changes of H3 acetylation occurs at submergence-inducible genes, *alcohol dehydrogenase 1* (*ADH1*) and *pyruvate decarboxylase 1* (*PDC1*) in rice (Tsuji et al., 2006). Forward genetic analysis has identified a rice mutant, *rice plasticity 1* (*rpl1*), which displays increased environment-dependent phenotypic variations and an elevation of overall H3K9 acetylation (Zhang et al., 2012a). Down-regulation of *OsHDT1/HDT701*, which encodes a histone H4 deacetylase, causes elevated levels of H4 acetylation and increased transcription of pattern recognition receptor (PRR) and defense-related genes (Ding et al., 2012a). Knockdown of *OsSRT1*, a member of SIR2-like HDAC family, results in an increase of H3K9 acetylation (H3K9ac), leading to DNA fragmentation and cell death, and the OsSRT1 protein binds to loci with relative low level of H3K9ac and regulates expression of many genes related to stress and metabolism as well as several families of transposable elements (Huang et al., 2007; Zhong et al., 2013).

READERS OF HISTONE MODIFICATIONS

Specific recognition of histone modifications by readers can recruit various components of the nuclear signaling network to chromatin, mediating fundamental processes such as gene transcription, DNA replication and recombination, DNA repair and chromatin remodeling (Musselman et al., 2012). Some readers are reported in *Arabidopsis* (reviewed in Berr et al., 2011), and more recent works have identified several novel chromodomain (CHD)- and/or plant homeodomain (PHD)-containing proteins as readers of H3K4me2/me3 and H3K36me3 (Bu et al., 2014; Lopez-Gonzalez et al., 2014; Molitor et al., 2014; Xu et al., 2014). Interestingly, the rice CHD3 protein can bind both the active mark H3K4me2 and the repressive mark H3K27me3 via its CHD and PHD domain, respectively (Hu et al., 2012). Knockdown of CHD3 caused reduction of H3K4me3 and H3K27me3 at many genes. It was thus suggested that the rice CHD3 may act as a bifunctional reader capable to recognize and modulate both H3K4 and H3K27 methylations (Hu et al., 2012).

SMALL AND LONG NON-CODING RNAs

Non-coding small RNAs (sRNA) of 21–24 nucleotides (nt) in length as well as long non-coding RNAs (lncRNAs, >200 nt in length) are known to be involved in chromatin modifications and thus epigenetic inheritance (reviewed in Castel and Martienssen, 2013; Bond and Baulcombe, 2014). Genome-wide profiling have identified several hundreds of different sRNAs, and differences exist at their expression levels between different rice subspecies, reciprocal hybrids, different plant tissues, and under different growth conditions (Chen et al., 2010; He et al., 2010; Jeong et al., 2010; Zhang et al., 2014). Remarkably, the most abundant sRNAs identified in rice panicles are 24 nt in length and mainly correspond to transposon-associated or repeat-associated small interfering RNAs (siRNAs; Jeong et al., 2011). The most intriguing role of siRNAs is in repression of transposons and repeat elements in reproductive tissues and epigenomic reprogramming during gametogenesis (Gutierrez-Marcos and Dickinson, 2012; Castel and Martienssen, 2013; Bond and Baulcombe, 2014). ARGONAUTE (AGO) proteins play important roles in microRNA-mediated post-transcriptional gene silencing (PTGS) and siRNA-mediated RdDM (Vaucheret, 2008). A germ line specific AGO-encoding gene, *MEIOSIS ARRESTED AT LEPTOTENE1* (*MEL1*), has been reported in rice, and the *mel1* mutant shows chromosome abortion during early meiotic stages, leading to impaired male and female fertilities (Nonomura et al., 2007). More recently, forward genetic analysis has identified a lncRNA, which could be subsequently processed to small RNAs, as a key regulator of male fertility in rice (Ding et al., 2012b,c). Meanwhile, Zhou et al. (2012) reported that a spontaneous mutation of a small RNA could cause male sterility in rice. Nevertheless, the precise role of lncRNA and sRNA, particularly at rice chromatin structure levels, requires future investigations.

EPIGENETIC REGULATION OF RICE FLOWERING

Flowering represents the transition from vegetative to reproductive growth, a key developmental switch during the plant life cycle. Flowering time is precisely controlled by complex gene network that integrates environmental signals, such as day length (photoperiod), light intensity and quality, and ambient temperature, as well as endogenous cues involving plant hormones (Albani and Coupland, 2010; Shrestha et al., 2014). Photoperiod is one of the most predictable cues in nature, and according to photoperiod responsiveness plants can be categorized into three groups: long-day (LD) plants, short-day (SD) plants, and day-neutral plants. *Arabidopsis* is a facultative LD plant whose flowering is accelerated when grown under LD photoperiods. Furthermore, flowering of most *Arabidopsis* ecotypes is promoted by a prolonged exposure to the cold of winter (a process known as vernalization), which has an epigenetic basis of competence memory (Ream et al., 2012; Song et al., 2012). During recent years, many chromatin modifiers have been shown as involved in *Arabidopsis* flowering time regulation, with majority of them acting *via* the transcriptional regulation of *FLOWERING LOCUS C* (*FLC*), a key flowering repressor at which vernalization and autonomous pathways converge (Berr et al., 2011; He, 2012; Ietswaart et al., 2012). In contrast to *Arabidopsis*, rice is a facultative SD plant and does not require vernalization to induce flowering and does not contain a *FLC* homolog. The complex gene network of rice flowering pathways primarily consists of flowering activators, and remarkably several chromatin modifiers have been shown recently as involved in rice flowering time control (**Figure 1**).

KEY TRANSCRIPTION FACTORS OF RICE FLOWERING PATHWAYS

Within the rice flowering pathways, the close paralogs *Heading date 3a* (*Hd3a*) and *RICE FLOWERING LOCUS T1* (*RFT1*) are specifically upregulated upon the inductive SD photoperiods in leaf phloem tissue and encode small globular proteins named florigens, which move to the shoot apex to promote flowering (Tsuji et al., 2013; Sun et al., 2014). There are at least two pathways that control the *Hd3a/RFT1* expression under either SD (**Figure 1A**) or LD (**Figure 1B**) photoperiods: the *Early heading date 1* (*Ehd1*) and the *Hd1* pathways (Tsuji et al., 2013; Sun et al., 2014). *Ehd1* encodes a B-type transcription factor that plays a key role in activation of both *Hd3a* and *RFT1* expression. The expression of *Ehd1* is modulated by at least three different types of function factors (Sun et al., 2014). The first type comprises day length-independent activators, including *Ehd2*, also known as *Rice Indeterminate1* (*RID1*) or *Os Indeterminate1* (*OsId1*), and *Ehd4*, which encode two different zinc-finger transcription factors and act in both SD and LD conditions in *Ehd1* induction (**Figure 1**). The second type comprises SD-preferential activators, including the PHD-finger factor Ehd3 and the MADS-box family transcription factor OsMADS51, which induce *Ehd1* expression specifically in SD conditions (**Figure 1A**). And the third type comprises LD-preferential repressors, including *Grain number, plant height, and heading date7* (*Ghd7*) that encodes a CCT-domain protein and *LEC2-FUSCA3-Like 1* (*OsLFL1*) that encodes a B3-type transcription factor, both repress *Ehd1* expression specifically in LD conditions (**Figure 1B**). Further upstream, the LD-preferential regulator *OsMADS50* promotes flowering *via* repression of *OsLFL1*. Interestingly, *Ehd3*, which acts as an activator of *Ehd1* to promote flowering in SD conditions (**Figure 1A**), displays a repressor function on *Ghd7* and thus also promotes flowering in LD conditions (**Figure 1B**). The rice circadian clock related protein GIGANTEA (OsGI) activates the *Ehd1* pathway partly *via* induction of *OsMAD51* expression (**Figure 1B**). While the *Ehd1* pathway is more unique to rice, the OsGI-Hd1-Hd3a pathway is very similar to the *Arabidopsis* GI-CO-FT pathway, composing of the respective orthologous proteins in the two plant species (Tsuji et al., 2013; Sun et al., 2014). An atypical helix-loop-helix (HLH) protein (OsLF) also is involved in the OsGI-Hd1-Hd3a pathway via *Hd1* repression. *Hd1* acts as an activator to promote rice flowering in SD conditions (**Figure 1A**) but as a suppressor of rice flowering in LD conditions (**Figure 1B**). Phytochrome signaling is crucial in conversion of Hd1 activity because mutation of *Phytochrome B* (*PHYB*) or phytochrome deficiency (e.g., in *photoperiod sensitivity5* mutant) maintains Hd1 as an activator independent of day length. Under LD conditions, the red-light photoreceptor PHYB pathway may convert and maintain Hd1 as a repressor possible *via* post-translational modification and/or protein complex formation. Because of space limitation, the one who is interested in more details about rice flowering pathways can read the two excellent review articles here cited (Tsuji et al., 2013; Sun et al., 2014) and the original research papers referred therein.

ACTIVE CHROMATIN MARKS ARE INVOLVED IN RICE FLOWERING TIME REGULATION

Understanding how the rice flowering pathway genes are regulated in the chromatin context has great importance. Recent studies have

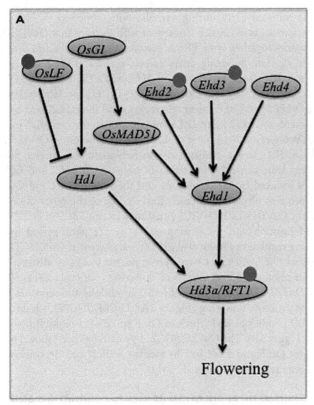 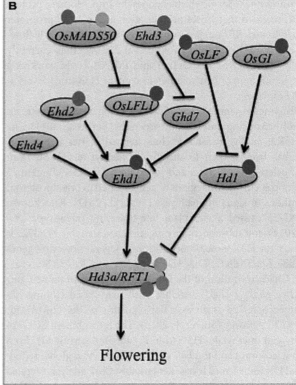

Flowering

Flowering

● H3K9 acetylation
● SDG725-mediated H3K36me2/3
● SDG724-mediated H3K36me2/3

● OsHDT1-mediated H4 deacetylation
● PRC2-mediated H3K27me3
● JMJ701-mediated H3K4me3 demethylation

FIGURE 1 | Regulatory networks of genetic and epigenetic control of rice flowering under short-day (A) and long-day (B) photoperiod conditions. Rice flowering network is integrated by two florigen genes *Hd3a* and *RFT1*, which are regulated by at least two pathways: the *Hd1*-dependent and the *Ehd1*-dependent pathways. Expressions of *Hd1* and *Ehd1* are further regulated by more upstream genes as indicated by different names in the circles. Arrows indicate for transcriptional activation, whereas bars indicate for transcriptional repression. Different color spheres surrounding the flowering gene circles indicate for different regulations by the indicated histone modifications at the gene locus, currently described in literatures.

found that histone acetylations, H3K4 and H3K36 methylations are involved in active transcription of several genes within the rice flowering pathways (**Figure 1**). It was reported that overexpression of the HD2-family HDAC gene *OsHDT1* in hybrid rice leads to early flowering under LD conditions, probably through transcriptional repression of *OsGI* and *Hd1* (Li et al., 2011a). Interestingly, the expression of *OsHDT1* displays a circadian rhythm under SD conditions, peaked at the end of day, which coincides with rhythmic expression of *OsGI* and advances that of *Hd1*. Ectopic *OsHDT1* expression in transgenic rice attenuates the overdominance rhythmic expression of *OsGI* and *Hd1* in hybrid rice, which may explains the early flowering phenotype specifically observed in hybrid but not parental rice lines (Li et al., 2011a). Histone H4 acetylation levels were observed to positively correlate with the rhythmic expression of *OsGI* and *Hd1*, and *OsHDT1* overexpression was shown to impair the acetylation increase at the peak time (Li et al., 2011a).

A positive DNA/histone methylation role in rice flowering promotion was first indicated by the study of the

S-adenosyl-L-methionine synthetase gene mutants (Li et al., 2011b). S-Adenosyl-L-methionine is a universal methyl group donor for both DNA and protein methylations. Its deficiency caused late-flowering of rice plants and reduction of *Ehd1*, *Hd3a*, and *RFT1* expression, which is associated with reduced levels of H3K4me3 and DNA CG/CHG-methylations at these flowering gene loci (Li et al., 2011b). More recently, it was reported that suppression of *OsTrx1*, an ortholog of the *Arabidopsis* H3K4-methyltransferase gene *ATX1*, delays rice flowering time under LD conditions (Choi et al., 2014). The *OsTrx1* suppression did not affect the *OsMADS50* and *Hd1* pathways, but elevated *Ghd7* expression and drastically reduced *Ehd1*, *Hd3a* and *RFT1* expression, which is consistent with the plant late-flowering phenotype (**Figure 1B**). The PHD domain of OsTrx1 can bind to native histone H3 and the SET domain of OsTrx1 can methylate histone H3 from oligonucleosomes *in vitro* (Choi et al., 2014). Yet the role of OsTrx1 in histone methylation *in vivo* remains undemonstrated. Because the OsTrx1 and Ehd3 proteins bind each other, the authors propose that OsTrx1 may promote rice flowering *via* interaction with Ehd3 (Choi et al., 2014). Mutant characterization of *Photoperiod sensitivity-14*(*Se14*), which encodes the JmjC-domain protein JMJ701, revealed that H3K4me3 elevation at the *RFT1* promoter region increases *RFT1* expression, leading to rice plant early flowering under LD conditions (Yokoo et al., 2014). It is currently unknown whether or not OsTrx1 and JMJ701 could work as a couple in an antagonistic manner to control H3K4me3 levels at the *RFT1* locus.

H3K36me3 is generally considered as acting more downstream of H3K4me3 during transcription processes (Berr et al., 2011). The first H3K36-methyltransferase characterized in rice is SDG725, which has been shown to specifically methylate H3K36 from mononucleosomes *in vitro* and is required for H3K36me2/me3 deposition at chromatin regions of genes related to brassinosteroid biosynthesis or signaling pathways (Sui et al., 2012). Knockdown of *SDG725* caused a rice plant late-flowering phenotype (Sui et al., 2012), and subsequent investigation revealed that SDG725 is necessary for H3K36me2/3 deposition at several flowering genes including *Ehd3*, *Ehd2*, *OsMADS50*, *Hd3a*, and *RFT1* (Sui et al., 2013). Characterization of the late-flowering mutant named *long vegetative phase 1* (*lvp1*) together with map-based cloning has uncovered SDG724 as an essential regulator of the *OsMAD50-Ehd1-RFT1* pathway (Sun et al., 2012). The recombinant SDG724 protein can methylate H3 (with K site undetermined) from oligonucleosomes and the *lvp1* mutant plants show global reduction of H3K36me2/me3 levels. Remarkably, ChIP analysis revealed specific reduction of H3K36me2/me3 at *OsMADS50* and *RFT1* but not at *Ehd1* and *Hd3a* in the *lvp1* mutant plants (Sun et al., 2012). Both the *lvp1* (*sdg724*) mutant and the SDG725-knockdown mutant exhibit late-flowering phenotypes under either SD or LD conditions (Sun et al., 2012; Sui et al., 2013), pointing to a crucial role of H3K36me2/me3 in promoting rice plant flowering irrespective of photoperiods. It is noteworthy that in *Arabidopsis* the SDG8-mediated H3K36me2/me3 also plays a major role in flowering time control, but in that case in prevention of early flowering (Shafiq et al., 2014). Future studies are necessary to investigate mechanisms underlying the overlap and specific targets of SDG724 and SDG725 in the rice flowering time control.

REPRESSIVE CHROMATIN MARKS ARE INVOLVED IN RICE FLOWERING TIME REGULATION

The repressive mark H3K27me3 is known to play a key role in *FLC* repression in vernalization-induced *Arabidopsis* plant flowering (He, 2012; Ietswaart et al., 2012). Interestingly, recent studies have shown that H3K27me3 deposited by PRC2-like complexes also plays an important role in vernalization-independent rice flowering time control (**Figure 1**). Loss-of-function of the PRC2 gene *OsEMF2b* causes late-flowering, which is associated with an increase of *OsLFL1* expression and a decrease of *Ehd1* expression (Yang et al., 2013). The OsEMF2b protein physically interacts with OsVIL3 (named as OsVIL2 in Yang et al., 2013, but here corrected to the first nomenclature used in Zhao et al., 2010; also called LC2), a PHD-domain protein showing homologies to the *Arabidopsis* VIN3-group proteins including VERNALIZATION INSENSITIVE 3 (VIN3), VIN3-LIKE 1 (VIL1)/VRN5, and VIL2/VEL1. The *Arabidopsis* VIN3-group proteins are know to be associated and to work together with the PRC2 core complex (constituting

the so-called PHD-PRC2 complexes) and the *VIN3* expression is induced early during vernalization (reviewed in He, 2012; Ietswaart et al., 2012). Consistent with the idea that OsVIL3/LC2 works together with PRC2, knockdown of *OsVIL3/LC2* results in rice late-flowering, increase of *OsLFL1* and *OsLF* expression, and decrease of *Ehd1* as well as *Hd3a* and *RFT1* expression (Wang et al., 2013; Yang et al., 2013). The OsVIL3/LC2 protein binds at the *OsLFL1* and *OsLF* chromatin regions and the H3K27me3 enrichments at *OsLFL1* and *OsLF* are impaired in the *osvil3*/lc2 mutant (Wang et al., 2013; Yang et al., 2013). In addition to *OsVIL3/LC2*, *OsVIL2* plays a similar but non-redundant role in rice flowering time control. Expression of both *OsVIL3/LC2* and *OsVIL2* is induced by SD conditions and the OsVIL3/LC2 and OsVIL2 proteins physically interact, thus leading to the proposition that the OsVIN3/LC2-OsVIL2 dimer may recruit PRC2 in H3K27me3 deposition and *OsLF* suppression in rice photoperiod flowering regulation (Wang et al., 2013). Very recently, *OsiEZ1/SDG718* and *OsCLF/SDG711* have been reported to display distinct roles in photoperiod regulation of flowering (Liu et al., 2014). While *OsiEZ1/SDG718* is induced in SD conditions and represses *OsLF* to promote flowering (**Figure 1A**), *OsCLF/SDG711* is induced in LD conditions and represses *OsLF* and *Ehd1* to inhibit flowering (**Figure 1B**). The OsCLF/SDG711 protein has been shown to target *OsLF* and *Ehd1* loci to mediate H3K27me3 deposition and gene repression (Liu et al., 2014).

EPIGENETIC REGULATION OF RICE REPRODUCTION AND SEED FORMATION

After flowering, plant sexual reproduction occurs in dedicated floral organs through sporogenesis, gametogenesis, embryo- and endosperm-genesis, resulting in seed formation. Studies in *Arabidopsis* have unraveled diverse epigenetic regulatory mechanisms as involved in different processes during floral organogenesis and plant sexual reproduction (Shen and Xu, 2009; Engelhorn et al., 2014; She and Baroux, 2014). Although more recent, studies in rice also have started to uncover multiple types of epigenetic modifiers involved in the regulation of plant reproduction (**Figure 2**).

EPIGENETIC REGULATION IN RICE REPRODUCTION

Compared to those of *Arabidopsis*, the rice inflorescence and flower have greatly diverged structures that are regulated by a conserved genetic framework together with rice specific genetic mechanisms (Yoshida and Nagato, 2011). Several epialleles are found to affect rice plant reproduction. The metastable epigenetic silencing of *DWARF1*, which is associated with DNA methylation and H3K9me2 at the gene promoter region, causes dwarf tillers, compact panicles (inflorescences) and small round rice grains (Miura et al., 2009). The *abnormal floral organ* (*afo*) epimutation causes increased DNA methylation and suppression of the transcription factor gene *OsMADS1*, leading to pseudovivipary, a specific asexual reproductive strategy (Wang et al., 2010). The transcription factor gene *SQUAMOSA PROMOTER BINDING PROTEIN-LIKE 14* (*SPL14*), also known as *IDEAL PLANT ARCHITECTURE 1* (*IPA1*) or *WEALTHY FARMER'S PANICLE* (*WFP*), promotes panicle branching and regulates a large number of genes, and differences in DNA methylation at the locus as well as the micro RNA 156 (OsmiR156) contribute to expression differences of

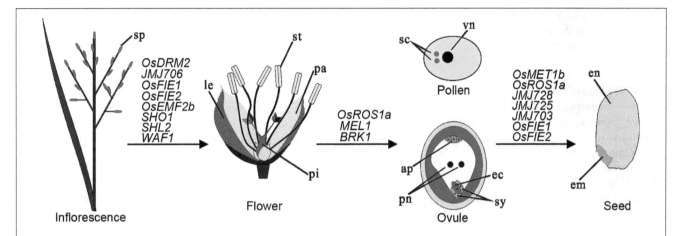

FIGURE 2 | Schematic representation of structures involved in rice reproduction together with chromatin modifier genes listed in regulation of three different steps. Inflorescence produces spikelets (sp) that generate numerous flowers. A mature flower contains different types of organs including lemma (le), palea (pa), stemen (st), and pistil (pi). The female gametophyte ovule is formed inside of ovary of the pistil and at maturation contains four different types of cells: antipodal (ap), polar nuclei (pn), synergid (sy), and egg cell (ec). The male gametophyte pollen is produced inside of anther of the stamen and at maturation contains two sperm cells (sc) and one vegetative cell (vn). Upon fertilization, one sperm cell fuses with egg cell to produce embryo (em) and the other sperm cell fuses with the two polar nuclei to produce endosperm (en), together forming a mature seed. Chromatin modifier genes playing important regulatory roles in floral organogenesis, gametophyte development, and fertilization/seed development are listed.

SPL14/IPA1/WFP in different rice varieties (Jiao et al., 2010; Miura et al., 2010; Lu et al., 2013). Important roles of sRNA (both miRNAs and siRNAs) in rice floral organ development are also evidenced by mutants of several sRNA-pathway genes, including *SHOOT ORGANIZATION 1* (*SHO1*) encoding a DICER-LIKE 4 homolog, *SHOOTLESS 2* (*SHL2*) encoding a RDR6 homolog), and *WAVY LEAF 1* (*WAF1*) encoding a HEN1 homolog (Abe et al., 2010; Toriba et al., 2010). lncRNAs are also reported as involved in plant reproductive process (Swiezewski et al., 2009; Heo and Sung, 2011). In rice, a point mutation that alter the secondary structure of the lncRNA called Long-Day-specific Male-fertility-Associated RNA (LDMAR) has been found to cause the photoperiod sensitive male sterility (Ding et al., 2012b).

Importance of DNA methylation in regulation of rice reproduction has been further supported by mutant studies. Targeted disruption of the DNA demethylase gene *ROS1a* leads to paternal allele transmission defect, presumably because of a male gametophytic defect prior to fertilization (Ono et al., 2012). Disruption of *OsDRM2* led to pleiotropic developmental defects in both vegetative and reproductive stages including semi-dwarfed stature, reductions in tiller number, and complete sterility (Moritoh et al., 2012). Consistently, transcriptome analysis of isolated rice gametes by deep sequencing indicates that *OsDRM2* is expressed in male cells but low in vegetative cells (Anderson et al., 2013).

Several modifiers of histone modifications are also critical for rice reproduction (**Figure 2**). Loss-of-function of the rice PRC2 gene *OsEMF2b* results in complete sterility, and severe floral organ defects and indeterminacy that resemble loss-of-function mutants in E-class floral organ specification genes (Conrad et al., 2014). The epimutation of *OsFIE1* (*Epi-df*) that is caused by DNA hypomethylation, reduced H3K9me2 and increased H3K4me3 at the gene locus, leads to ectopic expression of *OsFIE1*, resulting in a dwarf stature, diverse floral defects, and alteration of H3K27me3 levels

at hundreds of target genes (Zhang et al., 2012b). Mutation of the H3K27-demethylase gene *JMJ705* also causes partial sterility (Li et al., 2013). The *OsFIE2* RNAi lines display pleiotropic phenotypes including vegetative and reproductive organ formation, a decreased amount of pollen grains and a high proportion of male sterility (Li et al., 2014). These studies indicate that a balanced level of H3K27me3 is critical and that either its increase or decrease can cause rice reproduction defects. The other chromatin repressive mark H3K9me2/me3 is also important because mutation of the H3K9-demethylase gene *JMJ706* impairs spikelet development, including defective floral morphology and altered organ number (Sun and Zhou, 2008). Pleiotropic defective phenotypes including panicle morphology, rachis branch and spikelet numbers have also been described for mutants of the H3K36-methyltransferase gene *SDG725* and the H3K4-demethylase gene *JMJ703* (Sui et al., 2012; Cui et al., 2013), indicating that chromatin active marks also play important function during rice reproduction.

While precise reproduction processes affected by many of the above mentioned modifiers remain to be elucidated, meiosis is found to be regulated by several epigenetic factors. The rice germline-specific AGO-family protein MEL1 binds preferentially 21 nt siRNAs derived mostly from intergenic regions (Komiya et al., 2014), and its loss-of-function impairs both sporophytic germ-cell development and meiosis (Nonomura et al., 2007). The *mel1* mutant displays aberrant vacuolation of spore mother cells, and arrested chromosome condensation at early meiosis stages. H3K9me2 distribution as well as the localization of ZEP1, a component of transverse filaments of the rice synaptonemal complex, are affected in *mel1*, indicating for a role of *MEL1* in chromatin structure organization and homologous chromosome synapsis in early meiosis (Nonomura et al., 2007; Komiya et al., 2014). Histone phosphorylation is also involved in rice meiosis process. The rice Bub1-Related Kinase 1 (BRK1) is required for H2A phosphorylation and the centromeric recruitment of SHUGOSHIN 1 (SGO1), which is likely essential for generating proper tension between the homologous kinetochores at metaphase I to facilitate the accurate segregation of homologous chromosomes at anaphase I (Wang et al., 2012).

EPIGENETIC REGULATIONS IN SEED DEVELOPMENT

Like other angiosperms, sexual double fertilization initiates rice seed development, giving rise to two fertilization products, the embryo and the endosperm. Epigenetic mechanisms are thought to have important contribution to plant hybrid vigor (heterosis), a phenomenon referring to the increased yield and biomass of hybrid offspring relative to the parents (Chen and Zhou, 2013; Groszmann et al., 2013). In line with this idea, divers epigenetic pathways are found as involved in seed development and seed quality control (**Figure 2**).

Genome-wide analyses in rice have revealed that sRNA expression, DNA methylation, and histone modifications (e.g., H3K9ac, H3K4me3, and H3K27me3) significantly differ between hybrids and their parents (He et al., 2010; Chodavarapu et al., 2012; Zhang et al., 2014). Remarkably, the amount of 24 nt siRNAs, with most of them likely involved in regulation of the starch and sucrose biosynthesis pathway, declines with the process of rice grain-filling and this decline is to a lower degree in inferior grains then superior grains (Peng et al., 2013). The siRNAs may act *via* or together with DNA methylation in heterochromatin silencing. In line with this idea, the maternal loss of *ROS1a* causes failure of early stage endosperm development, leading to incomplete embryogenesis producing irregular but viable embryos that failed to complete seed dormancy (Ono et al., 2012). While the *met1a* null mutant displays a normal phenotype, the *met1b* mutant exhibits abnormal seed phenotypes, which is associated with either viviparous germination or early embryonic lethality (Hu et al., 2014; Yamauchi et al., 2014). Levels of DNA methylation in *met1b* are broadly reduced at genome-wide scale and in particular at repetitive centromeric and transposon sequences as well as at the *OsFIE1* gene locus in the embryos (Hu et al., 2014; Yamauchi et al., 2014).

OsFIE1 is an imprinted gene in rice endosperm but the *osfie1* mutant does not display any autonomous endosperm proliferation without fertilization, differing from the *Arabidopsis fie*, *mea* and *fis* mutants that are generally recognized with an autonomous endosperm proliferation phenotype (Luo et al., 2009). Nevertheless, over-expression of *OsFIE1* causes precocious cellularization and reduced seed size, and it has been proposed that that *OsFIE1* has a role in regulating seed enlargement under heat stress (Folsom et al., 2014). In addition, *OsFIE2* has a critical role in normal endosperm development and grain-filling. Down-regulation of *OsFIE2* results in small seeds and partial loss of seed dormancy, likely because of down-regulation of genes encoding the starch synthesis rate limiting step enzymes and multiple storage proteins (Nallamilli et al., 2013). Future studies are necessary to precise similarities and differences of PRC2-mediated H3K27me3 repression mechanisms involved in seed development between *Arabidopsis* and rice.

Involvement of other histone methylation marks in seed development are also evidenced from mutant studies (**Figure 2**). Down-regulation of the H3K9-methyltransferase gene *SDG728* reduces seed size and alters seed morphology (Qin et al., 2010).

Loss-of-function of the H3K4-demethylase gene *JMJ703* causes abnormal grain phenotypes, including reduced length, width, and thickness (Cui et al., 2013). Also, knockdown of the H3K36-methyltransferase gene SDG725 results in small seed size and reduced seed weight (Sui et al., 2012).

CONCLUSION REMARKS

The availability of full genome sequences and diverse improved powerful genomic and analytic tools have greatly advanced our knowledge about rice epigenetic modifiers and their biological roles. There are still a large number of modifiers uncharacterized, and molecular mechanisms of function of many chromatin modifiers remain to be investigated into details. It remains to be uncovered how the general histone modification and DNA methylation enzymes exert specific functions in plant growth and developmental processes and what effectors are involved. In particular, H3K27me3 is recognized as a crucial epigenetic mark associated with gene transcriptional repression, and the classical model proposes a sequential mode of action of the two Polycomb complexes: PRC2 is responsible H3K27me3 establishment, and PRC1 recognizes the H3K27me3 mark and further catalyzed downstream H2A monoubiquitination. While PRC1-like components and histone monoubiquitination have been recently studied in *Arabidopsis* (reviewed in Molitor and Shen, 2013; Feng and Shen, 2014), effectors acting together with H3K27me3 in rice remain unknown so far. Utilization of advanced technologies in proteomics, deep sequencing, and gene knockdown will facilitate future studies in functional characterization of interesting genes, investigation of protein complex composition and function, and gene networks controlling rice flowering and reproduction. The extensive agriculture breading has greatly enriched the rice germplasm resources with large collections of cultivated rice and their wild relatives. Comparative studies of different rice varieties and hybrids will likely impact on knowledge of genetics, epigenetics, and inheritance of agriculture traits as well as fundamental understanding of conservation and diversification of molecular mechanisms.

ACKNOWLEDGMENTS

The work in authors' laboratories were supported by National Basic Research Program of China (973 Program, grants no. 2012CB910500 and 2011CB944600), National Natural Science Foundation of China (grant no. 31300263), Science and Technology Commission of Shanghai Municipality (grant no. 13JC1401000), and the French Agence Nationale de la Recherche (ANR-12-BSV2-0013-02). The research was conducted within the context of the International Associated Laboratory Plant Epigenome Research, LIA PER.

REFERENCES

Abe, M., Yoshikawa, T., Nosaka, M., Sakakibara, H., Sato, Y., Nagato, Y., et al. (2010). WAVY LEAF1, an ortholog of *Arabidopsis* HEN1, regulates shoot development by maintaining microRNA and trans-acting small interfering RNA accumulation in rice. *Plant Physiol.* 154, 1335–1346. doi: 10.1104/pp.110.160234

Agius, F., Kapoor, A., and Zhu, J. K. (2006). Role of the *Arabidopsis* DNA glycosylase/lyase ROS1 in active DNA demethylation. *Proc. Natl. Acad. Sci. U.S.A.* 103, 11796–11801. doi: 10.1073/pnas.0603563103

Albani, M. C., and Coupland, G. (2010). Comparative analysis of flowering in annual and perennial plants. *Curr. Top. Dev. Biol.* 91, 323–348. doi: 10.1016/S0070-2153(10)91011-9

Anderson, S. N., Johnson, C. S., Jones, D. S., Conrad, L. J., Gou, X., Russell, S. D., et al. (2013). Transcriptomes of isolated *Oryza sativa* gametes characterized by deep sequencing: evidence for distinct sex-dependent chromatin and epigenetic states before fertilization. *Plant J.* 76, 729–741. doi: 10.1111/tpj.12336

Berr, A., Shafiq, S., and Shen, W. H. (2011). Histone modifications in transcriptional activation during plant development. *Biochim. Biophys. Acta* 1809, 567–576. doi: 10.1016/j.bbagrm.2011.07.001

Bond, D. M., and Baulcombe, D. C. (2014). Small RNAs and heritable epigenetic variation in plants. *Trends Cell Biol.* 24, 100–107. doi: 10.1016/j.tcb.2013.08.001

Bu, Z., Yu, Y., Li, Z., Liu, Y., Jiang, W., Huang, Y., et al. (2014). Regulation of *Arabidopsis* flowering by the histone mark readers MRG1/2 via interaction with CONSTANS to modulate FT expression. *PLoS Genet.* 10:e1004617. doi: 10.1371/journal.pgen.1004617

Cao, X., and Jacobsen, S. E. (2002). Role of the *Arabidopsis* DRM methyltransferases in de novo DNA methylation and gene silencing. *Curr. Biol.* 12, 1138–1144. doi: 10.1016/S0960-9822(02)00925-9

Castel, S. E., and Martienssen, R. A. (2013). RNA interference in the nucleus: roles for small RNAs in transcription, epigenetics and beyond. *Nat. Rev. Genet.* 14, 100–112. doi: 10.1038/nrg3355

Chen, F., He, G., He, H., Chen, W., Zhu, X., Liang, M., et al. (2010). Expression analysis of miRNAs and highly-expressed small RNAs in two rice subspecies and their reciprocal hybrids. *J. Integr. Plant Biol.* 52, 971–980. doi: 10.1111/j.1744-7909.2010.00985.x

Chen, Q., Chen, X., Wang, Q., Zhang, F., Lou, Z., Zhang, Q., et al. (2013). Structural basis of a histone H3 lysine 4 demethylase required for stem elongation in rice. *PLoS Genet.* 9:e1003239. doi: 10.1371/journal.pgen.1003239

Chen, S., and Rasmuson-Lestander, A. (2009). Regulation of the *Drosophila engrailed* gene by Polycomb repressor complex 2. *Mech. Dev.* 126, 443–448. doi: 10.1016/j.mod.2009.01.004

Chen, X., and Zhou, D. X. (2013). Rice epigenomics and epigenetics: challenges and opportunities. *Curr. Opin. Plant Biol.* 16, 164–169. doi: 10.1016/j.pbi.2013.03.004

Chen, Z. J., and Tian, L. (2007). Roles of dynamic and reversible histone acetylation in plant development and polyploidy. *Biochim. Biophys. Acta* 1769, 295–307. doi: 10.1016/j.bbaexp.2007.04.007

Chodavarapu, R. K., Feng, S., Ding, B., Simon, S. A., Lopez, D., Jia, Y., et al. (2012). Transcriptome and methylome interactions in rice hybrids. *Proc. Natl. Acad. Sci. U.S.A.* 109, 12040–12045. doi: 10.1073/pnas.1209297109

Choi, S. C., Lee, S., Kim, S. R., Lee, Y. S., Liu, C., Cao, X., et al. (2014). Trithorax group protein *Oryza sativa* TRITHORAX1 controls flowering time in rice via interaction with EARLY HEADING DATE 3. *Plant Physiol.* 164, 1326–1337. doi: 10.1104/pp.113.228049

Choi, Y., Gehring, M., Johnson, L., Hannon, M., Harada, J. J., Goldberg, R. B., et al. (2002). DEMETER, a DNA glycosylase domain protein, is required for endosperm gene imprinting and seed viability in *Arabidopsis*. *Cell* 110, 33–42. doi: 10.1016/S0092-8674(02)00807-3

Conrad, L. J., Khanday, I., Johnson, C., Guiderdoni, E., An, G., Vijayraghavan, U., et al. (2014). The Polycomb group gene EMF2B is essential for maintenance of floral meristem determinacy in rice. *Plant J.* 80, 883–894. doi: 10.1111/tpj.12688

Cui, X., Jin, P., Gu, L., Lu, Z., Xue, Y., Wei, L., et al. (2013). Control of transposon activity by a histone H3K4 demethylase in rice. *Proc. Natl. Acad. Sci. U.S.A.* 110, 1953–1958. doi: 10.1073/pnas.1217020110

Czermin, B., Melfi, R., Mccabe, D., Seitz, V., Imhof, A., and Pirrotta, V. (2002). *Drosophila* enhancer of Zeste/ESC complexes have a histone H3 methyltransferase activity that marks chromosomal polycomb sites. *Cell* 111, 185–196. doi: 10.1016/S0092-8674(02)00975-3

Dangwal, M., Malik, G., Kapoor, S., and Kapoor, M. (2013). De novo methyltransferase, OsDRM2, interacts with the ATP-dependent RNA helicase, OseIF4A, in rice. *J. Mol. Biol.* 425, 2853–2866. doi: 10.1016/j.jmb.2013.05.021

Ding, B., Bellizzi Mdel, R., Ning, Y., Meyers, B. C., and Wang, G. L. (2012a). HD1701, a histone H4 deacetylase, negatively regulates plant innate immunity by modulating histone H4 acetylation of defense-related genes in rice. *Plant Cell* 24, 3783–3794. doi: 10.1105/tpc.112.101972

Ding, J., Lu, Q., Ouyang, Y., Mao, H., Zhang, P., Yao, J., et al. (2012b). A long noncoding RNA regulates photoperiod-sensitive male sterility, an essential component of hybrid rice. *Proc. Natl. Acad. Sci. U.S.A.* 109, 2654–2659. doi: 10.1073/pnas.1121374109

Ding, J., Shen, J., Mao, H., Xie, W., Li, X., and Zhang, Q. (2012c). RNA-directed DNA methylation is involved in regulating photoperiod-sensitive male sterility in rice. *Mol. Plant* 5, 1210–1216. doi: 10.1093/mp/sss095

Ding, B., Zhu, Y., Bu, Z. Y., Shen, W. H., Yu, Y., and Dong, A. W. (2010). SDG714 regulates specific gene expression and consequently affects plant growth via H3K9 dimethylation. *J. Integr. Plant Biol.* 52, 420–430. doi: 10.1111/j.1744-7909.2010.00927.x

Ding, B., Zhu, Y., Gao, J., Yu, Y., Cao, K., Shen, W.-H., et al. (2007a). Molecular characterization of three rice SET-domain proteins. *Plant Sci.* 172, 1072–1078. doi: 10.1016/j.plantsci.2007.03.009

Ding, Y., Wang, X., Su, L., Zhai, J., Cao, S., Zhang, D., et al. (2007b). SDG714, a histone H3K9 methyltransferase, is involved in Tos17 DNA methylation and transposition in rice. *Plant Cell* 19, 9–22. doi: 10.1105/tpc.106.048124

Du, Z., Li, H., Wei, Q., Zhao, X., Wang, C., Zhu, Q., et al. (2013). Genome-wide analysis of histone modifications: H3K4me2, H3K4me3, H3K9ac, and H3K27ac in *Oryza sativa* L. Japonica. *Mol. Plant* 6, 1463–1472. doi: 10.1093/mp/sst018

Engelhorn, J., Blanvillain, R., and Carles, C. C. (2014). Gene activation and cell fate control in plants: a chromatin perspective. *Cell. Mol. Life Sci.* 71, 3119–3137. doi: 10.1007/s00018-014-1609-0

Feng, J., and Shen, W. H. (2014). Dynamic regulation and function of histone monoubiquitination in plants. *Front. Plant Sci.* 5:83. doi: 10.3389/fpls.2014.00083

Folsom, J. J., Begcy, K., Hao, X., Wang, D., and Walia, H. (2014). Rice FERTILIZATION-INDEPENDENT ENDOSPERM 1 regulates seed size under heat stress by controlling early endosperm development. *Plant Physiol.* 165, 238–248. doi: 10.1104/pp.113.232413

Furner, I. J., and Matzke, M. (2011). Methylation and demethylation of the *Arabidopsis* genome. *Curr. Opin. Plant Biol.* 14, 137–141. doi: 10.1016/j.pbi.2010.11.004

Gehring, M., Huh, J. H., Hsieh, T. F., Penterman, J., Choi, Y., Harada, J. J., et al. (2006). DEMETER DNA glycosylase establishes MEDEA polycomb gene self-imprinting by allele-specific demethylation. *Cell* 124, 495–506. doi: 10.1016/j.cell.2005.12.034

Gentry, M., and Hennig, L. (2014). Remodelling chromatin to shape development of plants. *Exp. Cell Res.* 321, 40–46. doi: 10.1016/j.yexcr.2013.11.010

Gong, Z., Morales-Ruiz, T., Ariza, R. R., Roldán-Arjona, T., David, L., and Zhu, J.-K. (2002). ROS1, a repressor of transcriptional gene silencing in *Arabidopsis*, encodes a DNA glycosylase/lyase. *Cell* 111, 803–814. doi: 10.1016/S0092-8674(02)01133-9

Groszmann, M., Greaves, I. K., Fujimoto, R., Peacock, W. J., and Dennis, E. S. (2013). The role of epigenetics in hybrid vigour. *Trends Genet.* 29, 684–690. doi: 10.1016/j.tig.2013.07.004

Gutierrez-Marcos, J. F., and Dickinson, H. G. (2012). Epigenetic reprogramming in plant reproductive lineages. *Plant Cell Physiol.* 53, 817–823. doi: 10.1093/pcp/pcs052

He, C., Huang, H., and Xu, L. (2013). Mechanisms guiding Polycomb activities during gene silencing in *Arabidopsis thaliana*. *Front. Plant Sci.* 4:454. doi: 10.3389/fpls.2013.00454

He, G., Zhu, X., Elling, A. A., Chen, L., Wang, X., Guo, L., et al. (2010). Global epigenetic and transcriptional trends among two rice subspecies and their reciprocal hybrids. *Plant Cell* 22, 17–33. doi: 10.1105/tpc.109.072041

He, Y. (2012). Chromatin regulation of flowering. *Trends Plant Sci.* 17, 556–562. doi: 10.1016/j.tplants.2012.05.001

Heo, J. B., and Sung, S. (2011). Vernalization-mediated epigenetic silencing by a long Intronic noncoding RNA. *Science* 331, 76–79. doi: 10.1126/science.1197349

Higo, H., Tahir, M., Takashima, K., Miura, A., Watanabe, K., Tagiri, A., et al. (2012). DDM1 (Decrease in DNA Methylation) genes in rice (*Oryza sativa*). *Mol. Genet. Genomics* 287, 785–792. doi: 10.1007/s00438-012-0717-5

Hu, L., Li, N., Xu, C., Zhong, S., Lin, X., Yang, J., et al. (2014). Mutation of a major CG methylase in rice causes genome-wide hypomethylation, dysregulated genome expression, and seedling lethality. *Proc. Natl. Acad. Sci. U.S.A.* 111, 10642–10647. doi: 10.1073/pnas.1410761111

Hu, Y., Liu, D., Zhong, X., Zhang, C., Zhang, Q., and Zhou, D.-X. (2012). CHD3 protein recognizes and regulates methylated histone H3 lysines 4 and 27 over a subset of targets in the rice genome. *Proc. Natl. Acad. Sci. U.S.A.* 109, 5773–5778. doi: 10.1073/pnas.1203148109

Hu, Y., Qin, F., Huang, L., Sun, Q., Li, C., Zhao, Y., et al. (2009). Rice histone deacety-lase genes display specific expression patterns and developmental functions. *Biochem. Biophys. Res. Commun.* 388, 266–271. doi: 10.1016/j.bbrc.2009.07.162

Huang, L., Sun, Q., Qin, F., Li, C., Zhao, Y., and Zhou, D. X. (2007). Down-regulation of a SILENT INFORMATION REGULATOR2-related histone deacetylase gene, OsSRT1, induces DNA fragmentation and cell death in rice. *Plant Physiol.* 144, 1508–1519. doi: 10.1104/pp.107.099473

Huang, Y., Liu, C., Shen, W.-H., and Ruan, Y. (2011). Phylogenetic analysis and classification of the *Brassica rapa* SET-domain protein family. *BMC Plant Biol.* 11:175. doi: 10.1186/1471-2229-11-175

Ietswaart, R., Wu, Z., and Dean, C. (2012). Flowering time control: another window to the connection between antisense RNA and chromatin. *Trends Genet.* 28, 445–453. doi: 10.1016/j.tig.2012.06.002

Jeong, D. H., German, M. A., Rymarquis, L. A., Thatcher, S. R., and Green, P. J. (2010). Abiotic stress-associated miRNAs: detection and functional analysis. *Methods Mol. Biol.* 592, 203–230. doi: 10.1007/978-1-60327-005-2_14

Jeong, D.-H., Park, S., Zhai, J., Gurazada, S. G. R., De Paoli, E., Meyers, B. C., et al. (2011). Massive analysis of rice small RNAs: mechanistic implications of regulated microRNAs and variants for differential target RNA cleavage. *Plant Cell* 23, 4185–4207. doi: 10.1105/tpc.111.089045

Jiang, D., Yang, W., He, Y., and Amasino, R. M. (2007). *Arabidopsis* relatives of the human lysine-specific Demethylase1 repress the expression of FWA and FLOWERING LOCUS C and thus promote the floral transition. *Plant Cell* 19, 2975–2987. doi: 10.1105/tpc.107.052373

Jiao, Y., Wang, Y., Xue, D., Wang, J., Yan, M., Liu, G., et al. (2010). Regulation of OsSPL14 by OsmiR156 defines ideal plant architecture in rice. *Nat. Genet.* 42, 541–544. doi: 10.1038/ng.591

Komiya, R., Ohyanagi, H., Niihama, M., Watanabe, T., Nakano, M., Kurata, N., et al. (2014). Rice germline-specific Argonaute MEL1 protein binds to phasiR-NAs generated from more than 700 lincRNAs. *Plant J.* 78, 385–397. doi: 10.1111/tpj.12483

La, H., Ding, B., Mishra, G. P., Zhou, B., Yang, H., Bellizzi, M. R., et al. (2011). A 5-methylcytosine DNA glycosylase/lyase demethylates the retrotransposon Tos17 and promotes its transposition in rice. *Proc. Natl. Acad. Sci. U.S.A.* 108, 15498–15503. doi: 10.1073/pnas.1112704108

Law, J. A., and Jacobsen, S. E. (2010). Establishing, maintaining and modifying DNA methylation patterns in plants and animals. *Nat. Rev. Genet.* 11, 204–220. doi: 10.1038/nrg2719

Li, C., Huang, L., Xu, C., Zhao, Y., and Zhou, D. X. (2011a). Altered levels of histone deacetylase OsHDT1 affect differential gene expression patterns in hybrid rice. *PLoS ONE* 6:e21789. doi: 10.1371/journal.pone.0021789

Li, W., Han, Y., Tao, F., and Chong, K. (2011b). Knockdown of SAMS genes encoding S-adenosyl-L-methionine synthetases causes methylation alterations of DNAs and histones and leads to late flowering in rice. *J. Plant Physiol.* 168, 1837–1843. doi: 10.1016/j.jplph.2011.05.020

Li, S., Zhou, B., Peng, X., Kuang, Q., Huang, X., Yao, J., et al. (2014). OsFIE2 plays an essential role in the regulation of rice vegetative and reproductive development. *New Phytol.* 201, 66–79. doi: 10.1111/nph.12472

Li, T., Chen, X., Zhong, X., Zhao, Y., Liu, X., Zhou, S., et al. (2013). Jumonji C domain protein JMJ705-mediated removal of histone H3 lysine 27 trimethylation is involved in defense-related gene activation in rice. *Plant Cell* 25, 4725–4736. doi: 10.1105/tpc.113.118802

Li, X., Zhu, J., Hu, F., Ge, S., Ye, M., Xiang, H., et al. (2012). Single-base resolution maps of cultivated and wild rice methylomes and regulatory roles of DNA methylation in plant gene expression. *BMC Genomics* 13:300. doi: 10.1186/1471-2164-13-300

Lindroth, A. M., Cao, X., Jackson, J. P., Zilberman, D., Mccallum, C. M., Henikoff, S., et al. (2001). Requirement of CHROMOMETHYLASE3 for main-tenance of CpXpG methylation. *Science* 292, 2077–2080. doi: 10.1126/science.1059745

Liu, C., Lu, F., Cui, X., and Cao, X. (2010). Histone methylation in higher plants. *Annu. Rev. Plant Biol.* 61, 395–420. doi: 10.1146/annurev.arplant.043008.091939

Liu, F., Quesada, V., Crevillen, P., Baurle, I., Swiezewski, S., and Dean, C. (2007). The Arabidopsis RNA-binding protein FCA requires a LYSINE-SPECIFIC DEMETHYLASE 1 homolog to downregulate FLC. *Mol. Cell* 28, 398–407. doi: 10.1016/j.molcel.2007.10.018

Liu, X., Luo, M., Zhang, W., Zhao, J., Zhang, J., Wu, K., et al. (2012). Histone acetyltransferases in rice (*Oryza sativa* L.): phylogenetic analysis, subcellular localization and expression. *BMC Plant Biol.* 12:145. doi: 10.1186/1471-2229-12-145

Liu, X., Zhou, C., Zhao, Y., Zhou, S., Wang, W., and Zhou, D.-X. (2014). The rice enhancer of zeste [E(z)] genes SDG711 and SDG718 are respectively involved in long day and short day signaling to mediate the accurate photoperiod control of flowering time. *Front. Plant Sci.* 5:591. doi: 10.3389/fpls.2014.00591

Lopez-Gonzalez, L., Mouriz, A., Narro-Diego, L., Bustos, R., Martinez-Zapater, J. M., Jarillo, J. A., et al. (2014). Chromatin-dependent repression of the *Arabidopsis* floral integrator genes involves plant specific PHD-containing proteins. *Plant Cell* 26, 3922–3938. doi: 10.1105/tpc.114.130781

Lu, Z., Yu, H., Xiong, G., Wang, J., Jiao, Y., Liu, G., et al. (2013). Genome-wide binding analysis of the transcription activator IDEAL PLANT ARCHITECTURE 1 reveals a complex network regulating rice plant architecture. *Plant Cell* 25, 3743–3759. doi: 10.1105/tpc.113.113639

Luo, M., Platten, D., Chaudhury, A., Peacock, W. J., and Dennis, E. S. (2009). Expression, imprinting, and evolution of rice homologs of the Polycomb group genes. *Mol. Plant* 2, 711–723. doi: 10.1093/mp/ssp036

Lusser, A., Kolle, D., and Loidl, P. (2001). Histone acetylation: lessons from the plant kingdom. *Trends Plant Sci.* 6, 59–65. doi: 10.1016/S1360-1385(00)01839-2

Ma, X., Lv, S., Zhang, C., and Yang, C. (2013). Histone deacetylases and their functions in plants. *Plant Cell Rep.* 32, 465–478. doi: 10.1007/s00299-013-1393-6

Malone, B. M., Tan, F., Bridges, S. M., and Peng, Z. (2011). Comparison of four ChIP-Seq analytical algorithms using rice endosperm H3K27 trimethylation profiling data. *PLoS ONE* 6:e25260. doi: 10.1371/journal.pone.0025260

McGinty, R., and Tan, S. (2014). "Histone, nucleosome, and chromatin structure," in *Fundamentals of Chromatin*, eds J. L. Workman and S. M. Abmayr (New York: Springer), 1–28.

Miura, K., Agetsuma, M., Kitano, H., Yoshimura, A., Matsuoka, M., Jacob-sen, S. E., et al. (2009). A metastable DWARF1 epigenetic mutant affecting plant stature in rice. *Proc. Natl. Acad. Sci. U.S.A.* 106, 11218–11223. doi: 10.1073/pnas.0901942106

Miura, K., Ikeda, M., Matsubara, A., Song, X. J., Ito, M., Asano, K., et al. (2010). OsSPL14 promotes panicle branching and higher grain productivity in rice. *Nat. Genet.* 42, 545–549. doi: 10.1038/ng.592

Molitor, A. M., Bu, Z., Yu, Y., and Shen, W. H. (2014). *Arabidopsis* AL PHD-PRC1 complexes promote seed germination through H3K4me3-to-H3K27me3 chromatin state switch in repression of seed developmental genes. *PLoS Genet.* 10:e1004091. doi: 10.1371/journal.pgen.1004091

Molitor, A., and Shen, W.-H. (2013). The Polycomb complex PRC1: com-position and function in plants. *J. Genet. Genomics* 40, 231–238. doi: 10.1016/j.jgg.2012.12.005

Moritoh, S., Eun, C. H., Ono, A., Asao, H., Okano, Y., Yamaguchi, K., et al. (2012). Targeted disruption of an orthologue of DOMAINS REARRANGED METHYLASE 2, OsDRM2, impairs the growth of rice plants by abnormal DNA methylation. *Plant J.* 71, 85–98. doi: 10.1111/j.1365-313X.2012.04974.x

Mosammaparast, N., and Shi, Y. (2010). Reversal of histone methylation: biochem-ical and molecular mechanisms of histone demethylases. *Annu. Rev. Biochem.* 79, 155–179. doi: 10.1146/annurev.biochem.78.070907.103946

Musselman, C. A., Lalonde, M. E., Cote, J., and Kutateladze, T. G. (2012). Perceiving the epigenetic landscape through histone readers. *Nat. Struct. Mol. Biol.* 19, 1218–1227. doi: 10.1038/nsmb.2436

Nallamilli, B. R., Edelmann, M. J., Zhong, X., Tan, F., Mujahid, H., Zhang, J., et al. (2014). Global analysis of lysine acetylation suggests the involvement of protein acetylation in diverse biological processes in rice (*Oryza sativa*). *PLoS ONE* 9:e89283. doi: 10.1371/journal.pone.0089283

Nallamilli, B. R. R., Zhang, J., Mujahid, H., Malone, B. M., Bridges, S. M., and Peng, Z. (2013). Polycomb group gene OsFIE2 regulates rice (*Oryza sativa*) seed development and grain filling via a mechanism distinct from *Arabidopsis*. *PLoS Genet.* 9:e1003322. doi: 10.1371/journal.pgen.1003322

Ng, D. W., Wang, T., Chandrasekharan, M. B., Aramayo, R., Kertbundit, S., and Hall, T. C. (2007). Plant SET domain-containing proteins: struc-ture, function and regulation. *Biochim. Biophys. Acta* 1769, 316–329. doi: 10.1016/j.bbaexp.2007.04.003

Nonomura, K.-I., Morohoshi, A., Nakano, M., Eiguchi, M., Miyao, A., Hirochika, H., et al. (2007). A germ cell-specific gene of the ARGONAUTE family is essential for the progression of premeiotic mitosis and meiosis during sporogenesis in rice. *Plant Cell* 19, 2583–2594. doi: 10.1105/tpc.107.053199

Ono, A., Yamaguchi, K., Fukada-Tanaka, S., Terada, R., Mitsui, T., and Iida, S. (2012). A null mutation of ROS1a for DNA demethylation in rice is not transmittable to progeny. *Plant J.* 71, 564–574. doi: 10.1111/j.1365-313X.2012.05009.x

Ortega-Galisteo, A. P., Morales-Ruiz, T., Ariza, R. R., and Roldan-Arjona, T. (2008). *Arabidopsis* DEMETER-LIKE proteins DML2 and DML3 are required for appropriate distribution of DNA methylation marks. *Plant Mol. Biol.* 67, 671–681. doi: 10.1007/s11103-008-9346-0

Pang, J., Dong, M., Li, N., Zhao, Y., and Liu, B. (2013). Functional characterization of a rice de novo DNA methyltransferase, OsDRM2, expressed in *Escherichia coli* and yeast. *Biochem. Biophys. Res. Commun.* 432, 157–162. doi: 10.1016/j.bbrc.2013.01.067

Peng, T., Du, Y., Zhang, J., Li, J., Liu, Y., Zhao, Y., et al. (2013). Genome-wide analysis of 24-nt siRNAs dynamic variations during rice superior and inferior grain filling. *PLoS ONE* 8:e61029. doi: 10.1371/journal.pone.0061029

Pikaard, C. S., and Scheid, O. M. (2013). Epigenetic regulation in plants. *Cold Spring Harb. Perspect. Biol.* 6:a019315. doi: 10.1101/cshperspect.a019315

Qin, F. J., Sun, Q. W., Huang, L. M., Chen, X. S., and Zhou, D. X. (2010). Rice SUVH histone methyltransferase genes display specific functions in chromatin modification and retrotransposon repression. *Mol. Plant* 3, 773–782. doi: 10.1093/mp/ssq030

Ream, T. S., Woods, D. P., and Amasino, R. M. (2012). The molecular basis of vernalization in different plant groups. *Cold Spring Harb. Symp. Quant. Biol.* 77, 105–115. doi: 10.1101/sqb.2013.77.014449

Saze, H., Mittelsten Scheid, O., and Paszkowski, J. (2003). Maintenance of CpG methylation is essential for epigenetic inheritance during plant gametogenesis. *Nat. Genet.* 34, 65–69. doi: 10.1038/ng1138

Schuettengruber, B., and Cavalli, G. (2009). Recruitment of Polycomb group complexes and their role in the dynamic regulation of cell fate choice. *Development* 136, 3531–3542. doi: 10.1242/dev.033902

Shafiq, S., Berr, A., and Shen, W. H. (2014). Combinatorial functions of diverse histone methylations in *Arabidopsis thaliana* flowering time regulation. *New Phytol.* 201, 312–322. doi: 10.1111/nph.12493

Sharma, R., Mohan Singh, R. K., Malik, G., Deveshwar, P., Tyagi, A. K., Kapoor, S., et al. (2009). Rice cytosine DNA methyltransferases – gene expression profiling during reproductive development and abiotic stress. *FEBS J.* 276, 6301–6311. doi: 10.1111/j.1742-4658.2009.07338.x

She, W., and Baroux, C. (2014). Chromatin dynamics during plant sexual reproduction. *Front. Plant Sci.* 5:354. doi: 10.3389/fpls.2014.00354

Shen, W. H., and Xu, L. (2009). Chromatin remodeling in stem cell maintenance in *Arabidopsis thaliana*. *Mol. Plant* 2, 600–609. doi: 10.1093/mp/ssp022

Shi, Y., Lan, F., Matson, C., Mulligan, P., Whetstine, J. R., Cole, P. A., et al. (2004). Histone demethylation mediated by the nuclear amine oxidase homolog LSD1. *Cell* 119, 941–953. doi: 10.1016/j.cell.2004.12.012

Shrestha, R., Gomez-Ariza, J., Brambilla, V., and Fornara, F. (2014). Molecular control of seasonal flowering in rice, arabidopsis and temperate cereals. *Ann. Bot.* 114, 1445–1458. doi: 10.1093/aob/mcu032

Song, J., Angel, A., Howard, M., and Dean, C. (2012). Vernalization – a cold-induced epigenetic switch. *J. Cell Sci.* 125, 3723–3731. doi: 10.1242/jcs.084764

Stroud, H., Do, T., Du, J., Zhong, X., Feng, S., Johnson, L., et al. (2014). Non-CG methylation patterns shape the epigenetic landscape in *Arabidopsis*. *Nat. Struct. Mol. Biol.* 21, 64–72. doi: 10.1038/nsmb.2735

Sui, P., Jin, J., Ye, S., Mu, C., Gao, J., Feng, H., et al. (2012). H3K36 methylation is critical for brassinosteroid-regulated plant growth and development in rice. *Plant J.* 70, 340–347. doi: 10.1111/j.1365-313X.2011.04873.x

Sui, P., Shi, J., Gao, X., Shen, W. H., and Dong, A. (2013). H3K36 methylation is involved in promoting rice flowering. *Mol. Plant* 6, 975–977. doi: 10.1093/mp/sss152

Sun, C., Chen, D., Fang, J., Wang, P., Deng, X., and Chu, C. (2014). Understanding the genetic and epigenetic architecture in complex network of rice flowering pathways. *Protein Cell* 5, 889–898. doi: 10.1007/s13238-014-0068-6

Sun, C., Fang, J., Zhao, T., Xu, B., Zhang, F., Liu, L., et al. (2012). The histone methyltransferase SDG724 mediates H3K36me2/3 deposition at MADS50 and RFT1 and promotes flowering in rice. *Plant Cell* 24, 3235–3247. doi: 10.1105/tpc.112.101436

Sun, Q., and Zhou, D. X. (2008). Rice jmjC domain-containing gene JMJ706 encodes H3K9 demethylase required for floral organ development. *Proc. Natl. Acad. Sci. U.S.A.* 105, 13679–13684. doi: 10.1073/pnas.0805901105

Swiezewski, S., Liu, F., Magusin, A., and Dean, C. (2009). Cold-induced silencing by long antisense transcripts of an *Arabidopsis* Polycomb target. *Nature* 462, 799–802. doi: 10.1038/nature08618

Teerawanichpan, P., Chandrasekharan, M. B., Jiang, Y., Narangajavana, J., and Hall, T. C. (2004). Characterization of two rice DNA methyltransferase genes and RNAi-mediated reactivation of a silenced transgene in rice callus. *Planta* 218, 337–349. doi: 10.1007/s00425-003-1112-6

Thorstensen, T., Grini, P. E., and Aalen, R. B. (2011). SET domain proteins in plant development. *Biochim. Biophys. Acta* 1809, 407–420. doi: 10.1016/j.bbagrm.2011.05.008

To, T. K., and Kim, J. M. (2014). Epigenetic regulation of gene responsiveness in *Arabidopsis*. *Front. Plant Sci.* 4:548. doi: 10.3389/fpls.2013.00548

Toriba, T., Suzaki, T., Yamaguchi, T., Ohmori, Y., Tsukaya, H., and Hirano, H. Y. (2010). Distinct regulation of adaxial-abaxial polarity in anther patterning in rice. *Plant Cell* 22, 1452–1462. doi: 10.1105/tpc.110.075291

Tsuji, H., Saika, H., Tsutsumi, N., Hirai, A., and Nakazono, M. (2006). Dynamic and reversible changes in histone H3-Lys4 methylation and H3 acetylation occurring at submergence-inducible genes in rice. *Plant Cell Physiol.* 47, 995–1003. doi: 10.1093/pcp/pcj072

Tsuji, H., Taoka, K.-I., and Shimamoto, K. (2013). Florigen in rice: complex gene network for florigen transcription, florigen activation complex, and multiple functions. *Curr. Opin. Plant Biol.* 16, 228–235. doi: 10.1016/j.pbi.2013.01.005

Tsukada, Y., Fang, J., Erdjument-Bromage, H., Warren, M. E., Borchers, C. H., Tempst, P., et al. (2006). Histone demethylation by a family of JmjC domain-containing proteins. *Nature* 439, 811–816. doi: 10.1038/nature04433

Van Lijsebettens, M., and Grasser, K. D. (2014). Transcript elongation factors: shaping transcriptomes after transcript initiation. *Trends Plant Sci.* 19, 717–726. doi: 10.1016/j.tplants.2014.07.002

Vaucheret, H. (2008). Plant ARGONAUTES. *Trends Plant Sci.* 13, 350–358. doi: 10.1016/j.tplants.2008.04.007

Wang, J., Hu, J., Qian, Q., and Xue, H.-W. (2013). LC2 and OsVIL2 promote rice flowering by photoperoid-induced epigenetic silencing of OsLF. *Mol. Plant* 6, 514–527. doi: 10.1093/mp/sss096

Wang, K., Tang, D., Hong, L., Xu, W., Huang, J., Li, M., et al. (2010). DEP and AFO regulate reproductive habit in rice. *PLoS Genet.* 6:e1000818. doi: 10.1371/journal.pgen.1000818

Wang, M., Tang, D., Luo, Q., Jin, Y., Shen, Y., Wang, K., et al. (2012). BRK1, a Bub1-related kinase, is essential for generating proper tension between homologous kinetochores at metaphase I of rice meiosis. *Plant Cell* 24, 4961–4973. doi: 10.1105/tpc.112.105874

Wu, H., and Zhang, Y. (2014). Reversing DNA methylation: mechanisms, genomics, and biological functions. *Cell* 156, 45–68. doi: 10.1016/j.cell.2013.12.019

Xu, Y., Gan, E. S., Zhou, J., Wee, W. Y., Zhang, X., and Ito, T. (2014). *Arabidopsis* MRG domain proteins bridge two histone modifications to elevate expression of flowering genes. *Nucleic Acids Res.* 42, 10960–10974. doi: 10.1093/nar/gku781

Yamauchi, T., Johzuka-Hisatomi, Y., Terada, R., Nakamura, I., and Iida, S. (2014). The MET1b gene encoding a maintenance DNA methyltransferase is indispensable for normal development in rice. *Plant Mol. Biol.* 85, 219–232. doi: 10.1007/s11103-014-0178-9

Yamauchi, T., Moritoh, S., Johzuka-Hisatomi, Y., Ono, A., Terada, R., Nakamura, I., et al. (2008). Alternative splicing of the rice OsMET1 genes encoding maintenance DNA methyltransferase. *J. Plant Physiol.* 165, 1774–1782. doi: 10.1016/j.jplph.2007.12.003

Yan, H., Kikuchi, S., Neumann, P., Zhang, W., Wu, Y., Chen, F., et al. (2010). Genome-wide mapping of cytosine methylation revealed dynamic DNA methylation patterns associated with genes and centromeres in rice. *Plant J.* 63, 353–365. doi: 10.1111/j.1365-313X.2010.04246.x

Yang, J., Lee, S., Hang, R., Kim, S. R., Lee, Y. S., Cao, X., et al. (2013). OsVIL2 functions with PRC2 to induce flowering by repressing OsLFL1 in rice. *Plant J.* 73, 566–578. doi: 10.1111/tpj.12057

Yao, X., and Shen, W. (2011). Crucial function of histone lysine methylation in plant reproduction. *Chin. Sci. Bull.* 56, 3493–3499. doi: 10.1007/s11434-011-4814-3

Yokoo, T., Saito, H., Yoshitake, Y., Xu, Q., Asami, T., Tsukiyama, T., et al. (2014). Se14, encoding a JmjC domain-containing protein, plays key roles in long-day suppression of rice flowering through the memethylation of H3K4me3 of RFT1. *PLoS ONE* 9:e96064. doi: 10.1371/journal.pone.0096064

Yoshida, H., and Nagato, Y. (2011). Flower development in rice. *J. Exp. Bot.* 62, 4719–4730. doi: 10.1093/jxb/err272

Yu, Y., Bu, Z., Shen, W.-H., and Dong, A. (2009). An update on histone lysine methylation in plants. *Progr. Nat. Sci.* 19, 407–413. doi: 10.1016/j.pnsc.2008.07.015

Zemach, A., Kim, M. Y., Silva, P., Rodrigues, J. A., Dotson, B., Brooks, M. D., et al. (2010). Local DNA hypomethylation activates genes in rice endosperm. *Proc. Natl. Acad. Sci. U.S.A.* 107, 18729–18734. doi: 10.1073/pnas.1009695107

Zhang, C. C., Yuan, W. Y., and Zhang, Q. F. (2012a). RPL1, a gene involved in epigenetic processes regulates phenotypic plasticity in rice. *Mol. Plant* 5, 482–493. doi: 10.1093/mp/ssr091

Zhang, L., Cheng, Z., Qin, R., Qiu, Y., Wang, J. L., Cui, X., et al. (2012b). Identification and characterization of an epi-allele of FIE1 reveals a regulatory linkage between two epigenetic marks in rice. *Plant Cell* 24, 4407–4421. doi: 10.1105/tpc.112.102269

Zhang, L., Peng, Y., Wei, X., Dai, Y., Yuan, D., Lu, Y., et al. (2014). Small RNAs as important regulators for the hybrid vigour of super-hybrid rice. *J. Exp. Bot.* 65, 5989–6002. doi: 10.1093/jxb/eru337

Zhao, S. Q., Hu, J., Guo, L. B., Qian, Q., and Xue, H. W. (2010). Rice LEAF INCLINATION 2, a VIN3-like protein, regulates leaf angle through modulating cell division of the collar. *Cell Res.* 20, 935–947. doi: 10.1038/cr.2010.109

Zhong, X., Zhang, H., Zhao, Y., Sun, Q., Hu, Y., Peng, H., et al. (2013). The rice NAD(+)-dependent histone deacetylase OsSRT1 targets preferentially to stress- and metabolism-related genes and transposable elements. *PLoS ONE* 8:e66807. doi: 10.1371/journal.pone.0066807

Zhou, H., Liu, Q., Li, J., Jiang, D., Zhou, L., Wu, P., et al. (2012). Photoperiod- and thermo-sensitive genic male sterility in rice are caused by a point mutation in a novel noncoding RNA that produces a small RNA. *Cell Res.* 22, 649–660. doi: 10.1038/cr.2012.28

Zhu, Y., Dong, A., and Shen, W. H. (2012). Histone variants and chromatin assembly in plant abiotic stress responses. *Biochim. Biophys. Acta* 1819, 343–348. doi: 10.1016/j.bbagrm.2011.07.012

12

Evolution and function of epigenetic processes in the endosperm

Claudia Köhler* and Clément Lafon-Placette

Department of Plant Biology, Uppsala BioCenter, Linnean Center of Plant Biology, Swedish University of Agricultural Sciences, Uppsala, Sweden

Edited by:
Ravishankar Palanivelu, University of Arizona, USA

Reviewed by:
Mary Gehring, Whitehead Institute for Biomedical Research, USA
Daniel Bouyer, Centre National de la Recherche Scientifique, France

*Correspondence:
Claudia Köhler, Department of Plant Biology, Uppsala BioCenter, Linnean Center of Plant Biology, Swedish University of Agricultural Sciences, P.O. Box 7080, Almas Allé 5, SE-75007 Uppsala, Sweden
e-mail: claudia.kohler@slu.se

The endosperm is an ephemeral tissue surrounding the embryo that is essential for its development. Aside from the embryo nourishing function, the endosperm serves as a battlefield for epigenetic processes that have been hypothesized to reinforce transposable element silencing in the embryo. Specifically, global DNA demethylation in the central cell may serve to produce small RNAs that migrate to egg cell and embryo to induce de novo DNA methylation. The Polycomb Repressive Complex 2 (PRC2) is particularly targeted to DNA hypomethylated regions, possibly alleviating the negative effects associated with loss of DNA methylation in the endosperm. The functional requirement of the PRC2 in the endosperm can be bypassed by increasing the maternal genome dosage in the endosperm, suggesting a main functional role of the endosperm PRC2 in reducing sexual conflict. We therefore propose that the functional requirement of an endosperm PRC2 was coupled to the evolution of a sexual endosperm and mechanisms enforcing transposon silencing in the embryo. The evolutionary consequences of this scenario for genome expansion will be discussed.

Keywords: endosperm, evolution, epigenetics, imprinting, Polycomb Repressive Complex 2

INTRODUCTION

The endosperm is a nutritive tissue formed in seeds of flowering plants that surrounds the embryo and is essential for its development. Embryo and endosperm are the products of two distinct fertilization events and enclosed by the maternally derived integuments that form the seed coat. While the embryo is derived from the fertilized egg cell, the endosperm is the descendent of the fertilized central cell (Li and Berger, 2012). The majority of angiosperms have an eight-nucleated, seven-celled Polygonum-type embryo sack that develops from the single surviving haploid megaspore after three rounds of nuclei divisions, followed by nuclei migrations and cellularization. The female gametophyte consists of the egg cell flanked by two synergid cells, a diploid central cell derived from fusion of two polar nuclei and three antipodal cells that will degenerate shortly before or after fertilization (Sprunck and Gross-Hardt, 2011). As a consequence of the central cell being diploid, the endosperm of most species is triploid. Ancestral to the seven-celled female gametophyte is the four celled female gametophyte that contains a haploid central cell that will form a diploid endosperm after fertilization (Friedman and Williams, 2003; Segal et al., 2003). There are also angiosperms (e.g., Piperaceae) where all four megaspores survive after meiosis and contribute to female gametophyte formation, giving rise to central cells with more than two nuclei that will form a high-ploidy ($>3n$) endosperm after fertilization (Friedman et al., 2008).

EVOLUTION OF PLOIDY SHIFTS IN THE ENDOSPERM

It has been hypothesized that the transition from a purely maternal embryo nourishing tissue to a biparental endosperm resulted

in two possible conflicts: (i) conflict of male and female parents over the allocation of nutrients to the developing progeny and (ii) conflict among the developing progeny for resources from the maternal sporophyte (Haig and Westoby, 1989a,b; Friedman, 1995). The kin-conflict theory provides a theoretical framework for both conflicts (Haig, 2013). This theory considers that the resources provided by the maternal sporophyte to provision the offspring are limited and that the relatedness of the endosperm to the maternal sporophyte is decisive for its ability to acquire nutrients for the developing embryo. In outcrossing species, the sexual endosperm containing maternal and paternal genomes is less related to the maternal sporophyte and sibling embryos compared to its own embryo (Charnov, 1979; Friedman et al., 2008) Therefore, paternally contributed alleles maximizing nutrient allocation to the endosperm will promote development of the embryo supported by this genetically related endosperm on the expense of the sibling embryos (Haig and Westoby, 1989a,b; Haig, 2013). Increasing the number of maternal genomes contributed to the central cell could have evolved as a mechanism to control resource provisioning to the developing progeny and to limit the selfish behavior of the endosperm. Therefore, transitions from the haploid central cell to the diploid and higher ploidy central cell can be viewed as evolutionary transitions to resolve conflict between maternal and paternal genomes on the provisioning of the progeny (Friedman et al., 2008).

EPIGENETIC PROCESSES IN THE ENDOSPERM

In *Arabidopsis*, DNA methylation occurs in CG, CHG, and CHH (where H = A, T, or C) sequence contexts and is controlled by three families of DNA methyltransferases that have different sequence preference. The DNA methyltransferase MET1 acts as a maintenance methyltransferase for symmetric CG residues, while non-CG methylation is maintained by the CHROMOMETHY-LASE3 (in CHG context) and the DOMAINS REARRANGED METHYLASES 1/2 (DRM1/2) and CMT2 (in CHH context). The small RNA (sRNA) pathway targets DRM1/2-mediated *de novo* methylation in all sequence contexts and is required for the maintenance of CHH methylation (Kim and Zilberman, 2014). Genome-wide methylation studies of embryo and endosperm in *Arabidopsis* seeds revealed that the endosperm is globally hypomethylated compared to the embryo (Gehring et al., 2009; Hsieh et al., 2009), signifying substantial epigenome differences of the two fertilization products. Hypomethylation in the endosperm is restricted to the maternally inherited alleles, suggesting that the hypomethylated status is established in the central cell and inherited to the endosperm (Ibarra et al., 2012). In *Arabidopsis*, demethylation of the maternal genome requires the DNA glycosylase DEMETER (DME) that excises 5-methylcytosine preferably at small transposable elements (TEs) and is expressed in the central cell of the female gametophyte before fertilization (Hsieh et al., 2009). Maternal demethylation is nearly fully reversed in *dme* mutant endosperm (Ibarra et al., 2012), indicating that DME is likely the only enzyme accounting for global DNA methylation differences between the maternal and paternal endosperm genomes in *Arabidopsis*. The process of extensive endosperm demethylation is likely conserved between monocots and dicots, but involves specific differences caused by divergent evolution of the DME-like family (Zemach et al., 2010). Importantly, DME function is not restricted to the female central cell but DME also acts in the vegetative cell in pollen, the companion cell to the sperm cells. Similar to its role in the central cell, DME is causing hypomethylation of distinct regions in the vegetative cell. Almost half of those hypomethylated regions in the vegetative cell overlap with hypomethylated regions identified in the maternal genomes of the endosperm that likely descend from the central cell (Ibarra et al., 2012). While *de novo* methylation in CHH context is generally depleted in sperm (Calarco et al., 2012; Ibarra et al., 2012), regions that become hypomethylated by DME in the vegetative cell have increased levels of CHH methylation in sperm (Ibarra et al., 2012), suggesting communication between vegetative cells and sperm cells. Thus, it seems likely that sRNAs that are formed from demethylated regions in the vegetative cell migrate to sperm cells and reinforce methylation at distinct target sites, a hypothesis that remains to be experimentally tested. In agreement with this notion, 21 nt and 24 nt sRNAs corresponding to differentially methylated regions accumulate in sperm cells (Slotkin et al., 2009; Calarco et al., 2012) and prominently target regions of imprinted genes (discussed below) that maintain the paternal allele silenced after fertilization (Calarco et al., 2012). While traveling of a micro-RNA from the vegetative cell to sperm cells was proposed (Slotkin et al., 2009), this study may have suffered from an unspecific promoter (Grant-Downton et al., 2013), biasing the conclusions. Nevertheless, expression of a micro-RNA in the central cell can silence a reporter in the egg cell (Ibarra et al., 2012); revealing that 21 nt sRNAs (that are preferentially formed from microRNAs)

can indeed travel from the companion cells to the neighboring gametes. While previously only 24 nt sRNAs were known to establish *de novo* methylation by the sRNA-dependent DNA methylation pathway (RdDM; Cao and Jacobsen, 2002; Cao et al., 2003), recent data revealed a role of 21 nt sRNAs in silencing of transcriptionally active TEs via the RdDM pathway (McCue et al., 2015).

De novo DNA methylation increases during embryo development, suggesting increased activity of the RdDM pathway during embryo development (Jullien et al., 2012). In agreement with this view, *de novo* DNA methylation in the embryo depends on the activity of DRM2 and in part as well on DRM1 (Jullien et al., 2012). After fertilization, increased production of sRNAs occurs in siliques, reaching maximum levels at 6 days after anthesis (Mosher et al., 2009). Increased sRNA production correlates with a steadily increasing *de novo* methylation in the embryo (Jullien et al., 2012), giving rise to the hypothesis that sRNAs migrate from the endosperm to the developing embryo and enforce TE silencing in the embryo by de novo DNA methylation. *De novo* DNA methyltransferases DRM1 and DRM2 seem not to be active in the early endosperm (Jullien et al., 2012) but are expressed around the time of endosperm cellularization (Belmonte et al., 2013), in agreement with the presence of substantial levels of CHH methylation levels in the cellularized endosperm (Hsieh et al., 2009; Ibarra et al., 2012).

GENOMIC IMPRINTING IN THE ENDOSPERM

As a consequence of DNA hypomethylation in the central cell, the parental genomes are differentially methylated in the endosperm, which can cause genes to become preferentially expressed from either the maternally or paternally inherited alleles. Parent-of-origin dependent gene expression as a consequence of epigenetic modification of maternal and paternal alleles in the gametes is a well-known phenomenon termed genomic imprinting (Gehring, 2013). Hypomethylation of TEs can cause either activation or silencing of the neighboring genes. What determines whether a gene will become activated or silenced in response to hypomethylation remains to be resolved, however, it seems likely that the distance of the TE to the gene is decisive. While many maternally and paternally expressed imprinted genes (MEGs and PEGs, respectively) have TEs in the vicinity of the 5′ region, many PEGs have TEs additionally in the coding and 3′ region of the gene (Wolff et al., 2011; Ibarra et al., 2012). Hypomethylation might expose binding sites for the repressive FERTILIZATION INDEPENDENT SEED (FIS)-Polycomb Repressive Complex 2 (PRC2), as it has been proposed for the differentially methylated region downstream of the *PHERES1* gene (Villar et al., 2009). The PRC2 is an evolutionary conserved repressive complex that modifies histones by applying histone trimethylation marks on histone H3 at lysine 27 (H3K27me3; Simon and Kingston, 2013). In *Arabidopsis*, there are at least three PRC2 complexes with different functional roles during plant development. The FIS-PRC2 is specifically expressed in the central cell and in the endosperm and consists of the subunits MEDEA (MEA), FIS2, FERTILIZATION INDEPENDENT ENDOSPERM (FIE), and MSI1 (Hennig and Derkacheva, 2009). Genome-wide profiling of H3K27me3 occupancy in the endosperm revealed that several

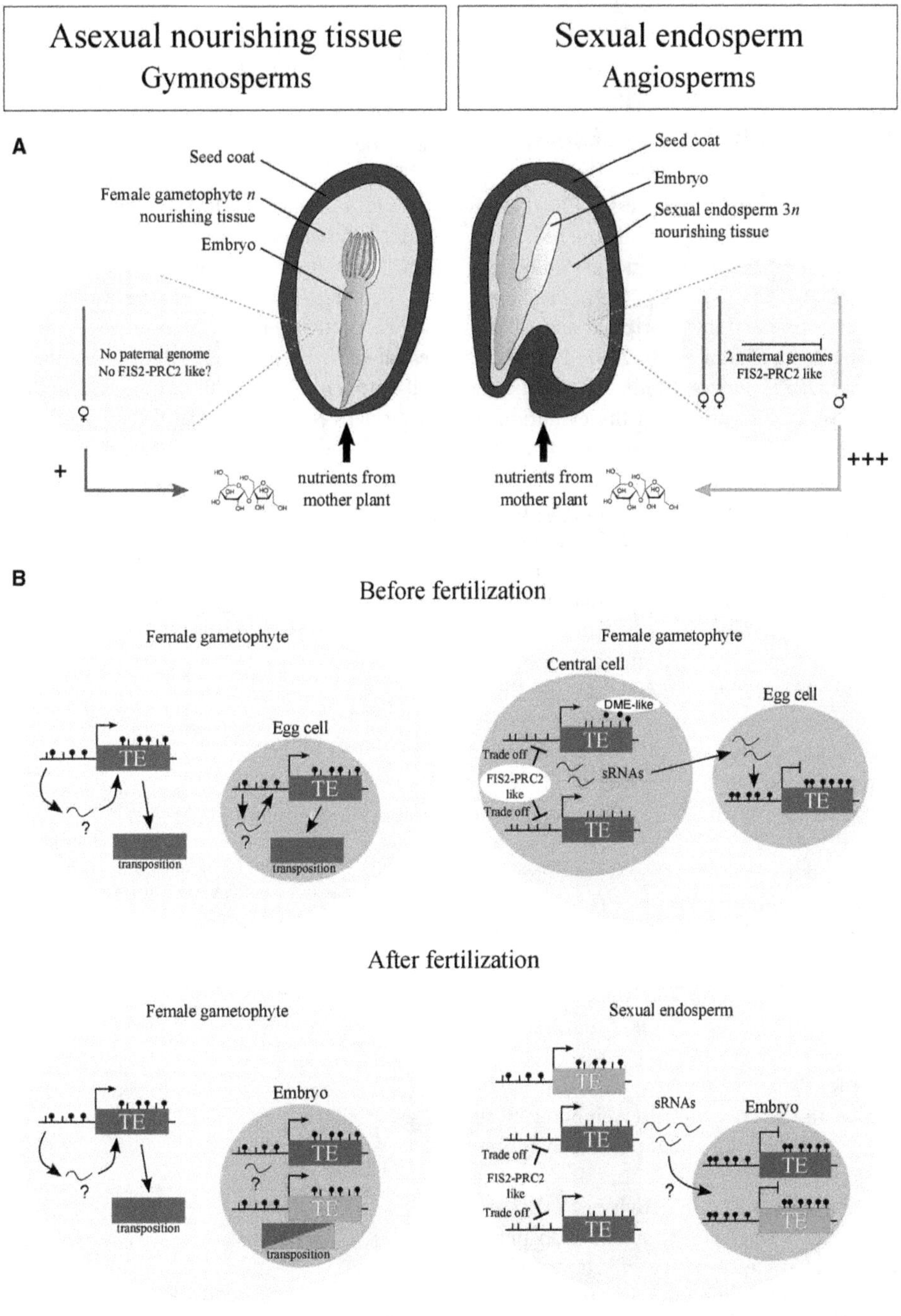

FIGURE 1 | Hypothesized emergence of FIS2-PRC2-like complex in angiosperms connected to its role in sexual endosperm. (A) In angiosperms, FIS2-PRC2-like function may have arisen to repress the paternal genome in the endosperm as predicted by the kin-conflict theory. The selfish behavior of the paternal genome, promoting nutrient allocation to the progeny, is symbolized by "+++". According to this hypothesis, in gymnosperms, where the nourishing tissue is purely maternal, FIS2-PRC2-like function is not required. **(B)** FIS2-PRC2-like function may have allowed the emergence of DNA hypomethylation via DME-like activity and small RNA (sRNA) production in central cell and endosperm. FIS2-PRC2 acts on hypomethylated regions, limiting the deleterious activity of transposable elements (TE). This tradeoff between TE silencing and activity allows the production of sRNAs traveling to the egg cell and embryo, reinforcing TE silencing. Such a demethylation process may not have emerged in gymnosperms, leading to a limited silencing of TEs and explaining the genome expansion in this taxon. Purple and green colors symbolize maternal and paternal genomes, respectively.

DNA hypomethylated TEs were targeted by the FIS-PRC2, revealing a redistribution of the FIS-PRC2 dependent on the location of DNA methylation (Weinhofer et al., 2010). A similar redistribution of H3K27me3 to TEs occurs in mutants deficient for MET1 (Deleris et al., 2012), as well as in mammalian cells depleted for DNA methylation (Reddington et al., 2013; Saksouk et al., 2014) revealing a general ability of PRC2 to target and possibly silence hypomethylated TEs.

EVOLUTION OF EPIGENETIC PROCESSES IN THE ENDOSPERM AS A MECHANISM TO PREVENT GENOME EXPANSION

Thus far, we do not know when endosperm-expressed PRC2 genes have emerged during angiosperm evolution. While the FIS-PRC2 subunit encoding genes *FIS2* and *MEA* are specific for *Arabidopsis thaliana* (Spillane et al., 2007; Chen et al., 2009), functional homologs of both genes are expressed in the triploid endosperm of monocots and lower eudicots (Haun et al., 2007; Luo et al., 2009; Gleason and Kramer, 2012) suggesting that the evolution of FIS-PRC2-like complexes is connected with the evolution of a sexual endosperm. Nevertheless, homologs but not orthologs fulfill the FIS2-PRC2 functional role in dicots and monocots and different PRC2 genes are regulated by genomic imprinting in monocots (Danilevskaya et al., 2003; Luo et al., 2009) and dicots (Grossniklaus et al., 1998; Luo et al., 2000), raising the hypothesis that FIS-PRC2-like complexes have evolved independently in both plant groups. The functional requirement of the FIS-PRC2 can be bypassed by increasing the maternal genome dosage in the endosperm (Kradolfer et al., 2013), suggesting that the FIS-PRC2 serves to suppress expression of paternally contributed genes. Supporting this view, *Arabidopsis* mutants lacking a FIS-PRC2 can form a functional diploid endosperm (Nowack et al., 2007); revealing that by reducing paternal genome dosage the functional requirement of the FIS-PRC2 can be bypassed. We therefore hypothesize that in gymnosperms, where the female gametophyte forms an endosperm-like nourishing structure, a FIS-PRC2-like complex would not be required. This notion is supported by the fact that the FIS-PRC2 prevents autonomous endosperm formation and thus couples fertilization to endosperm development (Guitton and Berger, 2005), a function that is not required in gymnosperms.

As outlined above, the PRC2 is targeted to regions with reduced DNA methylation in the endosperm (Weinhofer et al., 2010; Zhang et al., 2014) and could, therefore, potentially repress activity of TEs upon loss of DNA methylation. It can thus be envisioned that DNA hypomethylation activities in the central cell evolved concomitantly with a central cell/endosperm-expressed PRC2 that alleviated the negative effects associated with loss of DNA methylation. Following this logic, the evolution of hypomethylation mechanisms in the central cell to enforce silencing of TEs in the egg cell and descendent embryo were closely coupled to the evolution of FIS-PRC2-like complexes in the central cell and endosperm to ensure TE silencing after DNA hypomethylation. This scenario would imply that TE transcription upon DNA hypomethylation and PRC2-mediated TE repression are balanced to ensure sufficient TE-derived sRNAs being made to enforce silencing in the embryo but TE transposition remains suppressed (**Figure 1**).

If indeed a FIS-PRC2-like complex as well as DME-like enzymes evolved together with the sexual endosperm, neither of both should be present in the large female gametophytes of gymnosperms. The female gametophyte in gymnosperms serves an endosperm-like role in supporting embryo growth, but it is of pure maternal origin and forms a multicellular structure before fertilization. Consequently, the enormous genome expansion in gymnosperms may be a consequence of the lack of efficient TE silencing mechanisms that possibly have evolved after the female gametophyte became sexualized (**Figure 1**).

Testing this hypothesis requires to test whether orthologs of FIS-PRC2-like genes and DME-like genes are expressed in the female gametophyte of gymnosperms and basal angiosperms, which remains a challenge of future investigations.

Together, we propose that the evolution of mechanisms enforcing TE silencing in the embryo evolved concomitantly with the sexual endosperm. The evolution of PRC2 activity in the central cell and endosperm allowed DNA hypomethylation activities being active in the central cell to enforced TE silencing in the egg cell and embryo. Consequently, genome size restriction by efficient control of TE silencing may be directly coupled with the evolution of a sexual endosperm.

ACKNOWLEDGMENTS

We thank members of the Köhler laboratory for critical comments on this manuscript. This research was supported by a European Research Council Starting Independent Researcher grant (to CK).

REFERENCES

Belmonte, M. F., Kirkbride, R. C., Stone, S. L., Pelletier, J. M., Bui, A. Q., Yeung, E. C., et al. (2013). Comprehensive developmental profiles of gene activity in regions and subregions of the *Arabidopsis* seed. *Proc. Natl. Acad. Sci. U.S.A.* 110, 435–444. doi: 10.1073/pnas.1222061110

Calarco, J. P., Borges, F., Donoghue, M. T., Van Ex, F., Jullien, P. E., Lopes, T., et al. (2012). Reprogramming of DNA methylation in pollen guides epigenetic inheritance via small RNA. *Cell* 151, 194–205. doi: 10.1016/j.cell.2012.09.001

Cao, X., Aufsatz, W., Zilberman, D., Mette, M. F., Huang, M. S., Matzke, M., et al. (2003). Role of the *DRM* and *CMT3* methyltransferases in RNA-directed DNA methylation. *Curr. Biol.* 13, 2212–2217. doi: 10.1016/j.cub.2003.11.052

Cao, X., and Jacobsen, S. E. (2002). Role of the *Arabidopsis DRM* methyltransferases in de novo DNA methylation and gene silencing. *Curr. Biol.* 12, 1138–1144. doi: 10.1016/S0960-9822(02)00925-9

Charnov, E. L. (1979). Simultaneous hermaphroditism and sexual selection. *Proc. Natl. Acad. Sci. U.S.A.* 76, 2480–2484. doi: 10.1073/pnas.76.5.2480

Chen, L. J., Diao, Z. Y., Specht, C., and Sung, Z. R. (2009). Molecular evolution of *VEF*-domain-containing *PcG* genes in plants. *Mol. Plant* 2, 738–754. doi: 10.1093/mp/ssp032

Danilevskaya, O. N., Hermon, P., Hantke, S., Muszynski, M. G., Kollipara, K., and Ananiev, E. V. (2003). Duplicated fie genes in maize: expression pattern and imprinting suggest distinct functions. *Plant Cell* 15, 425–438. doi: 10.1105/tpc.006759

Deleris, A., Stroud, H., Bernatavichute, Y., Johnson, E., Klein, G., Schubert, D., et al. (2012). Loss of the DNA methyltransferase MET1 Induces H3K9 hyper-methylation at PcG target genes and redistribution of H3K27 trimethylation to transposons in *Arabidopsis thaliana*. *PLoS Genet.* 8:e1003062. doi: 10.1371/journal.pgen.1003062

Friedman, W. E. (1995). Organismal duplication, inclusive fitness theory, and altruism: understanding the evolution of endosperm and the angiosperm reproductive syndrome. *Proc. Natl. Acad. Sci. U.S.A.* 92, 3913–3917. doi: 10.1073/pnas.92.9.3913

Friedman, W. E., Madrid, E. N., and Williams, J. H. (2008). Origin of the fittest and survival of the fittest: relating female gametophyte development to endosperm genetics. *Int. J. Plant Sci.* 169, 79–92. doi: 10.1086/523354

Friedman, W. E., and Williams, J. H. (2003). Modularity of the angiosperm female gametophyte and its bearing on the early evolution of endosperm in flowering plants. *Evolution* 57, 216–230. doi: 10.1111/j.0014-3820.2003.tb00257.x

Gehring, M. (2013). Genomic imprinting: insights from plants. *Annu. Rev. Genet.* 47, 187–208. doi: 10.1146/annurev-genet-110711-155527

Gehring, M., Bubb, K. L., and Henikoff, S. (2009). Extensive demethylation of repetitive elements during seed development underlies gene imprinting. *Science* 324, 1447–1451. doi: 10.1126/science.1171609

Gleason, E. J., and Kramer, E. M. (2012). Characterization of Aquilegia Polycomb Repressive Complex 2 homologs reveals absence of imprinting. *Gene* 507, 54–60. doi: 10.1016/j.gene.2012.07.004

Grant-Downton, R., Kourmpetli, S., Hafidh, S., Khatab, H., Le Trionnaire, G., Dickinson, H., et al. (2013). Artificial microRNAs reveal cell-specific differences in small RNA activity in pollen. *Curr. Biol.* 23, R599–R601. doi: 10.1016/j.cub.2013.05.055

Grossniklaus, U., Vielle-Calzada, J. P., Hoeppner, M. A., and Gagliano, W. B. (1998). Maternal control of embryogenesis by MEDEA a polycomb group gene in *Arabidopsis*. *Science* 280, 446–450. doi: 10.1126/science.280.5362.446

Guitton, A. E., and Berger, F. (2005). Control of reproduction by polycomb group complexes in animals and plants. *Int. J. Dev. Biol.* 49, 707–716. doi: 10.1387/ijdb.051990ag

Haig, D. (2013). Kin conflict in seed development: an interdependent but fractious collective. *Annu. Rev. Cell Dev. Biol.* 29, 189–211. doi: 10.1146/annurev-cellbio-101512-122324

Haig, D., and Westoby, M. (1989a). Parent specific gene expression and the triploid endosperm. *Am. Nat.* 134, 147–155. doi: 10.1086/284971

Haig, D., and Westoby, M. (1989b). Selective forces in the emergence of the seed habit. *Biol. J. Linn. Soc.* 38, 215–238. doi: 10.1111/j.1095-8312.1989.tb01576.x

Haun, W. J., Laoueille-Duprat, S., O'Connell, M. J., Spillane, C., Grossniklaus, U., Phillips, A. R., et al. (2007). Genomic imprinting, methylation and molecular evolution of maize *Enhancer of zeste* (*Mez*) homologs. *Plant J.* 49, 325–337. doi: 10.1111/j.1365-313X.2006.02965.x

Hennig, L., and Derkacheva, M. (2009). Diversity of polycomb group complexes in plants: same rules, different players? *Trends Genet.* 25, 414–423. doi: 10.1016/j.tig.2009.07.002

Hsieh, T. F., Ibarra, C. A., Silva, P., Zemach, A., Eshed-Williams, L., Fischer, R. L., et al. (2009). Genome-wide demethylation of *Arabidopsis* endosperm. *Science* 324, 1451–1454. doi: 10.1126/science.1172417

Ibarra, C. A., Feng, X., Schoft, V. K., Hsieh, T. F., Uzawa, R., Rodrigues, J. A., et al. (2012). Active DNA demethylation in plant companion cells reinforces transposon methylation in gametes. *Science* 337, 1360–1364. doi: 10.1126/science.1224839

Jullien, P. E., Susaki, D., Yelagandula, R., Higashiyama, T., and Berger, F. (2012). DNA methylation dynamics during sexual reproduction in *Arabidopsis thaliana*. *Curr. Biol.* 22, 1825–1830. doi: 10.1016/j.cub.2012.07.061

Kim, M. Y., and Zilberman, D. (2014). DNA methylation as a system of plant genomic immunity. *Trends Plant Sci.* 19, 320–326. doi: 10.1016/j.tplants.2014.01.014

Kradolfer, D., Hennig, L., and Köhler, C. (2013). Increased maternal genome dosage bypasses the requirement of the FIS Polycomb Repressive Complex 2 in *Arabidopsis* seed development. *PLoS Genet.* 9:e1003163. doi: 10.1371/journal.pgen.1003163

Li, J., and Berger, F. (2012). Endosperm: food for humankind and fodder for scientific discoveries. *New Phytol.* 195, 290–305. doi: 10.1111/j.1469-8137.2012.04182.x

Luo, M., Bilodeau, P., Dennis, E. S., Peacock, W. J., and Chaudhury, A. (2000). Expression and parent-of-origin effects for FIS2, MEA, and FIE in the endosperm and embryo of developing *Arabidopsis* seeds. *Proc. Natl. Acad. Sci. U.S.A.* 97, 10637–10642. doi: 10.1073/pnas.170292997

Luo, M., Platten, D., Chaudhury, A., Peacock, W. J., and Dennis, E. S. (2009). Expression, imprinting, and evolution of rice homologs of the polycomb group genes. *Mol. Plant* 2, 711–723. doi: 10.1093/mp/ssp036

McCue, A. D., Panda, K., Nuthikattu, S., Choudury, S. G., Thomas, E. N., and Slotkin, R. K. (2015). ARGONAUTE 6 bridges transposable element mRNA-derived siRNAs to the establishment of DNA methylation *EMBO J.* 34, 20–35. doi: 10.15252/embj.201489499

Mosher, R. A., Melnyk, C. W., Kelly, K. A., Dunn, R. M., Studholme, D. J., and Baulcombe, D. C. (2009). Uniparental expression of PolIV-dependent siRNAs in developing endosperm of *Arabidopsis*. *Nature* 460, 283–286. doi: 10.1038/nature08084

Nowack, M. K., Shirzadi, R., Dissmeyer, N., Dolf, A., Endl, E., Grini, P. E., et al. (2007). Bypassing genomic imprinting allows seed development. *Nature* 447, 312–315. doi: 10.1038/nature05770

Reddington, J. P., Perricone, S. M., Nestor, C. E., Reichmann, J., Youngson, N. A., Suzuki, M., et al. (2013). Redistribution of H3K27me3 upon DNA hypomethylation results in de-repression of Polycomb target genes. *Genome Biol.* 14, R25. doi: 10.1186/gb-2013-14-3-r25

Saksouk, N., Barth, T. K., Ziegler-Birling, C., Olova, N., Nowak, A., Rey, E., et al. (2014). Redundant mechanisms to form silent chromatin at pericentromeric regions rely on BEND3 and DNA methylation. *Mol. Cell* 56, 580–594. doi: 10.1016/j.molcel.2014.10.001

Segal, E., Shapira, M., Regev, A., Pe'er, D., Botstein, D., Koller, D., et al. (2003). Module networks: identifying regulatory modules and their condition-specific regulators from gene expression data. *Nat. Genet.* 34, 166–176. doi: 10.1038/ng1165

Simon, J. A., and Kingston, R. E. (2013). Occupying chromatin: polycomb mechanisms for getting to genomic targets, stopping transcriptional traffic, and staying put. *Mol. Cell* 49, 808–824. doi: 10.1016/j.molcel.2013.02.013

Slotkin, R. K., Vaughn, M., Borges, F., Tanurdzic, M., Becker, J. D., Feijo, J. A., et al. (2009). Epigenetic reprogramming and small RNA silencing of transposable elements in pollen. *Cell* 136, 461–472. doi: 10.1016/j.cell.2008.12.038

Spillane, C., Schmid, K. J., Laoueille-Duprat, S., Pien, S., Escobar-Restrepo, J. M., Baroux, C., et al. (2007). Positive Darwinian selection at the imprinted MEDEA locus in plants. *Nature* 448, 349–352. doi: 10.1038/nature05984

Sprunck, S., and Gross-Hardt, R. (2011). Nuclear behavior, cell polarity, and cell specification in the female gametophyte. *Sex. Plant Reprod.* 24, 123–136. doi: 10.1007/s00497-011-0161-4

Villar, C. B., Erilova, A., Makarevich, G., Trösch, R., and Köhler, C. (2009). Control of PHERES1 imprinting in *Arabidopsis* by direct tandem repeats. *Mol. Plant* 2, 654–660. doi: 10.1093/mp/ssp014

Weinhofer, I., Hehenberger, E., Roszak, P., Hennig, L., and Köhler, C. (2010). H3K27me3 profiling of the endosperm implies exclusion of polycomb group protein targeting by DNA methylation. *PLoS Genet.* 6:e1001152. doi: 10.1371/journal.pgen.1001152

Wolff, P., Weinhofer, I., Seguin, J., Roszak, P., Beisel, C., Donoghue, M. T., et al. (2011). High-resolution analysis of parent-of-origin allelic expression in the *Arabidopsis* endosperm. *PLoS Genet.* 7:e1002126. doi: 10.1371/journal.pgen.1002126

Zemach, A., Kim, M. Y., Silva, P., Rodrigues, J. A., Dotson, B., Brooks, M. D., et al. (2010). Local DNA hypomethylation activates genes in rice endosperm. *Proc. Natl. Acad. Sci. U.S.A.* 107, 18729–18734. doi: 10.1073/pnas.1009695107

Zhang, M., Xie, S., Dong, X., Zhao, X., Zeng, B., Chen, J., et al. (2014). Genome-wide high resolution parental-specific DNA and histone methylation maps uncover patterns of imprinting regulation in maize. *Genome Res.* 24, 167–176. doi: 10.1101/gr.155879.113

13

Interpreting lemma and palea homologies: a point of view from rice floral mutants

Fabien Lombardo and Hitoshi Yoshida*

Rice Biotechnology Research Project, Rice Research Division, National Agriculture and Food Research Organization (NARO) Institute of Crop Science, Tsukuba, Ibaraki, Japan

Edited by:
Dazhong D. Zhao, University of Wisconsin at Milwaukee, USA

Reviewed by:
Clinton Whipple, Brigham Young University, USA
Elizabeth A. Kellogg, Donald Danforth Plant Science Center, USA

*Correspondence:
Hitoshi Yoshida, Rice Biotechnology Research Project, Rice Research Division, National Agriculture and Food Research Organization (NARO) Institute of Crop Science, 2-1-18 Kannondai, Tsukuba, Ibaraki 305-8518, Japan.
e-mail: yocida@affrc.go.jp

In contrast to eudicot flowers which typically exhibit sepals and petals at their periphery, the flowers of grasses are distinguished by the presence of characteristic outer organs. In place of sepals, grasses have evolved the lemma and the palea, two bract-like structures that partially or fully enclose the inner reproductive organs. With little morphological similarities to sepals, whether the lemma and palea are part of the perianth or non-floral organs has been a longstanding debate. In recent years, comparative studies of floral mutants as well as the availability of whole genome sequences in many plant species have provided strong arguments in favor of the hypothesis of lemma and palea being modified sepals. In rice, a feature of the palea is the bending of its lateral region into a hook-shaped marginal structure. This allows the palea to lock into the facing lemma region, forming a close-fitting lemma–palea enclosure. In this article, we focus on the rice lemma and palea and review some of the key transcription factors involved in their development and functional specialization. Alternative interpretations of these organs are also addressed.

Keywords: sepal, grass, MADS-box gene, perianth, bract, prophyll

EQUATING FLOWER ARCHITECTURES

Flowers are biological wonders. Flowering plants, or angiosperms, have evolved into an impressive number of species (the lowest estimations are well above 200,000; Scotland and Wortley, 2003) and are found in almost all ecological niches around the world. Flowers exist in a staggering variety of forms, colors and architectures and yet an exhaustive catalog is still a long way ahead (Endress, 2011). The ecological dominance and evolutionary success of the angiosperms is partly explained by the flexibility of their flower-based mode of reproduction which has allowed sustained species diversification over time (Crepet and Niklas, 2009).

At the molecular level, flowers are formed upon the action of numerous transcription factors, the majority belonging to the MIKC^c-type MADS-box family (Gramzow and Theissen, 2010; Wellmer and Riechmann, 2010). The current and widely-accepted model that describes how these transcription factors interact to direct the development of floral organs, the ABCDE model, is based on early mutant studies in two eudicot species, *Antirrhinum majus* (Plantaginaceae; Schwarz-Sommer et al., 1990) and *Arabidopsis thaliana* (Brassicaceae; Bowman et al., 1991). Consequently, conceptual thinking of flower development is rooted in the typical dicotyledonous, four-concentric whorl flower architecture in which each whorl is occupied by one type of organ with the following sequence: sepal, petal, stamen, and carpel (from the outermost to the innermost whorl). The model is flexible enough however to be extended to various floral architectures (Bowman, 1997; Erbar, 2007; Theissen and Melzer, 2007); and derived models, such as the "fading borders" model, have been

generated to describe the flowers of species as phylogenetically distant from *A. thaliana* as the basal angiosperms (Buzgo et al., 2004). In several monocot species for example, sepals and petals are not distinguishable and are collectively referred to as tepals. Nevertheless, the relation between tepals and sepals/petals can be accounted for in the ABCDE model by shifts in the domain of expression of B-function homeotic genes (Bowman, 1997).

There are several species however where the interpretation of the floral architecture itself, and most particularly the outer whorls and peripheral organs, is problematic to begin with. Within the monocots, this is the case for members of the grass family which bear characteristic flowers, termed florets, that differ substantially from the one described in the ABCDE model. The periphery of the grass flower is occupied by elongated and leafy organs, evocative of small bracts, in a striking contrast to a typical monocot perianth. The nature of these organs and the identity of their counterparts in non-grass related species, if any, have been subject to much debate for more than a century (Clifford, 1986). The identity of the bract-like organs closest to the inner flower, called lemma and palea, has been the most controversial. While various interpretations have been formulated (Clifford, 1986), the lemma and palea have been commonly interpreted as a bract and a prophyll, respectively (Linder, 1987; Rudall and Bateman, 2004). Alternatively, the palea has been interpreted as two fused sepals (adaxial tepals; Schuster, 1910; Stebbins, 1951) and the lemma has been rarely interpreted as a sepal (calyx; Francis, 1920).

Oryza sativa (common rice) is one of the best documented grass species and many rice mutants have been described in the literature. Focusing on *O. sativa*, in the following are reviewed some of the key pieces of data that have surfaced in the last 30 years or so which have shed light on the controversial nature of the lemma and palea.

LEMMA AS BRACT AND PALEA AS PROPHYLL EQUIVALENTS?

The structure of the rice flower is commonly described and organs designated as following (**Figure 1**): On a short axis, the rachilla, are proximally attached two cupule-shaped small outgrowths called rudimentary glumes. Above the rudimentary glumes are found a pair of scales called sterile lemmas or empty glumes depending on how they are interpreted. The floret is the unit above the sterile lemmas that comprises the lemma, the palea and the enclosed inner floral organs. The floret is commonly considered as the grass equivalent of the eudicot flower and the addition of the floret, the sterile lemmas and the rudimentary glumes forms the spikelet.

The above nomenclature stems from an early and common interpretation which is based on the observation that the lemma arises on the main axis, which is distinct from the floret axis. The lemma is thus regarded as a modified leaf which subtends the

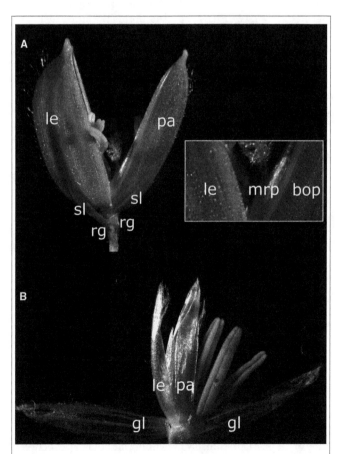

FIGURE 1 | Structure of the rice and maize spikelets. (A) Rice spikelet. Inset is a close-up view of the basal region of the lemma and palea. le, lemma; pa, palea; mrp, marginal region of palea; bop, body of palea; sl, sterile lemma; rg, rudimentary glume. **(B)** Maize tassel (male) spikelet. le, lemma; pa, palea; gl, glume.

floral meristem in its axil (Arber, 1934; Bell, 1991; Kellogg, 2001). The distinct origin of the lemma is illustrated in the *leafy lemma* barley mutant in which the lemma is specifically transformed into a leaf-like organ (Pozzi et al., 2000). While modified leaves growing near inflorescences, or bracts, can take petal-like vivid colors in some species (Buzgo and Endress, 2000), they do not belong to the perianth by definition and are therefore extra-floral organs. Facing the lemma, the palea originates on the floret axis and, since it is the first "leaf" arising from the meristem subtended by the lemma, it is commonly considered a prophyll. The basal bracts that subtend the spikelet are called glumes and in rice the term has been applied indiscriminately to both the rudimentary glumes and the empty glumes, bringing some confusion to which are the spikelet-subtending bracts. Based on serial sections, Arber (1934) concluded that the rudimentary glumes are the true basal bracts of the rice spikelet, only in an extremely reduced, vestigial form. Consequently the "empty glumes" are interpreted as sterile lemmas, since they do not bear any flowers in their axils.

LEMMA AND PALEA AS SEPAL EQUIVALENTS?

The interpretation of the lemma as bract and palea as prophyll equivalents relies for the most part on early morphological comparative studies (Arber, 1934; Stebbins, 1951; Clifford, 1986; Bell, 1991). However, more recent progress in the genetics of flower development highlighting the universal role of MADS-box genes as floral homeotic genes suggest that both lemma and palea are sepal equivalents, in the sense that they are outer perianth organs corresponding to the tepals/sepals of most other flowers. Such equivalency does not imply however that sepals and both lemma and palea are derived from the same ancestral organ. The sequencing of the genome of *Amborella trichopoda*, a species belonging to the sister lineage to all other extant flowering plants, has revealed that each of the eight major lineages of MADS-box genes were represented in the most recent common ancestor of the angiosperms (Amborella Genome Project, 2013). MADS-box genes are thus invaluable molecular markers toward determining floral organ identity. In the interpretation where the lemma and palea are floral organs, homeotic genes associated with floral identity are expected to be expressed in these structures. Conversely, such gene expression is expected to be lacking in bracts and other non-floral structures. In rice, inflorescence meristem identity is specified by *AP1/FUL*-like genes and a *SEP* gene (Kobayashi et al., 2012). Furthermore no significant expressions of floral MADS-box genes can be detected in the bracts of grasses, strongly suggesting that the lemma and palea are distinct from these structures (Kyozuka et al., 2000; Malcomber and Kellogg, 2004; Prasad et al., 2005; Preston and Kellogg, 2007). Expression analysis of key MADS-box genes in *Streptochaeta angustifolia*, a non-spikelet-bearing grass species, and in the grass outgroup monocot *Joinvillea ascendens* allowed Preston et al. (2009) to infer the putative floral architecture of the grass common ancestor: three categories of structure (glume-, sepal-, and petal-equivalents) would each express a different combination of *AP1/FUL*-like, *LHS1*-like and B-class genes. In any case, expression of any of these genes is neither expected in the bracts of the grass common ancestor nor detected in the bracts of any of the investigated monocot species. The authors suggest that the ancestral sepal-equivalent structures which express *AP1/FUL*-like and *LHS1*-like genes are the organs from which the lemma and palea are derived (Preston et al., 2009).

According to the ABCDE model, perianth whorls develop under the action of A-class genes (sepals) or cumulative action of A-and B-class genes (petals). Petals are therefore expected to homeotically transform into sepals or at least acquire some degree of sepal identity when B-class genes are disrupted, as documented in the *apetala3* (*ap3*) mutant of *A. thaliana* (Goto and Meyerowitz, 1994). The role of genes for B function has been shown to be conserved across the angiosperms (Whipple et al., 2007) and in maize, disruption of the B-class *SILKY1* gene leads to a homeotic conversion of the lodicules (organs commonly considered as petal equivalents) into lemma/palea-like structures (Ambrose et al., 2000). A similar homeotic conversion is observed in the loss-of-function alleles of the *SUPERWOMAN1* (*SPW1*) gene, the rice ortholog of *AP3* (Nagasawa, 2003). Following the ABCDE model, these results strongly suggest that the lemma and palea are equivalent to the sepals of most other flowers.

The phenotype of maize *branched silkless* (*bd1*), in which transition from the spikelet meristem to the floret meristem is blocked, supports that lemma and palea are floral organs. The mutant is able to produce glumes but neither lemma nor palea is formed (Colombo et al., 1998), indicating that the whorls holding the lemma and the palea originate from a floral meristem.

PALEA AS A DIFFERENTIATED LEMMA

Irrespective of the homology of the lemma and palea, the genetic mechanisms that control their development are distinct (summarized in **Figure 2**). There are mutants in which either the palea or the lemma is specifically affected, such as the *leafy lemma* mutant of barley or the *depressed palea1* (*dp1*) mutant of rice, which palea is dramatically reduced but its lemma remains unchanged (Pozzi et al., 2000; Luo et al., 2005). Ambrose et al. (2000) hypothesized that the lemma and the palea reside in two distinct whorls, which would account for some level of genetic independence and explains the asymmetrical phenotypes.

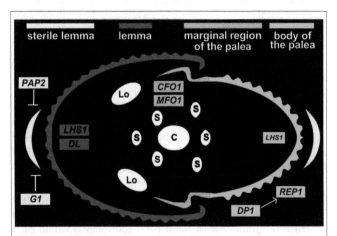

FIGURE 2 | Major transcription factors controlling the development of the lemma and palea in rice. Genes involved in development of lemma and palea are shown. lemma (dark green); marginal region of palea (mrp; blue); body of palea (bop; light green); sterile lemma (yellow); Lo, lodicule; S, stamen; C, carpel.

Depending on the grass species, the palea can be distinguished from the lemma by various morphological features, such as the number of vascular bundles, size, or surface structure. In *O. sativa*, the differentiation of the palea is particularly pronounced. Edges of the palea curl outwardly at its base in a hook-shaped marginal structure which fits together with the inwardly curled facing lemma. The marginal region of the palea is smooth and light colored, in contrast to the body of the palea which is populated with silicified cells bearing trichomes. Phenotypes of several mutants suggest that the rice palea can be considered as a composite of two types of domain: the body and the marginal region. In this hypothesis, the body is further interpreted as a structure with a lemma identity and the marginal regions as distinct structures with palea identity (Yoshida and Nagato, 2011). Phenotypes of the *mfo1* and *cfo1* mutants further support this idea (Ohmori et al., 2009; Sang et al., 2012; see below).

TRANSCRIPTION FACTORS INVOLVED IN PALEA DIFFERENTIATION

The *AGL6*-like MADS-box gene *MOSAIC FLORAL ORGANS1* (*MFO1*; *MADS6*) is a major determinant of the rice palea architecture. In *mfo1*, the palea acquires features of the lemma, namely inward curling, loss of the marginal region and ectopic expression of *DROOPING LEAF* (*DL*), a gene normally expressed in the lemma (Ohmori et al., 2009). In addition to its role in palea differentiation, *MFO1* has a central role in spikelet development and is involved in floral meristem determinacy. A phylogenetic analysis has revealed that expression of *MFO1* in the palea has appeared later in the evolution and correlated with the origin of the grass spikelet (Reinheimer and Kellogg, 2009). In maize, the *bearded-ear* (*bde*) gene is orthologous to *MFO1* and is also expressed in the palea but not in the lemma, suggesting a conserved role for *AGL6*-like genes in the palea across the grasses (Thompson et al., 2009). This hypothesis could be tested by investigating the role of *MFO1/bde* orthologs in other grass species.

Similarly to *mfo1*, *chimeric floral organs1* (*cfo1*; the mutant of rice *MADS32*) shows variable defects in the inner whorls but a rather consistent, somewhat similar phenotype to *mfo1* in the palea. The marginal region in *cfo1* mutants is enlarged and silicified and ectopic expression of *DL* is also observed. However, unlike in *mfo1* paleas, there is no lemma-like inward curling (Sang et al., 2012). *CFO1* was thought to be a grass-specific gene until the recent sequencing of *Amborella trichopoda* revealed the presence of an ortholog, implying that the gene has been lost outside of the grass group. The evolution of *CFO1* and its ancestral function remain to be elucidated and it would be particularly interesting to know if, similarly to *MFO1*, the gene was recruited in the palea to support its differentiation in grasses.

RETARDED PALEA1 (*REP1*) encodes a CYCLOIDEA (CYC)-like TCP transcription factor which promotes the growth of the body of the palea, presumably by defining the boundaries between the marginal region and the body (Yuan et al., 2009). In *rep1* the body is strongly reduced, resulting in a much smaller palea, whereas the marginal region is widened. Over-expression lines show the opposite phenotype, that is an overgrown body and narrower marginal region.

RETARDED PALEA1 is hypothesized to be downstream of the *DP1* gene which encodes an AT-hook transcription factor (Jin et al., 2011). The *dp1* mutant shows a more severe phenotype than *rep1*: The body is lost entirely, leaving two marginal leafy organs which are likely to be transformed marginal regions. The only putative ortholog to *DP1* described so far is the maize *BARREN STALK FASTIGIATE1* (*BAF1*) gene. The *BAF1/DP1* function is hypothesized to be conserved in all of the grasses (Gallavotti et al., 2011), and would contribute to the differentiation of the grass flower. The phenotypes of *rep1* and *dp1* mutants are consistent with the interpretation of the rice palea being composed of two types of domain: a lemma-identity structure (the body) and two differentiated lateral structures (the marginal regions; Jin et al., 2011).

TRANSCRIPTION FACTORS INVOLVED IN LEMMA DIFFERENTIATION

A common feature of both *mfo1* and *cfo1* mutants is the palea ectopic expression of *DL* in the abnormal paleas. Mutant alleles of *dl* have been well documented, mostly for the striking loss of carpel identity, a function which is conserved in *A. thaliana* via the *CRABS CLAW* (*CRC*) ortholog, and for the inability to maintain erect leaves (Bowman and Smyth, 1999; Yamaguchi et al., 2004). *DL* promotes cell proliferation in the leaf midrib structure and in the lemma, along its longitudinal axis. This is illustrated in the *dl-sup1* mutant which grows a shorter lemma; and for the requirement of the gene in awn development (Toriba and Hirano, 2014). In a *dl cfo1* double mutant, the altered marginal region phenotype of *cfo1* is rescued, suggesting that the defects observed in *cfo1* marginal regions are due to the ectopic activity of *DL*. The marginal region is not altered however in a *dl mfo1* double mutant, so the precise mechanisms by which ectopic *DL* expression disturbs palea development remain to be elucidated (Li et al., 2011).

Another gene involved in lemma differentiation is the rice *LHS1* (*MADS1*) gene. Ectopic expression of *LHS1* in the sterile lemma confers the organ lemma-like morphological and anatomical traits. Conversely, silenced lines of *LHS1* show transformation of their lemmas into sterile lemma-like organs with poor cellular differentiation (Prasad et al., 2005). The palea is only slightly affected in these mutants, suggesting that *LHS1* functions essentially as a lemma differentiation gene.

LEMMAS AND STERILE LEMMAS

Eighty years ago, Arber hypothesized that the sterile lemmas are the remaining organs of two additional spikelets, lost from an ancestral rice with a three-floret spikelet (Arber, 1934). The LONG STERILE LEMMA1 (G1) protein contains an ALOG domain and belongs to a recently described class of transcription factor. The *g1* mutant shows the striking phenotype of sterile lemmas transformed into lemmas, bringing genetic evidence to the long-standing hypothesis by Arber (1934; Yoshida et al., 2009). This idea is supported by similar phenotypes of *panicle phytomer2* (*pap2*; *mads34*; Lin et al., 2014).

The spikelet of the wild rice *O. grandiglumis* bears elongated sterile lemmas which are in a striking resemblance to the ones of *g1* or *pap2*. Nucleotide sequences of *O. grandiglumis G1* and *PAP2*

show some polymorphism in key functional domains, suggesting that the long sterile lemma phenotype of *O. grandiglumis* is the result of natural variations in the *G1* and/or *PAP2* sequences.

LEMMA AND PALEA ILLUSTRATE THE ANGIOSPERM FLOWER PLASTICITY

The large diversity in flower shape and architecture across the angiosperms makes unraveling the evolution of morphological features a laborious and challenging task. Identification and analysis of floral transcription factors have uncovered how subtle genetic alterations can result in dramatic morphological changes. Duplication, recruitment and/or sub-functionalization of the MADS-box transcription factors have been shown to correlate with floral diversification (Shan et al., 2009; Yockteng et al., 2013), and undoubtedly, the complexity and flexibility of floral feature evolution had been underestimated during the pre-molecular era (Endress and Matthews, 2012).

Before the advent of molecular biology, the lemma and palea of grasses have been arguably most commonly interpreted as a bract and prophyll, respectively, although a handful of authors over the last century have suggested that they might be modified perianth parts. While the lemma and palea of grasses show significant morphological variations depending on the observed species, expressions of *AP1/FUL*-like genes as well as *LHS1*-like genes are detected in these structures. This implies that the lemma and palea are emerging on a floral meristem and that they are very likely to be distinct from glumes since the expression of *LHS1*-like genes has not been observed in the glumes of any grasses yet (Preston et al., 2009). Some mutants affected in B function, which is likely to be conserved across angiosperms (Whipple et al., 2007), show a homeotic transformations of their second whorl organs into lemma/palea-like organs. Taken together, these data suggest that the lemma and palea of grasses are likely to be sepal equivalents.

Biotic-pollinated plants must accommodate for bud protection and attract pollinators at the same time, and their perianth has evolved under these constraints. In wind-pollinated grasses however, elongated and covering outer organs provide advantageous protection against pests and physical damage. Under the assumption that the grass lemma and palea are sepal equivalents, these organs, and most particularly in the case of rice, can be regarded as a remarkable illustration of the evolutionary potency of the angiosperms.

ACKNOWLEDGMENTS

This work was partly supported by a grant from the Ministry of Agriculture, Forestry, and Fishery of Japan (Research project for Genomics for Agricultural Innovation GRA-203-1-1) and by JSPS KAKENHI Grant Number 26292008.

REFERENCES

Amborella Genome Project. (2013). The *Amborella* genome and the evolution of flowering plants. *Science* 342, 1241089. doi: 10.1126/science.1241089

Ambrose, B. A., Lerner, D. R., Ciceri, P., Padilla, C. M., Yanofsky, M. F., and Schmidt, R. J. (2000). Molecular and genetic analyses of the *silky1* gene reveal conservation in floral organ specification between eudicots and monocots. *Mol. Cell* 5, 569–579. doi: 10.1016/S1097-2765(00)80450-5

Arber, A. (1934). *The Gramineae: A Study of Cereal, Bamboo, and Grass.* New York: Cambridge University Press.

Bell, A. D. (1991). *Plant Form: An Illustrated Guide to Flowering Plant Morphology.* Oxford: Oxford University Press.

Bowman, J. L. (1997). Evolutionary conservation of angiosperm flower development at the molecular and genetic levels. *J. Biosci.* 22, 515–527. doi: 10.1007/BF02703197

Bowman, J. L., and Smyth, D. R. (1999). CRABS CLAW, a gene that regulates carpel and nectary development in *Arabidopsis*, encodes a novel protein with zinc finger and helix-loop-helix domains. *Development* 126, 2387–2396.

Bowman, J. L., Smyth, D. R., and Meyerowitz, E. M. (1991). Genetic interactions among floral homeotic genes of *Arabidopsis*. *Development* 112, 1–20.

Buzgo, M., and Endress, P. K. (2000). Floral structure and development of Acoraceae and its systematic relationships with basal angiosperms. *Int. J. Plant Sci.* 161, 23–41. doi: 10.1086/314241

Buzgo, M., Soltis, P. S., and Soltis, D. E. (2004). Floral developmental morphology of *Amborella trichopoda* (Amborellaceae). *Int. J. Plant Sci.* 165, 925–947. doi: 10.1086/424024

Clifford, H. T. (1986). "Spikelet and floral morphology" in *Grass Systematics and Evolution*, eds T. R. Soderstrom, K. W. Hilu, C. S. Campbell, and M. E. Barkworth (Washington, DC, Smithsonian Institution Press), 21–30.

Colombo, L., Marziani, G., Masiero, S., Wittich, P. E., Schmidt, R. J., Gorla, M. S., et al. (1998). BRANCHED SILKLESS mediates the transition from spikelet to floral meristem during *Zea mays* ear development. *Plant J.* 16, 355–363. doi: 10.1046/j.1365-313x.1998.00300.x

Crepet, W. L., and Niklas, K. J. (2009). Darwin's second "abominable mystery": why are there so many angiosperm species? *Am. J. Bot.* 96, 366–381. doi: 10.3732/ajb.0800126

Endress, P. K. (2011). Evolutionary diversification of the flowers in angiosperms. *Am. J. Bot.* 98, 370–396. doi: 10.3732/ajb.1000299

Endress, P. K., and Matthews, M. L. (2012). Progress and problems in the assessment of flower morphology in higher-level systematics. *Plant Syst. Evol.* 298, 257–276. doi: 10.1007/s00606-011-0576-2

Erbar, C. (2007). Current opinions in flower development and the evo-devo approach in plant phylogeny. *Plant Syst. Evol.* 269, 107–132. doi: 10.1007/s00606-007-0579-1

Francis, M. E. (1920). *Book of Grasses*. Garden City, NY: Doubleday.

Gallavotti, A., Malcomber, S., Gaines, C., Stanfield, S., Whipple, C., Kellogg, E., et al. (2011). BARREN STALK FASTIGIATE1 is an AT-hook protein required for the formation of maize ears. *Plant Cell* 23, 1756–1771. doi: 10.1105/tpc.111.084590

Goto, K., and Meyerowitz, E. M. (1994). Function and regulation of the *Arabidopsis* floral homeotic gene PISTILLATA. *Genes Dev.* 8, 1548–1560. doi: 10.1101/gad.8.13.1548

Gramzow, L., and Theissen, G. (2010). A hitchhiker's guide to the MADS world of plants. *Genome Biol.* 11, 214. doi: 10.1186/gb-2010-11-6-214

Jin, Y., Luo, Q., Tong, H., Wang, A., Cheng, Z., Tang, J., et al. (2011). An AT-hook gene is required for palea formation and floral organ number control in rice. *Dev. Biol.* 359, 277–288. doi: 10.1016/j.ydbio.2011.08.023

Kellogg, E. A. (2001). Evolutionary history of the grasses. *Plant Physiol.* 125, 1198–1205. doi: 10.1104/pp.125.3.1198

Kobayashi, K., Yasuno, N., Sato, Y., Yoda, M., Yamazaki, R., Kimizu, M., et al. (2012). Inflorescence meristem identity in rice is specified by overlapping functions of three AP1/FUL-like MADS box genes and PAP2, a SEPALLATA MADS box gene. *Plant Cell* 24, 1848–1859. doi: 10.1105/tpc.112.097105

Kyozuka, J., Kobayashi, T., Morita, M., and Shimamoto, K. (2000). Spatially and temporally regulated expression of rice MADS box genes with similarity to *Arabidopsis* class A, B and C genes. *Plant Cell Physiol.* 41, 710–718. doi: 10.1093/pcp/41.6.710

Li, H., Liang, W., Hu, Y., Zhu, L., Yin, C., Xu, J., et al. (2011). Rice MADS6 interacts with the floral homeotic genes SUPERWOMAN1, MADS3, MADS58, MADS13, and DROOPING LEAF in specifying floral organ identities and meristem fate. *Plant Cell* 23, 2536–2552. doi: 10.1105/tpc.111.087262

Lin, X., Wu, F., Du, X., Shi, X., Liu, Y., Liu, S., et al. (2014). The pleiotropic SEPALLATA-like gene OsMADS34 reveals that the "empty glumes" of rice (*Oryza sativa*) spikelets are in fact rudimentary lemmas. *New Phytol.* 202, 689–702. doi: 10.1111/nph.12657

Linder, H. P. (1987). The evolutionary history of the Poales/Restionales: a hypothesis. *Kew Bull.* 42, 297–318. doi: 10.2307/4109686

Luo, Q., Zhou, K., Zhao, X., Zeng, Q., Xia, H., Zhai, W., et al. (2005). Identification and fine mapping of a mutant gene for paleless spikelet in rice. *Planta* 221, 222–230. doi: 10.1007/s00425-004-1438-8

Malcomber, S. T., and Kellogg, E. A. (2004). Heterogeneous expression patterns and separate roles of the SEPALLATA gene LEAFY HULL STERILE1 in grasses. *Gene* 16, 1692–1706. doi: 10.1105/tpc.021576.reduced

Nagasawa, N. (2003). SUPERWOMAN1 and DROOPING LEAF genes control floral organ identity in rice. *Development* 130, 705–718. doi: 10.1242/dev.00294

Ohmori, S., Kimizu, M., Sugita, M., Miyao, A., Hirochika, H., Uchida, E., et al. (2009). MOSAIC FLORAL ORGANS1, an AGL6-like MADS box gene, regulates floral organ identity and meristem fate in rice. *Plant Cell* 21, 3008–3025. doi: 10.1105/tpc.109.068742

Pozzi, C., Faccioli, P., Terzi, V., Stanca, A. M., Cerioli, S., Castiglioni, P., et al. (2000). Genetics of mutations affecting the development of a barley floral bract. *Genetics* 154, 1335–1346.

Prasad, K., Parameswaran, S., and Vijayraghavan, U. (2005). OsMADS1, a rice MADS-box factor, controls differentiation of specific cell types in the lemma and palea and is an early-acting regulator of inner floral organs. *Plant J.* 43, 915–928. doi: 10.1111/j.1365-313X.2005.02504.x

Preston, J. C., Christensen, A., Malcomber, S. T., and Kellogg, E. A. (2009). MADS-box gene expression and implications for developmental origins of the grass spikelet. *Am. J. Bot.* 96, 1419–1429. doi: 10.3732/ajb.0900062

Preston, J. C., and Kellogg, E. A. (2007). Conservation and divergence of APETALA1/FRUITFULL-like gene function in grasses: evidence from gene expression analyses. *Plant J.* 52, 69–81. doi: 10.1111/j.1365-313X.2007.03209.x

Reinheimer, R., and Kellogg, E. A. (2009). Evolution of AGL6-like MADS box genes in grasses (Poaceae): ovule expression is ancient and palea expression is new. *Plant Cell* 21, 2591–2605. doi: 10.1105/tpc.109.068239

Rudall, P. J., and Bateman, R. M. (2004). Evolution of zygomorphy in monocot flowers: iterative patterns and developmental constraints. *New Phytol.* 162, 25–44. doi: 10.1111/j.1469-8137.2004.01032.x

Sang, X., Li, Y., Luo, Z., Ren, D., Fang, L., Wang, N., et al. (2012). CHIMERIC FLORAL ORGANS1, encoding a monocot-specific MADS box protein, regulates floral organ identity in rice. *Plant Physiol.* 160, 788–807. doi: 10.1104/pp.112.200980

Schuster, J. (1910). Über die Morphologie der Grasblüte. *Flora* 100, 213–266.

Schwarz-Sommer, Z., Huijser, P., Nacken, W., Saedler, H., and Sommer, H. (1990). Genetic control of flower development by homeotic genes in *Antirrhinum majus*. *Science* 250, 931–936. doi: 10.1126/science.250.4983.931

Scotland, R. W., and Wortley, A. H. (2003). How many species of seed plants are there? *Taxon* 52, 101. doi: 10.2307/3647306

Shan, H., Zahn, L., Guindon, S., Wall, P. K., Kong, H., Ma, H., et al. (2009). Evolution of plant MADS box transcription factors: evidence for shifts in selection associated with early angiosperm diversification and concerted gene duplications. *Mol. Biol. Evol.* 26, 2229–2244. doi: 10.1093/molbev/msp129

Stebbins, G. L. (1951). Natural selection and the differentiation of angiosperm families. *Evolution* 5, 299. doi: 10.2307/2405676

Theissen, G., and Melzer, R. (2007). Molecular mechanisms underlying origin and diversification of the angiosperm flower. *Ann. Bot.* 100, 603–619. doi: 10.1093/aob/mcm143

Thompson, B. E., Bartling, L., Whipple, C., Hall, D. H., Sakai, H., Schmidt, R., et al. (2009). bearded-ear encodes a MADS box transcription factor critical for maize floral development. *Plant Cell* 21, 2578–2590. doi: 10.1105/tpc.109.067751

Toriba, T., and Hirano, H.-Y. (2014). The DROOPING LEAF and OsETTIN2 genes promote awn development in rice. *Plant J.* 77, 616–626. doi: 10.1111/tpj.12411

Wellmer, F., and Riechmann, J. L. (2010). Gene networks controlling the initiation of flower development. *Trends Genet.* 26, 519–527. doi: 10.1016/j.tig.2010.09.001

Whipple, C. J., Zanis, M. J., Kellogg, E. A., and Schmidt, R. J. (2007). Conservation of B class gene expression in the second whorl of a basal grass and outgroups links the origin of lodicules and petals. *Proc. Natl. Acad. Sci. U.S.A.* 104, 1081–1086. doi: 10.1073/pnas.0606434104

Yamaguchi, T., Nagasawa, N., Kawasaki, S., Matsuoka, M., Nagato, Y., and Hirano, H. (2004). The YABBY gene DROOPING LEAF regulates carpel specification and midrib development in *Oryza sativa*. *Plant Cell* 16, 500–509. doi: 10.1105/tpc.018044

Yockteng, R., Almeida, A. M. R., Morioka, K., Alvarez-Buylla, E. R., and Specht, C. D. (2013). Molecular evolution and patterns of duplication in the SEP/AGL6-like lineage of the Zingiberales: a proposed mechanism for floral diversification. *Mol. Biol. Evol.* 30, 2401–2422. doi: 10.1093/molbev/mst137

Yoshida, A., Suzaki, T., Tanaka, W., and Hirano, H.-Y. (2009). The homeotic gene long sterile lemma (G1) specifies sterile lemma identity in the rice spikelet. *Proc. Natl. Acad. Sci. U.S.A.* 106, 20103–20108. doi: 10.1073/pnas. 0907896106

Yoshida, H., and Nagato, Y. (2011). Flower development in rice. *J. Exp. Bot.* 62, 4719–4730. doi: 10.1093/jxb/err272

Yuan, Z., Gao, S., Xue, D.-W., Luo, D., Li, L.-T., Ding, S.-Y., et al. (2009). RETARDED PALEA1 controls palea development and floral zygomorphy in rice. *Plant Physiol.* 149, 235–244. doi: 10.1104/pp.108.128231

14

Comparative transcriptomic analysis of male and female flowers of monoecious *Quercus suber*

Margarida Rocheta[1†], Rómulo Sobral[2†], Joana Magalhães[2†], Maria I. Amorim[3], Teresa Ribeiro[1], Miguel Pinheiro[4], Conceição Egas[4], Leonor Morais-Cecílio[1* and Maria M. R. Costa[2]*

[1] *Departamento de Recursos Naturais Ambiente e Território, Instituto Superior de Agronomia, Universidade de Lisboa, Lisboa, Portugal*
[2] *Centre for Biodiversity, Functional & Integrative Genomics, Plant Functional Biology Centre, University of Minho, Braga, Portugal*
[3] *Departamento de Biologia, Faculdade de Ciências da Universidade do Porto, Porto, Portugal*
[4] *Biocant, Parque Tecnológico de Cantanhede, Cantanhede, Portugal*

Edited by:
Dazhong Dave Zhao, University of Wisconsin-Milwaukee, USA

Reviewed by:
Oliver Gailing, Michigan Technological University, USA
Wen-Wu Guo, Huazhong Agricultural University, China
Qingyi Yu, Texas A&M AgriLife Research, USA

*Correspondence:
*Leonor Morais-Cecílio, Departamento de Recursos Naturais Ambiente e Território, Instituto Superior de Agronomia, Universidade de Lisboa, Tapada da Ajuda, Lisboa, Portugal
e-mail: lmorais@isa.ulisboa.pt;
Maria M. R. Costa, Centre for Biodiversity, Functional & Integrative Genomics, Plant Functional Biology Centre, University of Minho, Campus de Gualtar, Braga 4710-057, Portugal
e-mail: manuela.costa@bio.uminho.pt*

[†] *These authors have contributed equally to this work.*

Monoecious species provide a comprehensive system to study the developmental programs underlying the establishment of female and male organs in unisexual flowers. However, molecular resources for most monoecious non-model species are limited, hampering our ability to study the molecular mechanisms involved in flower development of these species. The objective of this study was to identify differentially expressed genes during the development of male and female flowers of the monoecious species *Quercus suber*, an economically important Mediterranean tree. Total RNA was extracted from different developmental stages of *Q. suber* flowers. Non-normalized cDNA libraries of male and female flowers were generated using 454 pyrosequencing technology producing a total of 962,172 high-quality reads with an average length of 264 nucleotides. The assembly of the reads resulted in 14,488 contigs for female libraries and 10,438 contigs for male libraries. Comparative analysis of the transcriptomes revealed genes differentially expressed in early and late stages of development of female and male flowers, some of which have been shown to be involved in pollen development, in ovule formation and in flower development of other species with a monoecious, dioecious, or hermaphroditic sexual system. Moreover, we found differentially expressed genes that have not yet been characterized and others that have not been previously shown to be implicated in flower development. This transcriptomic analysis constitutes a major step toward the characterization of the molecular mechanisms involved in flower development in a monoecious tree with a potential contribution toward the knowledge of conserved developmental mechanisms in other species.

Keywords: flower development, monoecious, pyrosequencing, *Quercus suber*, RNA-seq, transcriptomics, cork oak, EST

INTRODUCTION

Quercus suber (L.) is one of the most important forest species in Portugal, being the dominant tree of the oak woodlands (Aronson et al., 2009). Due to its ecological and socio-economic significance, the cork oak forest is a unique resource. There is a growing interest in the management of woods for the production of acorns destined either for nursery production or for animal feed stocks. Therefore, the knowledge of the molecular mechanisms that control flower induction and fertilization is crucial to fully understand the reproductive success of this species.

Quercus suber is a monoecious tree species with a protandrous system and a long progamic phase (period between pollination and fertilization). Male flowers are organized in catkins that emerge in reproductive buds of the previous growth season or at the base of the branches of the current season. Each individual catkin contain 15–25 staminate flowers that are radially set around the catkin's axis (Natividade, 1950). The staminate flowers present a perianth with four to six tepals with an equal or double number of anthers that do not burst simultaneously

(Boavida et al., 1999). Female inflorescences arise in spikes, with three to five flowers, on the axil of the new leaves. Female flowers are included in a cupule and contain three carpels, with two ovules each (Boavida et al., 1999). Male flowering buds occur in early spring and sometimes also in autumn, whereas female flowers appear in spring and only get fully developed a few months later, if pollinated. During spike elongation, three to five styles emerge from the cupule and the stigma becomes receptive (Ducousso et al., 1993). At the time of pollination the ovary is still undifferentiated and the transmitting tissue extends only to the base of the styles. The wind driven pollen lays on the receptive stigmatic surface, germinates and the pollen tube grows throughout the transmitting tissue, until it reaches the base of the style. Usually, the pollen tube growth is arrested for 6 weeks, overlapping with ovule differentiation (Boavida et al., 1999; Kanazashi and Kanazashi, 2003). After fertilization, only one of the six ovules develops into a monospermic seed, which matures during autumn (Ducousso et al., 1993; Boavida et al., 1999).

Flower development is a complex and dynamic process that requires the tight coordination of gene expression and environmental cues (Fornara et al., 2010). During the past several years, a significant progress has been made in elucidating the genetic networks involved in flower organ specification in hermaphroditic model (reviewed in Wellmer et al., 2014) and non-model species (Wu et al., 2010; Yoo et al., 2010; Zahn et al., 2010; Logacheva et al., 2011; Varkonyi-Gasic et al., 2011; Zhang et al., 2012). Unisexual flower specification requires developmentally regulated processes that initiate male and female organ primordia in separate parts of the plant (Dellaporta and Calderon-Urrea, 1993). Studies focusing on mutant isolation revealed that several genes affect the key steps of sex determination in a variety of species. For example, in maize, unisexuality is controlled by *TASSELSEED2* that is expressed in the male structure (tassel) and is involved in pistil primordia abortion (DeLong et al., 1993). Also, in melon, a single nucleotide change in the *1-AMINOCYCLOPROPANE-1-CARBOXYLIC ACID SYNTHASE* gene is responsible for the specific inhibition of the male reproductive organs (Boualem et al., 2008). With the advent of next generation sequencing (NGS) technology, the previous limitation of mutant isolation in important model and non-model species was surpassed (Rowan et al., 2011). In another Cucurbitaceae, prior knowledge established a link between the de-regulation of the homeotic ABC model genes and sex determinacy (Kater et al., 2001). Using NGS technology, Guo et al. (2010) further helped to understand the molecular mechanisms underlying sex determinacy in cucumber by comparing the transcriptomes of the two types of flowers (gynoecious and hermaphroditic). In *Quercus* spp., many studies have been conducted focusing on the morphology of reproductive organs (Kaul, 1985), life cycle (Ducousso et al., 1993; Elena-Rossello et al., 1993), flowering process (Varela and Valdiviesso, 1996), and embryogenesis (Stairs, 1964). However, molecular information regarding these mechanisms is still scarce. Ueno et al. (2010) described the first large-scale study of bud transcriptomes of the two main European white oak species (*Q. petraea* and *Q. robur*). Ueno et al. (2013) used the same pyrosequencing technology to characterize the bud transcriptomes of endo- and ecodormant sessile oak (*Q. petraea*). Recently, the transcriptome of *Q. suber* has been reported using 21 normalized cDNA libraries derived from multiple *Q. suber* tissues and organs, developmental stages and physiological conditions (Pereira-Leal et al., 2014). This work included two normalized libraries of *Q. suber* (male and female) flowers that could serve as a tool to mine genes in each flower type. However, data concerning differentially expressed genes during different developmental stages of each type of flower were still missing.

In the present work, with the aim of capturing the diversity of transcripts differentially expressed in male and female *Q. suber* flowers, inflorescences in different developmental stages were separately collected and non-normalized cDNA libraries were generated and sequenced using the 454 GS-FLX Titanium technology. This study provides a unique set of databases, invaluable for gene discovery, which might reveal the regulatory networks of sex-specific flower development of a non-model monoecious tree species.

MATERIALS AND METHODS
PLANT MATERIAL

Six developmental stages of male and female cork oak flowers were collected from different trees in three different locations in Portugal (Lisbon, Porto, and Braga). The classification of the different phenological phases was based on visual observation, according to Varela and Valdiviesso (1996) (**Figure 1**). Samples were harvested between the end of March and the beginning of June and were frozen in liquid nitrogen immediately after collection.

RNA EXTRACTION AND cDNA PREPARATION

RNA was extracted from each sample using the RNAqueous® Kit (Ambion), following the manufacturer's instructions. The same amount of RNA was combined to create four specific RNA pools, two for female flowers (1F and 2F) and two for male flowers (1M and 2M), covering either early (1F or 1M) or late (2F or 2M) developmental stages. Pool 1F (**Figure 1A**) contained RNA from female buds enclosed by protective scales (Cf), female reddish buds with open scales (Df) and buds showing the elongation of the spike axe and the emergency of the first pair of flowers (Ef). The 2F pool (**Figure 1B**) included RNA

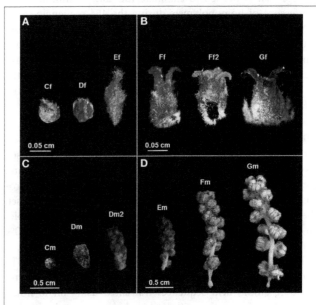

FIGURE 1 | *Quercus suber* **female and male flowers in different developmental stages used in RNA-seq. (A)** Early and **(B)** late stages of female flower development used in pools 1F and 2F, respectively. **(C)** Early and **(D)** late stages of male flower development used in pools 1M and 2M, respectively. (Cf) female bud enclosed by protective scales; (Df) female reddish bud with open scales; (Ef) elongation of the spike axe and the emergency of the first pair of flowers; (Ff) female flower showing distinct, erect, yellow stigmas with curved pinkish/brownish tips; (Ff2) flower with shining yellow and viscous pattern stigmas in clear divergent position; (Gf) female flower with closed stigmas that lost the receptivity, exhibiting a dark brown color. (Cm) catkin with red round shape due to the tight clustering of the flowers; (Dm) elongated cluster of male flowers; (Dm2) pendent catkin with some individualized flowers; (Em) male flowers with the anthers individualized; (Fm) flowers with individualized green/yellow anthers where pollen shedding begins; (Gm) catkin with male flowers in full anthesis.

from female flowers showing distinct, erect, yellow stigmas with curved pinkish/brownish tips (Ff), flowers with shining yellow and viscous pattern stigmas in clear divergent position (Ff2) and flowers with closed stigmas that lost the receptivity, exhibiting a dark brown color (Gf). The 1M pool (**Figure 1C**) comprised RNA from catkins with red round shape (Cm), elongated clustered male flowers (Dm), and pendent catkins with some individualized flowers (Dm2). The 2M pool (**Figure 1D**) included male flowers in which the anthers were becoming individualized (Em), flowers with individualized green/yellow anthers, prior to pollen shedding (Fm) and catkins with male flowers in full anthesis (pollen shedding occurring in half of the flowers with some anthers eventually empty) (Gm). RNA integrity was verified on an Agilent 2100 Bioanalyzer with the RNA 6000 Pico kit (Agilent Technologies) and the quantity assessed by fluorometry with the Quant-iT RiboGreen RNA kit (Invitrogen). A fraction of 2.0 μg of each pool of total RNA was used as starting material for cDNA synthesis using the MINT cDNA synthesis kit (Evrogen), where a strategy based on SMART double-stranded cDNA synthesis (Zhu et al., 2001) was applied. During the amplification of the poly RNA molecules, a known adapter sequence was introduced to both ends of the first strand of cDNA. The synthesis was also performed using a modified oligo-dT, containing a restriction site for *Bsg*I that is needed to eliminate the tails, to minimize the interference of homopolymers during the 454-sequencing run. cDNA was quantified by fluorescence and sequenced in a full plate of 454 GS FLX Titanium system, according to the standard manufacturer's instructions (Roche-454 Life Sciences) at Biocant (Cantanhede, Portugal). Sequence reads were deposited in the NCBI Sequence Read Archive (SRA) under the accession number SRP044882.

SEQUENCE PROCESSING ASSEMBLY AND ANNOTATION

Prior to the assembly of sequences, the raw reads were processed in order to remove sequences with less than 100 nucleotides and low-quality regions. The ribosomal, mitochondrial and chloroplast reads were also identified and removed from the data set. The reads were then assembled into contigs using 454 Newbler 2.6 (Roche) with the default parameters (40 bp overlap and 90% identity).

The translation frame of contigs was assessed through BLASTx searches against Swissprot (e-value = 1e–6), and the corresponding amino acid sequences translated using an in-house script. Next, the contigs without translation were submitted to FrameDP software (Gouzy et al., 2009) and the remaining contigs were analyzed with ESTScan (Lottaz et al., 2003). Transcripts resulting from these two last sequence identification steps (FrameDP and ESTScan) were searched using BLASTp against the non-redundant NBCI (National Center for Biotechnology Information) database in order to translate the putative proteins.

The deduced aminoacid sequences were annotated using InterProScan (Hunter et al., 2009) and each was given the Gene ontology terms (GOs) (Ashburner et al., 2000).

To identify the differential gene expression between samples, the contigs were clustered using the CD-Hit 454 (Niu et al., 2010) application (90% similarity) in order to eliminate redundant sequences and generate reference contigs. After this step,

the contigs that codify non-redundant proteins were used as reference to map the reads. The mapping process was made using 454 Newbler Mapping 2.6 (Roche). The mapping results were quantified to obtain the number of reads from different samples and a contingency table with contig names was created using the number of reads per reference contig per sample. The contingency table was normalized at a 95 percentile using the MyRNA (Langmead et al., 2010) statistical analysis package and the differential gene expression was evaluated using a linear regression model based on a Gaussian distribution and taking into account only contigs with a minimum of eight mapped reads and FDR < 0.05. Differentially expressed were clustered using the Self-organizing Trees algorithm (SOTA), euclidean distance (Dopazo and Carazo, 1997; Herrero et al., 2001) and the default settings of the MeV: MultiExperiment Viewer program (http://www.tm4.org/mev.html).

qRT-PCR ANALYSIS

cDNA was synthesized from the same RNA samples used for the 454 sequencing, according to the manufacturer's instructions. cDNA was amplified using SsoFast™ EvaGreen® Supermix (Bio-Rad), 250 nM of each gene-specific primer (listed in **Supplementary Table S1**) and 1 μL of cDNA (1:100 dilution). Quantitative real-time PCR (qRT-PCR) reactions were performed in triplicates on the CFX96 Touch™ Real-Time PCR Detection System (Bio-Rad). After an initial period of 3 min at 95°C, each of the 40 PCR cycles consisted of a denaturation step of 10 s at 95°C and an annealing/extension step of 10 s at the gene specific primer temperature. With each PCR reaction, a melting curve was obtained to check for amplification specificity and reaction contaminations, by heating the amplification products from 60°C to 95°C in 5 s intervals. Primer efficiency was analyzed with CFX Manager™ Software v3.1 (Bio-Rad), using the Livak calculation method for normalized expression (Livak and Schmittgen, 2001). Gene expression analysis was established based on three technical and biological replicates, and normalized with the reference gene *QsPP2AA3* (Marum et al., 2012).

RESULTS AND DISCUSSION

Due to the large number of reads attainable, the 454 DNA sequencing technology has great potential for discovering transcripts in non-model organisms. A prior study by Pereira-Leal et al. (2014) provided the first step toward the assembly of the monoecious tree *Q. suber* transcriptome using normalized libraries. In order to capture the diversity of transcripts differentially expressed during the development of female (F) and male (M) flowers, different developmental stages of flowers were collected covering either early (1F and 1M) or late (2F and 2M) developmental stages (**Figure 1**).

SEQUENCING AND ASSEMBLY OF *Q. SUBER* FLOWER TRANSCRIPTOME

Pyrosequencing resulted in 332,607 (1F), 312,282 (2F), 255,962 (1M), 270,871 (2M) raw reads for each library. After trimming, a total of 280,092 (1F), 252,024 (2F), 205,781 (1M), 224,275 (2M) high-quality reads were available with an average length of 264, 253, 269, and 270 bp, respectively (**Table 1**). Reads were

Table 1 | Sequencing and annotation statistics of *Quercus suber* flower libraries.

	1F	2F	1M	2M	1F_1M_2F_2M
Number of raw reads	332,607	312,282	255,962	270,871	
Number of reads after trimming	280,092	252,024	205,781	224,275	837,163
Average read length after trimming	264	253	269	270	263
Number of contigs	7565	6923	5267	5171	16,832
Average contig length	773	714	779	812	914
Range of contig length	[60.. 3489]	[52.. 3394]	[36.. 3384]	[15.. 3392]	[16.. 3848]
Number of translated contigs	7289	6600	5090	4981	16,152
Peptide sequences with BLASTx matches	5723	5154	4083	3997	11,956
Peptide sequences translated by FrameDP	2211	1907	1386	1398	6621
Peptide sequences translated by ESTScan	108	117	116	114	297
Total of amino acid sequences	8042	7178	5585	5509	18,874
Peptide sequences with BLASTP matches	1294	1086	809	825	3372
Amino acid sequence assigned InterPro terms	5940	5312	4251	4217	12,698
Amino acid sequence assigned to GO terms	4536	4056	3272	3269	9459

Individual libraries were generated from four specific RNA pools, two for female flowers (1F and 2F) and two for male flowers (1M and 2M), covering either early (1F and 1M) or late (2F and 2M) stages of flower development. The four individually EST projects were assembled into the 1F_1M_2F_2M library. The libraries were assembled using 454 Newbler 2.6 (Roche) and annotated in a three-step process using BLASTx search, FrameDP and ESTScan using default parameters.

assembled into 7565, 6923, 5267, and 5171 contigs in 1F, 2F, 1M, and 2M, respectively, with an average contig length of 714 to 812 bp (**Table 1**).

High-quality reads from the four individual EST libraries were assembled together into a single library (1F_2F_1M_2M), generating 837,163 high-quality reads with an average length of 263 bp that were assembled into 16,832 contigs with an average contig length of 914 bp (**Table 1**).

To annotate the *Q. suber* flower transcriptome, a three-step process (BLASTx search, FrameDP and ESTScan) was performed resulting in 16,152 (95.96%) translated contigs for the 1F_2F_1M_2M library (**Table 1**). GO terms were then assigned, indicating a total of 9459 aminoacid sequences (50%) with at least one GO term (**Table 1**). Based on the GO annotations, cell, metabolic process and binding were the most abundant GO slims within the cellular component, molecular function, and biological process categories, respectively (**Figure 2**). Metabolic process (41.34%) and cellular process (30.40%) were the most highly represented groups within the biological process category, indicating that the floral tissues were undergoing extensive physiological activity in accordance with what was observed in *Arabidopsis thaliana* reproductive tissues (Hennig et al., 2004).

VALIDATION OF THE *Q. SUBER* FLOWER TRANSCRIPTOMES

The *in silico* analysis of the transcriptomes allowed the identification of differences between the distinct developmental stages of male and female flowers. In order to validate the differences observed between male and female flower libraries, several contigs were identified by homology with functional important genes known to be involved in carpel or stamen development in model organisms. Homologs for *ABORTED MICROSPORES* (*AMS*, Xu et al., 2010), *LESS ADHERENT POLLEN3* (*LAP3*, Dobritsa et al., 2009), *LESS ADHESIVE POLLEN5* (*LAP5*, Dobritsa et al., 2010), and *LESS ADHESIVE POLLEN6* (*LAP6*, Dobritsa et al., 2010) were chosen as the male candidate genes

due to their involvement in pollen development. Homologs for the female candidate genes, *At4g27290* (Pagnussat et al., 2005), *CYTOCHROME P450 78A9* (*CYP78A9*, Ito and Meyerowitz, 2000), *POLYGALACTURONASE-1* (*PG1*, Tacken et al., 2010), and *STIGMA SPECIFIC1* (*STIG1*, Verhoeven et al., 2005) were selected based on their relevance in pollen recognition, stigma and transmitting tract development.

As expected, the *Q. suber* homologous genes presented differential expression ratios between male and female libraries (**Table 2**), and thus were considerate good candidates for qRT-PCR analysis. The qRT-PCR results confirmed that genes involved in pollen exine formation (*LAP3*, *LAP5*, and *LAP6*) and in the tapetum cell development (*AMS*) were more expressed in the early stages of male flower development (**Figure 3**), whereas genes involved in stigma-specific recognition (*STIG1*), in the recognition of pollen (*At4g27290*) and in fruit growth and development (*CYP78A9*, *PG1*) were more expressed in the female flowers. These results were in close agreement with the RNAseq data (**Table 2**) suggesting the reliability of the transcriptomic profiling data.

DIFFERENTIAL GENE EXPRESSION BETWEEN *Q. SUBER* FLOWER-TYPE LIBRARIES

In order to identify exclusive transcripts of early and late developmental stages of female and male flower development, the assembly of the four non-normalized libraries was analyzed. The analysis showed that there were 230 unique contigs for the early (1F) and 214 contigs unique for the late (2F) stages of female flower development (**Figure 4A**). The 1F unique contigs might correspond to genes controlling early flower development, whereas the 2F unique contigs might be associated with stigma maturation, ovule development and fertilization. Accordingly, there were 198 contigs unique in the early stages of male flower development (1M), most probably involved in early stages of anther development and 327 contigs specific for the late stages

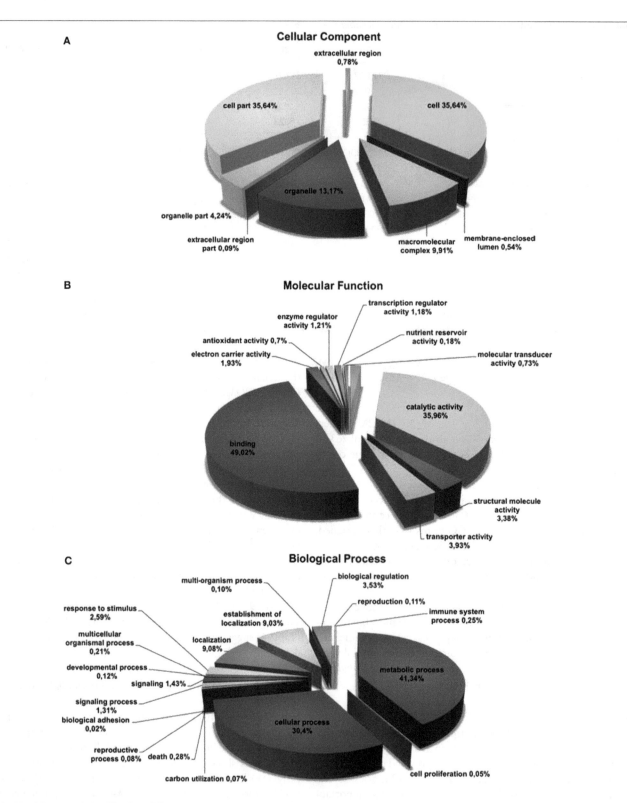

FIGURE 2 | Functional classification of *Quercus suber* unigenes. Four EST projects were generated from four-specific RNA pools, two for female flowers (1F and 2F) and two for male flowers (1M and 2M), covering either early (1F and 1M) or the late (2F and 2M) developmental stages. The four individually EST projects were assembled into the 1F_1M_2F_2M library and the deduced aminoacid sequences of this library were annotated using InterProScan. The Gene Ontology terms (GOs) for each translated amino acid sequence were used to classify the transcript products within the category of **(A)** cellular component, **(B)** molecular function, and **(C)** biological process sub-ontologies.

Table 2 | Candidate genes that were selected to validate the transcriptional levels determined by RNAseq results.

Candidate genes	Gene accession	1F	2F	1M	2M
• At4g27290	QSP078589.0	29	14	0	1
• CYTOCHROME P450 78A9	QSP091316.0	70	70	0	0
• POLYGALACTURO-NASE1	QS094531.0	19	337	0	0
• STIGMA SPECIFIC1	QS121989.0	5	34	0	0
* ABORTED MICROSPORES	QS049646.0	0	0	38	4
* LESS ADHERENT POLLEN3	QS049611.0	1	2	170	46
* LESS ADHESIVE POLLEN5	QS003695.0	0	0	566	6
* LESS ADHERENT POLLEN6	QS039918.0	0	0	924	49
PROTEIN PHOSPHATASE 2A SUBUNIT A3	QS092015.0	55	71	57	65

The number of ESTs represents their distribution in the female and male Quercus suber combined flower library, covering either early (1F and 1M) or late (2F and 2M) stages of flower development. The selected genes were identified by homology with functional important genes known to be involved in carpel or stamen development.
• Female candidate genes.
* Male candidate genes.
PROTEIN PHOSPHATASE 2A SUBUNIT A3 was chosen as the reference gene. Gene accessions according to CorkOak database (www.corkoakdb.org).

(2M) that could be indicative of genes controlling pollen development and maturation (**Figure 4A**). A normalization cut-off of eight reads at the 95th percentile was applied, resulting in 3760 differentially expressed genes (19.9%) for the 1F_2F_1M_2M transcriptome. Differentially expressed genes were then clustered into different groups according to their expression profile similarity (**Figure 4**). Groups of genes that were either unique (**Figure 4B**) or significantly more expressed in the male samples (**Figures 4C,D**) were identified. At least 430 differentially expressed genes were predominantly expressed in the last stages of male flower development (**Figures 4E,F**), whereas 239 genes were absent from this stage and present in all the other libraries (**Figure 4G**). We also found genes that were more expressed in the early stages of both male and female flower development (**Figure 4H**). A group of genes (115) was more expressed in both female libraries (**Figure 4I**), whereas 217 genes appear to be preferentially expressed in late stages of female flower development (**Figure 4J**).

Out of the 3760 differentially expressed genes, a GO term was assigned to 1797 female and to 745 male transcripts. No significant differences were found between male and female GO categories apart from the molecular and cellular functions. In the former, 16% of the female GO terms were assigned to protein binding in contrast with the male GO terms (7%), whereas in the latter, 8% of the female GO terms were assigned to the nucleus and just 1% of the male were appointed to the nucleus (**Figure S1**).

POTENTIAL ALLERGEN GENES PRESENT IN *Q. SUBER* LIBRARIES

During spring, Fagales tree species produce and release large amounts of pollen. In Southern Europe, pollen from these plants and other anemophilous trees, like *Platanus acerifolia* and *Olea europaea*, has been proved to elicit allergic diseases, such as pollinosis rhinitis/rhino conjunctivitis (D'Amato et al., 2007; Esteve et al., 2012). The official site for the systematic allergen nomenclature (http://www.allergen.org), that was approved by the World Health Organization and International Union of Immunological Societies (WHO/IUIS) Allergen Nomenclature Sub-committee, lists 263 allergenic proteins to the taxonomic group Plantae Magnoliopsida. Among these 263 allergens, 34 were associated with the Fagales order and only one (Quea1) was related to the genus *Quercus*.

In order to identify transcripts encoding potential allergens in cork oak, blast searches were carried on the 1F_1M_2F_2M transcriptome against the proteins reported as allergenic and included in the WHO–IUIS list. This analysis revealed several potential orthologs for genes coding for potential allergens in *Q. alba*, *Betula pendula*, *Corylus avellana*, *O. europaea*, *Hevea brasiliensis*, and *P. acerifolia* (**Table 3**). Of major interest was the identification of a potential ortholog of Quea1, which is the major allergen of *Quercus alba* (Wallner et al., 2009). As expected, almost all the potential orthologs of allergen genes were highly expressed in the male libraries (**Table 3**).

Q. SUBER MOST DIFFERENTIALLY EXPRESSED GENES BETWEEN FEMALE AND MALE TISSUES

The ten genes most differentially expressed in both male and female tissues were identified by establishing a ratio between male and female EST counts (**Table 4**). Concerning the differentially expressed genes more represented in female flowers, we identified a homolog for *POLYGALACTURONASE-1* that is comparatively 356 times more expressed in female tissues. Interestingly, several studies report the involvement of polygalacturonases associated genes to both carpel (Ogawa et al., 2009) and pollen development (Allen and Lonsdale, 1993; Tebbutt et al., 1994; Rhee et al., 2003). *QsENDO-BETA-1,3-1,4 GLUCANASE*, a member of the glycoside hydrolase family, is 199 times more expressed in female samples. In *Populus trichocarpa*, a member of this family, *PtrCel9A6*, is tightly involved in sexual determinism (Yu et al., 2013). Overexpression of *PtrCel9A6* in *A. thaliana* resulted in male sterility due to defects in anther dehiscence (Yu et al., 2013). It is possible that the *QsENDO-BETA-1,3-1,4 GLUCANASE* might have a similar function by inhibiting the development of male structures in female flowers.

A *CYTOCHOME P450* transcript (*QsCYTOCHROME P450 78A3*) was also highly represented in the female samples with a possible role in carpel gametophyte and sporophyte development as it was shown for homologous genes in other species (Ito and Meyerowitz, 2000; Chakrabarti et al., 2013). Ito and Meyerowitz (2000) identified *AtCYP450 78A9*, a gene that when overexpressed in *A. thaliana* results in altered fruit and seed,

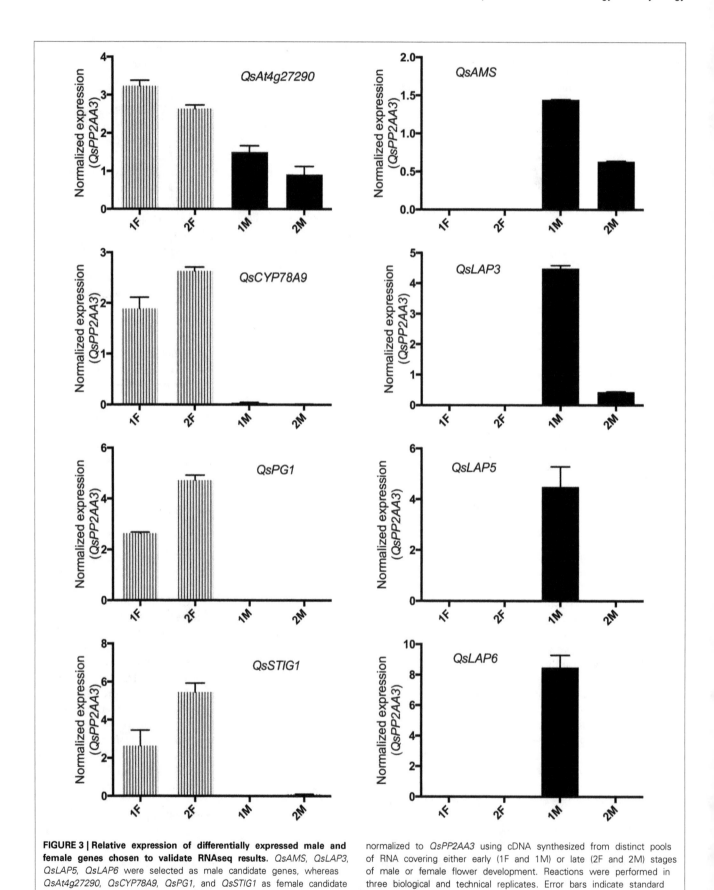

FIGURE 3 | Relative expression of differentially expressed male and female genes chosen to validate RNAseq results. *QsAMS, QsLAP3, QsLAP5, QsLAP6* were selected as male candidate genes, whereas *QsAt4g27290, QsCYP78A9, QsPG1,* and *QsSTIG1* as female candidate genes. Transcript abundance was determined using qPCR, and normalized to *QsPP2AA3* using cDNA synthesized from distinct pools of RNA covering either early (1F and 1M) or late (2F and 2M) stages of male or female flower development. Reactions were performed in three biological and technical replicates. Error bars indicate standard deviation (SD).

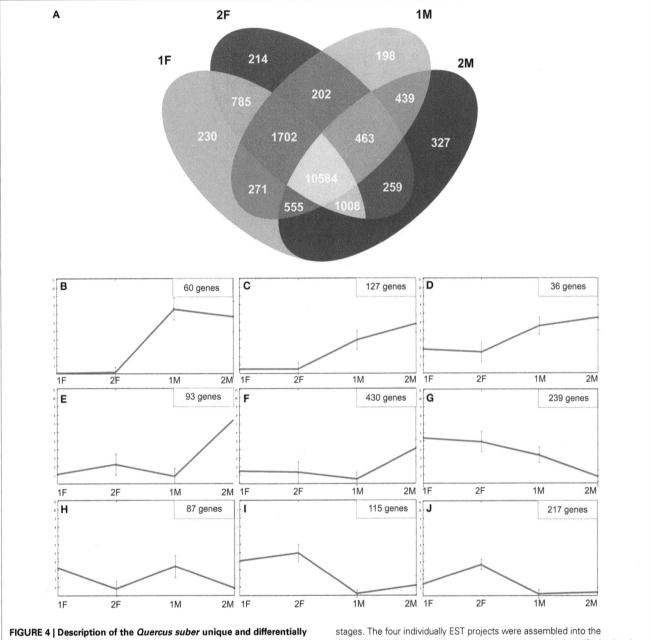

FIGURE 4 | Description of the *Quercus suber* unique and differentially expressed genes. (A) Venn diagram indicating the number of exclusive and shared transcripts of early and late developmental stages of *Quercus suber* flower. Four EST projects were generated from four-specific RNA pools, two for female flowers (1F and 2F) and two for male flowers (1M and 2M), covering either early (1F and 1M) or the late (2F and 2M) developmental stages. The four individually EST projects were assembled into the 1F_1M_2F_2M library and the exclusive transcripts were identified using the Venny application (Oliveros, 2007). **(B–J)** Differentially expressed genes were clustered using the Self-organizing Trees algorithm (SOTA), euclidean distance (Dopazo and Carazo, 1997; Herrero et al., 2001) and the default settings of the MeV, MultiExperiment Viewer program (http://www.tm4.org/mev.html).

Another member of this family is *SlKLUH*, which controls not only plant architecture but also fruit mass and ripening in tomato (Chakrabarti et al., 2013). We also identified a homolog for *1-AMINOCYCLOPROPANE-1-CARBOXYLATE OXIDASE 66*, that belongs to a family of genes that has been associated to ethylene biosynthesis (Barry and Giovannoni, 2007). Several *RECEPTOR-LIKE PROTEIN KINASES (RLPK)* have been linked to key aspects of plant development: the brassinosteroids signaling pathway (Schumacher and Chory, 2000), meristem maintenance

(Clark, 2001), or pollen-pistil interaction (McCubbin and Kao, 2000). Thus, a 65 times more expressed *QsRLPK* transcript in female flowers might have an important role in carpel development.

Within the group of genes with the highest differential expression in male flowers, there were three unknown genes without a significant BLAST hit. These three genes may be specific to *Q. suber* and pivotal to anther differentiation and development in this species. Two highly represented *QsCHALCONE SYNTHASE*

Table 3 | *Quercus suber* potential allergens in flower transcriptome.

Gene designation	Gene accession	ESTs 1F+2F	ESTs 1M+2M	Allergen	Closest species homolog	InterProScan description
QsQUEA1	QS091157.0	11	367	Quea1	*Quercus alba*	Pollen allergen Bet v 1
QsBETV2	QS034447.0	2	22	Betv2	*Betula pendula*	Profilin
QsBETV3	QSP068142.0	4	26	Betv3	*Betula pendula*	Calcium-binding allergen Bet v 3
QsBETV6	QSP015348.0	40	100	Betv6	*Betula pendula*	Isoflavone reductase homolog
QsCyP	QS017405.0	274	183	Betv7	*Betula pendula*	Cyclophilin
QsBETV4	QS115544.0	0	65	Betv4	*Betula pendula*	Polcalcin Bet v 4
QsBIP	QS126407.0	380	703	Cora10	*Corylus avellana*	Luminal binding protein
QsOLE1	QS069793.0	1	89	Olee1	*Olea europae*	Ole e1-like protein
QsOLE9	QS095617.0	9	248	Olee9	*Olea europae*	β-1.3 Glucanase
QsPME	QS101292.0	1	72	Olee11	*Olea europea*	Pectin methyl esterase
QsSOD_CU_ZN	QS057690.0	115	96	Olee5	*Olea europea*	Superoxide dismutase. copper/zinc binding
QsHEV1	QS150670.0	12	11	Hevb6	*Hevea brasiliensis*	Pro-hevein
QsPLAA2	QS060812.0	5	311	Plaa2	*Platanus acerifolia*	Exopolygalacturonase
QsnsLTP	QS106891.0	1097	3077	Plaa3	*Platanus acerifolia*	Non-specific lipid-transfer protein

The number of ESTs represents their distribution in the female (1F+2F) and male (1M+2M) Q. suber combined flower library. Selected genes were identified by homology with functional important allergen genes included in the WHO-IUIS allergen list. The deduced aminoacid sequences of these genes were annotated using InterProScan. Gene accessions according to CorkOak database (www.corkoakdb.org).

A genes were identified as male unique transcripts. A suppressor mutant of the *CHALCONE SYNTHASE A* homolog gene in *Petunia* generates viable pollen, however, pollen germination and tube growth are severely affected (Taylor and Jorgensen, 1992), which indicates that this gene is essential for proper anther development. *CHALCONE SYNTHASE A* associated genes are also known to be involved in the metabolic pathway that leads to the production of flavonoids and anthocyanin pigments in several species (Winkel-Shirley, 2002). Thus, it is possible that the coloration on the anthers might be due to the action of the highly expressed *QsCHALCONE SYNTHASE A* genes. A *Qs4-COURAMATE–CoA LIGASE-LIKE 1* gene that is related to the *ACOS* genes was also identified. These genes have been associated with proper pollen development (De Azevedo Souza et al., 2009). The *DEFECTIVE IN ANTHER DEHISCENCE1* (*DAD1*) is a lipase-like gene involved in pollen development (Ishiguro et al., 2001). In *A. thaliana dad1* shows defects in anther dehiscence, pollen maturation, and flower opening (Ishiguro et al., 2001). *QsLIPASE-LIKE* might also have a similar function.

To investigate whether these genes were flower specific, available root, leaf, bud and fruit libraries (Pereira-Leal et al., 2014) were used. Interestingly, all the male-associated genes are exclusive to the male flower except for *QsLIPASE-like*, which is expressed in almost all the other tissues with the exception of the female flower and leaves (**Table 4**). The majority of the female-associated genes analyzed are not flower specific. Interestingly, there is one gene that is present only in the buds, flowers and fruits (*QsNON-SPECIFIC LIPID-TRANSFER PROTEIN*), suggesting a putative role in the female reproductive determinism. At least two genes are expressed in all the organs (*QsPOLYGALACTURONASE-1* and *QsRECEPTOR-LIKE PROTEIN KINASE*) except for the male flowers. Considering that these genes are expressed in all *Q. suber* tissues analyzed

and absent from the male flowers might be indicative that these flowers go through a very distinctive developmental programme, or that the mentioned genes expression could be detrimental to proper male flower development. It will be very interesting to perform functional studies to analyse the involvement of the aforementioned genes in plant reproduction or flower development in *Q. suber* and other flowering species.

TRANSCRIPTION FACTORS DIFFERENTIALLY EXPRESSED IN FEMALE AND MALE FLOWERS

Differential expression of transcription factors (TF) has a pivotal role in the control of mechanisms that direct organ development (Latchman, 1997). Based on the analysis of the different TF groups, a group of biologically interesting genes that are sex-specific or differentially expressed in each library was identified (**Table 5**).

Zinc-finger TF family

The zinc-finger family of genes is an example of diversification in the Plant Kingdom and consists of a large number of proteins that are further classified into distinct subfamilies (Takatsuji, 1999). Among these, the C2H2-type and B-box zinc finger proteins constitute one of the largest families of transcriptional regulators in plants (Ciftci-Yilmaz and Mittler, 2008). Proteins containing zinc finger domains which play important roles in eukaryotic cells, regulating different signal transduction pathways and controlling processes such as development (Colasanti et al., 1998), homeostasis (Devaiah et al., 2007) and abiotic stress responses (Rizhsky et al., 2004; Sakamoto et al., 2004; Davletova et al., 2005). Some floral regulators contain a zinc-finger domain such as *CONSTANS* (*CO*), which has been linked to floral induction in several species by integrating the circadian clock and light signals (Putterill et al., 1995;

Table 4 | *Quercus suber* most differentially expressed genes in the flower transcriptome.

	Gene designation	Gene accession	Organ	ESTs 1F+2F	ESTs 1M+2M	InterProScan description
More expressed in Female Flowers	*QsPOLYGALACTURONASE-1*	QS094531.0	R, L, B, F, Fr	356	1	BURP
	QsENDO-BETA-1,3-1,4 GLUCANASE	QS112778.0	R, L, F, Fr	397	2	Glycoside hydrolase, family 17
	QsNON-SPECIFIC LIPID-TRANSFER PROTEIN	QS158755.0	B, F, Fr	162	0	Plant lipid transfer protein/Par allergen
	QsCYTOCHROME P450 78A3	QS073368.0	R, F, Fr	140	0	Cytochrome P450
	QsLEUCINE-RICH REPEAT EXTENSIN-LIKE PROTEIN3	QS109035.0	R, F, Fr	273	2	Unintegrated
	QsENDOCHITINASE PR4	QS078296.0	R, B, F, Fr	2553	22	Glycoside hydrolase, family 19, catalytic
	QsISOFLAVONE REDUCTASE HOMOLOG P3	QS124429.0	R, L, F, Fr	86	1	NmrA-like
	Qs1-AMINOCYCLOPROPANE-1-CARBOXYLATE OX. HOMOLOG 66	QS023535.0	L, F, Fr	78	1	Oxoglutarate/iron-dependent oxygenase
	QsACID PHOSPHATASE 1	QS083994.0	L, B, F, Fr	203	3	Acid phosphatase (Class B)
	QsRECEPTOR-LIKE PROTEIN KINASE	QS006331.0	R, L, B, F, Fr	65	1	Protein kinase, core
More expressed in Male Flowers	*QsUNKNOWN1*	QS082953.0	F	0	1376	PLT protein/seed storage/trypsin-α-amylase inhibitor
	QsKDEL-TAILED CYSTEINE ENDOPEPTIDASE	QS035274.0	F	1	1122	Peptidase, cysteine peptidase active site
	QsCHALCONE SYNTHASE A1	QS039918.0	F	0	973	Chalcone/stilbene synthase, N-terminal
	QsUNKNOWN2	QS047555.0	F	0	759	Bifunctional inhibitor/PLT protein/seed storage
	Qs4-COURAMATE–CoA LIGASE-LIKE 1	QS024768.0	F	0	712	AMP-dependent synthetase/ligase
	QsBETA-1,3-GALACTOSYLTRANSFERASE 8	QS087345.0	F	0	580	Glycosyl transferase, family 31
	QsCHALCONE SYNTHASE A2	QS003695.0	F	0	572	Chalcone/stilbene synthase, N-terminal
	QsANTHRANILATE N-BENZOYLTRANSFERASE PROTEIN 2	QS009481.0	F	0	533	Transferase
	QsUNKNOWN3	(?)	F	0	522	Unintegrated
	QsLIPASE-LIKE	QS011358.0	R, B, F, Fr	0	441	Lipase, GDXG, active site

The 10 genes most differentially expressed in male or female tissues were identified by establishing a ratio between male and female EST count in female (1F+2F) and male (1M+2M) Q. suber combined flower library. The deduced aminoacid sequences of these genes were annotated using InterProScan. Gene accessions according to the CorkOak database [www.corkoakdb.org; (?) not present]. Transcript distribution of the most differentially expressed genes in the Q. suber organs was obtained by blasting the genes sequences against the organ libraries made available by Pereira-Leal et al. (2014). R, root; L, leaf; B, bud; F, flower; Fr, Fruit.

Böhlenius et al., 2006). A clear *CO* homolog was not identified, as expected for the type of biological sample (flowers) used in the RNAseq. However, four *CO-like* transcripts differentially expressed in female flowers as compared to male flowers were identified. Of these, the *QsCONSTANS-LIKE 9* (*QsCOL9*) homolog was five times more expressed in female flowers. In *A. thaliana*, *COL9* delays flowering by reducing expression of *CO* and *FLOWERING LOCUS T* in leaves (Cheng and Wang, 2005). Its high level of expression in female tissues, particularly in early stages of the reproductive program, could suggest a novel function yet undisclosed. A homolog for the *A. thaliana* zinc-finger protein *TRANSPARENT TESTA 1* (*TT1*) was also seven times more expressed in female than in male *Q. suber* flowers. In melon, *CmWIP1* (a homolog of *TT1*) has a masculinizing effect by indirect repression of the ethylene driven *CmACS-7* gene (Boualem et al., 2008). *CmWIP1* needs to be epigenetically silenced to generate a fully functional female flower (Martin et al., 2009). The expression of *QsTT1* in the female flowers might point out the differences between developmental programs that give rise to sexual dimorphism in monoecious and dioecious/hermaphrodite species.

Basic helix-loop-helix (bHLH) TF family

The bHLH family encloses one of the largest groups of plants TF (Heim et al., 2003). These TF are involved in, among others, wound and stress responses (De Pater et al., 1997; Smolen et al., 2002; Chinnusamy et al., 2003; Kiribuchi et al., 2004), hormonal regulation (Abe et al., 1997; Friedrichsen et al., 2002) stigma and anther development, and fruit development and differentiation (Rajani and Sundaresan, 2001; Liljegren et al., 2004; Szécsi et al., 2006; Gremski et al., 2007). From the differentially expressed bHLH genes in *Q. suber* floral libraries, three were up-regulated in male flowers. One was the homolog of *ABORTED MICROSPORES*, a gene essential to the development of pollen (Sorensen et al., 2003; Xu et al., 2010). The other two *QsbHLH* are homologs to genes associated with iron deficiency (Wang et al., 2013). In female flowers, nine *Q. suber* transcripts were significantly more expressed, with one transcript being exclusive to the female samples (*QsBR ENHANCED EXPRESSION 1*). In *A. thaliana*, *BEE1* is involved in the brassinosteroids signaling and associated with the development of the reproductive tract (Crawford and Yanofsky, 2011). Homologs for *GLABRA3*, *MYC2*, *INDUCER OF CBF*

Table 5 | Transcription factors differentially expressed in male and female flowers of *Quercus suber*.

Gene designation	Gene accession	Organ	ESTs 1F+2F	ESTs 1M+2M	InterProScan description
QsCONSTANS LIKE 9	QS124225.0	F,L	34	7	Zinc finger, B-box
QsCONSTANS LIKE 4	QS050989.0	R, L, B, F, Fr	270	68	Zinc finger, B-box
QsCONSTANS LIKE 5	QS116713.0	R, L, B, F, Fr	82	35	Zinc finger, B-box
QsTRANSPARENT TESTA 1	QS064811.0	R, L, B, F, Fr	38	4	Zinc finger, C2H2-type
QsABORTED MICROSPORES	QS049646.0	F	0	42	bHLH dimerization region
QsBRASSINOSTEROID ENHANCED EXPRESSION1	QS048556.0	F	39	0	bHLH dimerization region
QsINDUCER OF CBF EXPRESSION1	QS129356.0	R, L, B, F, Fr	38	4	bHLH dimerization region
QsGLABRA3	QS150704.0	B, F Fr	31	5	bHLH dimerization region
QsMYC2	QS073124.0	R, L, B, F, Fr	114	26	bHLH dimerization region
QsIAA-LEUCINE RESISTANT3	QS154009.0	R, L, B, F, Fr	83	39	bHLH dimerization region
QsPERIANTHIA	QS095157.0	F	39	15	Basic-leucine zipper (bZIP)
QsCUC1	QS061789.0	R, F, Fr	33	6	No apical meristem (NAM) protein
QsCUC2	QS009784.0	R, F, Fr	33	7	No apical meristem (NAM) protein
QsIAA27	QS075617.0	R, L, B, F, Fr	110	17	AUX/IAA protein
QsIAA9	QS117343.0	R, L, B, F, Fr	160	97	AUX/IAA protein
QsPIN1	QS117199.0	R, B, F, Fr	20	1	Auxin efflux carrier
QsEIN3	QS119163.0	R, F	18	1	Ethylene insensitive 3
QsMYB33	QS020061.0	F	0	13	Homeodomain-like
QsSUPRESSOR OF CONSTANS1	QS149164.0	R, L, B, F Fr	25	2	Transcription factor, MADS-box
QsAP1	QS003005.0	F Fr	68	8	Transcription factor, MADS-box
QsFRUITFULL	QS029922.0	F Fr	53	7	Transcription factor, MADS-box
QsSHORT VEGETATIVE PHASE 1	QS116365.0	B, F Fr	36	6	Transcription factor, MADS-box
QsSHORT VEGETATIVE PHASE 2	QS055926.0	B, F Fr	9	23	Transcription factor, MADS-box

Based on the analysis of the different transcription factor groups, a number of biological interesting genes were identified and their expression level was determined according to the number of ESTs in female (1F+2F) and male (1M+2M) Q. suber combined flower library Gene accessions according to CorkOak database (www. corkoakdb.org). Transcript distribution of the most differentially expressed genes in the Q. suber organs was obtained by blasting the genes sequences against the organ libraries made available by Pereira-Leal et al. (2014). R, root; L, leaf; B, bud; F, flower; Fr, Fruit.

EXPRESSION1, and *IAA LEUCINE RESISTANT3* were also up-regulated in female flowers. These genes are involved in hormonal regulation, cold acclimation, cell fate and double fertilization (Bernhardt et al., 2003; Chinnusamy et al., 2003; Yadav et al., 2005; Rampey et al., 2006; Dombrecht et al., 2007). However, there were three bHLH transcripts differentially expressed whose function is yet to be characterized in other species, making them good candidates at least to be involved in carpel development.

Basic Leucine Zipper (bZIP) TF Family

The bZIP TFs regulate diverse biological processes in plants including flower development (Jakoby et al., 2002). Eight *Q. suber* bZIP associated genes were differentially expressed in female samples. Among them is the homolog of *VIP1*, an *A. thaliana* bZIP TF that regulates pathogen responses and rehydration responses (Tzfira et al., 2001; Tsugama et al., 2012). Another bZIP TF up-regulated in female samples is the homolog of *PERIANTHIA* (*PAN*), a gene involved in flower development in *A. thaliana* by altering floral organ number and initiation pattern (Running and Meyerowitz, 1996; Wynn et al., 2014). *PAN* is also involved in the activation of the C-class MADS box protein *AGAMOUS* (*AG*), a gene essential for carpel development (Das et al., 2009; Maier et al., 2009).

CUC/NAM TF Family

Data also showed several *CUP-SHAPED COTYLEDON/NO APICAL MERISTEM* (*CUC/NAM*) genes highly represented in female flowers and the majority of them is differentially expressed. The *CUC/NAM* family encloses genes that control boundary formation and lateral organ separation, which are critical for proper leaf and floral patterning (Aida et al., 1997; Vroemen et al., 2003). A *CUC/NAM* gene in *Medicago trunculata* is needed for proper regulation of floral organ identity (Cheng et al., 2012). Also, in *A. thaliana*, *Petunia* and rice, mutants for *CUC/NAM* genes lead to the fusion of the cotyledons and some floral organs, as well as severe defects of the primary apical meristem (Souer et al., 1996; Aida et al., 1997; Mao et al., 2007). Out of *QsCUC/NAM* genes that are differentially expressed, homologs for *CUC1* and *CUC2* genes were up-regulated in female flowers.

MADS-Box TF Family

The MADS family of TF include a group of genes that play prominent roles in plant development (Smaczniak et al., 2012). Particularly, MADS TFs were found to be crucial for proper flower development in several species across the angiosperm lineage (reviewed in Theissen and Melzer, 2007). According to the canonical ABC model, which explains how homeotic genes control flower identity, stamens are formed by the activity of the

B-Class and a C-class gene, whereas the same C-class is responsible for carpel development (Coen and Meyerowitz, 1991). As expected, B-class genes were differentially expressed in the male flowers (*QsAPETALA3* and *QsPISTILLATA*), and a similar level of expression of *QsAGAMOUS* (C-class gene) in both male and female libraries. The E-function genes (*SEPALLATA1-4*) that act as cadastral genes for proper organ development and identity (Pelaz et al., 2000) were also identified in both libraries but there was no sex differential expression. Several other homologs for MADS genes (Qs*APETALA1*, Qs*FRUITFUL*, Qs*FLOWERING LOCUS*, or Qs*SUPRESSOR OF CONSTANS1*) that influence flowering in model and non-model species were also identified both in female and male libraries. Two transcripts similar to the *SHORT VEGETATIVE PHASE* gene were identified. It was interesting to detect a *QsSVP* gene differentially expressed in the female libraries and another in the male libraries. Genes of the *SVP* lineage in peach (*dormancy-associated genes, DAM*) are involved in growth cessation, bud set and break (Li et al., 2009).

HORMONE RELATED GENES DIFFERENTIALLY EXPRESSED IN FEMALE AND MALE FLOWERS

Flower development is strongly affected by hormonal regulation (reviewed in Chandler, 2011). Auxin is tightly linked to the initiation of floral organ primordia and the disruption of auxin biosynthesis, polar auxin transport or auxin signaling leads to failure of flower formation (reviewed in Aloni et al., 2006). In agreement, the floral meristem identity gene *LEAFY* was recently shown to act through the regulation of the auxin response pathway (Li et al., 2013). Aux/IAA proteins, Auxin Efflux Carriers, and AUXIN RESPONSIVE FACTORS (ARF) are core components of the auxin-signaling cascade (Guilfoyle and Hagen, 2007). Several genes associated with the auxin regulatory network are highly represented and several are differentially expressed in female flowers. Particularly, homologs for the *A. thaliana IAA27* and *IAA9* genes were differentially expressed. The latter is a gene involved in fruit development and leaf embryogenesis in tomato (Wang et al., 2005) while mutants for *IAA27* showed altered fruit and flower development. A homolog of the *ARF4* gene that in *A. thaliana*, together with *ARF3*, control perianth organ number and spacing, as well as organ borders (Sessions et al., 1997) was identified in female flowers. Mutants for these genes showed defects in the stamens and gynoecium, as well as in the perianth organs, indicating an involvement in regional identity determination (Sessions et al., 1997). An auxin efflux carrier *QsPIN1* was also found to be up-regulated 20 fold in female flowers. Loss of *PIN1* function severely affects organ initiation; *pin1* mutants are characterized by an inflorescence meristem that does not initiate any flowers, resulting in the formation of a naked inflorescence stem (Okada et al., 1991). The abundance of auxin related machinery in female tissues leads to the possibility that female tissue determination might be under strong control of this hormone. Another hormone strongly correlated with sex determination is ethylene (Byers et al., 1972; Rudich et al., 1972). In cucumber, ethylene signaling is important in the inhibition of stamen development (Yamasaki et al., 2001). Interestingly, all the differentially expressed genes containing an ETHYLENE RESPONSIVE

FACTOR (ERF) domain were detected in the female flowers, in agreement with the aforementioned role of ethylene in the feminization of the flower meristem. Another gene linked to ethylene signaling pathway is *EIN3*, a nuclear TF that initiates downstream transcriptional cascades for ethylene responses (Potuschak et al., 2003; Yanagisawa et al., 2003). *QsEIN3* was unique to the female samples. Interestingly in *Arabidopsis*, activated ethylene signaling reduces bioactive GA levels, thus enhancing the accumulation of DELLAs (repressors of the gibberellins pathway) and this most likely happens downstream of the transcriptional regulator *EIN3* (Achard et al., 2007). This is very interesting because the gibberellin hormone is thought to be essential for the developmental of a fully functional stamen in several species like *A. thaliana*, *Oryza sativa* or *Cucurbita maxima* (Pimenta Lange and Lange, 2006). This goes in agreement with our RNAseq results, in which several gibberellin related genes are found exclusively in the male flowers equally expressed in early and late stages of the developmental program, indicating a role in floral primordia and anther differentiation. Also, a *GAMYB* firmly involved in anther development, *QsMYB33*, was only expressed in male samples (Millar and Gubler, 2005). Interestingly, all the GRAS associated transcripts (known to be important regulators in GA signaling) that includes the gibberellin repressors DELLA, were only present in the female database.

CONCLUSION

Monoecious and dioecious species have been long considered unique tools to study the developmental programs involved in the formation of separate male and female flowers. However, for the majority of these species, insufficient or inexistent genomic and transcriptomic data availability has hampered functional studies. Advances in NGS technologies have made possible to perform a rapid and cost-effective compilation of large RNA sequence data sets in non-model organisms with no or little prior genomic data available. Here, a broad flowering transcriptome composed of four independent libraries was obtained for early and late developmental stages of male and female flowers of *Q. suber*, a monoecious tree. In the future, to further enhance our knowledge on sex-specific genetic networks, individual EST libraries could be obtained for each phenological stage to fine map male and female flower specific regulators. Comparative studies revealed a subset of transcripts that were differentially expressed in the different libraries, many of which have a known role in flower and/or plant development. Transcriptome analysis also revealed a group of genes expressed exclusively in each type of flower gender that may have a functional role in male and female flower organ development or in sex specification. Some of the genes that showed differential expression have not been previously characterized in other species and others have not, to our knowledge, been implicated in flower development. Thus, it would be very interesting to perform functional studies using the above mentioned genes to identify its roles in plant reproduction or flower development in *Q. suber* and other flowering species. The analysis of *Q. suber* flower transcriptome may therefore contribute to uncover sex-specific regulatory networks hidden by hermaphroditism and serve as a platform to future studies in model and non-model species independently of their sexual habit.

AUTHOR CONTRIBUTIONS

Leonor Morais-Cecílio, Maria M. R. Costa—Conceived and designed the experiments. Margarida Rocheta, Rómulo Sobral, Maria I. Amorim, Teresa Ribeiro, Leonor Morais-Cecílio, Maria M. R. Costa—Preparation of plant material and RNA. Margarida Rocheta, Rómulo Sobral, Joana Magalhães, Maria I. Amorim, Leonor Morais-Cecílio, Maria M. R. Costa—Performed the experiments. Miguel Pinheiro, Conceição Egas—Transcriptome sequencing and Bioinformatics. Margarida Rocheta, Rómulo Sobral, Joana Magalhães, Maria I. Amorim, Miguel Pinheiro, Leonor Morais-Cecílio, Maria M. R. Costa—Data analysis. Margarida Rocheta, Rómulo Sobral, Joana Magalhães, Maria I. Amorim, Miguel Pinheiro, Leonor Morais-Cecílio, Maria M. R. Costa—Paper writing and discussion. All authors read and approved the final manuscript.

ACKNOWLEDGMENTS

This work was funded by FEDER funds through the Operational Competitiveness Programme—COMPETE and by National Funds through FCT—Fundação para a Ciência e a Tecnologia under the project FCOMP—01—0124—FEDER—019461 (PTDC/AGR-GPL/118508/2010) and the sub-project SOBREIRO/0019/2009 within the National Consortium (COEC—Cork Oak ESTs Consortium). Rómulo Sobral was supported by funding from FCT with a PhD grant (ref. SFRH/BD/84365/2012). Margarida Rocheta was supported by funding from FCT with a Post-Doc grant (ref. SFRH/BPD/64905/2009). Teresa Ribeiro was supported by funding from FCT with a Post-Doc grant (SFRH/BPD/64618/2009). We are grateful to Alexandre Magalhães for the kind help given in bioinformatic analysis.

REFERENCES

Abe, H., Yamaguchi-Shinozaki, K., Urao, T., Iwasaki, T., Hosokawa, D., and Shinozaki, K. (1997). Role of Arabidopsis MYC and MYB homologs in drought- and abscisic acid-regulated gene expression. Plant Cell 9, 1859–1868.

Achard, P., Baghour, M., Chapple, A., Hedden, P., Van Der Straeten, D., Genschik, P., et al. (2007). The plant stress hormone ethylene controls floral transition via DELLA-dependent regulation of floral meristem-identity genes. Proc. Natl. Acad. Sci. U.S.A. 104, 6484–6489. doi: 10.1073/pnas.0610717104

Aida, M., Ishida, T., Fukaki, H., Fujisawa, H., and Tasaka, M. (1997). Genes involved in organ separation in Arabidopsis: an analysis of the cup-shaped cotyledon mutant. Plant Cell 9, 841–857. doi: 10.1105/tpc.9.6.841

Allen, R. L., and Lonsdale, D. M. (1993). Molecular characterization of one of the maize polygalacturonase gene family members which are expressed during late pollen development. Plant J. 3, 261–271. doi: 10.1111/j.1365-313X.1993.tb00177.x

Aloni, R., Aloni, E., Langhans, M., and Ullrich, C. I. (2006). Role of auxin in regulating Arabidopsis flower development. Planta 223, 315–328. doi: 10.1007/s00425-005-0088-9

Aronson, J., Pereira, J., and Pausas, J. (2009). Cork Oak Woodlands on the Edge: Ecology, Adaptive Management, and Restoration. Washington; Covelo; London: Island Press.

Ashburner, M., Ball, C. A., Blake, J. A., Botstein, D., Butler, H., Cherry, J. M., et al. (2000). Gene ontology: tool for the unification of biology. The Gene Ontology Consortium. Nat. Genet. 25, 25–29. doi: 10.1038/75556

Barry, C. S., and Giovannoni, J. J. (2007). Ethylene and fruit ripening. J. Plant Growth Regul. 26, 143–159. doi: 10.1007/s00344-007-9002-y

Bernhardt, C., Lee, M. M., Gonzalez, A., Zhang, F., Lloyd, A., and Schiefelbein, J. (2003). The bHLH genes GLABRA3 (GL3) and ENHANCER OF GLABRA3 (EGL3) specify epidermal cell fate in the Arabidopsis root. Development 130, 6431–6439. doi: 10.1242/dev.00880

Boavida, L. C., Varela, M. C., and Feijo, J. A. (1999). Sexual reproduction in the cork oak (Quercus suber L.). I. The progamic phase. Sex. Plant Reprod. 11, 347–353. doi: 10.1007/s004970050162

Böhlenius, H., Huang, T., Charbonnel-Campaa, L., Brunner, A. M., Jansson, S., Strauss, S. H., et al. (2006). CO/FT regulatory module controls timing of flowering and seasonal growth cessation in trees. Science 312, 1040–1043. doi: 10.1126/science.1126038

Boualem, A., Fergany, M., Fernandez, R., Troadec, C., Martin, A., Morin, H., et al. (2008). A conserved mutation in an ethylene biosynthesis enzyme leads to andromonoecy in melons. Science 321, 836–838. doi: 10.1126/science.1159023

Byers, R. E., Baker, L. R., Sell, H. M., Herner, R. C., and Dilley, D. R. (1972). Ethylene: a natural regulator of sex expression of Cucumis melo L. Proc. Natl. Acad. Sci. U.S.A. 69, 717–720. doi: 10.1073/pnas.69.3.717

Chakrabarti, M., Zhang, N., Sauvage, C., Muños, S., Blanca, J., Cañizares, J., et al. (2013). A cytochrome P450 regulates a domestication trait in cultivated tomato. Proc. Natl. Acad. Sci. U.S.A. 110, 17125–17130. doi: 10.1073/pnas.1307313110

Chandler, J. (2011). The hormonal regulation of flower development. J. Plant Growth Regul. 30, 242–254. doi: 10.1007/s00344-010-9180-x

Cheng, X.-F., and Wang, Z.-Y. (2005). Overexpression of COL9, a CONSTANS-LIKE gene, delays flowering by reducing expression of CO and FT in Arabidopsis thaliana. Plant J. 43, 758–768. doi: 10.1111/j.1365-313X.2005.02491.x

Cheng, X., Peng, J., Ma, J., Tang, Y., Chen, R., Mysore, K. S., et al. (2012). NO APICAL MERISTEM (MtNAM) regulates floral organ identity and lateral organ separation in Medicago truncatula. New Phytol. 195, 71–84. doi: 10.1111/j.1469-8137.2012.04147.x

Chinnusamy, V., Ohta, M., Kanrar, S., Lee, B.-H., Hong, X., Agarwal, M., et al. (2003). ICE1: a regulator of cold-induced transcriptome and freezing tolerance in Arabidopsis. Genes Dev. 17, 1043–1054. doi: 10.1101/gad.1077503

Ciftci-Yilmaz, S., and Mittler, R. (2008). The zinc finger network of plants. Cell. Mol. Life Sci. 65, 1150–1160. doi: 10.1007/s00018-007-7473-4

Clark, S. E. (2001). Cell signalling at the shoot meristem. Nat. Rev. Mol. Cell Biol. 2, 276–284. doi: 10.1038/35067079

Coen, E. S., and Meyerowitz, E. M. (1991). The war of the whorls: genetic interactions controlling flower development. Nature 353, 31–37. doi: 10.1038/353031a0

Colasanti, J., Yuan, Z., and Sundaresan, V. (1998). The indeterminate gene encodes a zinc finger protein and regulates a leaf-generated signal required for the transition to flowering in maize. Cell 93, 593–603. doi: 10.1016/S0092-8674(00)81188-5

Crawford, B. C. W., and Yanofsky, M. F. (2011). HALF FILLED promotes reproductive tract development and fertilization efficiency in Arabidopsis thaliana. Development 138, 2999–3009. doi: 10.1242/dev.067793

D'Amato, G., Cecchi, L., Bonini, S., Nunes, C., Annesi-Maesano, I., Behrendt, H., et al. (2007). Allergenic pollen and pollen allergy in Europe. Allergy Eur. J. Allergy Clin. Immunol. 62, 976–990. doi: 10.1111/j.1398-9995.2007.01393.x

Das, P., Ito, T., Wellmer, F., Vernoux, T., Dedieu, A., Traas, J., et al. (2009). Floral stem cell termination involves the direct regulation of AGAMOUS by PERIANTHIA. Development 136, 1605–1611. doi: 10.1242/dev.035436

Davletova, S., Rizhsky, L., Liang, H., Shengqiang, Z., Oliver, D. J., Coutu, J., et al. (2005). Cytosolic ascorbate peroxidase 1 is a central component of the reactive oxygen gene network of Arabidopsis. Plant Cell 17, 268–281. doi: 10.1105/tpc.104.026971

De Azevedo Souza, C., Kim, S. S., Koch, S., Kienow, L., Schneider, K., McKim, S. M., et al. (2009). A novel fatty Acyl-CoA Synthetase is required for pollen development and sporopollenin biosynthesis in Arabidopsis. Plant Cell 21, 507–525. doi: 10.1105/tpc.108.062513

Dellaporta, S. L., and Calderon-Urrea, A. (1993). Sex determination in flowering plants. Plant Cell 5, 1241–1251. doi: 10.1105/tpc.5.10.1241

DeLong, A., Calderon-Urrea, A., and Dellaporta, S. L. (1993). Sex determination gene TASSELSEED2 of maize encodes a short-chain alcohol dehydrogenase required for stage-specific floral organ abortion. Cell 74, 757–768. doi: 10.1016/0092-8674(93)90522-R

De Pater, S., Pham, K., Memelink, J., and Kijne, J. (1997). RAP-1 is an Arabidopsis MYC-like R protein homologue, that binds to G-box sequence motifs. Plant Mol. Biol. 34, 169–174. doi: 10.1023/A:1005898823105

Devaiah, B. N., Karthikeyan, A. S., and Raghothama, K. G. (2007). WRKY75 transcription factor is a modulator of phosphate acquisition and root development in *Arabidopsis*. *Plant Physiol.* 143, 1789–1801. doi: 10.1104/pp.106. 093971

Dobritsa, A. A., Lei, Z., Nishikawa, S.-I., Urbanczyk-Wochniak, E., Huhman, D. V., Preuss, D., et al. (2010). *LAP5* and *LAP6* encode anther-specific proteins with similarity to chalcone synthase essential for pollen exine development in *Arabidopsis*. *Plant Physiol.* 153, 937–955. doi: 10.1104/pp.110.157446

Dobritsa, A. A., Nishikawa, S. I., Preuss, D., Urbanczyk-Wochniak, E., Sumner, L. W., Hammond, A., et al. (2009). *LAP3*, a novel plant protein required for pollen development, is essential for proper exine formation. *Sex. Plant Reprod.* 22, 167–177. doi: 10.1007/s00497-009-0101-8

Dombrecht, B., Xue, G. P., Sprague, S. J., Kirkegaard, J. A., Ross, J. J., Reid, J. B., et al. (2007). *MYC2* differentially modulates diverse jasmonate-dependent functions in *Arabidopsis*. *Plant Cell* 19, 2225–2245. doi: 10.1105/tpc.106.048017

Dopazo, J., and Carazo, J. M. (1997). Phylogenetic reconstruction using an unsupervised growing neural network that adopts the topology of a phylogenetic tree. *J. Mol. Evol.* 44, 226–233. doi: 10.1007/PL00006139

Ducousso, A., Michaud, H., and Lumaret, R. (1993). Reproduction and gene flow in the genus *Quercus* L. *Ann. For. Sci.* 50, 91–106. doi: 10.1051/forest:19930708

Elena-Rossello, J. A., de Rio, J. M., Valdecantos Garcia, J. L., and Santamaria, I. G. (1993). Ecological aspects of the floral phenology of the cork-oak (*Q suber* L): why do annual and biennial biotypes appear? *Ann. For. Sci.* 50, 114s–121s.

Esteve, C., Montealegre, C., Marina, M. L., and García, M. C. (2012). Analysis of olive allergens. *Talanta* 92, 1–14. doi: 10.1016/j.talanta.2012.01.016

Fornara, F., Montaigu, A., and Coupland, G. (2010). SnapShot: control of flowering in *Arabidopsis*. *Cell* 141, 550, 550.e1-2. doi: 10.1016/j.cell.2010.04.024

Friedrichsen, D. M., Nemhauser, J., Muramitsu, T., Maloof, J. N., Alonso, J., Ecker, J. R., et al. (2002). Three redundant brassinosteroid early response genes encode putative bHLH transcription factors required for normal growth. *Genetics* 162, 1445–1456.

Gouzy, J., Carrere, S., and Schiex, T. (2009). FrameDP: sensitive peptide detection on noisy matured sequences. *Bioinformatics* 25, 670–671. doi: 10.1093/bioinformatics/btp024

Gremski, K., Ditta, G., and Yanofsky, M. F. (2007). The *HECATE* genes regulate female reproductive tract development in *Arabidopsis thaliana*. *Development* 134, 3593–3601. doi: 10.1242/dev.011510

Guilfoyle, T. J., and Hagen, G. (2007). Auxin response factors. *Curr. Opin. Plant Biol.* 10, 453–460. doi: 10.1016/j.pbi.2007.08.014

Guo, S., Zheng, Y., Joung, J.-G., Liu, S., Zhang, Z., Crasta, O. R., et al. (2010). Transcriptome sequencing and comparative analysis of cucumber flowers with different sex types. *BMC Genomics* 11:384. doi: 10.1186/1471-2164-11-384

Heim, M. A., Jakoby, M., Werber, M., Martin, C., Weisshaar, B., and Bailey, P. C. (2003). The basic helix-loop-helix transcription factor family in plants: a genome-wide study of protein structure and functional diversity. *Mol. Biol. Evol.* 20, 735–747. doi: 10.1093/molbev/msg088

Hennig, L., Gruissem, W., Grossniklaus, U., and Köhler, C. (2004). Transcriptional programs of early reproductive stages in *Arabidopsis*. *Plant Physiol.* 135, 1765–1775. doi: 10.1104/pp.104.043182

Herrero, J., Valencia, A., and Dopazo, J. (2001). A hierarchical unsupervised growing neural network for clustering gene expression patterns. *Bioinformatics* 17, 126–136. doi: 10.1093/bioinformatics/17.2.126

Hunter, S., Apweiler, R., Attwood, T. K., Bairoch, A., Bateman, A., Binns, D., et al. (2009). InterPro: the integrative protein signature database. *Nucleic Acids Res.* 37, 211–215. doi: 10.1093/nar/gkn785

Ishiguro, S., Kawai-Oda, A., Ueda, J., Nishida, I., and Okada, K. (2001). The *DEFECTIVE IN ANTHER DEHISCENCE* gene encodes a novel phospholipase A1 catalyzing the initial step of jasmonic acid biosynthesis, which synchronizes pollen maturation, anther dehiscence, and flower opening in *Arabidopsis*. *Plant Cell* 13, 2191–2209. doi: 10.1105/tpc.13.10.2191

Ito, T., and Meyerowitz, E. M. (2000). Overexpression of a gene encoding a cytochrome P450, *CYP78A9*, induces large and seedless fruit in *Arabidopsis*. *Plant Cell* 12, 1541–1550. doi: 10.1105/tpc.12.9.1541

Jakoby, M., Weisshaar, B., Droge-Laser, W., Vicente-Carbajosa, J., Tiedemann, J., Kroj, T., et al. (2002). bZIP transcription factors in *Arabidopsis*. *Trends Plant Sci.* 7, 106–111. doi: 10.1016/S1360-1385(01)02223-3

Kanazashi, T., and Kanazashi, A. (2003). Estimation for the timing of the internal developmental processes of acorns from fruit size in *Quercus serrata* Thunb. *ex Murray*. *J. For. Res.* 8, 261–266. doi: 10.1007/s10310-003-0035-1

Kater, M. M., Franken, J., Carney, K. J., Colombo, L., and Angenent, G. C. (2001).

Sex determination in the monoecious species cucumber is confined to specific floral whorls. *Plant Cell* 13, 481–493. doi: 10.1105/tpc.13.3.481

Kaul, R. B. (1985). Reproductive morphology of *Quercus* (Fagaceae). *Am. J. Bot.* 72, 1962. doi: 10.2307/2443613

Kiribuchi, K., Sugimori, M., Takeda, M., Otani, T., Okada, K., Onodera, H., et al. (2004). *RERJ1*, a jasmonic acid-responsive gene from rice, encodes a basic helix-loop-helix protein. *Biochem. Biophys. Res. Commun.* 325, 857–863. doi: 10.1016/j.bbrc.2004.10.126

Langmead, B., Hansen, K. D., and Leek, J. T. (2010). Cloud-scale RNA-sequencing differential expression analysis with Myrna. *Genome Biol.* 11:R83. doi: 10.1186/gb-2010-11-8-r83

Latchman, D. S. (1997). Transcription factors: an overview. *Int. J. Biochem. Cell Biol.* 29, 1305–1312. doi: 10.1016/S1357-2725(97)00085-X

Li, W., Zhou, Y., Liu, X., Yu, P., Cohen, J. D., and Meyerowitz, E. M. (2013). *LEAFY* controls auxin response pathways in floral primordium formation. *Sci. Signal.* 6:ra23. doi: 10.1126/scisignal.2003937

Li, Z., Reighard, G. L., Abbott, A. G., and Bielenberg, D. G. (2009). Dormancy-associated MADS genes from the *EVG* locus of peach [*Prunus persica* (L.) Batsch] have distinct seasonal and photoperiodic expression patterns. *J. Exp. Bot.* 60, 3521–3530. doi: 10.1093/jxb/erp195

Liljegren, S. J., Roeder, A. H. K., Kempin, S. A., Gremski, K., Ostergaard, L., Guimil, S., et al. (2004). Control of fruit patterning in *Arabidopsis* by INDEHISCENT. *Cell* 116, 843–853. doi: 10.1016/S0092-8674(04)00217-X

Livak, K. J., and Schmittgen, T. D. (2001). Analysis of relative gene expression data using real-time quantitative PCR and the 2(-Delta Delta C(T)) Method. *Methods* 25, 402–408. doi: 10.1006/meth.2001.1262

Logacheva, M. D., Kasianov, A. S., Vinogradov, D. V., Samigullin, T. H., Gelfand, M. S., Makeev, V. J., et al. (2011). *De novo* sequencing and characterization of floral transcriptome in two species of buckwheat (*Fagopyrum*). *BMC Genomics* 12:30. doi: 10.1186/1471-2164-12-30

Lottaz, C., Iseli, C., Jongeneel, C. V., and Bucher, P. (2003). Modeling sequencing errors by combining Hidden Markov models. *Bioinformatics* 19, 103–112. doi: 10.1093/bioinformatics/btg1067

Maier, A. T., Stehling-Sun, S., Wollmann, H., Demar, M., Hong, R. L., Haubeiss, S., et al. (2009). Dual roles of the bZIP transcription factor PERIANTHIA in the control of floral architecture and homeotic gene expression. *Development* 136, 1613–1620. doi: 10.1242/dev.033647

Mao, C., Ding, W., Wu, Y., Yu, J., He, X., Shou, H., et al. (2007). Overexpression of a NAC-domain protein promotes shoot branching in rice. *New Phytol.* 176, 288–298. doi: 10.1111/j.1469-8137.2007.02177.x

Martin, A., Troadec, C., Boualem, A., Rajab, M., Fernandez, R., Morin, H., et al. (2009). A transposon-induced epigenetic change leads to sex determination in melon. *Nature* 461, 1135–1138. doi: 10.1038/nature08498

Marum, L., Miguel, A., Ricardo, C. P., and Miguel, C. (2012). Reference gene selection for quantitative real-time PCR normalization in *Quercus suber*. *PLoS ONE* 7:e35113. doi: 10.1371/journal.pone.0035113

McCubbin, A. G., and Kao, T. (2000). Molecular recognition and response in pollen and pistil interactions. *Annu. Rev. Cell Dev. Biol.* 16, 333–364. doi: 10.1146/annurev.cellbio.16.1.333

Millar, A. A., and Gubler, F. (2005). The Arabidopsis GAMYB-like genes, *MYB33* and *MYB65*, are microRNA-regulated genes that redundantly facilitate anther development. *Plant Cell* 17, 705–721. doi: 10.1105/tpc.104. 027920

Natividade, J. (1950). *Subericultura*. Lisboa: DGSFA.

Niu, B., Fu, L., Sun, S., and Li, W. (2010). Artificial and natural duplicates in pyrosequencing reads of metagenomic data. *BMC Bioinformatics* 11:187. doi: 10.1186/1471-2105-11-187

Ogawa, M., Kay, P., Wilson, S., and Swain, S. M. (2009). *ARABIDOPSIS DEHISCENCE ZONE POLYGALACTURONASE1* (ADPG1), ADPG2, and QUARTET2 are Polygalacturonases required for cell separation during reproductive development in *Arabidopsis*. *Plant Cell* 21, 216–233. doi: 10.1105/tpc.108.063768

Okada, K., Ueda, J., Komaki, M., Bell, C., and Shimura, Y. (1991). Requirement of the auxin polar transport system in early stages of *Arabidopsis* floral bud formation. *Plant Cell* 3, 677–684. doi: 10.1105/tpc.3.7.677

Oliveros, J. C. (2007). "VENNY. An interactive tool for comparing lists with Venn diagrams," in *BioinfoGP of CNB-CSIC*. Available online at: http://bioinfogp. cnnb.csic.es/tools/venny/index.html

Pagnussat, G. C., Yu, H.-J., Ngo, Q. A., Rajani, S., Mayalagu, S., Johnson, C. S., et al.

(2005). Genetic and molecular identification of genes required for female gametophyte development and function in *Arabidopsis*. *Development* 132, 603–614. doi: 10.1242/dev.01595

Pelaz, S., Ditta, G. S., Baumann, E., Wisman, E., and Yanofsky, M. F. (2000). B and C floral organ identity functions require *SEPALLATA* MADS-box genes. *Nature* 405, 200–203. doi: 10.1038/35012103

Pereira-Leal, J. B., Abreu, I. A., Alabaça, C. S., Almeida, M. H., Almeida, P., Almeida, T., et al. (2014). A comprehensive assessment of the transcriptome of cork oak (*Quercus suber*) through EST sequencing. *BMC Genomics* 15:371. doi: 10.1186/1471-2164-15-371

Pimenta Lange, M. J., and Lange, T. (2006). Gibberellin biosynthesis and the regulation of plant development. *Plant Biol.* 8, 281–290. doi: 10.1055/s-2006-923882

Potuschak, T., Lechner, E., Parmentier, Y., Yanagisawa, S., Grava, S., Koncz, C., et al. (2003). *EIN3*-dependent regulation of plant ethylene hormone signaling by two *Arabidopsis* F box proteins: EBF1 and EBF2. *Cell* 115, 679–689. doi: 10.1016/S0092-8674(03)00968-1

Putterill, J., Robson, F., Lee, K., Simon, R., and Coupland, G. (1995). The *CONSTANS* gene of Arabidopsis promotes flowering and encodes a protein showing similarities to zinc finger transcription factors. *Cell* 80, 847–857. doi: 10.1016/0092-8674(95)90288-0

Rajani, S., and Sundaresan, V. (2001). The *Arabidopsis* myc/bHLH gene *ALCATRAZ* enables cell separation in fruit dehiscence. *Curr. Biol.* 11, 1914–1922. doi: 10.1016/S0960-9822(01)00593-0

Rampey, R. A., Woodward, A. W., Hobbs, B. N., Tierney, M. P., Lahner, B., Salt, D. E., et al. (2006). An *Arabidopsis* basic helix-loop-helix leucine zipper protein modulates metal homeostasis and auxin conjugate responsiveness. *Genetics* 174, 1841–1857. doi: 10.1534/genetics.106.061044

Rhee, S. Y., Osborne, E., Poindexter, P. D., and Somerville, C. R. (2003). Microspore separation in the *QUARTET3* mutants of *Arabidopsis* is impaired by a defect in a developmentally regulated polygalacturonase required for pollen mother cell wall degradation. *Plant Physiol.* 133, 1170–1180. doi: 10.1104/pp.103.028266

Rizhsky, L., Davletova, S., Liang, H., and Mittler, R. (2004). The zinc finger protein Zat12 is required for cytosolic ascorbate peroxidase 1 expression during oxidative stress in *Arabidopsis*. *J. Biol. Chem.* 279, 11736–11743. doi: 10.1074/jbc.M313350200

Rowan, B. A., Weigel, D., and Koenig, D. (2011). Developmental genetics and new sequencing technologies: the rise of nonmodel organisms. *Dev. Cell* 21, 65–76. doi: 10.1016/j.devcel.2011.05.021

Rudich, J., Halevy, A. H., and Kedar, N. (1972). Ethylene evolution from cucumber plants as related to sex expression. *Plant Physiol.* 49, 998–999. doi: 10.1104/pp.49.6.998

Running, M. P., and Meyerowitz, E. M. (1996). Mutations in the *PERIANTHIA* gene of *Arabidopsis* specifically alter floral organ number and initiation pattern. *Development* 124, 1261–1269.

Sakamoto, H., Maruyama, K., Sakuma, Y., Meshi, T., Iwabuchi, M., Shinozaki, K., et al. (2004). *Arabidopsis* Cys2/His2-type zinc-finger proteins function as transcription repressors under drought, cold, and high-salinity stress conditions. *Plant Physiol.* 136, 2734–2746. doi: 10.1104/pp.104.046599

Schumacher, K., and Chory, J. (2000). Brassinosteroid signal transduction: Still casting the actors. *Curr. Opin. Plant Biol.* 3, 79–84. doi: 10.1016/S1369-5266(99)00038-2

Sessions, A., Nemhauser, J. L., McColl, A., Roe, J. L., Feldmann, K. A., and Zambryski, P. C. (1997). *ETTIN* patterns the *Arabidopsis* floral meristem and reproductive organs. *Development* 124, 4481–4491.

Smaczniak, C., Immink, R. G. H., Muino, J. M., Blanvillain, R., Busscher, M., Busscher-Lange, J., et al. (2012). Characterization of MADS-domain transcription factor complexes in *Arabidopsis* flower development. *Proc. Natl. Acad. Sci. U.S.A.* 109, 1560–1565. doi: 10.1073/pnas.1112871109

Smolen, G. A., Pawlowski, L., Wilensky, S. E., and Bender, J. (2002). Dominant alleles of the basic helix-loop-helix transcription factor *ATR2* activate stress-responsive genes in *Arabidopsis*. *Genetics* 161, 1235–1246.

Sorensen, A. M., Krober, S., Unte, U. S., Huijser, P., Dekker, K., and Saedler, H. (2003). The *Arabidopsis ABORTED MICROSPORES* (*AMS*) gene encodes a MYC class transcription factor. *Plant J.* 33, 413–423. doi: 10.1046/j.1365-313X.2003.01644.x

Souer, E., van Houwelingen, A., Kloos, D., Mol, J., and Koes, R. (1996). The *NO APICAL MERISTEM* gene of *Petunia* is required for pattern formation in embryos and flowers and is expressed at meristem and primordia boundaries. *Cell* 85, 159–170. doi: 10.1016/S0092-8674(00)81093-4

Stairs, G. R. (1964). Microsporogenesis and embryogenesis in *Quercus*. *Bot. Gaz.* 125, 115–121. doi: 10.1086/336255

Szécsi, J., Joly, C., Bordji, K., Varaud, E., Cock, J. M., Dumas, C., et al. (2006). *BIGPETALp*, a bHLH transcription factor is involved in the control of *Arabidopsis* petal size. *EMBO J.* 25, 3912–3920. doi: 10.1038/sj.emboj.7601270

Tacken, E., Ireland, H., Gunaseelan, K., Karunairetnam, S., Wang, D., Schultz, K., et al. (2010). The role of ethylene and cold temperature in the regulation of the apple *POLYGALACTURONASE1* gene and fruit softening. *Plant Physiol.* 153, 294–305. doi: 10.1104/pp.109.151092

Takatsuji, H. (1999). Zinc-finger proteins: the classical zinc finger emerges in contemporary plant science. *Plant Mol. Biol.* 39, 1073–1078. doi: 10.1023/A:1006184519697

Taylor, L. P., and Jorgensen, R. (1992). Conditional male fertility in Chalcone Synthase-deficient *Petunia*. *J. Hered.* 83, 11–17.

Tebbutt, S. J., Rogers, H. J., and Lonsdale, D. M. (1994). Characterization of a tobacco gene encoding a pollen-specific polygalacturonase. *Plant Mol. Biol.* 25, 283–297. doi: 10.1007/BF00023244

Theissen, G., and Melzer, R. (2007). Molecular mechanisms underlying origin and diversification of the angiosperm flower. *Ann. Bot.* 100, 603–619. doi: 10.1093/aob/mcm143

Tsugama, D., Liu, S., and Takano, T. (2012). A bZIP protein, VIP1, is a regulator of osmosensory signaling in *Arabidopsis*. *Plant Physiol.* 159, 144–155. doi: 10.1104/pp.112.197020

Tzfira, T., Vaidya, M., and Citovsky, V. (2001). VIP1, an Arabidopsis protein that interacts with *Agrobacterium* VirE2, is involved in VirE2 nuclear import and *Agrobacterium* infectivity. *EMBO J.* 20, 3596–3607. doi: 10.1093/emboj/20.13.3596

Ueno, S., Klopp, C., Leplé, J. C., Derory, J., Noirot, C., Léger, V., et al. (2013). Transcriptional profiling of bud dormancy induction and release in oak by next-generation sequencing. *BMC Genomics* 14:236. doi: 10.1186/1471-2164-14-236

Ueno, S., Le Provost, G., Léger, V., Klopp, C., Noirot, C., Frigerio, J.-M., et al. (2010). Bioinformatic analysis of ESTs collected by Sanger and pyrosequencing methods for a keystone forest tree species: oak. *BMC Genomics* 11:650. doi: 10.1186/1471-2164-11-650

Varela, M. C., and Valdiviesso, T. (1996). Phenological phases of *Quercus suber* L. flowering. *For. Genet.* 3, 93–102.

Varkonyi-Gasic, E., Moss, S. M., Voogd, C., Wu, R., Lough, R. H., Wang, Y.-Y., et al. (2011). Identification and characterization of flowering genes in kiwifruit: sequence conservation and role in kiwifruit flower development. *BMC Plant Biol.* 11:72. doi: 10.1186/1471-2229-11-72

Verhoeven, T., Feron, R., Wolters-Arts, M., Edqvist, J., Gerats, T., Derksen, J., et al. (2005). *STIG1* controls exudate secretion in the pistil of petunia and tobacco. *Plant Physiol.* 138, 153–160. doi: 10.1104/pp.104.054809

Vroemen, C. W., Mordhorst, A. P., Albrecht, C., Kwaaitaal, M. A. C. J., and de Vries, S. C. (2003). The *CUP-SHAPED COTYLEDON3* gene is required for boundary and shoot meristem formation in *Arabidopsis*. *Plant Cell* 15, 1563–1577. doi: 10.1105/tpc.012203

Wallner, M., Erler, A., Hauser, M., Klinglmayr, E., Gadermaier, G., Vogel, L., et al. (2009). Immunologic characterization of isoforms of Car b 1 and Que a 1, the major hornbeam and oak pollen allergens. *Allergy Eur. J. Allergy Clin. Immunol.* 64, 452–460. doi: 10.1111/j.1398-9995.2008.01788.x

Wang, H., Jones, B., Li, Z., Frasse, P., Delalande, C., Regad, F., et al. (2005). The tomato Aux/IAA transcription factor *IAA9* is involved in fruit development and leaf morphogenesis. *Plant Cell* 17, 2676–2692. doi: 10.1105/tpc.105.033415

Wang, L., Ying, Y., Narsai, R., Ye, L., Zheng, L., Tian, J., et al. (2013). Identification of *OsbHLH133* as a regulator of iron distribution between roots and shoots in *Oryza sativa*. *Plant Cell Environ.* 36, 224–236. doi: 10.1111/j.1365-3040.2012.02569.x

Wellmer, F., Graciet, E., and Riechmann, J. L. (2014). Specification of floral organs in *Arabidopsis*. *J. Exp. Bot.* 65, 1–9. doi: 10.1093/jxb/ert385

Winkel-Shirley, B. (2002). Biosynthesis of flavonoids and effects of stress. *Curr. Opin. Plant Biol.* 5, 218–223. doi: 10.1016/S1369-5266(02)00256-X

Wu, T., Qin, Z., Zhou, X., Feng, Z., and Du, Y. (2010). Transcriptome profile analysis of floral sex determination in cucumber. *J. Plant Physiol.* 167, 905–913. doi: 10.1016/j.jplph.2010.02.004

Wynn, A. N., Seaman, A. A., Jones, A. L., and Franks, R. G. (2014). Novel functional roles for *PERIANTHIA* and *SEUSS* during floral organ identity specification, floral meristem termination, and gynoecial development. *Front. Plant Sci.* 5:130. doi: 10.3389/fpls.2014.00130

Xu, J., Yang, C., Yuan, Z., Zhang, D., Gondwe, M. Y., Ding, Z., et al. (2010). The *ABORTED MICROSPORES* regulatory network is required for postmeiotic male reproductive development in *Arabidopsis thaliana*. *Plant Cell* 22, 91–107. doi: 10.1105/tpc.109.071803

Yadav, V., Mallappa, C., Gangappa, S. N., Bhatia, S., and Chattopadhyay, S. (2005). A basic helix-loop-helix transcription factor in *Arabidopsis*, *MYC2*, acts as a repressor of blue light-mediated photomorphogenic growth. *Plant Cell* 17, 1953–1966. doi: 10.1105/tpc.105.032060

Yamasaki, S., Fujii, N., Matsuura, S., Mizusawa, H., and Takahashi, H. (2001). The M locus and ethylene-controlled sex determination in andromonoecious cucumber plants. *Plant Cell Physiol.* 42, 608–619. doi: 10.1093/pcp/pce076

Yanagisawa, S., Yoo, S.-D., and Sheen, J. (2003). Differential regulation of *EIN3* stability by glucose and ethylene signalling in plants. *Nature* 425, 521–525. doi: 10.1038/nature01984

Yoo, M. J., Chanderbali, A. S., Altman, N. S., Soltis, P. S., and Soltis, D. E. (2010). Evolutionary trends in the floral transcriptome: insights from one of the basalmost angiosperms, the water lily *Nuphar advena* (Nymphaeaceae). *Plant J.* 64, 687–698. doi: 10.1111/j.1365-313X.2010.04357.x

Yu, L., Sun, J., and Li, L. (2013). *PtrCel9A6*, an endo-1,4-β-glucanase, is required for cell wall formation during xylem differentiation in *Populus*. *Mol. Plant* 6, 1904–1917. doi: 10.1093/mp/sst104

Zahn, L. M., Ma, X., Altman, N. S., Zhang, Q., Wall, P. K., Tian, D., et al. (2010). Comparative transcriptomics among floral organs of the basal eudicot *Eschscholzia californica* as reference for floral evolutionary developmental studies. *Genome Biol.* 11:R101. doi: 10.1186/gb-2010-11-10-r101

Zhang, X. M., Zhao, L., Larson-Rabin, Z., Li, D. Z., and Guo, Z. H. (2012). *De novo* sequencing and characterization of the floral transcriptome of *Dendrocalamus latiflorus* (poaceae: Bambusoideae). *PLoS ONE* 7:e42082. doi: 10.1371/journal.pone.0042082

Zhu, Y. Y., Machleder, E. M., Chenchik, A., Li, R., and Siebert, P. D. (2001). Reverse transcriptase template switching: a SMART approach for full-length cDNA library construction. *Biotechniques* 30, 892–897.

15

Reproduction and the pheromonal regulation of sex type in fern gametophytes

Nadia M. Atallah and Jo Ann Banks*

Department of Botany and Plant Pathology, Purdue University, West Lafayette, IN, USA

Edited by:
Dazhong D. Zhao, University of Wisconsin-Milwaukee, USA
Reviewed by:
Elena M. Kramer, Harvard University, USA
Keiko Sakakibara, University of Tokyo, Japan

*Correspondence:
Jo Ann Banks, Department of Botany and Plant Pathology, Purdue University, 915 West State Street, West Lafayette, IN 47906, USA
e-mail: banksj@purdue.edu

The fern life cycle includes a haploid gametophyte that is independent of the sporophyte and functions to produce the gametes. In homosporous ferns, the sex of the gametophyte is not fixed but can vary depending on its social environment. In many species, the sexual phenotype of the gametophyte is determined by the pheromone antheridiogen. Antheridiogen induces male development and is secreted by hermaphrodites once they become insensitive to its male-inducing effect. Recent genetic and biochemical studies of the antheridiogen response and sex-determination pathway in ferns, which are highlighted here, reveal many similarities and interesting differences to GA signaling and biosynthetic pathways in angiosperms.

Keywords: antheridiogen, sex determination, ferns, GA signaling, GA biosynthesis

INTRODUCTION

The fern life cycle, illustrated in **Figure 1**, features two distinct body types: the large diploid sporophyte and the tiny haploid gametophyte. From a reproduction point of view, the sole function of the sporophyte is to produce then release haploid spores, while the gametophyte, which grows from a spore, functions to produce the gametes. Some ferns, like all angiosperms, are heterosporous and produce both mega- and microspores that are destined to develop as female and male gametophytes, respectively. Most ferns species are homosporous and produce only one type of spore. While textbook drawings of homosporous fern gametophytes typically show a heart-shaped hermaphrodite, fern gametophytes can be male, female, male then female, female then male, hermaphroditic or asexual, depending on the species. In this review we highlight old and recent studies that have revealed the fascinating cross-talk that occurs between neighboring gametophytes in determining what their sexual phenotype will be.

ASEXUAL REPRODUCTION IN FERN GAMETOPHYTES

In addition to reproducing sexually, there are many examples of fern gametophytes that circumvent sex and reproduce asexually. The most common type of asexual reproduction is apogamy, whereby a sporophyte plant develops from a gametophyte without fertilization, similar to apomixis in angiosperms. In naturally occurring apogamous species, the viable spores produced by the sporophyte have the same chromosome number as the sporophyte (Walker, 1962, 1979). Obligate apogamy often occurs naturally in species of ferns that produce no or only one type of gametangia. Because water is required for the flagellated sperm to swim to the egg in ferns, apogamous species are typically found in dry habitats where water is limiting (White, 1979). Apogamy also can be artificially induced in many ferns by adding sucrose to the culture media in which gametophytes are grown (Whittier and Steeves, 1962; White,

1979). By optimizing the conditions for inducing apospory in *Ceratopteris richardii* gametophytes, a recent study has established C. richardii as a useful experimental system for studying this phenomenon (Cordle et al., 2007). Induced apogamous sporophytes of *C. richardii* have features typical of the sporophyte, including stomata, vascular tissue and scale-like ramenta; however, they are abnormal compared to sexually-derived diploid sporophytes, which could be a consequence of being haploid. To better understand how sucrose promotes the development of a sporophyte from cells of the gametophyte, the same researchers identified 170 genes whose expression is up-regulated during the period of apogamy commitment. Many of them are associated with stress and metabolism or are homologs of genes preferentially expressed in seed and flower tissues (Cordle et al., 2012). Understanding apogamy, coupled with studies of apospory in *C. richardii*, where diploid gametophytes develop from cells of sporophyte leaves without meiosis (DeYoung et al., 1997), should provide useful insights into genes and molecular mechanisms that regulate the alternation of gametophyte and sporophyte generations in ferns in the absence of meiosis and fertilization.

A second form of asexual reproduction in homosporous ferns involves vegetative propagation of the gametophyte. While relatively rare, such gametophytes typically do not produce sex organs. The fern *Vittaria appalachiana*, for example, is only known from its gametophytes (Farrar and Mickel, 1991). Each gametophyte forms vegetative buds, or gammae, that allow gametophytes to multiply and form mats in dark, moist cavities and rock shelters in the Appalachian Mountains. While the origin of *V. appalachiana* (is it a recent hybrid or ancient relict?) and why it is unable to form sporophytes are unknown at this time, its persistent gametophyte suggest that fern gametophytes, like bryophyte gametophytes, can persist and thrive for very long periods of time.

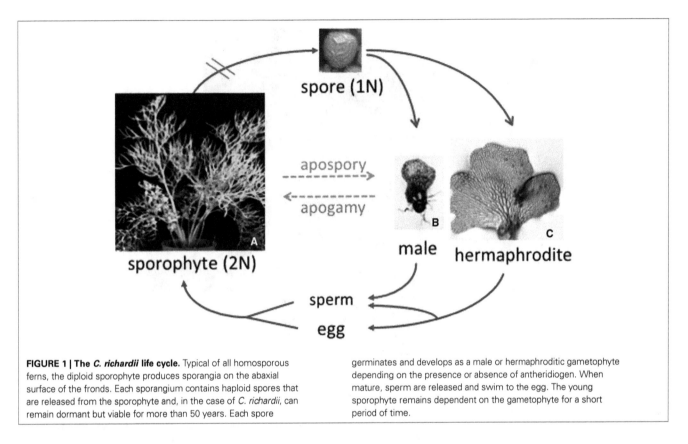

FIGURE 1 | The *C. richardii* life cycle. Typical of all homosporous ferns, the diploid sporophyte produces sporangia on the abaxial surface of the fronds. Each sporangium contains haploid spores that are released from the sporophyte and, in the case of *C. richardii*, can remain dormant but viable for more than 50 years. Each spore germinates and develops as a male or hermaphroditic gametophyte depending on the presence or absence of antheridiogen. When mature, sperm are released and swim to the egg. The young sporophyte remains dependent on the gametophyte for a short period of time.

SEXUAL REPRODUCTION

Most homosporous ferns that reproduce sexually ultimately form hermaphroditic gametophytes that have antheridia and archegonia. While hermaphroditism increases the probability that a single gametophyte will reproduce, self-fertilization of a hermaphrodite (which is genetically similar to a doubled haploid in angiosperms) results in a completely homozygous sporophyte. Given that this absolute inbreeding could have negative consequences to the individual and reduce genetic variation in populations, it is not surprising that homosporous ferns have evolved mechanisms to promote outcrossing. One such mechanism that is common to many species of ferns involves the pheromonal regulation of sexual identity, where the sexual phenotype of an individual gametophyte depends on its social environment.

ONE GENOTYPE—TWO OR MORE PHENOTYPES

In the late 1800's, botanists began noting that fern gametophytes are often sexually dimorphic, with larger gametophytes bearing archegonia and smaller gametophytes bearing antheridia (Prantl, 1881; Yin and Quinn, 1995). The size difference between them was attributed to the presence or absence of a meristem, with females or hermaphrodites being "meristic" (with a meristem) and males "ameristic" (without a meristem). In a major discovery, Döpp noted that the medium harvested from cultures of *Pteridium aquilinum* gametophytes contained a pheromone that promoted the development of males in juvenile gametophytes (Döpp, 1950); this pheromone is referred to as antheridiogen. Antheridiogens or antheridiogen responses have since been identified in over 20 species of

ferns (Yamane, 1998; Kurumatani et al., 2001; Jimenez et al., 2008).

Much of what is known about the biology of antheridiogen responses can be attributed to studies by Näf and Schraudolf during the 1950s and 1960s (reviewed in Näf, 1959, 1979). This response is illustrated here for the fern *C. richardii*, originally characterized by Hickok et al. (1995). In this species, an individual spore always develops as a relatively large hermaphrodite (**Figure 2A**) that produces egg-forming archegonia (**Figure 2B**), sperm-forming antheridia and multicellular lateral meristem. The hermaphrodite also secretes antheridiogen, or A_{CE} (for antheridiogen *Ceratopteris*) into its surroundings. If the hermaphrodite is removed then replaced with a genetically identical spore, the new spore will develop as an ameristic male gametophyte (**Figure 2C**) with many antheridia (**Figure 2D**) in response to A_{CE} secreted by the hermaphrodite. In a population of spores, spores that germinate first become hermaphrodites that secrete A_{CE}, while slower-growing members of the population become male in response to the secreted A_{CE}. In comparison to chromosomal based sex determination, this mechanism of sex-determination is unusual because it allows the ratio of males to hermaphrodites to vary depending on population size and density and it is inherently flexible rather than fixed.

Typical of other ferns, a *C. richardii* gametophyte is able to respond to A_{CE} for a limited period of time, prior to the establishment of a lateral meristem. The lateral meristem not only confers indeterminate growth to the gametophyte, but its formation coincides with a loss in ability to respond to A_{CE} as well as the secretion of A_{CE}. Archegonia invariably initiate

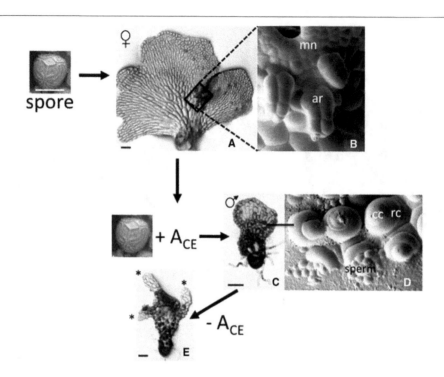

FIGURE 2 | The antheridiogen response in *C. richardii*. A single spore always develops as a hermaphrodite when grown in the absence of A$_{CE}$. The hermaphrodite consists of a single sheet of cells with a distinct multicellular meristem that forms a meristem notch and multiple archegonia that develop adjacent to the meristem notch, which are highlighted in the SEM (boxed area of the hermaphrodite). Hermaphrodites secrete A$_{CE}$; in the presence of A$_{CE}$, spores develop as males. The male lacks a meristem and almost all cells differentiate as antheridia. The SEM shows six antheridia, each having a ring cell and a cap cell that pops open to release sperm. When a male gametophyte is transferred to media lacking A$_{CE}$, some cells divide and begin to form a hermaphroditic prothallus. The "switched" male shown is forming three such prothalli. mn: meristem notch; ar: archegonia; cc: cap cell; rc: ring cell.

close to the meristem notch of the hermaphrodite, well after the lateral meristem is well developed. While the hermaphroditic program of expression cannot be reversed, the male program of expression is reversible. Cells of the male gametophyte prothallus, when transferred to media lacking A$_{CE}$, will divide to ultimately form one or more new hermaphroditic prothalli (**Figure 2E**). Antheridiogen thus serves multiple functions in male gametophyte development: it represses divisions of the prothallus that establish the lateral meristem; it promotes the rapid differentiation of antheridia; it represses its own biosynthesis; and it serves to maintain in the gametophyte an ability to respond to itself.

All of the antheridiogens that have been structurally characterized from ferns are gibberellins (GAs) (Yamane et al., 1979; Furber et al., 1989; Takeno et al., 1989; Yamane, 1998). Although the structure of ACE is unknown, GA biosynthetic inhibitors reduce the proportion of males in a population of *C. richardii* gametophytes suggesting that ACE and GA share a common biosynthetic pathway (Warne and Hickok, 1989). ABA, a known antagonist of GA responses in angiosperms, completely blocks the ACE response in *C. richardii*, also indicating that ACE is likely a GA (Hickok, 1983).

THE SEX-DETERMINING PATHWAY IN *Ceratopteris*

Most recent studies aimed at understanding how antheridiogen determines the sex of the gametophyte have focused on two species of homosporous ferns: *C. richardii* and *Lygodium japonicum*. *Ceratopteris richardii* is a semi-tropical, annual species and is useful as a genetic system for many reasons. Large numbers of single-celled, haploid spores (typically 10^6) can be mutagenized and mutants identified within 2 weeks after mutagenesis. Gametophytes can be dissected and regrown, making it possible to simultaneously self-fertilize and out-cross a single mutant gametophyte. Because self-fertilization of a gametophyte results in a completely homozygous sporophyte that produces $>10^7$ spores within a 6-month period, suppressor mutants are also easy to generate. Because *C. richardii* gametophytes are sexually dimorphic, mutations affecting the sex of the gametophyte are especially easy to identify (Hickok, 1977, 1985; Hickok et al., 1985, 1991; Warne and Hickok, 1986; Warne et al., 1988; Hickok and Schwarz, 1989; Vaughn et al., 1990; Scott and Hickok, 1991; Chun and Hickok, 1992; Banks, 1994, 1997a,b; Eberle and Banks, 1996; Strain et al., 2001; Renzaglia et al., 2004). Over 70 mutants affecting sex determination have been characterized, most falling into three major phenotypic groups: the *hermaphroditic* (*her*) mutants, which are hermaphroditic in the presence or absence of A$_{CE}$, the *transformer* (*tra*) mutants, which are male in the presence or absence of A$_{CE}$, and the *feminization* (*fem*) mutants, which are female in the presence or absence of A$_{CE}$ and produce no antheridia. Through test of epistasis (i.e., comparing mutant phenotypes of single and various combinations of double and triple mutants), a genetic model of the sex determination pathway has been developed and

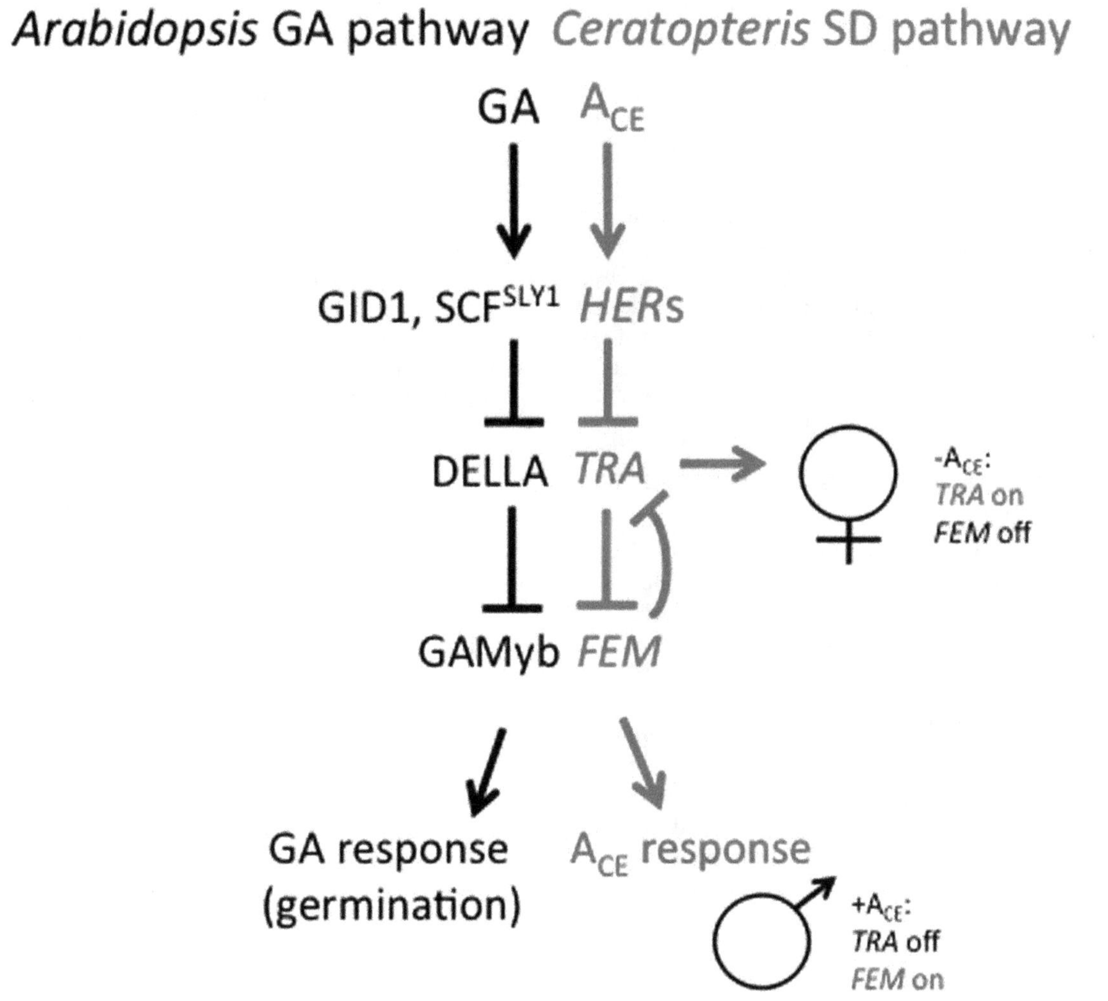

FIGURE 3 | A comparison of the GA signaling pathway in angiosperms and the sex-determining (SD) pathway in *C. richardii*. The SD pathway in *C. richardii* is based solely on the epistatic interactions among sex-determining mutants but it is consistent with recent molecular and biochemical studies in the fern *L. japonicum*. T bars represent repressive events whereas arrows indicate activating events.

is illustrated in **Figure 3** (Eberle and Banks, 1996; Banks, 1997a,b; Strain et al., 2001). This pathway reveals that there are two major regulators of sex: *TRA*, which is necessary for lateral meristem and archegonia development (female traits), and *FEM*, which is necessary for antheridia development (the male trait). *FEM* and *TRA* negatively regulate each other such that only one can be expressed in the gametophyte. What determines whether *FEM* or *TRA* is expressed in the gametophyte is A_{CE}. A_{CE} activates the *HERs*, which, in turn, repress *TRA*. Because *TRA* cannot repress *FEM*, *FEM* is expressed and the gametophyte develops as a male. In the absence of A_{CE}, *HER* is not active and is thus unable to repress *TRA*. *TRA* promotes the development of a gametophyte with female traits and represses the development of antheridia by repressing the *FEM* gene that promotes male development. Additional genetic experiments have revealed that the repression of *FEM* by *TRA* and of *TRA* by *FEM* is indirect and involves other genes (Strain et al., 2001). What is remarkable about this pathway is that it is inherently flexible, which is consistent with what is understood about sex determination in this species by

A_{CE}. This "battle of the sexes"—deciding whether to be male or female—depends on which of the two major regulatory sex genes prevails in the young gametophyte, a decision that is ultimately determined by the presence or absence A_{CE}.

While this model explains how male and female gametophyte identities are determined, it does not explain the hermaphrodite. One possibility is that in certain cells of the hermaphrodite, the activities of *FEM* and *TRA* are reversed, allowing *FEM* to be expressed in cells that will eventually differentiate as antheridia. Testing this and other possibilities will require the cloning of the sex-determining genes and assessing their temporal and spatial patterns of expression in the developing hermaphrodite.

The sex-determining pathway in *C. richardii* is remarkable in its resemblance to the GA signaling pathway in angiosperms (Sun, 2011), as illustrated in **Figure 3**. In *Arabidopsis*, GA is bound by its receptor GIBBERELLIN INSENSITIVE DWARF1 (GID1). The GA-GID1 complex triggers the rapid proteolysis of one or more DELLA proteins that are ultimately responsible for repressing GA responses. Proteolysis of DELLA requires

GID1 and the specific F-box protein SLEEPY1 (SLY1), which promotes poly-ubiquitination of DELLA by the SCR$^{SLY1/GID2}$ complex and results in its degradation by the 26S proteasome. Since DELLA acts as a repressor of GA responses, its GA-induced degradation results in a GA response. While targets of DELLA repression have been identified (Fleet and Sun, 2005), in the case of barley seed germination (which requires GA), DELLA directly or indirectly represses *GAMYB*, a transcription factor that promotes α-amylase expression in germinating barley seeds (Gubler et al., 1995, 1999). Based on the similarities between the GA signaling pathway in angiosperms and the sex determination pathway in *C. richardii*, it is tempting to speculate that the *HER* genes in *C. richardii* encode GID1 and SLY1, that *TRA* encodes a DELLA protein, and that *FEM* encodes a GAMYB-like protein.

ANTHERIDIOGEN BIOSYNTHESIS IS SPLIT BETWEEN YOUNG AND OLDER GAMETOPHYTES IN *Lygodium japonicum*

Lygodium japonicum is another homosporous fern species with an antheridiogen response. This species has the distinct advantage of having its antheridiogens structurally well characterized. Two different GAs have been identified as antheridiogens in this species, including GA$_9$ methyl ester (Yamane et al., 1979) and GA$_{73}$ methyl ester (Yamane et al., 1988). GA$_{73}$ methyl ester is the most active antheridiogen and is able to induce antheridia formation at the incredibly low concentration of 10^{-15} M. To test the hypothesis that antheridiogen is synthesized through the GA biosynthetic pathway, *L. japonicum* genes related to five different GA synthesis genes, including *ent-copalyl diphosphate/ent-kaurene synthase* (*CPS/KS*), *ent-kaurenoic acid oxidase* (*KAO*), *kaurene oxidase* (*KO*), *GA 20-oxidase* (*GA20ox*), and *GA3-oxidase* (*GA3ox*), were identified and their expression patterns in developing gametophytes investigated (Tanaka et al., 2014). Their expression patterns revealed that all but *GA3ox* were more highly expressed in older gametophytes that secrete antheridiogen, consistent with the expectation that antheridiogen biosynthesis genes are up-regulated in gametophytes that secrete it. *GA3ox* expression showed the opposite pattern of expression; i.e., it was more highly expressed in young gametophytes that did not secrete antheridiogen but were capable of responding to antheridiogen. To explore this further, the same authors assayed the effects of prohexadione, a GA3ox inhibitor, on antheridia formation in the presence of GA$_4$ (which has an OH group at the C3 position) or GA$_9$ methyl ester (which lacks the OH group at C3); both GA$_9$ and GA$_4$ induce antheridia formation by themselves. Whereas prohexadione plus GA$_9$ methyl ester inhibited antheridia formation, prohexadione plus GA$_4$ did not, demonstrating that C3 hydroxylation of antheridiogen is essential for inducing antheridia formation. In another series of experiments, the authors found that GA$_9$ methyl ester was converted to GA$_9$ in young gametophytes. Based on these and other results, a model was proposed whereby antheridiogen (GA$_9$ methyl ester) is synthesized via a GA biosynthetic pathway and secreted by older gametophytes. When it is taken up by younger gametophytes, the methyl ester is removed by a possible methyl esterase then hydroxylated at the C3 position by GA3ox to GA$_4$, where it is perceived and transduced by the GA signaling pathway in young

gametophyte. Because GA$_9$ methyl ester is more hydrophobic and more efficiently taken up by gametophytes than GA$_9$, splitting the GA biosynthetic pathway between young and older gametophytes was proposed to enhance the sensitivity of young gametophytes to the secreted antheridiogen by their neighbors and, at the same time, promote the activation of male traits once inside the young gametophyte (Tanaka et al., 2014).

In addition to characterizing antheridiogen biosynthesis in *L. japonicum*, Tanaka et al. (2014) also made two other important discoveries. They found that a *L. japonicum* DELLA protein was degraded in GA$_4$ and GA$_9$ methyl ester treated gametophytes, and that the *L. japonicum* GID1 and DELLA proteins could interact in a yeast–two hybrid assay, but only in the presence of GA$_4$ (and not GA$_4$ methyl ester or GA$_9$ methyl ester). All told, the results of these experiments were used to define a model of the antheridiogen response in *L. japonicum* that is remarkably similar to the pathways illustrated in **Figure 3**.

FUTURE DIRECTIONS

The elucidation of the antheridiogen biosynthetic and signaling pathways in ferns has only just begun and many questions regarding sex determination and sexual reproduction remain, many of which can be resolved by cloning all of the sex determining genes. Some of these questions are: To what extent are other hormones involved in sex determination? Is the split GA biosynthetic pathway in *L. japonicum* typical of other ferns? What is the relationship between the antheridiogen response in the gametophyte to GA responses in the sporophyte? Knowing that some mutations in *C. richardii* (e.g., *her* mutations) have no effect on the sporophyte while other mutations (e.g., *tra* mutations) severely affect the sporophyte suggest that at least some, but not all, genes are necessary in both generations. Is antheridiogen also involved in the developmental decision to produce mega- and micro-sporangia in heterosporous ferns? From an evolutionary perspective, was the antheridiogen signaling and responses in the gametophyte co-opted during or important for the evolution of heterospory from homospory in ferns? Addressing these and other questions will lead to a more comprehensive understanding of sex determination in ferns, including an understanding of the molecular mechanisms at play.

REFERENCES

Banks, J. A. (1994). Sex-determining genes in the homosporous fern *Ceratopteris*. *Development* 120, 1949–1958.

Banks, J. A. (1997a). Sex determination in the fern *Ceratopteris*. *Trends Plant Sci.* 2, 175–180. doi: 10.1016/S1360-1385(97)85223-5

Banks, J. A. (1997b). The TRANSFORMER genes of the fern *Ceratopteris* simultaneously promote meristem and archegonia development and repress antheridia development in the developing gametophyte. *Genetics* 147, 1885–1897.

Chun, P. T., and Hickok, L. G. (1992). Inheritance of two mutations conferring glyphosate tolerance in the fern *Ceratopteris richardii*. *Can. J. Bot. Rev. Can. Bot.* 70, 1097–1099. doi: 10.1139/b92-135

Cordle, A. R., Irish, E. E., and Cheng, C. L. (2007). Apogamy induction in *Ceratopteris richardii*. *Int. J. Plant Sci.* 168, 361–369. doi: 10.1086/511049

Cordle, A. R., Irish, E. E., and Cheng, C. L. (2012). Gene expression associated with apogamy commitment in *Ceratopteris richardii*. *Sex. Plant Reprod.* 25, 293–304. doi: 10.1007/s00497-012-0198-z

DeYoung, B., Weber, T., Hass, B., and Banks, J. A. (1997). Generating autotetraploid sporophytes and their use in analyzing mutations affecting gametophyte development in the fern *Ceratopteris*. *Genetics* 147, 809–814.

Döpp, W. (1950). Eine die antheridienbildung bei farnen fördernde substanz in den prothallien von *Pteridium aquilinum* (L.). Kuhn. *Ber. Deut. Botan. Ges.* 63, 139–146.

Eberle, J. R., and Banks, J. A. (1996). Genetic interactions among sex-determining genes in the fern *Ceratopteris richardii*. *Genetics* 142, 973–985.

Farrar, D. R., and Mickel, J. T. (1991). Society *Vittaria appalachiana*: a name for the "Appalachian Gametophyte." *Am. Fern J.* 81, 69–75. doi: 10.2307/1547574

Fleet, C. M., and Sun, T. P. (2005). A DELLAcate balance: the role of gibberellin in plant morphogenesis. *Curr. Opin. Plant Biol.* 8, 77–85. doi: 10.1016/j.pbi.2004.11.015

Furber, M., Mander, L., Nester, J., Takahashi, N., and Yamane, H. (1989). Structure of a novel antheridiogen from the fern Anemia mexicana. *Phytochem* 28, 63–66. doi: 10.1016/0031-9422(89)85008-3

Gubler, F., Kalla, R., Roberts, J. K., and Jacobsen, J. V. (1995). Gibberellin-regulated expression of a myb gene in barley aleurone cells: evidence for Myb transactivation of a high-pI alpha-amylase gene promoter. *Plant Cell* 7, 1879–1891.

Gubler, F., Raventos, D., Keys, M., Watts, R., Mundy, J., and Jacobsen, J. V. (1999). Target genes and regulatory domains of the GAMYB transcriptional activator in cereal aleurone. *Plant J.* 17, 1–9. doi: 10.1046/j.1365-313X.1999.00346.x

Hickok, L. G. (1977). Apomictic mutant for sticky chromosomes in fern *Ceratopteris*. *Can. J. Bot.* 55, 2186–2195. doi: 10.1139/b77-247

Hickok, L. G. (1983). Abscisic acid blocks antheridiogen-induced antheridium formation in gametophytes of the fern *Ceratopteris*. *Can. J. Bot.* 63, 888–892.

Hickok, L. G. (1985). Abscisic acid resistant mutants in the fern *Ceratopteris*: characterization and genetic analysis. *Can. J. Bot.* 63, 1582–1585. doi: 10.1139/b85-220

Hickok, L. G., and Schwarz, O. J. (1989). Genetic characterization of a mutation that enhances paraquat tolerance in the fern *Ceratopteris richardii*. *Theor. Appl. Genet.* 77, 200–204. doi: 10.1007/BF00266187

Hickok, L. G., Scott, R. J., and Warne, T. R. (1985). Isolation and characterization of antheridiogen-resistant mutants in the fern *Ceratopteris*. *Am. J. Bot.* 72, 922–922.

Hickok, L. G., Vogelien, D. L., and Warne, T. R. (1991). Selection of a mutation conferring high nacl tolerance to gametophytes of ceratopteris. *Theor. Appl. Genet.* 81, 293–300. doi: 10.1007/BF00228666

Hickok, L. G., Warne, T. R., and Fribourg, R. S. (1995). The biology of the fern *Ceratopteris* and its use as a model system. *Int. J. Plant Sci.* 156, 332–345. doi: 10.1086/297255

Jimenez, A., Quintanilla, L. G., Pajaron, S., and Pangua, E. (2008). Reproductive and competitive interactions among gametophytes of the allotetraploid fern *Dryopteris corleyi* and its two diploid parents. *Ann. Bot.* 102, 353–359. doi: 10.1093/aob/mcn099

Kurumatani, M., Yagi, K., Murata, T., Tezuka, M., Mander, L. N., Nishiyama, M., et al. (2001). Isolation and identification of antheridiogens in the ferns, *Lygodium microphyllum* and *Lygodium reticulatum*. *Biosci. Biotechnol. Biochem.* 65, 2311–2314. doi: 10.1271/bbb.65.2311

Näf, U. (1959). Control of antheridium formation in the fern species anemia phyllitides. *Nature* 184, 798–800. doi: 10.1038/184798a0

Näf, U. (1979). "Antheridiogens and antheridial development," in *The Experimental Biology of Ferns*, ed. A. F. Dyer (New York: Academic Press), 436–470.

Prantl, K. A. E. (1881). Beobachtungen üer die ernährung der farnprothallien und die verteilung der sexual organe: apogamie. *Just's Bot. Jahrb.* 15, 553–574.

Renzaglia, K. S., Wood, K. D., Rupp, G., and Hickok, L. G. (2004). Characterization of the sleepy sperm mutant in the fern *Ceratopteris richardii*: a new model for the study of axonemal function. *Can. J. Bot.* 82, 1602–1617. doi: 10.1139/b04-125

Scott, R. J., and Hickok, L. G. (1991). Inheritance and characterization of a dark-germinating, light-inhibited mutant in the fern *Ceratopteris richardii*. *Can. J. Bot.* 69, 2616–2619. doi: 10.1139/b91-326

Strain, E., Hass, B., and Banks, J. A. (2001). Characterization of mutations that feminize gametophytes of the fern *Ceratopteris*. *Genetics* 159, 1271–1281.

Sun, T. P. (2011). The molecular mechanism and evolution of the GA-GID1-DELLA signaling module in plants. *Curr. Biol.* 21, R338–R345. doi: 10.1016/j.cub.2011.02.036

Takeno, K., Yamane, H., Yamauchi, T., Takahashi, N., Furber, M., and Mander, L. (1989). Biological activities of the methyl ester of gibberellin a73, a novel and principal antheridiogen in Lygodium japonicum. *Plant Cell Physiol.* 30, 201–215.

Tanaka, J., Yano, K., Aya, K., Hirano, K., Takehara, S., Koketsu, E., et al. (2014). Antheridiogen determines sex in ferns via a spatiotemporally split gibberellin synthesis pathway. *Science* 346, 469–473. doi: 10.1126/science.1259923

Vaughn, K. C., Hickok, L. G., Warne, T. R., and Farrow, A. C. (1990). Structural analysis and inheritance of a clumped-chloroplast mutant in the fern *Ceratopteris*. *J. Heredity* 81, 146–151.

Walker, T. G. (1962). Cytology and evolution in the fern genus Pteris L. *Evolution* 16, 27–43.

Walker, T. G. (1979). "The cytogenetics of ferns," in *The Experimental Biology of Ferns*, ed. A. F. Dyer (London: Academic Press), 87–132.

Warne, T. R., and Hickok, L. G. (1986). Selection and characterization of sodium-chloride tolerant mutants in the fern *Ceratopteris richardii*. *Am. J. Bot.* 73, 741–741.

Warne, T. R., and Hickok, L. G. (1989). Evidence for a gibberellin biosynthetic origin of *Ceratopteris antheridiogen*. *Plant Physiol.* 89, 535–538. doi: 10.1104/Pp.89.2.535

Warne, T. R., Hickok, L. G., and Scott, R. J. (1988). Characterization and genetic Analysis of antheridiogen-insensitive mutants in the fern *Ceratopteris*. *Bot. J. Linn. Soc.* 96, 371–379. doi: 10.1111/j.1095-8339.1988.tb00692.x

White, R. A. (1979). "Experimental investigations of fern sporophyte development," in *The Experimental Biology of Ferns*, ed. A. F. Dyer (New York: Academic Press), 505–549.

Whittier, D. P., and Steeves, T. A. (1962). Further studies on induced apogamy in ferns. *Can. J. Bot.* 40, 1525–1531. doi: 10.1139/b62-148

Yamane, H. (1998). Fern antheridiogens. *Int. Rev. Cytol.* 184, 1–32. doi: 10.1016/S0074-7696(08)62177-4

Yamane, H., Satoh, Y., Nohara, K., Nakayama, M., Murofushi, N., Takahashi, N., et al. (1988). The methyl ester of a new gibberellin, GA73: the principal antheridiogen in *Lygodium japonicum*. *Tetrahedron Letters* 29, 3959–3962. doi: 10.1016/S0040-4039(00)80393-7

Yamane, H., Takahashi, N., Takeno, K., and Furuya, M. (1979). Identification of gibberellin A9 methyl ester as a natural substance regulating formation of reproductive organs in *Lygodium japonicum*. *Planta* 147, 251–256. doi: 10.1007/BF00388747

Yin, T., and Quinn, J. A. (1995). Tests of a mechanistic model of one hormone regulating both sexes in *Cucumis sativus* (Cucurbitaceae). *Am. J. Bot.* 82, 1537–1546. doi: 10.2307/2446182

16

MeioBase: a comprehensive database for meiosis

Hao Li[1,2], Fanrui Meng[1,2], Chunce Guo[1], Yingxiang Wang[3], Xiaojing Xie[4], Tiansheng Zhu[4], Shuigeng Zhou[4], Hong Ma[3,5], Hongyan Shan[1]* and Hongzhi Kong[1]*

[1] *State Key Laboratory of Systematic and Evolutionary Botany, Institute of Botany, Chinese Academy of Sciences, Beijing, China*
[2] *University of Chinese Academy of Sciences, Beijing, China*
[3] *State Key Laboratory of Genetic Engineering and Collaborative Innovation Center for Genetics and Development, Ministry of Education Key Laboratory of Biodiversity Sciences and Ecological Engineering, Institute of Plant Biology, Institute of Biodiversity Science, Center for Evolutionary Biology, School of Life Sciences, Fudan University, Shanghai, China*
[4] *Shanghai Key Lab of Intelligent Information Processing and School of Computer Science, Fudan University, Shanghai, China*
[5] *Institutes of Biomedical Sciences, Fudan University, Shanghai, China*

Edited by:
Dazhong Dave Zhao, University of Wisconsin–Milwaukee, USA

Reviewed by:
Raphael Mercier, Institut National de la Recherche Agronomique, France
Fangpu Han, Chinese Academy of Sciences, China

*Correspondence:
Hongyan Shan and Hongzhi Kong, State Key Laboratory of Systematic and Evolutionary Botany, Institute of Botany, Chinese Academy of Sciences, Nanxincun 20, Beijing 100093, China
e-mail: shanhongyan@ibcas.ac.cn; hzkong@ibcas.ac.cn

Meiosis is a special type of cell division process necessary for the sexual reproduction of all eukaryotes. The ever expanding meiosis research calls for an effective and specialized database that is not readily available yet. To fill this gap, we have developed a knowledge database MeioBase (http://meiosis.ibcas.ac.cn), which is comprised of two core parts, *Resources* and *Tools*. In the *Resources* part, a wealth of meiosis data collected by curation and manual review from published literatures and biological databases are integrated and organized into various sections, such as *Cytology*, *Pathway*, *Species*, *Interaction*, and *Expression*. In the *Tools* part, some useful tools have been integrated into MeioBase, such as *Search*, *Download*, *Blast*, *Comparison*, *My Favorites*, *Submission*, and *Advice*. With a simplified and efficient web interface, users are able to search against the database with gene model IDs or keywords, and batch download the data for local investigation. We believe that MeioBase can greatly facilitate the researches related to meiosis.

Keywords: meiosis, MeioBase, knowledge database, meiotic genes, protein-protein interaction, eukaryotes, sexual reproduction

INTRODUCTION

Meiosis is a specialized cell division process essential for all sexually reproducing organisms. During meiosis, a single round of DNA replication is followed by two successive rounds of nuclear division, meiosis I and meiosis II. Meiosis I is unique and involves the segregation of homologous chromosomes (homologs), whereas meiosis II is similar to mitosis and results in the segregation of sister chromatids. The function of meiosis is to generate four haploid gametes, which are able to develop into germ cells. Fertilization of the germ cells, then, can restore the offspring to the chromosome number and complexity level of their parents (Zickler and Kleckner, 1998; Hamant et al., 2006). Meiosis not only ensures the stability of chromosome numbers between generations, but also provides genetic materials for biodiversity.

Studies of meiosis have been carried out extensively for over 100 years (Hamant et al., 2006). Chromosome behaviors in some species have been examined in detail by using cytological approaches (Orr-Weaver, 1995; Zickler and Kleckner, 1999; Ma, 2005; Birchler and Han, 2013). Over the last two decades, much efforts have been devoted to understanding the genetic basis and molecular mechanisms of meiosis in model species, such as nematode (*Caenorhabditis elegans*), budding yeast (*Saccharomyces cerevisiae*), *Arabidopsis* (*Arabidopsis thaliana*), rice (*Oryza sativa*), and maize (*Zea mays*). Genes regulating meiosis, especially those involved in homologous chromosome paring, synapsis, recombination and separation in prophase I, have also been cloned and characterized in terms of their functions (Hollingsworth et al., 1990; Sym et al., 1993; Keeney et al., 1997; Klimyuk and Jones, 1997; Yang et al., 1999; Li et al., 2004; Tang et al., 2014). Recent advances in transcriptomics, protein-protein interactions (PPIs), and phylogenetic analyses of genes and gene families related to meiosis have improved our understanding of this complex process dramatically (Kee and Keeney, 2002; Wang et al., 2004; Lin et al., 2006; Vignard et al., 2007; Chen et al., 2010; Tang et al., 2010; Yang et al., 2011; Dukowic-Schulze et al., 2014).

The ever expanding studies of meiosis and the data accumulated, which are scattered in tremendously diverse literatures and a few of large databases, such as NCBI, Ensemble, and TAIR, call for an integrative and encyclopedia-like platform for meiosis (Hamant et al., 2006; Handel and Schimenti, 2010; Luo et al., 2014). Recently, two databases related to reproductive development, GermOnline and Plant Male Reproduction Database (PMRD), have been established. GermOnline is a cross-species microarray expression database focusing on germline development, reproduction and cancer (Lardenois et al., 2010). PMRD is a comprehensive resource for browsing and retrieving knowledge about genes and mutants related to plant male reproduction (Cui et al., 2012). Notwithstanding, neither databases provide comprehensive information about meiosis, because data related to meiosis is heavily fragmented. Therefore, researchers are in great need of effective tools or databases to quickly obtain precise meiotic data from the exponentially increasing amount of information.

FIGURE 1 | Screenshot of MeioBase homepage.

Here, by collecting and integrating all sorts of information related to meiosis, as well as including and developing powerful tools for search, comparison and analysis, we established a comprehensive and specialized database for meiosis, MeioBase. It will not only serve the meiosis research community, but also help any users who need an easy and efficient access to various kinds of information related to meiosis.

DATABASE STRUCTURE AND WEB INTERFACE

The database intends to provide all necessary resources and tools for meiosis-related researches (**Figure 1**). In the *Resources* part, information related to meiosis are categorized and integrated into five major sections (i.e., *Cytology, Pathway, Species, Interaction,* and *Expression*). In the *Tools* part, useful functions for searching, analyzing, uploading, and downloading data are included in seven sections (i.e., *Search, Download, Blast, Comparison, My Favorites, Submission,* and *Advice*). This structure can provide a centralized and user-friendly web portal for meiosis-related studies.

The website is built on a Linux, Apache, MySQL, and PHP (LAMP) stack. MySQL is used for storage, maintenance, and operation of the database, and 44 data sheets have been designed to form a network storing all the data (**Figure 2**). The front–end interface is implemented in PHP, which is a popular scripting language for dynamic web page. A well-defined and packaged JavaScript called jQuery is used to enhance the interface of the website and improve user experience. A navigation tool bar containing search box and links, such as *Home, Blast, Submission, Download,* and *Help*, are also included in each web page.

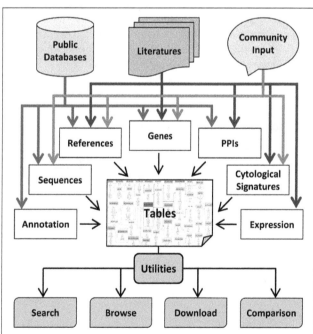

FIGURE 2 | The flowchart for construction of MeioBase. MeioBase is a comprehensive database for browsing and retrieving knowledge about meiosis, meiotic genes, and related data. MeioBase brings together three main sources of knowledge: (1) basic information about genes from public databases; (2) detailed curation of meiosis-related studies from the literatures; (3) public contributions from research community and other users. All of the information is stored in relational database tables that could be accessed by the utilities of MeioBase through any web browser.

RESOURCES

Cytological features of meiosis have been described in detail for many species, which laid a solid foundation for the study of the molecular basis of meiosis (Zickler and Kleckner, 1999; Ma, 2005). Here, in the *Cytology* section, we provide an overview of the cytological process of meiosis, with special emphasis on the chromosome behaviors at different stages. To help understand the conservation and variability of the meiotic process during evolution, cytological features in model species are summarized and compared. Important advances related to meiosis can also be retrieved through clicking the *Updated Advances* links from the overview page.

During meiotic prophase I, several critical events related to meiotic chromosome structure and interaction occur, including pairing, synapsis, recombination, and segregation. Consistent with this, genes regulating meiosis have also been grouped into pathways or networks, each of which corresponds to one of the meiotic events (Gerton and Hawley, 2005; Chang et al., 2011). For this reason, we have established the *Pathway* section, in which genes with similar or related functions, as well as the complex regulatory relationships between them, are visualized by diagrams.

During the last 20 years, many genes involved in meiosis have been discovered and functionally characterized. By literature mining, 483 meiotic genes have been collected (**Table 1**). In the *Species* section, users can find the genes of a certain species and go into the details. In the *Gene Detailed Information* page, we have integrated useful information from references, databases and web servers. Data are organized by different aspects, such as *General Information*, *Featured Domains*, *Protein Signatures*, *Gene Ontology*, *Protein Sequence*, and *References*. *General Information* contains species name, gene name, gene family name, a brief description, gene model ID, and the coding sequence (CDS) of the gene.

Featured Domains provides protein domains predicted by Pfam. *Protein Signatures* and *Gene Ontology* provide more information of the functions of the proteins. *References* includes literatures with PMIDs in PubMed (**Figure 3**).

We also include 11,201 pieces of PPI data of the collected meiotic genes of *Arabidopsis*, nematode, and budding yeast in MeioBase, and display them in the *Interaction* section (**Table 1**). The PPI data are retrieved from 283 literatures and 11 databases, such as BIND, BioGRID, DIP, IntAct, and STRING. Users can search for partners of the proteins of interest, for which corresponding sources and references are provided.

To provide more comprehensive information on meiosis, references related to gene expression patterns are collected and displayed in the *Expression* section. Until now, important references of various species, such as *Arabidopsis*, rice, maize, petunia, wheat, mouse, rat, budding yeast, and fission yeast, have been listed in this section and more expression data of meiosis are being collected.

In addition to the aforementioned five sections, MeioBase provides many other resources in the *Links* section, such as the commonly used databases, powerful web servers for molecular and genomic analysis, and experimental protocols for meiosis research, etc. An introduction to our "Plant Meiosis Project" can be found in the *Project* section. Important progresses and events on meiosis research are available in the *News* section. Release notes of MeioBase are announced in the *Notice* section. To provide an overview of our database, we also include a detailed introduction in the *Help* section.

TOOLS

MeioBase provides various ways to retrieve the data that users are interested in. By using the search box, users can search genes by gene names, gene model IDs or MeioBase IDs. In the *Search*

Table 1 | Data status in the current MeioBase.

Species	Common name	Number of meiotic genes[1]						PPIs[2]
		Total	Initiation & Pairing	Synaptonemal complex	Chromosomal recombination	Chromosome segregation	Unclassified	
Arabidopsis thaliana	Arabidopsis	88	17	8	37	11	15	445
Oryza sativa	Rice	32	8	4	18	2	0	NA
Zea mays	Maize	8	4	1	2	1	0	NA
Caenorhabditis elegans	Nematode	178	16	11	31	62	58	675
Mus musculus	Mouse	10	0	6	4	0	0	NA
Saccharomyces cerevisiae	Budding yeast	162	18	9	71	22	42	10081
Schizosaccharomyces pombe	Fission yeast	5	1	0	3	1	0	NA
Total		483	64	39	166	99	115	11201

[1]*The number of functionally characterized meiotic genes that have been integrated into MeioBase.*
[2]*The number of PPI data that have been integrated into MeioBase.*

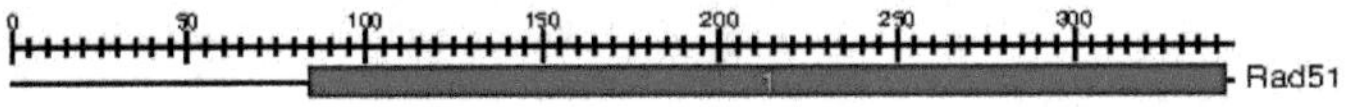

A

General Information

Species	*Arabidopsis thaliana*
MDB ID	Ath000138
Gene Family	*RAD51*
Gene Name	DMC1
Protein Properties	Length: 344 aa MW: 37512.8 Da PI: 5.4905
Description	Functionally-known meiotic gene. DISRUPTION OF MEIOTIC CONTROL 1 (DMC1); FUNCTIONS IN: in 6 functions; INVOLVED IN: DNA repair, meiosis, chiasma assembly, reciprocal meiotic recombination, DNA metabolic process; LOCATED IN: nucleus; EXPRESSED IN: 25 plant structures; EXPRESSED DURING: 12 growth stages; CONTAINS InterPro DOMAIN/s: DNA recombination/repair protein RecA/RadB, ATP-binding domain (InterPro:IPR020588), DNA repair Rad51/transcription factor NusA, alpha-helical (InterPro:IPR010995), ATPase, AAA+ type, core (InterPro:IPR003593), Meiotic recombinase Dmc1 (InterPro:IPR011940), DNA recombination and repair protein, RecA-like (InterPro:IPR016467), DNA recombination/repair protein RecA, monomer-monomer interface (InterPro:IPR020587), DNA recombination and repair protein Rad51, C-terminal (InterPro:IPR013632); BEST Arabidopsis thaliana protein match is: RAS associated with diabetes protein 51 (TAIR:AT5G20850.1); Has 11998 Blast hits to 11924 proteins in 3797 species: Archae - 689; Bacteria - 8026; Metazoa - 742; Fungi - 447; Plants - 503; Viruses - 22; Other Eukaryotes - 1569 (source: NCBI BLink).

Gene Model	Gene Model ID	Type	Source	Coding Sequence
	AT3G22880.1	genome	TAIR	View CDS
	NM_113188.2	refseq	Refseq	View CDS

B

Featured Domains

No.	Domain	Score	E-value	Start	End	HMM Start	HMM End
1	Rad51	416.9	1.8E-125	89	343	2	256

C

Protein Signatures

Database	Entry ID	E-value	Start	End	InterPro ID	Description
PIR-PSD	PIRSF005856	0	1	344	IPR016467	Rad51
PANTHER	PTHR22942:SF13	0	9	344	noIPR	PTHR22942:SF13
PANTHER	PTHR22942	0	9	344	noIPR	PTHR22942

D

Gene Ontology

GO Term	GO Category	GO Description
GO:0006259	Biological Process	DNA metabolic process
GO:0007131	Biological Process	reciprocal meiotic recombination

E

Protein Sequence

Length: 344 aa   Download sequence   Send to blast

```
MMASLKAEET SQMQLVEREE NDEDEDLFEM IDKLIAQGIN AGDVKKLQEA GIHTCNGLMM HTKKNLTGIK GLSEAKVDKI 80
CEAAEKIVNF GYMTGSDALI KRKSVVKITT GCQALDDLLG GGIETSAITE AFGEFRSGKT QLAHTLCVTT QLPTNMKGGN 160
GKVAYIDTEG TFRPDRIVPI AERFGMDPGA VLDNIIYARA YTYEHQYNLL LGLAAKMSEE PFRILIVDSI IALFRVDFTG 240
RGELADRQQK LAQMLSRLIK IAEEFNVAVY MTNQVIADPG GGMFISDPKK PAGGHVLAHA ATIRLLFRKG KGDTRVCKVY 320
DAPNLAEAEA SFQITQGGIA DAKD
```

F

References

1 Klimyuk VI, Jones JD.

AtDMC1, the Arabidopsis homologue of the yeast DMC1 gene: characterization, transposon-induced allelic variation and meiosis-associated expression

PMID: 9025299

FIGURE 2 | Screenshot of gene information. The *Gene Detailed Information* page of a meiotic gene consists of six aspects. **(A)** *General Information*; **(B)** *Featured Domains*; **(C)** *Protein Signatures*, **(D)** *Gene Ontology*, **(E)** *Protein Sequence*; **(F)** *References*.

section, users can search genes not only by inputting their model IDs but also by providing keywords describing them. PPI data can also be searched with keywords. All data can be downloaded in the *Download* section, which links to the FTP site.

Blast search against all the data in MeioBase is provided to facilitate users of finding similar sequence of a given sequence. In the *Comparison* section, users can add any two genes in the gene list and compare them. Users can also add any ten genes into the *My Favorites* section for fast checking afterward.

Moreover, we have integrated sections specifically for contributing meiosis data or suggestions to MeioBase. In the *Submission* section, users can first download and fill in a customized excel file as directions with meiotic genes, PPI data, and other data not yet included in the database, and then upload it to the database. After checking the uploaded data, we will add the qualified ones into MeioBase timely. In *Advice* section in homepage and at the foot of every page, users could ask any questions or give us comments or suggestions about this database. We will appreciate every user for improving MeioBase and reply as soon as possible.

FUTURE PLANS

MeioBase is the first web database providing comprehensive information on meiosis. It is only a start of establishing a large and well-known database on meiosis. We will reiterate the process of database structure and user interface development to enhance the data content and functionality. The major data content enhancement will come from elaboration of the gene annotation and incorporation of more meiotic genes in other species, various expression data from references and other databases, PPI networks of different species, and other vital pathways during meiosis.

ACKNOWLEDGMENTS

We thank Yongjie Sui for assistance in developing web interface, Dr. Zhihao Cheng and Prof. Zhukuan Cheng for help in checking meiosis data, and Dr. Hong Luo, Dr. Zhe Li, Prof. Jingchu Luo, Prof. Fangpu Han, and Prof. Ji Yang for helpful discussions, and two anonymous reviewers for valuable suggestions. This work was supported by grants from the Ministry of Sciences and Technology of China (2011CB944604).

REFERENCES

Birchler, J. A., and Han, F. (2013). Meiotic behavior of small chromosomes in maize. *Front. Plant Sci.* 4:505. doi: 10.3389/fpls.2013.00505

Chang, F., Wang, Y., Wang, S., and Ma, H. (2011). Molecular control of microsporogenesis in *Arabidopsis*. *Curr. Opin. Plant Biol.* 14, 66–73. doi: 10.1016/j.pbi.2010.11.001

Chen, C., Farmer, A. D., Langley, R. J., Mudge, J., Crow, J. A., May, G. D., et al. (2010). Meiosis-specific gene discovery in plants: RNA-Seq applied to isolated *Arabidopsis* male meiocytes. *BMC Plant Biol.* 10:280. doi: 10.1186/1471-2229-10-280

Cui, X., Wang, Q., Yin, W., Xu, H., Wilson, Z. A., Wei, C., et al. (2012). PMRD: a curated database for genes and mutants involved in plant male reproduction. *BMC Plant Biol.* 12:215. doi: 10.1186/1471-2229-12-215

Dukowic-Schulze, S., Sundararajan, A., Mudge, J., Ramaraj, T., Farmer, A. D., Wang, M., et al. (2014). The transcriptome landscape of early maize meiosis. *BMC Plant Biol.* 14:118. doi: 10.1186/1471-2229-14-118

Gerton, J. L., and Hawley, R. S. (2005). Homologous chromosome interactions in meiosis: diversity amidst conservation. *Nat. Rev. Genet.* 6, 477–487. doi: 10.1038/nrg1614

Hamant, O., Ma, H., and Cande, W. Z. (2006). Genetics of meiotic prophase I in plants. *Annu. Rev. Plant Biol.* 57, 267–302. doi: 10.1146/annurev.arplant.57.032905.105255

Handel, M. A., and Schimenti, J. C. (2010). Genetics of mammalian meiosis: regulation, dynamics and impact on fertility. *Nat. Rev. Genet.* 11, 124–136. doi: 10.1038/nrg2723

Hollingsworth, N. M., Goetsch, L., and Byers, B. (1990). The *HOP1* gene encodes a meiosis-specific component of yeast chromosomes. *Cell* 61, 73–84. doi: 10.1016/0092-8674(90)90216-2

Kee, K., and Keeney, S. (2002). Functional interactions between SPO11 and REC102 during initiation of meiotic recombination in *Saccharomyces cerevisiae*. *Genetics* 160, 111–122.

Keeney, S., Giroux, C. N., and Kleckner, N. (1997). Meiosis-specific DNA double-strand breaks are catalyzed by Spo11, a member of a widely conserved protein family. *Cell* 88, 375–384. doi: 10.1016/S0092-8674(00)81876-0

Klimyuk, V. I., and Jones, J. D. (1997). *AtDMC1*, the *Arabidopsis* homologue of the yeast *DMC1* gene: characterization, transposon-induced allelic variation and meiosis-associated expression. *Plant J.* 11, 1–14. doi: 10.1046/j.1365-313X.1997.11010001.x

Lardenois, A., Gattiker, A., Collin, O., Chalmel, F., and Primig, M. (2010). GermOnline 4.0 is a genomics gateway for germline development, meiosis and the mitotic cell cycle. *Database* 2010:baq030. doi: 10.1093/database/baq030

Li, W., Chen, C., Markmann-Mulisch, U., Timofejeva, L., Schmelzer, E., Ma, H., et al. (2004). The *Arabidopsis AtRAD51* gene is dispensable for vegetative development but required for meiosis. *Proc. Natl. Acad. Sci. U.S.A.* 101, 10596–10601. doi: 10.1073/pnas.0404110101

Lin, Z., Kong, H., Nei, M., and Ma, H. (2006). Origins and evolution of the *recA/RAD51* gene family: evidence for ancient gene duplication and endosymbiotic gene transfer. *Proc. Natl. Acad. Sci. U.S.A.* 103, 10328–10333. doi: 10.1073/pnas.0604232103

Luo, Q., Li, Y., Shen, Y., and Cheng, Z. (2014). Ten years of gene discovery for meiotic event control in rice. *J. Genet. Genomics* 41, 125–137. doi: 10.1016/j.jgg.2014.02.002

Ma, H. (2005). Molecular genetic analyses of microsporogenesis and microgametogenesis in flowering plants. *Annu. Rev. Plant Biol.* 56, 393–434. doi: 10.1146/annurev.arplant.55.031903.141717

Orr-Weaver, T. L. (1995). Meiosis in *Drosophila*: seeing is believing. *Proc. Natl. Acad. Sci. U.S.A.* 92, 10443–10449. doi: 10.1073/pnas.92.23.10443

Sym, M., Engebrecht, J. A., and Roeder, G. S. (1993). ZIP1 is a synaptonemal complex protein required for meiotic chromosome synapsis. *Cell* 72, 365–378. doi: 10.1016/0092-8674(93)90114-6

Tang, D., Miao, C., Li, Y., Wang, H., Liu, X., Yu, H., et al. (2014). OsRAD51C is essential for double-strand break repair in rice meiosis. *Front. Plant Sci.* 5:167. doi: 10.3389/fpls.2014.00167

Tang, X., Zhang, Z., Zhang, W., Zhao, X., Li, X., Zhang, D., et al. (2010). Global gene profiling of laser-captured pollen mother cells indicates molecular pathways and gene subfamilies involved in rice meiosis. *Plant Physiol.* 154, 1855–1870. doi: 10.1104/pp.110.161661

Vignard, J., Siwiec, T., Chelysheva, L., Vrielynck, N., Gonord, F., Armstrong, S. J., et al. (2007). The interplay of RecA-related proteins and the MND1-HOP2 complex during meiosis in *Arabidopsis thaliana*. *PLoS Genet.* 3:1894–1906. doi: 10.1371/journal.pgen.0030176

Wang, G., Kong, H., Sun, Y., Zhang, X., Zhang, W., Altman, N., et al. (2004). Genome-wide analysis of the cyclin family in *Arabidopsis* and comparative phylogenetic analysis of plant cyclin-like proteins. *Plant Physiol.* 135, 1084–1099. doi: 10.1104/pp.104.040436

Yang, H., Lu, P., Wang, Y., and Ma, H. (2011). The transcriptome landscape of *Arabidopsis* male meiocytes from high-throughput sequencing: the complexity and evolution of the meiotic process. *Plant J.* 65, 503–516. doi: 10.1111/j.1365-313X.2010.04439.x

Yang, M., Hu, Y., Lodhi, M., Mccombie, W. R., and Ma, H. (1999). The *Arabidopsis SKP1-LIKE1* gene is essential for male meiosis and may control homologue separation. *Proc. Natl. Acad. Sci. U.S.A.* 96, 11416–11421. doi: 10.1073/pnas.96.20.11416

17

Organization and regulation of the actin cytoskeleton in the pollen tube

Xiaolu Qu¹, Yuxiang Jiang², Ming Chang², Xiaonan Liu², Ruihui Zhang² and Shanjin Huang²*

¹ *Center for Plant Biology, School of Life Sciences, Tsinghua University, Beijing, China*
² *Key Laboratory of Plant Molecular Physiology, Institute of Botany – Chinese Academy of Sciences, Beijing, China*

Edited by:
Dazhong Dave Zhao, University of Wisconsin-Milwaukee, USA

Reviewed by:
David G. Oppenheimer, University of Florida, USA
Alice Y. Cheung, University of Massachusetts, USA

*Correspondence:
Shanjin Huang, Key Laboratory of Plant Molecular Physiology, Institute of Botany – Chinese Academy of Sciences, Beijing 100093, China
e-mail: sjhuang@ibcas.ac.cn

Proper organization of the actin cytoskeleton is crucial for pollen tube growth. However, the precise mechanisms by which the actin cytoskeleton regulates pollen tube growth remain to be further elucidated. The functions of the actin cytoskeleton are dictated by its spatial organization and dynamics. However, early observations of the distribution of actin filaments at the pollen tube apex were quite perplexing, resulting in decades of controversial debate. Fortunately, due to improvements in fixation regimens for staining actin filaments in fixed pollen tubes, as well as the adoption of appropriate markers for visualizing actin filaments in living pollen tubes, this issue has been resolved and has given rise to the consensus view of the spatial distribution of actin filaments throughout the entire pollen tube. Importantly, recent descriptions of the dynamics of individual actin filaments in the apical region have expanded our understanding of the function of actin in regulation of pollen tube growth. Furthermore, careful documentation of the function and mode of action of several actin-binding proteins expressed in pollen have provided novel insights into the regulation of actin spatial distribution and dynamics. In the current review, we summarize our understanding of the organization, dynamics, and regulation of the actin cytoskeleton in the pollen tube.

Keywords: actin, pollen tube, actin-binding protein, formin, villin, ADF, fimbrin

INTRODUCTION

Pollen represents a critical stage of the plant life cycle and is essential for the production of seeds in flowering plants (McCormick, 2013). Upon landing on the surface of the stigma, pollen begins to hydrate and germinate, protruding outgrowth to form a tubular structure that extends rapidly in the style. This structure provides the passage for two non-motile sperm cells to be delivered to the female gametophyte and finally effect the double fertilization (Hepler et al., 2001; Lord and Russell, 2002; Cheung and Wu, 2008). Pollen tube growth is very rapid; the growth rate for maize pollen tubes in the style can reach up to 1 cm/h (Bedinger, 1992). During the journey of fertilization, the pollen tube normally traverses a distance 1000s of times the diameter of its grain. However, growth of the pollen tube is restricted to the tip region, which is therefore called "tip growth". This type of tip growth found in the pollen tube is shared by several other cell types, including root hairs in plants, protonemal cells in moss, hyphae in fungi, and neurites in animals (Cheung and Wu, 2008; Rounds and Bezanilla, 2013). Among these systems, pollen tube growth is particularly analogous to neurite growth. Surprisingly, however, despite its rigid cell wall, the pollen tube extends very fast, e.g., the lily pollen tubes even grow one order of magnitude faster than neurite (Hepler et al., 2001). The rapidity of growth implies that the underlying cellular activities may be amplified in the pollen tube. Pollen is an excellent cellular model for study of tip growth, as it is easy to culture, and most of the features associated with *in vivo* growth of pollen tubes are also observed *in vitro*. Additionally, essential mutations associated with pollen function can be maintained under a heterozygous state, which makes pollen tube a very nice genetic system to study polarized cell growth. For these reasons, over the past several decades, the pollen tube has served a very important model cellular system for intensive study of the mechanisms underlying polarized cell growth.

The actin cytoskeleton has been shown to be crucial for pollen tube growth (Taylor and Hepler, 1997; Gibbon et al., 1999; Hepler et al., 2001; Vidali et al., 2001; Smith and Oppenheimer, 2005; Hussey et al., 2006; Chen et al., 2009; Fu, 2010; Staiger et al., 2010). The precise molecular mechanisms underlying the function of the actin cytoskeleton in the pollen tube, however, remain poorly understood. Different models have been proposed regarding the function of the actin cytoskeleton during pollen tube growth. One of the more common ideas is that the actin cytoskeleton drives the intracellular transport system that carries Golgi-derived vesicles containing the materials necessary for cell wall synthesis and membrane fusion to the tip (Pierson and Cresti, 1992; Hepler et al., 2001; Vidali and Hepler, 2001). The actin cytoskeleton has also been viewed as a structural element that supports the turgor pressure needed to drive and maintain rapid pollen tube growth (Picton and Steer, 1982; Steer and Steer, 1989; Derksen et al., 1995). Additionally, actin polymerization itself has also been shown to be important for pollen tube growth (Gibbon et al., 1999; Vidali et al., 2001). Irrespective of the particular mechanism underlying the function of the actin cytoskeleton in polarized growth, it is important to precisely describe its spatial distribution and dynamics in the pollen tube. Since, the spatial distribution and dynamics of the actin cytoskeleton are modified by the presence of various actin-binding proteins (ABPs) in cells (Staiger et al., 2010; Huang et al., 2014), it is also important to characterize the function and mode of action of these ABPs. Indeed, recent characterization of the mode of action of several pollen-expressed ABPs has provided exceptional insights into the regulation of actin organization and dynamics in the pollen tube. Thus, the purpose of this review is to summarize our current understanding of the organization and dynamics of the actin cytoskeleton, as well as its regulation, in the pollen tube.

SPATIAL DISTRIBUTION OF ACTIN FILAMENTS IN THE POLLEN TUBE

The distributions of the actin cytoskeleton in fixed pollen tubes from different species have been characterized using immunostaining with anti-actin antibodies and staining with fluorescent phalloidin (Tang et al., 1989; Rutten and Derksen, 1990; Gibbon et al., 1999; Geitmann et al., 2000; Li et al., 2001; Ye et al., 2009). Historically, reaching a consensus with respect to the distribution of actin filaments in the apical and subapical regions has been quite problematic. Different results have been reported regarding the distribution of actin filaments in the apical and subapical regions (Derksen et al., 1995; Miller et al., 1996). The variation could be due to differences between species or due to the use of different staining methods. The variation in actin structures in the apical region of pollen tubes stained using different methods most likely results from alterations in the fixation step, in which apical actin filaments may not be well-preserved. This problem is presumably due to two factors: one is that the pollen tube grows too rapidly and cannot be fixed instantaneously, and another is that apical actin filaments are highly dynamic and fragile. The original observations from experiments using conventional fixation procedures showed that dense actin filaments are present in the tip (Tiwari and Polito, 1988; Tang et al., 1989; Derksen et al., 1995). However, injection of rhodamine phalloidin into pollen tubes failed to label actin filaments in the extreme apex (Miller et al., 1996), and several other studies showed that the extreme apex is depleted of actin filaments in fixed pollen tubes (Li et al., 2001; Raudaskoski et al., 2001; Vidali et al., 2001). Later, using improved fixation regimens more likely to better preserve the cellular structures, such as rapid freeze fixation, the distribution of the actin cytoskeleton in the apical region was reproducibly revealed in the pollen tube (Lovy-Wheeler et al., 2005). This actin distribution pattern was further confirmed by the results of actin filament labeling using live-cell actin markers (Kost et al., 1998; Fu et al., 2001; Qu et al., 2013; see also the description below).

Thus, the current consensus view of the spatial distribution of actin filaments is that they are arrayed into at least three distinct structures in the pollen tube, consistent with the zonation of cytoplasm (**Figure 1A**; Lovy-Wheeler et al., 2005; Ren and Xiang, 2007; Chen et al., 2009; Staiger et al., 2010). In the shank, actin filaments are arranged axially into bundles with uniform polarity, which allows the transport of organelles or vesicles from the base to the tip along the cell cortex. At the subapex, actin filaments form regular structures referred to as the collar (Gibbon et al., 1999; Fu et al., 2001), fringe (Lovy-Wheeler et al., 2005), mesh (Geitmann et al., 2000; Chen et al., 2002), or funnel (Vidali et al., 2001) in pollen tubes from different species. In this region,

cytoplasmic streaming reverses direction and turns back toward the base along the axial actin cables in the center of the tube, giving rise to the reverse-fountain cytoplasmic streaming pattern (Hepler et al., 2001; Ye et al., 2009). Though large organelles do not enter the apical region, small vesicles enter into and become accumulated in the apical region (**Figures 1A,B**). In the apical region, actin filaments are less abundant, but are highly dynamic. The dynamics of the tip-localized population of actin filaments have been investigated in tobacco, lily, and *Arabidopsis* pollen tubes (Fu et al., 2001; Qu et al., 2013; Rounds et al., 2014 see also the detailed description below), which has expanded our understanding of the function of actin filaments. However, the means by which those actin filaments precisely regulate underlying cellular events, like vesicle targeting and fusion, remains to be explored.

ACTIN MARKERS USED TO DECORATE ACTIN FILAMENTS IN GROWING POLLEN TUBES

Staining of actin filaments in fixed pollen tubes has yielded a great deal of useful information regarding the spatial distribution of actin filaments in the pollen tube (Gibbon et al., 1999; Geitmann et al., 2000; Lovy-Wheeler et al., 2005). However, the use of fixed tissues provides only a static image and does not reveal how the actin cytoskeleton is remodeled during pollen tube growth. Therefore, development of markers to label actin filaments in living pollen tubes was much needed, in order to allow tracing of actin filament dynamics during pollen tube growth. The introduction of green fluorescent protein (GFP) has revolutionized the way in which cellular dynamics are visualized (Brandizzi et al., 2002). Numerous useful GFP fusion markers have been developed in order to visualize cytoskeletal dynamics and have revolutionized research in the cytoskeleton field (Stringham et al., 2012). In plants, to date, GFP-actin has never been demonstrated to be a useful marker for decorating filamentous actin, presumably due to one or more of the following reasons. One possible issue is that the GFP tag alters the function of actin and prevents its incorporation into filamentous actin or alters the conformation of filamentous actin after its incorporation. Another possible reason is that filamentous actin only represents a small population of the total actin in plant cells (Staiger and Blanchoin, 2006); therefore, the filamentous signal might be masked by the overwhelming amount of monomeric signal.

Tagging of some ABPs or the actin-binding domains derived from them with GFP has provided a non-invasive way to image actin filaments in plants. Several of these actin markers have been used to image actin filaments in growing pollen tubes (**Table 1**). The earliest actin marker used for decorating actin filaments in pollen tubes was GFP-mTalin (**Table 1**; Kost et al., 1998). However, nowadays, GFP-mTalin is rarely used because it causes excessive filament bundling (Ketelaar et al., 2004b). GFP-ADF was shown to decorate actin filaments and most prominently at the subapical region of the pollen tube (**Table 1**; Chen et al., 2002; Cheung et al., 2008, 2010), and LIM-GFP and GFP-fimbrin/ABD2-GFP decorate longitudinal actin cables in the shanks of pollen tubes (**Table 1**; **Figure 1C**; Wilsen et al., 2006; Cheung et al., 2008). However, these actin markers each have distinct disadvantages

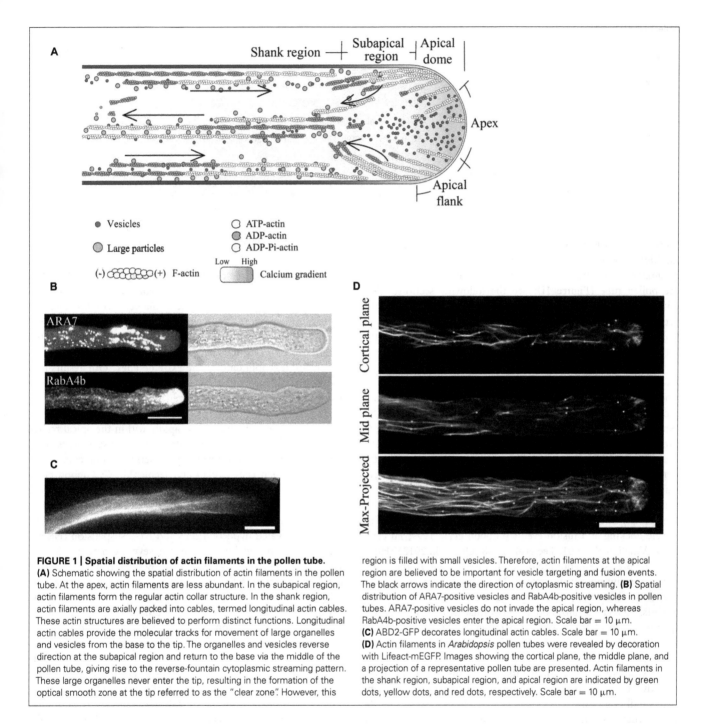

FIGURE 1 | Spatial distribution of actin filaments in the pollen tube.
(A) Schematic showing the spatial distribution of actin filaments in the pollen tube. At the apex, actin filaments are less abundant. In the subapical region, actin filaments form the regular actin collar structure. In the shank region, actin filaments are axially packed into cables, termed longitudinal actin cables. These actin structures are believed to perform distinct functions. Longitudinal actin cables provide the molecular tracks for movement of large organelles and vesicles from the base to the tip. The organelles and vesicles reverse direction at the subapical region and return to the base via the middle of the pollen tube, giving rise to the reverse-fountain cytoplasmic streaming pattern. These large organelles never enter the tip, resulting in the formation of the optical smooth zone at the tip referred to as the "clear zone". However, this region is filled with small vesicles. Therefore, actin filaments at the apical region are believed to be important for vesicle targeting and fusion events. The black arrows indicate the direction of cytoplasmic streaming. **(B)** Spatial distribution of ARA7-positive vesicles and RabA4b-positive vesicles in pollen tubes. ARA7-positive vesicles do not invade the apical region, whereas RabA4b-positive vesicles enter the apical region. Scale bar = 10 μm. **(C)** ABD2-GFP decorates longitudinal actin cables. Scale bar = 10 μm. **(D)** Actin filaments in *Arabidopsis* pollen tubes were revealed by decoration with Lifeact-mEGFP. Images showing the cortical plane, the middle plane, and a projection of a representative pollen tube are presented. Actin filaments in the shank region, subapical region, and apical region are indicated by green dots, yellow dots, and red dots, respectively. Scale bar = 10 μm.

in revealing actin structures in the pollen tube; for example, GFP-fimbrin/ABD2-GFP does not label actin filaments well in the apical and subapical regions (**Figure 1C**; Cheung et al., 2008). Despite these issues, these markers are useful for labeling different aspects of the actin cytoskeleton. Thus, the use of different actin markers can be effective for study of the distribution and changes in the actin cytoskeleton in the pollen tube. However, the ideal actin marker would be able to detect all arrays of actin filaments present in the growing pollen tube, and it would be even better if it had minimal effect on normal actin dynamics. In this regard, Lifeact-mEGFP has become the actin marker of choice

in the pollen tube. Lifeact-mEGFP contains an actin-binding site consisting of 17 amino acids derived from yeast ABP-140 fused with mEGFP. This protein decorates actin filaments in animal cells (Riedl et al., 2008) and has been used to detect actin filaments in growing tobacco and lily pollen tubes (Vidali et al., 2009; Dong et al., 2012). Recently, it has been employed to detect actin filaments in *Arabidopsis* pollen tubes (Qu et al., 2013; Zhu et al., 2013; Qin et al., 2014). Lifeact-mEGFP reveals actin structures nicely within different regions of the *Arabidopsis* pollen tube (**Figure 1D**; Qu et al., 2013), and results using this marker are reminiscent of results from actin staining of fixed pollen tubes

Table 1 | Actin markers used to decorate actin filaments in living pollen tubes.

Actin markers	Pollen tubes	Reference
GFP-mTalin	Tobacco and *Arabidopsis*	Kost et al. (1998), Fu et al. (2001), Ketelaar et al. (2004b), Wilsen et al. (2006), Wang et al. (2008a), Zhang et al. (2009, 2010b), Gui et al. (2014)
GFP-FIMBRIN/ABD2-GFP	Tobacco and lily	Wilsen et al. (2006), Liao et al. (2010)
GFP-ADF	Tobacco and lily	Chen et al. (2002), Cheung and Wu (2004), Wilsen et al. (2006), Cheung et al. (2008, 2010)
LIM-GFP	Tobacco	Cheung et al. (2008)
Lifeact-mEGFP	Tobacco, lily, and *Arabidopsis*	Vidali et al. (2009), Daher and Geitmann (2011), Dong et al. (2012), Qu et al. (2013), Zhu et al. (2013), Gui et al. (2014), Qin et al. (2014), Rounds et al. (2014)

(Gibbon et al., 1999; Lovy-Wheeler et al., 2005). Therefore, Lifeact-mEGFP represents an ideal actin marker for visualization of the organization and tracing of the dynamics of actin filaments in the pollen tube (**Figure 1D**; see the following section). Certainly, careful analysis of the organization and dynamics of the actin cytoskeleton in the entire pollen tube using a combination of standard and new actin markers will be useful in the future.

ACTIN FILAMENT DYNAMICS IN THE POLLEN TUBE

The actin cytoskeleton plays an integral role during pollen tube growth. It is well-appreciated that the relatively stable longitudinal actin cables drive intracellular movement in the shank to propel pollen tube growth (Hepler et al., 2001; Ye et al., 2009; Wu et al., 2010). Meanwhile, several studies have indicated that the dynamic state of actin filaments in the tip region is also crucial for growth (Gibbon et al., 1999; Vidali et al., 2001). Apical actin filaments are thought to play an important role in regulating the velocity and direction of pollen tube growth by controlling vesicle docking and fusion events (Gibbon et al., 1999; Fu et al., 2001; Vidali et al., 2001; Lee et al., 2008; Qu et al., 2013). However, the precise functioning of actin filaments within the apical region remains poorly understood. This is partly because we know very little about the precise organization of the apical actin filaments. Direct visualization of individual actin filaments and quantification of the associated parameters are needed to provide insights into the organizational nature of these apical actin filaments.

Previous studies have shown that pollen tube growth is more sensitive than cytoplasmic streaming to actin depolymerizing drugs (Gibbon et al., 1999; Vidali et al., 2001), implying that the actin structures at the tip are highly dynamic. Direct visualization of GFP-mTalin-decorated actin filaments showed that the tip-localized actin filaments, termed short actin bundles, are indeed highly dynamic, and that their dynamics are connected to the formation of actin structures in the subapical region (Fu et al., 2001; Hwang et al., 2005). Our recent visualization and quantification of individual actin filaments within the apical dome of the *Arabidopsis thaliana* pollen tubes has provided further insight into this system. Through the use of the advanced imaging technology of spinning disk confocal microscopy, we traced the dynamics of individual actin filaments and quantified their associated parameters, such as filament elongation and shortening rates, severing

frequency, and other factors (Qu et al., 2013). Our observations revealed that actin filaments are constantly generated from the apical membrane of the pollen tube (**Figure 2A**; Qu et al., 2013), and that this process is most likely mediated by formins, such as AtFH5 (Cheung et al., 2010). Actin filaments originating from the extreme apex are highly dynamic. They are either turned over locally or moved to the apical flank, presumably with the membrane flow (**Figure 2B**), partially explaining why actin filaments are less abundant at the extreme apex. Our results also provide convincing evidence that exocytosis occurs at the extreme apex, supporting previous findings (Lee et al., 2008; Wang et al., 2013).

Actin filaments elongate very rapidly within the apical region (**Table 2**; Qu et al., 2013), consistent with measurements indicating that a high concentration of actin/profilin complex is present in pollen cells (Vidali and Hepler, 1997; Gibbon et al., 1999; Snowman et al., 2002). Actin filaments are also severed frequently within the apical region (**Figure 2B**; Qu et al., 2013), similar to findings showing that actin filaments in the cortical region of etiolated hypocotyl cells and BY-2 suspension cells are primarily eliminated by the filament severing activity (Staiger et al., 2009; Smertenko et al., 2010). Besides driving the local turnover of apical actin filaments, this severing activity may facilitate the departure of the severed actin filaments from the apical region, providing a pool of actin filaments that can be used for the construction of actin structures in the subapical region (**Figure 2A**; Qu et al., 2013). Thus, to some extent, actin filaments in the apical and subapical regions appear to be inherently connected, consistent with previous findings (Fu et al., 2001; Hwang et al., 2005; Cheung et al., 2010). Based on these observations, we present a simple model describing the organization and dynamics of actin filaments within the apical dome of the pollen tube (**Figure 2C**; Qu et al., 2013). A highly dynamic pool of actin filaments are constantly generated from the apical membrane. They are either turned over locally by filament severing and depolymerization activities, or they can move from the extreme apex to the apical flank, leading to decreased abundance of actin filaments in the apical region of the pollen tube (**Figure 2C**).

The dynamics of actin filaments within the shank region were also traced and quantified. By comparison, maximal filament length substantially increased, and severing frequency substantially decreased in the shank region compared to that

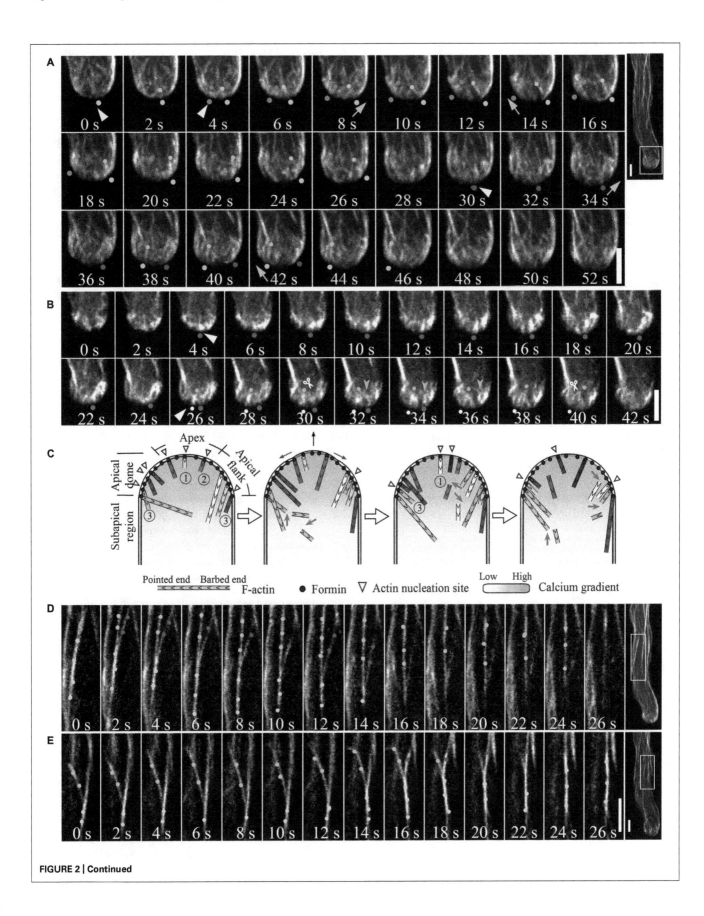

FIGURE 2 | Continued

FIGURE 2 | Continued

Actin dynamics in the pollen tube. (A) Actin filaments are constantly generated from the apical membrane within the apical dome. The images presented are maximally projected time-lapse images. Emerging individual actin filaments are marked by two dots of the same color. Yellow triangles indicate the origination of actin polymerization events, and movement of the filaments from the apex to the apical flank is indicated by orange arrows. Images are a higher magnification of the boxed region of the whole pollen tube shown in the far right panel. Scale bars = 4 μm. **(B)** Corresponding single optical slices of images shown in **(A)** allowing clear visualization of single actin filament dynamics. Actin filaments are highlighted by two dots of the same color. Red arrows indicate filament elongation events, green arrows indicate filament shrinking events, and white scissors indicate severing events. Scale bar = 4 μm. **(C)** Schematic describing the dynamics of actin filaments within the apical dome. Figure adapted from Qu et al. (2013). With the permission from American Society of Plant Biologists (www.plantcell.org). For a detailed description, see the associated text and Qu et al. (2013). 1, 2, and 3 mark actin filaments that were nucleated from the membrane at the extreme apex, that moved from apex to the apical flank, and that were nucleated from the membrane at the apical flank, respectively. **(D,E)** Dynamic formation of actin bundles in the shank region. **(D)** Filament debundling events. Yellow dots highlight actin bundles that split into two bundles highlighted with red dots and green dots. The bundle marked by red dots is subjected to severing (indicted by scissors) and depolymerization. Images are a higher magnification of the boxed region shown in the far right panel. **(E)** Bundling event. Actin filaments marked by green dots and red dots were brought together via "zipping" to form the larger bundle indicated by yellow dots. Scale bars = 4 μm.

in the apical region (**Table 2**), suggesting that shank-oriented actin filaments are relatively stable. This finding is consistent with previous observations showing that cytoplasmic streaming is more resistant than pollen tube growth to actin depolymerizing drugs (Gibbon et al., 1999; Vidali et al., 2001). Given that most actin filaments are packed into bundles, presumably by "zippering" individual actin filaments together via actin bundling factors, such as villin, fimbrin, and others, actin bundling activities were analyzed in *Arabidopsis* pollen tubes (**Figures 2D,E**). Two metrics describing bundle dynamics during live-cell imaging were analyzed: the bundling and debundling frequencies (**Table 2**). Bundling was observed at a frequency of 2.3×10^{-4} events/μm^2/s, whereas unbundling occurred at a frequency of 5.4×10^{-5} events/μm^2/s (Zheng et al., 2013). By comparison, the bundling frequency in hypocotyl epidermal cells was measured to be 6.9×10^{-5} events/μm^2/s (Hoffmann et al., 2014). These findings may explain, to some extent, why most actin filaments exist

in longitudinal bundles in the shank of pollen tubes. Future documentation of the role and mechanism of action of several major actin bundling factors in these processes will shed light on the regulation of the equilibrium between individual actin filaments and bundles.

REGULATION OF ACTIN DYNAMICS IN THE POLLEN TUBE: THE ROLES AND MECHANISMS OF ACTION OF SEVERAL POLLEN-EXPRESSED ABPs

Considering the fact that the organization and dynamics of the actin cytoskeleton are directly regulated by various ABPs, studying their functions and mechanisms of action should provide insights into the regulation of the organization and dynamics of the actin cytoskeleton in the pollen tube. Indeed, genetic manipulation of pollen-expressed ABPs has increasingly enriched our knowledge in this area. Given that actin filaments in the apical and subapical regions are highly dynamic and inherently connected (Fu et al., 2001; Hwang et al., 2005; Cheung et al., 2010; Qu et al., 2013) and that shank-oriented longitudinal actin cables are relatively stable and, to some extent, functionally distinct, in the next sections, we will review the current state of knowledge regarding several ABPs that have been implicated in regulation of actin structures either in the apical and subapical regions or in the shank region.

SEVERAL ABPs HAVE BEEN IMPLICATED IN REGULATION OF THE CONSTRUCTION AND REMODELING OF ACTIN STRUCTURES IN THE APICAL AND SUBAPICAL REGIONS

It is now generally accepted that a population of highly dynamic actin filaments are present in the apical region of the pollen tube. However, the organizational nature of the actin structures within the apical region remains poorly understood. Recent characterization of several ABPs has provided unique insights into this question and has also substantially expanded our understanding of the functions of these actin filaments, leading us to consider how they regulate cellular processes, such as exocytosis.

FORMIN NUCLEATES ACTIN FILAMENTS FROM THE APICAL MEMBRANE

Live-cell imaging of actin filament dynamics has shown that actin filaments are constantly generated from the apical plasma membrane (**Figure 2A**), suggesting that membrane-anchored actin

Table 2 | Dynamic parameters associated with actin filaments in different regions of the pollen tube.

	Apical region	Subapical region	Shank
Maximal filament length (μm)	2.5 ± 0.2[1]	3.277 ± 0.322[2]	4.63 ± 0.25[2]
Filament lifetime (s)	20.2 ± 2.9[1]	25.4 ± 1.73[2]	25.7 ± 1.2[2]
Severing frequency (breaks/μm/s)	0.034 ± 0.009[1]	0.024 ± 0.005[2]	0.020 ± 0.002[2]
Elongation rate (μm/s)	0.25 ± 0.02[1]	0.245 ± 0.02[2]	0.430 ± 0.021[2]
Depolymerization rate (μm/s)	0.22 ± 0.01[1]	0.204 ± 0.01[2]	0.334 ± 0.022[2]
Bundling frequency (events/μm^2/s, $\times 10^{-4}$)	—	—	2.3 ± 0.54[2]
Debundling frequency (events/μm^2/s, $\times 10^{-4}$)	—	—	5.4 ± 0.65[2]

The quantitative parameters associated with the dynamics of individual actin filaments were reported in Qu et al. (2013)[1] and Zheng et al. (2013)[2].

nucleation factors may be required for this role. The Arp2/3 complex and formin proteins are arguably the best characterized actin nucleation factors in plants (Deeks et al., 2002; Blanchoin and Staiger, 2010; Zheng et al., 2012; Yanagisawa et al., 2013). Historically, the Arp2/3 complex was assumed to play a role in regulating the nucleation of actin filaments in the apical region of the pollen tube (Mathur and Hulskamp, 2001). However, considering the observations that actin filaments grow outward quite linearly from the apical membrane and that loss of function of the Arp2/3 complex yields no obvious phenotype in pollen (unpublished observation), it is very unlikely that the Arp2/3 complex acts as the major nucleation factor in that region. In contrast, the formins might be more reasonable candidates for actin nucleation factors in this region. Overexpression of *Arabidopsis* Formin1 (AtFH1) has been shown to result in excessive formation of actin cables and to induce membrane curvature at the pollen tube tip (Cheung and Wu, 2004), implicating this formin in the generation of actin filaments from the apical membrane. Quite recently, it was demonstrated that AtFH5 is a major regulator of the nucleation of actin filaments growing from the apical membrane of the pollen tube (Cheung et al., 2010). AtFH5 localized at the apical membrane, and knockdown of AtFH5 diminished the abundance of actin structures in the apical and subapical regions of pollen tubes (Cheung et al., 2010), suggesting that apical membrane-anchored AtFH5 nucleates actin assembly for the construction of actin structures in the apical and subapical regions. In support of this hypothesis, biochemical data revealed that AtFH5 is a bona fide actin nucleation factor and is capable of nucleating actin assembly from actin monomers or actin monomers bound to profilin (Ingouff et al., 2005). Further studies are needed to dissect how the activity of AtFH5 is regulated during pollen tube growth, as well as how it may coordinate its activity with that of other formins. Additionally, considering the fact that the formin proteins can utilize the profilin/actin complex for assembly and that the actin monomer pool in plant cells is predicted to be buffered by profilin (Staiger and Blanchoin, 2006), future analysis of the relationship between AtFH5 and profilin in the pollen tube is also necessary.

CALCIUM-DEPENDENT FILAMENT SEVERING PROTEINS DRIVE THE TURNOVER OF APICAL ACTIN FILAMENTS

Functional characterization of villin proteins has recently provided unique insights into how the rapid turnover of actin filaments in the apical region of the pollen tube is achieved. Considering the fact that a tip-focused calcium gradient is present in the pollen tube with calcium concentrations that can reach 1–3 μM in the apical region (Pierson et al., 1994; Holdaway-Clarke et al., 1997; Messerli et al., 2000), villin, a calcium-responsive actin depolymerization promoting factor, was suggested to be an important player (Hepler et al., 2001; Vidali and Hepler, 2001; Yokota et al., 2005; Staiger et al., 2010). Villin was originally isolated from lily (*Lilium longiflorum*) pollen biochemically (Nakayasu et al., 1998; Yokota et al., 1998) and was shown to reduce the length of actin filaments in the presence of calcium/calmodulin (Yokota et al., 2005). Since then, several members of the villin/gelsolin/fragmin superfamily of proteins have been implicated in regulation of actin dynamics in the pollen

tube (Xiang et al., 2007; Wang et al., 2008b). Importantly, by taking advantage of the power of *A. thaliana* genetic approaches, characterization of villins in *Arabidopsis* has provided exceptional insights into the roles of these proteins in regulating actin dynamics in the pollen tube and during its growth (Huang et al., 2014).

We have previously shown that pollen-expressed *Arabidopsis* villin2 (VLN2) and VLN5 are able to sever actin filaments in the presence of micromolar concentrations of free calcium (Zhang et al., 2010a; Bao et al., 2012). Remarkably, we also found that VLN5 promotes actin depolymerization in the presence of profilin under similar conditions (Zhang et al., 2010a), leading to the hypothesis that villins may regulate actin dynamics by promoting actin depolymerization in the apical region. Consistent with this idea, actin filaments accumulated in the apical region of *vln2 vln5* double mutant pollen tubes (Qu et al., 2013). Visualization of the dynamics of actin filaments at a single filament resolution showed that the average filament severing frequency decreased and the average maximum filament lifetime increased in the apical region of *vln2 vln5* double mutant pollen tubes (Qu et al., 2013). These results suggest that villins promote actin turnover via their calcium-dependent filament severing activity at pollen tube tips.

Consistent with the idea that actin filaments nucleated from the apical membrane are required for the construction of actin structures at the subapical region (Cheung et al., 2010), the accumulation of actin filaments in the apical region of *vln2 vln5* double mutant pollen tubes was accompanied by disorganization of actin filaments at the subapical region (Qu et al., 2013). Certainly, as villins are known to be versatile actin regulatory proteins (Huang et al., 2014), the filament bundling and stabilizing activity of VLN2 and VLN5 may also contribute to the construction of actin structures in the subapical region of the pollen tube. In support of this idea, actin filaments in the subapical region of *vln2 vln5* double mutant pollen tubes are more wavy and are thinner than those in their wild-type counterparts (Qu et al., 2013). Additionally, villin-mediated filament severing activity has also been implicated in regulation of the construction of the subapical region by eliminating actin filaments that do not align longitudinally (Qu et al., 2013).

Several recently characterized calcium-responsive filament severing proteins, such as MAP18 (Zhu et al., 2013) and MDP25 (Qin et al., 2014), may coordinate with villins to regulate actin dynamics in the pollen tube. Both MAP18 and MDP25 were shown to function as microtubule-associated proteins in vegetative cells (Wang et al., 2007; Li et al., 2011). Surprisingly, these proteins have been shown to act as regulators of actin dynamics in the pollen tube (Zhu et al., 2013; Qin et al., 2014). *In vitro* biochemical characterization revealed that both proteins are able to sever actin filaments in a calcium-dependent manner (Zhu et al., 2013; Qin et al., 2014). Direct visualization of the actin cytoskeleton showed that more actin bundles were present in the apical and subapical region of *map18* pollen tubes compared to the wild-type (Zhu et al., 2013), suggesting that MAP18 drives the turnover of actin filaments by severing. Similarly, filament severing frequency was decreased, and actin filaments were more abundant in the subapical region of *mdp25* pollen tubes (Qin et al., 2014). How these calcium-responsive filament severing proteins coordinate to regulate the turnover of actin filaments at the pollen tube apex remains to be determined. Additionally, actin-depolymerizing factor (ADF) and actin-interacting protein (AIP1) have been implicated in the regulation of actin structures at the subapical region of the pollen tube (Chen et al., 2002; Lovy-Wheeler et al., 2006), although how they coordinate with calcium-responsive filament severing proteins to drive the turnover of actin filaments remains to be addressed.

SEVERAL ABPs HAVE BEEN IMPLICATED IN THE GENERATION AND REMODELING OF LONGITUDINAL ACTIN CABLES IN THE SHANK

In the shank region, actin filaments are arranged into longitudinal cables (Lancelle and Hepler, 1992; Gibbon et al., 1999; Fu et al., 2001; Lovy-Wheeler et al., 2005). It is clear that these longitudinal actin cables regulate cytoplasmic streaming in pollen tubes by providing molecular tracks for myosins, similar to observations in other plant cells (Shimmen and Yokota, 2004). Previous results have shown that cytoplasmic streaming is relatively more resistant than pollen tube growth to actin depolymerizing drugs (Gibbon et al., 1999; Vidali et al., 2001), suggesting that longitudinal actin cables are relatively stable. However, the mechanisms by which these longitudinal actin cables are generated and maintained, as well as how they are remodeled, remain largely unknown. Characterization of several pollen-expressed ABPs, including several recently characterized in *Arabidopsis*, has shed new light on these mechanisms.

FORMINS NUCLEATE ACTIN ASSEMBLY FOR THE CONSTRUCTION OF LONGITUDINAL ACTIN CABLES

Characterization of the pollen-specific Class I formin, AtFH3, has provided insight into the nucleation step required for the generation of longitudinal actin cables in the shank of the pollen tube (Ye et al., 2009). AtFH3 is a bona fide actin nucleation factor, capable of using the actin/profilin complex to nucleate actin assembly. It is also able to interact with the barbed end of actin filaments. Knockdown of *AtFH3* using RNAi affects the formation of longitudinal actin cables, resulting in depolarized growth of the pollen tube (Ye et al., 2009). Given that AtFH3 is an important regulator of actin nucleation, it will be very interesting to better understand exactly how its activity is regulated. Furthermore, characterization of how AtFH3 coordinates with other formins to nucleate actin assembly in the shank region should be a subject of future investigation.

ADF/COFILIN REGULATES THE REMODELING OF LONGITUDINAL ACTIN CABLES

Actin-depolymerizing factor/cofilin is important for driving the rapid turnover of actin filaments in cells. Members of the ADF family have been implicated in regulation of actin turnover in pollen. For example, lily ADFs were shown to accumulate at the germination aperture during tube protrusion, but distribute evenly in the pollen tube (Smertenko et al., 2001). In contrast to this observation, tagging of tobacco pollen-specific ADF/cofilin (NtADF1) with GFP revealed that this protein decorates filamentous actin, particularly subapical actin structures and longitudinal actin cables in the shank (Chen et al., 2002). Overexpression of NtADF1 reduces the number of axially arranged fine actin cables

(Chen et al., 2002), implicating ADF as a driver in the turnover of longitudinal actin cables.

In order to probe the intracellular localization of pollen-specific *Arabidopsis* actin-depolymerizing factor 7 (ADF7) in the pollen tube, we generated several ADF7-EGFP fusion constructs containing EGFP inserted in different locations within the ADF7 molecule in hopes of minimizing the interference of the EGFP fusion on the function of ADF7. Among which, one ADF7-EGFP fusion protein (ADF7-EGFP$_{V10}$) is fully functional (Zheng et al., 2013) and decorates filamentous actin throughout the pollen tube (Daher et al., 2011; Zheng et al., 2013). *In vitro* biochemical characterization showed that ADF7 is a typical ADF; it prefers ADP-loaded actin and inhibits nucleotide exchange, and it is able to promote actin depolymerization and sever actin filaments. However, by comparison, its actin depolymerizing and severing activity are lower than that of the vegetative ADF1 (Zheng et al., 2013). This observation is consistent with a previous report that the lily pollen ADF1 (LIADF1) has weak actin depolymerizing activity compared to the vegetative *Zea maize* ADF, ZmADF3 (Smertenko et al., 2001). These data suggest that reproductive ADFs may have evolved to play specific roles in the regulation of actin dynamics in the context of pollen.

Interestingly, ADF7-EGFP$_{V10}$ was found to fully retain both the monomer actin (G-actin) binding and filament severing activities and are fully functional *in vivo*, whereas another ADF7-EGFP fusion protein (ADF7-EGFP$_{D75}$) retained G-actin binding, but was deficient in severing actin filaments, was non-functional in the pollen tube (Zheng et al., 2013). Results from analysis of these proteins suggested that the severing activity of ADF7 is crucial for its functions *in vivo*. Consistent with this hypothesis, specific abolishment of the severing activity of yeast cofilin has been shown to affect its *in vivo* function (Chen and Pollard, 2011). These data suggest that the associated severing activity is important for the function of ADF/cofilin protein family members *in vivo*. Additional observations showed that the turnover rate of actin filaments was decreased in *adf7* pollen tubes, consistent with a role in promotion of actin depolymerization. Consequently, the amount of filamentous actin and the extent of filament bundling were increased in *adf7* pollen tubes. These data suggest that ADF7 is an important player in driving the turnover of actin filaments in the shanks of pollen tubes (Zheng et al., 2013). Detailed documentation of the functional coordination of ADF7 with other ADF isovariants, such as ADF10 (Daher et al., 2011), will provide further insight into the regulation of actin turnover in the pollen tube. Additionally, determination of the functional relationship between ADF7 and other players, such as cyclase-associated protein (CAP1) (Chaudhry et al., 2007; Deeks et al., 2007) and AIP1 (Ketelaar et al., 2004a; Shi et al., 2013) will also shed light on the regulation of the dynamic turnover of longitudinal actin cables.

ACTIN FILAMENT BUNDLING PROTEINS GENERATE LONGITUDINAL CABLES AND MAINTAIN THEIR LONGITUDINAL ARRANGEMENT

Several actin filament bundling proteins have been implicated in organization of actin filaments into bundles, as well as maintenance of the longitudinal arrangement of actin bundles in the shanks of pollen tubes. *In vitro* biochemical studies have shown that *Arabidopsis* FIMBRIN5 (FIM5) is a bona fide actin bundling factor that stabilizes actin filaments (Wu et al., 2010). Loss of function of FIM5 affects pollen germination and polarized tube growth. FIM5 decorates actin filaments throughout the pollen tube, and loss of function of FIM5 results in disorganization of actin filaments in the pollen tube and alters the longitudinal arrangement of actin cables (Wu et al., 2010; Su et al., 2012). As a result, the pattern of cytoplasmic streaming is altered, exhibiting decreased velocity and altered direction (Wu et al., 2010). Unexpectedly, actin bundles were found to be thicker in *fim5* pollen tubes compared to wild-type tubes (Wu et al., 2010), suggesting that loss of function of FIM5 may upregulate the activity some other actin bundling factors. This question might be worthwhile to examine in the future. Furthermore, the activity of LI-FIM1 was shown to be sensitive to pH (Su et al., 2012), suggesting that fimbrin might act as a sensor that regulates actin dynamics in response to pH. Further study is needed to characterize the mechanisms by which fimbrin regulates actin dynamics in the pollen tube in response to oscillations in intracellular pH.

The bundling factor villin has also been implicated in regulation of longitudinal actin bundle formation in the shank of the pollen tube. Two pollen-expressed *Arabidopsis* villin isovariants, VLN2 and VLN5, were demonstrated to be bona fide actin filament bundling proteins (Zhang et al., 2010a; Bao et al., 2012). Though loss of function of VLN5 alone did not have an overt effect on the generation and formation of longitudinal actin cables (Zhang et al., 2010a), loss of function of both VLN2 and VLN5 decreased the amount of actin filaments, suggesting that these proteins stabilize actin filaments in the pollen tube. Additionally, actin cables became thinner and more disorganized in the shanks of *vln2 vln5* pollen tubes (Qu et al., 2013), suggesting that VLN2 and VLN5 function as actin bundling factors that regulate the formation of shank-oriented longitudinal actin bundles.

Several other actin filament bundling factors may also be involved in regulating the construction and maintenance of longitudinal actin cables, such as LIMs (Papuga et al., 2010) and the recently identified, novel, plant actin-crosslinking protein, CROLIN1 (Jia et al., 2013). For instance, LI-LIM1 was shown to promote actin filament bundling and stabilize actin filaments in the pollen tube (Wang et al., 2008a), suggesting that LI-LIM1 is involved in regulating the formation of longitudinal actin cables. Loss of function of CROLIN1 led to instability of actin filaments in the shanks of pollen tubes (Jia et al., 2013), implicating this protein in regulation of the construction of shank-oriented, longitudinal actin cables. However, the means by which CROLIN1 regulates the construction and dynamics of longitudinal actin cables needs to be carefully examined.

SCHEMATIC DESCRIBING THE REGULATION OF THE CONSTRUCTION AND REMODELING OF DISTINCT ACTIN STRUCTURES IN THE POLLEN TUBE

As described above, our knowledge regarding the organization and regulation of the actin cytoskeleton in the pollen tube has grown substantially. Functional characterization of several pollen-expressed ABPs has enriched our understanding of the relevant mechanisms. In particular, direct visualization and quantitative analysis of the dynamics of individual actin filaments in pollen tubes with loss of function of specific ABPs, as well as

careful comparisons with wild-type pollen tubes have yielded substantial insight. Based on these data, we propose a simple model describing the role of various ABPs in regulating the organization of the actin cytoskeleton in the pollen tube (**Figure 3**).

Two formins, AtFH5 and AtFH3, regulate the construction of actin structures in the apical and subapical regions and shank region, respectively. AtFH5 is localized on the apical membrane within the apical dome, where it nucleates actin filaments from the apical membrane that are used for the construction of actin structures in the apical and subapical regions (**Figure 3B**; Cheung et al., 2010). In contrast, AtFH3 nucleates actin filaments within the cytoplasm or from the membrane to generate longitudinal actin cables in the shank (**Figure 3B**; Ye et al., 2009). Certainly, transport of actin filaments from the apical and subapical regions to the shank could represent another potential mechanism leading to the construction of longitudinal actin cables. Additionally, other pollen-expressed formin proteins may play yet undiscovered roles in construction of distinct actin arrays in the pollen tube.

To build actin structures in the apical and subapical regions, AtFH5-generated actin filaments are instantaneously bundled by actin bundling factors, such as villins (Qu et al., 2013), fimbrins (Wu et al., 2010; Su et al., 2012), and/or LIMs (Papuga et al., 2010), allowing them to grow outward linearly from the membrane. These actin filaments are subjected to rapid turnover due to filament severing and depolymerizing activities. With respect to these activities, several calcium-dependent filament severing and depolymerizing proteins, including villins (Qu et al., 2013), MAP18 (Zhu et al., 2013), and MDP25 (Qin et al., 2014), are reasonable candidates for these roles. Together, these mechanisms presumably lead to the decreased abundance of actin filaments at the pollen tube apex. Certainly, a potential role for ADF and its cofactors in promoting the turnover of actin filaments in the apical region needs to be examined in the future.

Filament severing activity mediated by calcium-responsive severing proteins may allow the departure of actin filaments originating from the apical membrane away from the apical region. These filaments can then be used for the construction of subapical actin structures. This idea is partially supported by the observation that loss of function of villins affects the formation of actin structures at the subapex (Qu et al., 2013). Furthermore, actin filaments originating from the extreme apical membrane can shift toward the apical flank via membrane flow and can further elongate to directly participate in the construction of actin structures at the subapex. At the subapical region, villin plays a major role in the formation of regular actin collars through bundling and stabilization of longitudinally-aligned actin filaments (**Figure 3B**; Qu et al., 2013). Additionally, villin may facilitate the formation of regular actin collars by eliminating actin filaments that do not align longitudinally via its filament severing activity (**Figure 3B**; Qu et al., 2013). MDP25 may also have a similar function in severing of actin filaments that do not align longitudinally in the subapical region (Qin et al., 2014). Furthermore, previous studies suggest that ADF and AIP1 may also be involved in regulating the turnover of actin structures at the subapical region (Chen et al., 2002; Lovy-Wheeler et al., 2006).

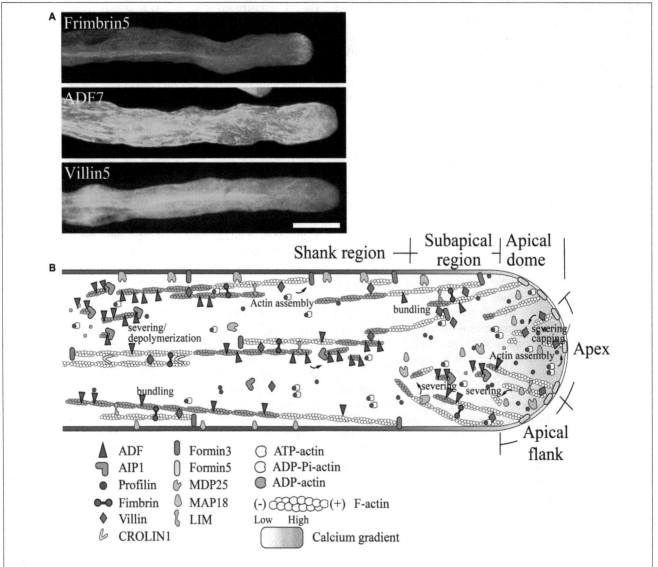

FIGURE 3 | Schematic describing the regulation of actin dynamics in the pollen tube based on the functional characterization of ABPs derived mainly from *Arabidopsis*. (A) Intracellular localization of several ABPs in the pollen tube. For methods used to determine the localization of FIMBRIN5, ADF7, and VILLIN5, see descriptions in previous studies (Wu et al., 2010; Qu et al., 2013; Zheng et al., 2013). Scale bar = 10 μm. **(B)** Schematic describing the intracellular localization and function of various ABPs in the pollen tube. For detailed information regarding the intracellular localization and function of each ABP, see the description in the text.

In the shank, AtFH3-generated actin filaments (Ye et al., 2009) initially undergo dynamic turnover due to the activity of ADFs, such as ADF7 (Zheng et al., 2013), along with several ADF cofactors, like AIP1 (Ketelaar et al., 2004a; Shi et al., 2013) and CAP1 (Chaudhry et al., 2007; Deeks et al., 2007). Subsequently, actin bundling factors, such as fimbrins, villins, LIMs, and CROLIN1, may participate in packing these filaments into longitudinal actin cables in addition to maintaining their longitudinal arrangement and stabilizing them (**Figure 3B**).

CONCLUDING REMARKS

There is no doubt that our knowledge regarding the organization, dynamics, and regulation of the actin cytoskeleton in the pollen tube has grown substantially in recent years, although many issues remain to be resolved. Importantly, adoption of the *A. thaliana* pollen tube as a cellular system for the study of these processes has greatly facilitated progress in this field due to the powerful combination of *Arabidopsis* genomic and genetic approaches, as well as the introduction of complex spatiotemporal imaging technology and the development of appropriate actin markers that have allowed real-time visualization of individual actin filaments. Recent improvements in imaging technology have expanded our view of actin filament dynamics, as well as our understanding of the underlying organization of actin structures in the pollen tube. A detailed description of the dynamic properties of actin filaments in the apical region represents one of the exciting achievements resulting from these technologies (Qu et al., 2013), and this data has greatly enhanced our understanding of the function of the actin cytoskeleton in regulating polarized pollen tube growth. Careful characterization of the mode of action of several pollen-expressed ABPs has provided additional insights into regulation of the actin cytoskeleton during pollen tube growth. As some ABPs act as direct sensors for various signals, these studies have also shed light on how various signals converge on these ABPs to control actin dynamics. For example, Ca^{2+} signaling has been shown to regulate actin dynamics by controlling the activity of villins in the pollen tube. An important future challenge will be to delineate the roles of various signaling transduction pathways in order to determine how various signals converge on ABPs to regulate actin dynamics in the pollen tube. Certainly, in this field, the eternal and most challenging question is still to understand precisely how the actin cytoskeleton functions to regulate the pollen tube growth.

ACKNOWLEDGMENTS

This work was supported by grants from the Ministry of Science and Technology of China (2013CB945100 and 2011CB944600) and the National Natural Science Foundation of China (31125004).

REFERENCES

Bao, C. C., Wang, J., Zhang, R. H., Zhang, B. C., Zhang, H., Zhou, Y. H., et al. (2012). *Arabidopsis* VILLIN2 and VILLIN3 act redundantly in sclerenchyma development via bundling of actin filaments. *Plant J.* 71, 962–975. doi: 10.1111/j.1365-313x.2012.05044.x

Bedinger, P. (1992). The remarkable biology of pollen. *Plant Cell* 4, 879–887. doi: 10.1105/tpc.4.8.879

Blanchoin, L., and Staiger, C. J. (2010). Plant formins: diverse isoforms and unique molecular mechanism. *Biochim. Biophys. Acta* 1803, 201–206. doi: 10.1016/j.bbamcr.2008.09.015

Brandizzi, F., Fricker, M., and Hawes, C. (2002). A greener world: the revolution in plant bioimaging. *Nat. Rev. Mol. Cell Biol.* 3, 520–530. doi: 10.1038/nrm861

Chaudhry, F., Guerin, C., Von Witsch, M., Blanchoin, L., and Staiger, C. J. (2007). Identification of *Arabidopsis* cyclase-associated protein 1 as the first nucleotide exchange factor for plant actin. *Mol. Biol. Cell* 18, 3002–3014. doi: 10.1091/mbc.e06-11-1041

Chen, C. Y., Wong, E. I., Vidali, L., Estavillo, A., Hepler, P. K., Wu, H. M., et al. (2002). The regulation of actin organization by actin-depolymerizing factor in elongating pollen tubes. *Plant Cell* 14, 2175–2190. doi: 10.1105/tpc.003038

Chen, N. Z., Qu, X. L., Wu, Y. J., and Huang, S. J. (2009). Regulation of actin dynamics in pollen tubes: control of actin polymer level. *J. Integr. Plant Biol.* 51, 740–750. doi: 10.1111/j.1744-7909.2009.00850.x

Chen, Q., and Pollard, T. D. (2011). Actin filament severing by cofilin is more important for assembly than constriction of the cytokinetic contractile ring. *J. Cell Biol.* 195, 485–498. doi: 10.1083/jcb.201103067

Cheung, A. Y., Duan, Q. H., Costa, S. S., De Graaf, B. H. J., Di Stilio, V. S., Feijo, J., et al. (2008). The dynamic pollen tube cytoskeleton: live cell studies using actin-binding and microtubule-binding reporter proteins. *Mol. Plant* 1, 686–702. doi: 10.1093/mp/ssn026

Cheung, A. Y., Niroomand, S., Zou, Y. J., and Wu, H. M. (2010). A transmembrane formin nucleates subapical actin assembly and controls tip-focused growth in pollen tubes. *Proc. Natl. Acad. Sci. U.S.A.* 107, 16390–16395. doi: 10.1073/pnas.1008527107

Cheung, A. Y., and Wu, H. M. (2004). Overexpression of an *Arabidopsis* formin stimulates supernumerary actin cable formation from pollen tube cell membrane. *Plant Cell* 16, 257–269. doi: 10.1105/tpc.016550

Cheung, A. Y., and Wu, H. M. (2008). Structural and signaling networks for the polar cell growth machinery in pollen tubes. *Annu. Rev. Plant Biol.* 59, 547–572. doi: 10.1146/annurev.arplant.59.032607.092921

Daher, F. B., and Geitmann, A. (2011). Actin is involved in pollen tube tropism through redefining the spatial targeting of secretory vesicles. *Traffic* 12, 1537–1551. doi: 10.1111/j.1600-0854.2011.01256.x

Daher, F. B., Van Oostende, C., and Geitmann, A. (2011). Spatial and temporal expression of actin depolymerizing factors ADF7 and ADF10 during male gametophyte development in *Arabidopsis thaliana*. *Plant Cell Physiol.* 52, 1177–1192. doi: 10.1093/pcp/pcr068

Deeks, M. J., Hussey, P. J., and Davies, B. (2002). Formins: intermediates in signal-transduction cascades that affect cytoskeletal reorganization. *Trends Plant Sci.* 7, 492–498. doi: 10.1016/s1360-1385(02)02341-5

Deeks, M. J., Rodrigues, C., Dimmock, S., Ketelaar, T., Maciver, S. K., Malho, R., et al. (2007). *Arabidopsis* CAP1 – a key regulator of actin organisation and development. *J. Cell Sci.* 120, 2609–2618. doi: 10.1242/jcs.007302

Derksen, J., Rutten, T., Vanamstel, T., Dewin, A., Doris, F., and Steer, M. (1995). Regulation of pollen tube growth. *Acta Bot. Neerl.* 44, 93–119. doi: 10.1111/j.1438-8677.1995.tb00773.x

Dong, H. J., Pei, W. K., and Ren, H. Y. (2012). Actin fringe is correlated with tip growth velocity of pollen tubes. *Mol. Plant* 5, 1160–1162. doi: 10.1093/mp/sss073

Fu, Y. (2010). The actin cytoskeleton and signaling network during pollen tube tip growth. *J. Integr. Plant Biol.* 52, 131–137. doi: 10.1111/j.1744-7909.2010.00922.x

Fu, Y., Wu, G., and Yang, Z. B. (2001). Rop GTPase-dependent dynamics of tip-localized F-actin controls tip growth in pollen tubes. *J. Cell Biol.* 152, 1019–1032. doi: 10.1083/jcb.152.5.1019

Geitmann, A., Snowman, B. N., Emons, A. M. C., and Franklin-Tong, V. E. (2000). Alterations in the actin cytoskeleton of pollen tubes are induced by the self-incompatibility reaction in *Papaver rhoeas*. *Plant Cell* 12, 1239–1251. doi: 10.1105/tpc.12.7.1239

Gibbon, B. C., Kovar, D. R., and Staiger, C. J. (1999). Latrunculin B has different effects on pollen germination and tube growth. *Plant Cell* 11, 2349–2363. doi: 10.2307/3870960

Gui, C. P., Dong, X., Liu, H. K., Huang, W. J., Zhang, D., Wang, S. J., et al. (2014). Overexpression of the tomato pollen receptor kinase LePRK1 rewires pollen tube growth to a blebbing mode. *Plant Cell* 26, 3538–3555. doi: 10.1105/tpc.114.127381

Hepler, P. K., Vidali, L., and Cheung, A. Y. (2001). Polarized cell growth in higher plants. *Annu. Rev. Cell Dev. Biol.* 17, 159–187. doi: 10.1146/annurev.cellbio.17.1.159

Hoffmann, C., Moes, D., Dieterle, M., Neumann, K., Moreau, F., Dumas, D., et al. (2014). Live cell imaging approaches reveal actin cytoskeleton-induced self-association of the actin-bundling protein WLIM1. *J. Cell Sci.* 127, 583–598. doi: 10.1242/jcs.134536

Holdaway-Clarke, T. L., Feijo, J. A., Hackett, G. R., Kunkel, J. G., and Hepler, P. K. (1997). Pollen tube growth and the intracellular cytosolic calcium gradient oscillate in phase while extracellular calcium influx is delayed. *Plant Cell* 9, 1999–2010. doi: 10.1105/tpc.9.11.1999

Huang, S., Qu, X., and Zhang, R. (2014). Plant villins: versatile actin regulatory proteins. *J. Integr. Plant Biol.* doi: 10.1111/jipb.12293 [Epub ahead of print].

Hussey, P. J., Ketelaar, T., and Deeks, M. J. (2006). Control of the actin cytoskeleton in plant cell growth. *Annu. Rev. Plant Biol.* 57, 109–125. doi: 10.1146/annurev.arplant.57.032905.105206

Hwang, J. U., Gu, Y., Lee, Y. J., and Yang, Z. B. (2005). Oscillatory ROP GTPase activation leads the oscillatory polarized growth of pollen tubes. *Mol. Biol. Cell* 16, 5385–5399. doi: 10.1091/mbc.E05-05-0409

Ingouff, M., Gerald, J. N. F., Guerin, C., Robert, H., Sorensen, M. B., Van Damme, D., et al. (2005). Plant formin AtFH5 is an evolutionarily conserved actin nucleator involved in cytokinesis. *Nat. Cell Biol.* 7, 374–380. doi: 10.1038/ncb1238

Jia, H. L., Li, J. S., Zhu, J. G., Fan, T. T., Qian, D., Zhou, Y. L., et al. (2013). *Arabidopsis* CROLIN1, a novel plant actin-binding protein, functions in cross-linking and stabilizing actin filaments. *J. Biol. Chem.* 288, 32277–32288. doi: 10.1074/jbc.m113.483594

Ketelaar, T., Allwood, E. G., Anthony, R., Voigt, B., Menzel, D., and Hussey, P. J. (2004a). The actin-interacting protein AIP1 is essential for actin organization and plant development. *Curr. Biol.* 14, 145–149. doi: 10.1016/j.cub.2004.01.004

Ketelaar, T., Anthony, R. G., and Hussey, P. J. (2004b). Green fluorescent protein-mTalin causes defects in actin organization and cell expansion in *Arabidopsis* and inhibits actin depolymerizing factor's actin depolymerizing activity in vitro. *Plant Physiol.* 136, 3990–3998. doi: 10.1104/pp.104.050799

Kost, B., Spielhofer, P., and Chua, N. H. (1998). A GFP-mouse talin fusion protein labels plant actin filaments in vivo and visualizes the actin cytoskeleton in growing pollen tubes. *Plant J.* 16, 393–401. doi: 10.1046/j.1365-313x.1998.00304.x

Lancelle, S. A., and Hepler, P. K. (1992). Ultrastructure of freeze-substituted pollen tubes of *Lilium longiflorum*. *Protoplasma* 167, 215–230. doi: 10.1007/bf01403385

Lee, Y. J., Szumlanski, A., Nielsen, E., and Yang, Z. B. (2008). Rho-GTPase-dependent filamentous actin dynamics coordinate vesicle targeting and exocytosis during tip growth. *J. Cell Biol.* 181, 1155–1168. doi: 10.1083/jcb.2008 01086

Li, J. J., Wang, X. L., Qin, T., Zhang, Y., Liu, X. M., Sun, J. B., et al. (2011). MDP25, a novel calcium regulatory protein, mediates hypocotyl cell elongation by destabilizing cortical microtubules in *Arabidopsis*. *Plant Cell* 23, 4411–4427. doi: 10.1105/tpc.111.092684

Li, Y., Zee, S. Y., Liu, Y. M., Huang, B. Q., and Yen, L. F. (2001). Circular F-actin bundles and a G-actin gradient in pollen and pollen tubes of *Lilium davidii*. *Planta* 213, 722–730. doi: 10.1007/s004250100543

Liao, F. L., Wang, L., Yang, L. B., Peng, X. B., and Sun, M. X. (2010). NtGNL1 plays an essential role in pollen tube tip growth and orientation likely via regulation of post-golgi trafficking. *PLoS ONE* 5:e13401. doi: 10.1371/journal.pone.00 13401

Lord, E. M., and Russell, S. D. (2002). The mechanisms of pollination and fertilization in plants. *Annu. Rev. Cell Dev. Biol.* 18, 81–105. doi: 10.1146/annurev.cellbio.18.012502.083438

Lovy-Wheeler, A., Kunkel, J. G., Allwood, E. G., Hussey, P. J., and Hepler, P. K. (2006). Oscillatory increases in alkalinity anticipate growth and may regulate actin dynamics in pollen tubes of lily. *Plant Cell* 18, 2182–2193. doi: 10.1105/tpc.106.044867

Lovy-Wheeler, A., Wilsen, K. L., Baskin, T. I., and Hepler, P. K. (2005). Enhanced fixation reveals the apical cortical fringe of actin filaments as a consistent feature of the pollen tube. *Planta* 221, 95–104. doi: 10.1007/s00425-004-1423-2

Mathur, J., and Hulskamp, M. (2001). Cell growth: how to grow and where to grow. *Curr. Biol.* 11, R402–R404. doi: 10.1016/s0960-9822(01)00219-6

McCormick, S. (2013). Pollen. *Curr. Biol.* 23, R988–R990. doi: 10.1016/j.cub.2013.08.016

Messerli, M. A., Creton, R., Jaffe, L. F., and Robinson, K. R. (2000). Periodic increases in elongation rate precede increases in cytosolic Ca^{2+} during pollen tube growth. *Dev. Biol.* 222, 84–98. doi: 10.1006/dbio.2000.9709

Miller, D. D., Lancelle, S. A., and Hepler, P. K. (1996). Actin microfilaments do not form a dense meshwork in *Lilium longiflorum* pollen tube tips. *Protoplasma* 195, 123–132. doi: 10.1007/bf01279191

Nakayasu, T., Yokota, E., and Shimmen, T. (1998). Purification of an actin-binding protein composed of 115-kDa polypeptide from pollen tubes of lily. *Biochem. Biophys. Res. Commun.* 249, 61–65. doi: 10.1006/bbrc.1998.9088

Papuga, J., Hoffmann, C., Dieterle, M., Moes, D., Moreau, F., Tholl, S., et al. (2010). *Arabidopsis* LIM proteins: a family of actin bundlers with distinct expression patterns and modes of regulation. *Plant Cell* 22, 3034–3052. doi: 10.1105/tpc.110.075960

Picton, J. M., and Steer, M. W. (1982). A model for the mechanism of tip extension in pollen tubes. *J. Theor. Biol.* 98, 15–20. doi: 10.1016/0022-5193(82)90054-6

Pierson, E. S., and Cresti, M. (1992). Cytoskeleton and cytoplasmic organization of pollen and pollen tubes. *Int. Rev. Cytol.* 140, 73–125. doi: 10.1016/S0074-7696(08)61094-3

Pierson, E. S., Miller, D. D., Callaham, D. A., Shipley, A. M., Rivers, B. A., Cresti, M., et al. (1994). Pollen tube growth is coupled to the extracellular calcium ion flux and the intracellular calcium gradient: effect of BAPTA-type buffers and hypertonic media. *Plant Cell* 6, 1815–1828. doi: 10.1105/tpc.6.12.1815

Qin, T., Liu, X. M., Li, J. J., Sun, J. B., Song, L. N., and Mao, T. L. (2014). *Arabidopsis* microtubule-destabilizing protein 25 functions in pollen tube growth by severing actin filaments. *Plant Cell* 26, 325–339. doi: 10.1105/tpc.113.119768

Qu, X. L., Zhang, H., Xie, Y. R., Wang, J., Chen, N. Z., and Huang, S. J. (2013). *Arabidopsis* villins promote actin turnover at pollen tube tips and facilitate the construction of actin collars. *Plant Cell* 25, 1803–1817. doi: 10.1105/tpc.113.110940

Raudaskoski, M., Astrom, H., and Laitiainen, E. (2001). Pollen tube cytoskeleton: structure and function. *J. Plant Growth Regul.* 20, 113–130. doi: 10.1007/s003440010015

Ren, H. Y., and Xiang, Y. (2007). The function of actin-binding proteins in pollen tube growth. *Protoplasma* 230, 171–182. doi: 10.1007/s00709-006-0231-x

Riedl, J., Crevenna, A. H., Kessenbrock, K., Yu, J. H., Neukirchen, D., Bista, M., et al. (2008). Lifeact: a versatile marker to visualize F-actin. *Nat. Methods* 5, 605–607. doi: 10.1038/nmeth.1220

Rounds, C. M., and Bezanilla, M. (2013). Growth mechanisms in tip-growing plant cells. *Annu. Rev. Plant Biol.* 64, 243–265. doi: 10.1146/annurev-arplant-050312-120150

Rounds, C. M., Hepler, P. K., and Winship, L. J. (2014). The apical actin fringe contributes to localized cell wall deposition and polarized growth in the lily pollen tube. *Plant Physiol.* 166, 139–151. doi: 10.1104/pp.114.242974

Rutten, T. L. M., and Derksen, J. (1990). Organization of actin filaments in regenerating and outgrowing subprotoplasts from pollen tubes of *Nicotiana tabacum* L. *Planta* 180, 471–479. doi: 10.1007/bf02411443

Shi, M., Xie, Y. R., Zheng, Y. Y., Wang, J. M., Su, Y., Yang, Q. Y., et al. (2013). *Oryza sativa* actin-interacting protein1 is required for rice growth by promoting actin turnover. *Plant J.* 73, 747–760. doi: 10.1111/tpj.12065

Shimmen, T., and Yokota, E. (2004). Cytoplasmic streaming in plants. *Curr. Opin. Cell Biol.* 16, 68–72. doi: 10.1016/j.ceb.2003.11.009

Smertenko, A. P., Allwood, E. G., Khan, S., Jiang, C. J., Maciver, S. K., Weeds, A. G., et al. (2001). Interaction of pollen-specific actin-depolymerizing factor with actin. *Plant J.* 25, 203–212. doi: 10.1046/j.1365-313x.2001.00954.x

Smertenko, A. P., Deeks, M. J., and Hussey, P. J. (2010). Strategies of actin reorganisation in plant cells. *J. Cell Sci.* 123, 3019–3029. doi: 10.1242/jcs.0 79749

Smith, L. G., and Oppenheimer, D. G. (2005). Spatial control of cell expansion by the plant cytoskeleton. *Annu. Rev. Cell Dev. Biol.* 21, 271–295. doi: 10.1146/annurev.cellbio.21.122303.114901

Snowman, B. N., Kovar, D. R., Shevchenko, G., Franklin-Tong, V. E., and Staiger, C. J. (2002). Signal-mediated depolymerization of actin in pollen during the self-incompatibility response. *Plant Cell* 14, 2613–2626. doi: 10.1105/tpc.0 02998

Staiger, C. J., and Blanchoin, L. (2006). Actin dynamics: old friends with new stories. *Curr. Opin. Plant Biol.* 9, 554–562. doi: 10.1016/j.pbi.2006.09.013

Staiger, C. J., Poulter, N. S., Henty, J. L., Franklin-Tong, V. E., and Blanchoin, L. (2010). Regulation of actin dynamics by actin-binding proteins in pollen. *J. Exp. Bot.* 61, 1969–1986. doi: 10.1093/jxb/erq012

Staiger, C. J., Sheahan, M. B., Khurana, P., Wang, X., Mccurdy, D. W., and Blanchoin, L. (2009). Actin filament dynamics are dominated by rapid growth and severing activity in the *Arabidopsis* cortical array. *J. Cell Biol.* 184, 269–280. doi: 10.1083/jcb.200806185

Steer, M. W., and Steer, J. M. (1989). Tansley review No. 16 pollen tube tip growth. *New Phytol.* 111, 323–358. doi: 10.1111/j.1469-8137.1989.tb00697.x

Stringham, E. G., Marcus-Gueret, N., Ramsay, L., and Schmidt, K. L. (2012). Live cell imaging of the cytoskeleton. *Methods Enzymol.* 505, 203–217. doi: 10.1016/b978-0-12-388448-0.00019-x

Su, H., Zhu, J. S., Cai, C., Pei, W. K., Wang, J. J., Dong, H. J., et al. (2012). FIMBRIN1 is involved in lily pollen tube growth by stabilizing the actin fringe. *Plant Cell* 24, 4539–4554. doi: 10.1105/tpc.112.099358

Tang, X. J., Lancelle, S. A., and Hepler, P. K. (1989). Fluorescence microscopic localization of actin in pollen tubes: comparison of actin antibody and phalloidin staining. *Cell Motil. Cytoskeleton* 12, 216–224. doi: 10.1002/cm.9701 20404

Taylor, L. P., and Hepler, P. K. (1997). Pollen germination and tube growth. *Annu. Rev. Plant Phys.* 48, 461–491. doi: 10.1146/annurev.arplant.48.1.461

Tiwari, S. C., and Polito, V. S. (1988). Organization of the cytoskeleton in pollen tubes of *Pyrus communis*: a study employing conventional and freeze-substitution electron microscopy, immunofluorescence, and rhodamine-phalloidin. *Protoplasma* 147, 100–112. doi: 10.1007/bf01403337

Vidali, L., and Hepler, P. K. (1997). Characterization and localization of profilin in pollen grains and tubes of *Lilium longiflorum*. *Cell Motil. Cytoskeleton* 36, 323–338. doi: 10.1002/(SICI)1097-0169(1997)36:43.0.CO;2-6

Vidali, L., and Hepler, P. K. (2001). Actin and pollen tube growth. *Protoplasma* 215, 64–76. doi: 10.1007/bf01280304

Vidali, L., Mckenna, S. T., and Hepler, P. K. (2001). Actin polymerization is essential for pollen tube growth. *Mol. Biol. Cell* 12, 2534–2545. doi: 10.1091/mbc.12. 8.2534

Vidali, L., Rounds, C. M., Hepler, P. K., and Bezanilla, M. (2009). Lifeact-mEGFP reveals a dynamic apical F-actin network in tip growing plant cells. *PLoS ONE* 4:e5744. doi: 10.1371/journal.pone.0005744

Wang, H. J., Wan, A. R., and Jauh, G. Y. (2008a). An actin-binding protein, LlLIM1, mediates calcium and hydrogen regulation of actin dynamics in pollen tubes. *Plant Physiol.* 147, 1619–1636. doi: 10.1104/pp.108.118604

Wang, T., Xiang, Y., Hou, J., and Ren, H. Y. (2008b). ABP41 is involved in the pollen tube development via fragmenting actin filaments. *Mol. Plant* 1, 1048–1055. doi: 10.1093/mp/ssn073

Wang, H., Zhuang, X. H., Cai, Y., Cheung, A. Y., and Jiang, L. W. (2013). Apical F-actin-regulated exocytic targeting of NtPPME1 is essential for construction and rigidity of the pollen tube cell wall. *Plant J.* 76, 367–379. doi: 10.1111/tpj.12300

Wang, X., Zhu, L., Liu, B. Q., Wang, C., Jin, L. F., Zhao, Q., et al. (2007). *Arabidopsis* MICROTUBULE-ASSOCIATED PROTEIN18 functions in directional cell growth by destabilizing cortical microtubules. *Plant Cell* 19, 877–889. doi: 10.1105/tpc.106.048579

Wilsen, K. L., Lovy-Wheeler, A., Voigt, B., Menzel, D., Kunkel, J. G., and Hepler, P. K. (2006). Imaging the actin cytoskeleton in growing pollen tubes. *Sex. Plant Reprod.* 19, 51–62. doi: 10.1007/s00497-006-0021-9

Wu, Y. J., Yan, J., Zhang, R. H., Qu, X. L., Ren, S. L., Chen, N. Z., et al. (2010). *Arabidopsis* FIMBRIN5, an actin bundling factor, is required for pollen germination and pollen tube growth. *Plant Cell* 22, 3745–3763. doi: 10.1105/tpc.110.080283

Xiang, Y., Huang, X., Wang, T., Zhang, Y., Liu, Q. W., Hussey, P. J., et al. (2007). ACTIN BINDING PROTEIN29 from *Lilium* pollen plays an important role in dynamic actin remodeling. *Plant Cell* 19, 1930–1946. doi: 10.1105/tpc.106.048413

Yanagisawa, M., Zhang, C. H., and Szymanski, D. B. (2013). ARP2/3-dependent growth in the plant kingdom: SCARs for life. *Front. Plant Sci.* 4:166. doi: 10.3389/fpls.2013.00166

Ye, J. R., Zheng, Y. Y., Yan, A., Chen, N. Z., Wang, Z. K., Huang, S. J., et al. (2009). *Arabidopsis* Formin3 directs the formation of actin cables and polarized growth in pollen tubes. *Plant Cell* 21, 3868–3884. doi: 10.1105/tpc.109.068700

Yokota, E., Takahara, K., and Shimmen, T. (1998). Actin-bundling protein isolated from pollen tubes of lily – Biochemical and immunocytochemical characterization. *Plant Physiol.* 116, 1421–1429. doi: 10.1104/pp.116.4.1421

Yokota, E., Tominaga, M., Mabuchi, I., Tsuji, Y., Staiger, C. J., Oiwa, K., et al. (2005). Plant villin, lily P-135-ABP, possesses G-actin binding activity and accelerates the polymerization and depolymerization of actin in a Ca2+-sensitive manner. *Plant Cell Physiol.* 46, 1690–1703. doi: 10.1093/pcp/pci185

Zhang, H., Qu, X. L., Bao, C. C., Khurana, P., Wang, Q. N., Xie, Y. R., et al. (2010a). *Arabidopsis* VILLIN5, an actin filament bundling and severing protein, is necessary for normal pollen tube growth. *Plant Cell* 22, 2749–2767. doi: 10.1105/tpc.110.076257

Zhang, Y., He, J. M., Lee, D., and Mccormick, S. (2010b). Interdependence of endomembrane trafficking and actin dynamics during polarized growth of *Arabidopsis* pollen tubes. *Plant Physiol.* 152, 2200–2210. doi: 10.1104/pp.109.142349

Zhang, Y., He, J. M., and Mccormick, S. (2009). Two *Arabidopsis* AGC kinases are critical for the polarized growth of pollen tubes. *Plant J.* 58, 474–484. doi: 10.1111/j.1365-313X.2009.03792.x

Zheng, Y. Y., Xie, Y. R., Jiang, Y. X., Qu, X. L., and Huang, S. J. (2013). *Arabidopsis* ACTIN-DEPOLYMERIZING FACTOR7 severs actin filaments and regulates actin cable turnover to promote normal pollen tube growth. *Plant Cell* 25, 3405–3423. doi: 10.1105/tpc.113.117820

Zheng, Y. Y., Xin, H. B., Lin, J. X., Liu, C. M., and Huang, S. J. (2012). An *Arabidopsis* class II formin, AtFH19, nucleates actin assembly, binds to the barbed end of actin filaments, and antagonizes the effect of AtFH1 on actin dynamics. *J. Integr. Plant Biol.* 54, 800–813. doi: 10.1111/j.1744-7909.2012.01160.x

Zhu, L., Zhang, Y., Kang, E., Xu, Q., Wang, M., Rui, Y., et al. (2013). MAP18 regulates the direction of pollen tube growth in *Arabidopsis* by modulating F-actin organization. *Plant Cell* 25, 851–867. doi: 10.1105/tpc.113.110528

18

The expression and roles of parent-of-origin genes in early embryogenesis of angiosperms

An Luo[1,2], Ce Shi[1], Liyao Zhang[1] and Meng-Xiang Sun[1]*

[1] State Key Laboratory of Hybrid Rice, Department of Cell and Developmental Biology, College of Life Sciences, Wuhan University, Wuhan, China
[2] College of Life Sciences, Yangtze University, Jingzhou, China

Edited by:
Dazhong Dave Zhao, University of Wisconsin-Milwaukee, USA

Reviewed by:
Scott D. Russell, University of Oklahoma, USA
Dolf Weijers, Wageningen University, Netherlands
Wolfgang Lukowitz, University of Georgia, USA

*Correspondence:
Meng-Xiang Sun, State Key Laboratory of Hybrid Rice, Department of Cell and Developmental Biology, College of Life Sciences, Wuhan University, Wuhan 430072, Hubei, China
e-mail: mxsun@whu.edu.cn

Uniparental transcripts during embryogenesis may arise due to gamete delivery during fertilization or genomic imprinting. Such transcripts have been found in a number of plant species and appear critical for the early development of embryo or endosperm in seeds. Although the regulatory expression mechanism and function of these genes in embryogenesis require further elucidation, recent studies suggest stage-specific and highly dynamic features that might be essential for critical developmental events such as zygotic division and cell fate determination during embryogenesis. Here, we summarize the current work in this field and discuss future research directions.

Keywords: uniparental transcripts, gamete-delivered transcript, maternal control, paternal allele, genomic imprinting, embryo

INTRODUCTION

During sexual reproduction of flowering plants, male and female gametes are formed in the haploid gametophytic generation (Walbot and Evans, 2003; Chang et al., 2011). In angiosperm, the typical female gametophyte contains two kinds of female gametes, a haploid egg cell and a diploid central cell with two identical copies of the maternal genome. The male gametophyte is found in pollen, which carries one generative cell or two sperm cells. During pollen generation, sperm cells are transported though the pollen tube to the female gametophyte. Upon double fertilization, two sperm cells enter the embryo sac. One sperm cell fuses with the egg cell, and the other fuses with the central cell. This integration of the two gamete genomes results in the formation of a diploid embryo and a triploid endosperm, respectively. After fertilization, the embryo forms basic morphological and physiological structures (Le et al., 2007), during which the endosperm plays a nutritive role, similar to the placenta of mammals, to support embryonic development (Lopes and Larkins, 1993; Olsen, 2004). During this process, both paternal and maternal genetic information may contribute to the fertilization and development of the embryo, which leads to generation of the sporophyte. These parental information includes RNA that transcribed in sperm and/or egg cells, proteins that synthesized and deposited in gametes, paternal or maternal genome, and mitochondria and plastid genome. After fertilization they are brought into and integrated in zygote (**Figure 1**). Due to technical limitations the contribution of gamete-delivered proteins, mitochondria and plastid genome to zygote development and early embryogenesis are hardly investigated. Current studies as pioneer works mainly focus on *de novo* expression of imprinted genes and gamete-delivered transcripts.

The molecular mechanisms of fertilization and early embryogenesis, especially the role of parent-of-origin genes, have been well studied in animals. However, little is known about these processes in plants due to technical limitations. Gametogenesis, fertilization and embryogenesis occur deep in the plant saprophytic tissues, thus rendering it difficult to observe the developmental events and investigate the molecular mechanisms of these processes directly. Modern technological advances have allowed the isolation and analysis of gametes, zygotes, and early embryos in a wide variety of plants including maize, tobacco, *Arabidopsis*, rice, and wheat (Engel et al., 2003; Zhao et al., 2011; Nodine and Bartel, 2012; Anderson et al., 2013; Domoki et al., 2013). Therefore, great advances have been made toward understanding the role of uniparental transcripts in plant embryogenesis.

In animals, maternal allele products synthesized during gametogenesis exert control in all aspects of embryonic development prior to the global activation of the zygotic genome (Tadros and Lipshitz, 2009). However, in plants, the parental contribution in early embryogenesis has not yet been fully understood. Early reports indicated that the transcripts in early embryos were mainly originated from the maternally inherited alleles and the transcription of paternal alleles was delayed (Vielle-Calzada et al., 2000; Baroux et al., 2001; Golden et al., 2002; Grimanelli et al., 2005). Baroux et al. (2008) further suggested that early embryogenesis in plants was maternally controlled similar to that in animals, as early studies indicated that maternal transcripts could support embryonic development until the proembryo stage. At the same time, some other researches presented evidences of early activated paternal genome (Weijers et al., 2001; Scholten et al., 2002; Lukowitz et al., 2004; Sheldon et al., 2008). Recently, paternal transcripts were proved to be critical for the normal development of early

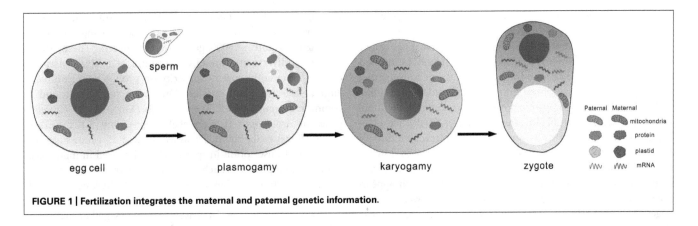

FIGURE 1 | Fertilization integrates the maternal and paternal genetic information.

embryo (Ueda et al., 2011; Babu et al., 2013). More impressively, interleukin-1 receptor-associated kinase (IRAK)/Pelle-like kinase gene, *SHORT SUSPENSOR (SSP)* transcripts were found to be produced in mature pollen and were believed to be carried into the egg cell via fertilization in *Arabidopsis thaliana* (Bayer et al., 2009). *SSP* functioned during the asymmetric first division in the zygote, indicating that the paternal transcripts from sperm cells may be involved in many aspects of zygotic development and early embryogenesis in plants. Our previous work also confirmed that paternal transcripts in sperm cells could be found in zygotes soon after fertilization (Xin et al., 2011), suggesting the possibility that sperm-delivered paternal transcripts may be involved in zygotic development.

Meyer and Scholten (2007) reported the relative expression levels of parental transcripts in zygotes, suggesting equivalent parental contribution in maize zygotic development. Maternally expressed in embryo 1 *(mee1)* in maize was the first reported imprinted gene in a plant embryo, although its function is unclear (Jahnke and Scholten, 2009). Using deep sequencing in a genome-wide analysis, Autran et al. (2011) assessed the parental contributions in early embryogenesis and found that the maternal transcripts predominated at early embryonic stages in *Arabidopsis*. With development, the relative paternal contribution arose due to the gradual activation of the embryonic genome. Subsequently, Nodine and Bartel (2012) found that a majority of genes were expressed equally from both parents at the beginning of embryogenesis in *Arabidopsis*. Interestingly, some of these works focused on the quantitative ratio of maternal and paternal transcripts, some mainly analyzed the regulatory roles of these genes. It is not surprised to see various conclusions. Even more, a latest report indicated that the different results might be due to the different material (e.g., ecotypes) they used in their experiments (Del Toro-De León et al., 2014). Despite all these discussions, it is believed that some transcripts are derived primarily from one parent or from imprinted genes in embryos soon after fertilization (Autran et al., 2011; Nodine and Bartel, 2012). These studies indicate that the parent-of-origin gene transcripts indeed exist in the zygote or early embryo. Such transcripts could arise from both the gamete-delivered and *de novo* expression of imprinted genes. Each type of uniparental transcript may play specific roles in plant development, since they are regulated by different molecular mechanisms. This review highlights the

characteristics of uniparental transcripts during early embryogenesis.

GAMETE-CARRIED MATERNAL OR PATERNAL TRANSCRIPTS INVOLVED IN EARLY EMBRYOGENESIS
MATERNAL TRANSCRIPTS
The embryo originates from a fertilized egg cell, termed a zygote. Two sequential events occur during the integration of a sperm and an egg cell: plasmogamy and karyogamy. Not only do the two genomes integrate, but also various components of the cytoplasm mix during the fertilization process. For example, sperm mitochondria could be found in fertilized egg cells of tobacco (Yu and Russell, 1994) although mitochondria is usually inherited maternally.

During early embryogenesis in most animal species, maternal transcripts deposited in the egg cells are involved in various developmental processes before activation of the zygotic genome, such as formation of embryonic axes, cell differentiation, and morphogenesis (Johnston, 1995; Wylie et al., 1996; Nishida, 1997; Angerer and Angerer, 2000; Mohr et al., 2001; Pellettieri and Seydoux, 2002). Although various experimental data support the hypothesis that maternal control may also exist during early embryogenesis in plants (Baroux et al., 2008), little is known about transcripts stored in egg cells and their role in early embryogenesis (Xin et al., 2012).

Using microdissection, Sprunck et al. (2005) constructed a cDNA library from wheat egg cells, and a total of 404 clusters were found to function in metabolic activity, mRNA translation and protein turnover. Subsequently, another 226 expressed sequence tags (ESTs) were studied in wheat egg cells (Domoki et al., 2013). In a similar analysis carried out in tobacco, thousands of ESTs were detected, which may be involved in a variety of developmental processes (Ning et al., 2006; Zhao et al., 2011). In addition, microarray technology combined with laser-assisted microdissection (LAM) was used to analyze the expression profile in *Arabidopsis* egg cells (Wuest et al., 2010). Transcriptomic analysis of egg cells isolated by manual manipulation was performed in rice (Ohnishi et al., 2011; Abiko et al., 2013), and genome-wide deep sequencing was used to characterize the gene expression profile in rice egg cells (Anderson et al., 2013). The functional categories of approximately 27,000 genes detected proved to be comprehensive. However, a comparison of the egg-specific expression of transcriptomes in rice and

Arabidopsis revealed relatively different sets of genes in egg cells of rice and *Arabidopsis* (Ohnishi et al., 2011).

The role of mRNA stored in egg cells has been investigated. Although downregulation of RNA polymerase II by RNA interference (RNAi) impeded *de novo* transcription, the development of *Arabidopsis* embryos continued until the preglobular stage (Pillot et al., 2010). In tobacco, zygotic development continued without *de novo* transcription until 72 h after pollination (HAP). The cytological observation of developmental events in transcriptionally inhibited zygotes showed that maternal transcripts stored in egg cells were functionally competent in gamete fusion, zygote volume reduction, complete cell wall formation, large vacuole disappearance, and limited cell enlargement during early developmental stages. However, *de novo* transcripts would then seize control of embryogenesis to trigger subsequent developmental processes (Zhao et al., 2011).

Interestingly, small RNA-mediated transposon silencing is thought to be an essential regulatory mechanism in male and female gametes (Slotkin et al., 2009; Martínez and Slotkin, 2012). Anderson et al. (2013) evaluated the expression of genes involved in the miRNA and siRNA pathways in transcriptomes of rice gametes and showed that all important components involved in these pathways were active in egg cells rather than in sperm cells. Thus, transposon silencing is mediated by small RNAs produced in egg cells; moreover, it is regulated in the zygote by small RNAs inherited from the egg cells (Anderson et al., 2013).

Currently, the roles of female gamete transcripts in zygotes and early embryogenesis are unclear. Although *de novo* transcription in the zygotic genome is activated within hours after fertilization in maize, tobacco and *Arabidopsis*, the maternal transcripts deposited in the egg cells still play a key role in the initial stages of zygotic development (Meyer and Scholten, 2007; Zhao et al., 2011; Nodine and Bartel, 2012).

PATERNAL TRANSCRIPTS

The sperm cell, the other contributor to zygote, has a simple structure including the karyoplasm and very little cytoplasm. Due to the condensed chromatin observed in sperm cells, it was generally thought that inactive male transcription made no contribution to early embryogenesis prior to zygotic genome activation. This view might be supported in animals, since almost all mRNAs in zygotes are inherited from egg cells (Ostermeier et al., 2004; Krawetz, 2005). However, the cytoplasm of sperm cells may play an important role during early embryogenesis after fertilization in plants, as the extracted sperm nuclei in maize was insufficient to achieve successful fertilization *in vitro* (Matthys-Rochon et al., 1994).

Recently, increasing evidence has confirmed the presence of a number of transcripts in the sperm cell, refuting the hypothesis that highly condensed chromatin in sperm cells impede activation of transcription. Various cDNA libraries have been constructed based on isolated sperm cells from rice (Gou et al., 2001), tobacco (Xin et al., 2011), maize (*Zea mays*; Engel et al., 2003), and Plumbago dimorphic (Gou et al., 2009). Additionally, genome-wide expression has been detected in different plants. Using microarray analysis, the transcriptomic profile in sperm cells was investigated in rice and *Arabidopsis* (Borges et al., 2008;

Russell et al., 2012; Abiko et al., 2013). The transcriptomes of rice sperm cells were studied by deep sequencing (Anderson et al., 2013), and ~25000 genes were analyzed. These studies revealed a diverse and broad constitution of mRNAs in sperm cells. Subsequently, sperm transcription profiles were compared among different plant species. Only 35 genes were found in common among 1,048 ESTs in tobacco (Xin et al., 2011), 5,829 genes in *Arabidopsis* (Borges et al., 2008), and 5,174 ESTs in maize (Engel et al., 2003). Analysis of these 35 genes suggested that active transcription in sperm cells is involved in many basic pathways and processes such as metabolism, transcription, translation, signal transduction and intercellular trafficking (Xin et al., 2011).

For years, plant scientists have questioned whether male transcripts are delivered to the zygote during the fertilization process, and if so, whether these sperm-carrying transcripts have a role in zygote development or early embryogenesis. In our previous work, we identified sperm-specific transcripts in zygotes at 96 HAP (Ning et al., 2006). Subsequently, two kinds of sperm transcripts with unknown function were revealed in zygotes ~10 h after fertilization (HAF). These results strongly suggested that paternal transcripts could be delivered into zygotes, where they might play a role in zygote activation and/or early embryogenesis (Xin et al., 2011). Similarly, Ohnishi et al. (2014) found abundant expression of the Os07g0182900 rice gene in sperm cells (Abiko et al., 2013), but not in unfertilized egg cells. The fact that its transcripts could be detected in the zygote ~10–20 min after fertilization indicated that transcripts in plant zygotes could be delivered from the sperm cells by plasmogamy. The Os07g0182900 gene encoding cytosine-5 DNA methyltransferase 1 (MET1) may be involved in the transition from the zygote to two-celled pre-embryo stage, as the process could be partially inhibited by a specific inhibitor of MET1 (Abiko et al., 2013). In *Arabidopsis*, the polarity of elongated zygotes contributed substantially to regular embryonic development. Corrected asymmetric cell division led to normal formation of the initial apical–basal axis and the embryo and suspensor ancestors in plants (Jeong et al., 2011; Zhang and Laux, 2011; Ueda and Laux, 2012). Sperm transcripts are now believed to be essential in this critical developmental process. Bayer et al. (2009) reported that transcripts of the IRAK/Pelle-like kinase gene, *SSP*, were produced in mature sperm cells and translated in zygotes after fertilization. Defective *SSP* influenced the elongation of the zygote and the formation of suspensor through the YODA-dependent MAPKKK signaling pathway. These examples suggest that sperm mRNAs might have vital functions in normal developmental embryogenesis.

IMPRINTED GENES IN EARLY EMBRYOGENESIS

In mammals and flowering plants, genomic imprinting is a general epigenetic mechanism associated with the differential expression of parental alleles (Feil and Berger, 2007). The differential *de novo* transcription of parental alleles is caused by different epigenetic influences established in the germ line, rather than the nucleotide changes or uniparental transcripts caused by gamete delivery. Maternally expressed imprinted genes (MEGs) are expressed maternally but silenced paternally, whereas maternally expressed imprinted genes (PEGs) are expressed paternally but silenced maternally.

IMPRINTED GENES IN PLANT EMBRYOS

Imprinting is another cause of unequal contributions from parental transcripts in the early embryo. A minority of imprinted genes has been identified in endosperm using conventional methods, such as sequence homologies, small-scale transcriptional surveys, assays for reduced DNA methylation and mutant identification. The *mee1* gene in maize provided the evidence confirming the presence of imprinted genes in embryos (Raissig et al., 2011). The differential methylation status between paternal and maternal alleles regulates the maternal expression of *mee1* in the embryo and endosperm. Dynamic expression of *mee1* was found in the early embryo, but its function remains unclear (Jahnke and Scholten, 2009).

Recently, genome-wide approaches have been used to identify imprinted genes in *Arabidopsis*, maize and rice (Gehring et al., 2011; Hsieh et al., 2011; Luo et al., 2011; Waters et al., 2011; Wolff et al., 2011; Zhang et al., 2011). Several 100 endosperm-specific imprinted genes were newly detected in these species. However, the presence of imprinted genes in the embryo remains controversial. For example, Hsieh et al. (2011) identified 116 MEGs and 10 PEGs in *Arabidopsis* endosperm 7–8 days after pollination (DAP), while 37 MEGs and one PEG were found in the embryo during the same period. However, the imprinted genes in the embryo were considered to be false positives due to contamination with endosperm or maternal tissue (Hsieh et al., 2011). Similarly, Gehring et al. (2011) identified 165 MEGs and 43 PEGs in *Arabidopsis* endosperm at 6–7 DAP; additionally, 17 MEGs and one PEG were found in embryos during the same period. However, the imprinted genes in the embryo could have been due to endosperm contamination or biased expression dependent on an unchangeable allele (Gehring et al., 2011). In monocots, Luo et al. (2011) found 262 imprinted loci in rice endosperm at 5 DAF. An imprinted gene was detected in both the embryo and endosperm; however, this candidate requires further confirmation by confirming its expression in gametes (Luo et al., 2011). Waters et al. (2011) found 54 MEGs and 46 PEGs in maize endosperm at 14 DAP, with 29 MEGs and nine PEGs in embryos during the same period. However, these imprinted genes in embryos might be due to contamination, trafficking of transcripts produced in the endosperm to the embryo, or relatively stable transcripts inherited from the gametes (Waters et al., 2011).

Currently, genomic imprinting in *Arabidopsis* embryos has not been validated conclusively. Raissig et al. (2013) constructed cDNA libraries using 2 to 4-cells embryos and globular embryos isolated from the reciprocal cross of the Col-0 and the Ler accessions. Imprinted gene candidates were then chosen, and their relative expression levels between parental alleles were assessed by reverse transcription polymerase chain reaction (RT-PCR) and Sanger sequencing (Raissig et al., 2013). A total of 11 MEGs were expressed at the 2 to 4-cells and globular embryo stages, and one PEG was expressed at the 2 to 4-cells embryo stage. No transcripts in the one PEG or in nine of the MEGs were detected in the gametes, indicating that their imprinted expression in the embryo was derived from *de novo* transcription and was reliable.

To avoid contamination, strict procedures were adopted in constructing the cDNA libraries. In addition, an independent assay was used to confirm the genomic imprinting in embryos by fusing the promoters of seven MEGs and one PEG with the reporter gene β-glucuronidase (GUS). Promoter-GUS reporter lines (Col-0 background) were crossed reciprocally with wild-type plants (Col-0), and the analysis of stained F1 embryos showed that six MEG reporter lines were either imprinted fully or showed a strong bias for maternal expression (Raissig et al., 2013). Furthermore, Raissig et al. (2013) detected imprinted expression of all embryonic MEGs and the PEG in other samples, as early Col-0 × Cvi embryos (different accession, similar stage; Nodine and Bartel, 2012) and late torpedo-stage Col-0 × Ler embryos (same accession, but later stage; Gehring et al., 2011). The results confirmed that the expression of most imprinted genes during early embryogenesis was maintained regardless of the different accessions or later developmental stage (Raissig et al., 2013). Therefore, these results indicated that genomic imprinting may not be restricted to the endosperm and may be more extensive in embryos than thought previously.

FUNCTION OF IMPRINTED GENES INVOLVED IN EMBRYOGENESIS

In mammals, 100s of imprinted genes have been identified that are connected to the location of nutrient transfer from mother to offspring, embryogenesis, and postnatal development (Constância et al., 2004; Gregg et al., 2010). Abnormal imprinting can harm fetal growth, hormone systems after birth, and adult brain function. Whereas genome-wide approaches have revealed many imprinted genes involved in transcriptional regulation, chromatin modification, hormone signaling, ubiquitin degradation, small RNA pathways and metabolism (Gehring et al., 2011; Hsieh et al., 2011; Luo et al., 2011; Raissig et al., 2011, 2013) in plants, little is known regarding the involvement of imprinted genes in plant development (**Table 1**).

To date, only four imprinted genes in endosperm are known to be involved in embryogenesis (Raissig et al., 2011; Costa et al., 2012). In *Arabidopsis*, the FERTILIZATION-INDEPENDENT SEED *(FIS)* genes *MEA* *(FIS1)* and *FIS2* belong to the Polycomb group family (PcG). *MEDEA (MEA)* is expressed in both the embryo and endosperm; however, maternal imprinting has been confirmed only in the latter, and it remains to be determined whether *MEA* is imprinted in the embryo (Raissig et al., 2013). *FIS2* is a maternally imprinted gene, and its expression was detected both in the central cell before fertilization and endosperm after fertilization (Luo et al., 2000). Double fertilization products that contained maternal alleles of *mea* and *fis2* resulted in failure of endosperm cellularization. Moreover, embryogenesis ceased at the heart/torpedo stage, resulting in seed abortion (Luo et al., 2000). Another novel maternal imprinted gene MPC was found to be active in the central cell before fertilization and in the endosperm from fertilization to 4 DAP. Knockdown of *MPC* through RNAi resulted in defective seed development, with delayed embryogenesis and abnormal embryo and endosperm morphology (Tiwari et al., 2008). In maize, maternally expressed gene 1 *(Meg1)* encodes a new kind of signaling peptide located in endosperm nutrient transfer cells, where it regulates their establishment and differentiation. *Meg1* is the first identified imprinted gene in plants that

Table 1 | Uniparental genes in sexual plant reproduction.

	Genes	Epigenetic mark	Ecotype/inbred line for detection	Period of detection	Mutant/RNAi phenotype	Reference
Endosperm						
Arabidopsis	MEA (MEG)[1,2]	H3K27me3 DNA-me	Ler&RLD	6,7,8 DAP	Seed abortion	Kinoshita et al. (1999), Vielle-Calzada et al. (1999)
	FWA (MEG)	DNA-me	Col-0, Ler, WS	6,7,8 DAP	—	Kinoshita et al. (2004), Jullien et al. (2006)
	FIS2 (MEG)[1]	DNA-me	C24&Col	0.5-5 DAP	Seed abortion	Jullien et al. (2006, 2008), Luo et al. (2000)
	MPC (MEG)[1]	DNA-me	Col&Ler	3,5,7 DAP	Abnormal seed	Tiwari et al. (2008)
	PHE1 (PEG)[1,2]	H3K27me3 DNA-me	Col&C24	1-4 DAP	—	Köhler et al. (2003, 2005), Makarevich et al. (2008)
	FH5 (MEG)[1,2]	H3K27me3	Ler&C24	5 DAP	Endosperm defects	Ingouff et al. (2005), Fitz Gerald et al. (2009)
	AGL36 (MEG)[1]	H3K27me3 DNA-me	Col&Ler	3 DAP	—	Shirzadi et al. (2011), Wuest et al. (2010)
	3 MEGs&2 PEGs	DNA-me	Col-gl&Ler	Torpedo-stage	—	Gehring et al. (2009)
	116 MEGs&10 PEGs	—	Col&Ler	7–8 DAP	—	Hsieh et al. (2011)
	165 MEGs&43 PEGs	—	Col-0&Ler	6 or 7 DAP	—	Gehring et al. (2011)
	39 MEGs&27 PEGs	—	Col-0&Bur-0	4DAP	—	Wolff et al. (2011)
Maize	Fie1 (MEG)[2]	DNA-me, H3K27me3, H3/H4-Ac	B73&Mo17; SSS1&NSS1	2-15 DAP	—	Danilevskaya et al. (2003), Gutiérrez-Marcos et al. (2006), Hermon et al. (2007)
	Fie2 (MEG)	DNA-me	B73&Mo17	2,5 DAP	—	Danilevskaya et al. (2003), Gutiérrez-Marcos et al. (2006), Hermon et al. (2007),
	Nrp1 (MEG)[2]	DNA-me, H3K27me3, H3/H4-Ac	B73&Mo17; SSS1&NSS1	10,14,21 DAP	—	Guo et al. (2003), Haun and Springer (2008)
	Peg1 (PEG)	—	W228Tx303	12 DAP	—	Gutiérrez-Marcos et al. (2003)

(Continued)

Table 1 | Continued

	Genes	Epigenetic mark	Ecotype/inbred line for detection	Period of detection	Mutant/RNAi phenotype	Reference
	Meg1 (MEG)[1,2]	DNA-me	F2, A69Y, W23	4 DAP	Reduced-size seeds	Gutiérrez-Marcos et al. (2004), Costa et al. (2012)
	Mez1 (MEG)	DNA-me, H3K27me3, H3/H4-Ac	B73&Mo17	8-27 DAP	—	Haun et al. (2007), Haun and Springer (2008)
	Mee1 (MEG)	DNA-me	UH005&UH301	6 DAP	—	Jahnke and Scholten (2009)
	α-Tubulin (MEG)	DNA-me	W64A&A69Y	20 DAP	—	Lund et al. (1995b)
	Zein (MEG)	DNA-me	W64A&A69Y	19 DAP	—	Lund et al. (1995a)
	R gene (MEG)	—	—	—	—	Kermicle (1970), Ludwig et al. (1989)
	Dzr-1 (MEG)	—	BSSS53&Mo17	15-27 DAP	—	Chaudhuri and Messing (1994)
	54 MEGs&46 PEGs	—	B73&Mo17	14 DAP	—	Waters et al. (2011)
	93 MNCs&124 PNCs[3]	—	B73&Mo17	10DAP	—	Zhang et al. (2011)
Rice	177 MEGs&85 PEGs	—	Nip&93-11	5 DAF	—	Luo et al. (2011)
Embryo						
Arabidopsis	11 MEGs&1 PEGs[1,2]	H3K27me3[4]	Col-0&Ler	2.5,4 DAP	—	Raissig et al. (2013)
Maize	Mee1 (MEG)[2]	DNA-me	UH005&UH301	3,6,8 DAP	—	Jahnke and Scholten (2009)

	Gamete-carried transcripts		Origin of transcripts		Defective in embryogenesis	Reference
Zygote						
Arabidopsis	SSP		Sperm-specific		Abnormal division of zygote	Bayer et al. (2009)
Rice	Os07g0182900		Sperm-specific		Abnormal division of zygote	Ohnishi et al. (2014)
Tobacco	Ntsp0002&Ntsp0003		Sperm-specific		—	Xin et al. (2011)

The table lists the known information of imprinted and potentially imprinted genes discovered in plants to date. Phenotypes in seed development are also displayed. These data are partially adapted from Raissig et al. (2011). —: not shown or not known.

[1] Reporter activity of parent-of-origin-expression was identified (part of imprinted genes in Arabidopsis embryo).
[2] De novo transcription was identified (part of imprinted genes in Arabidopsis embryo).
[3] MNCs, maternal expressed transcripts; PNCs, paternal expressed transcripts; MNCs and PNCs include protein-coding genes and long non-coding RNAs.
[4] Epigenetic marks of a part of imprinted genes in Arabidopsis embryo.

participates in nutrient distribution to the embryo. Interestingly, in contrast to imprinted genes in mammals, *Meg1* promotes rather than restricts the transfer of nutrient flow from the mother to fetus (Costa et al., 2012).

The imprinted genes in the endosperm mentioned above have defined roles in the endosperm; however, the role of the imprinted genes in the embryo remain unknown. To identify the contribution of imprinted genes in the embryo during embryogenesis, T-DNA gene insertions were used to search for deviant phenotypes relative to embryonic development, but no obvious phenotypes were observed (Raissig et al., 2013). Interestingly, all the maternally imprinted genes in the *Arabidopsis* embryo were expressed in the seed coat, and some even showed a slightly biased expression toward the basal embryo and the suspensor (Raissig et al., 2013). Notably, some maternally imprinted genes were involved in metabolism (Raissig et al., 2013). Therefore, maternally imprinted genes in the embryo might function at the interface between the embryo and maternal tissue, possibly by linking seed coat metabolism and embryo metabolism, and rendering the genes in the embryo under maternal control (Raissig et al., 2013). This result may support the maternal–offspring **coadaptation** theory, which posits that maternally imprinted genes are critical for the events during mother–offspring interactions (Bateson, 1994; Wolf and Hager, 2006). Further research on the roles of imprinted genes in the embryo will lead to a better understanding of the function and evolution of genomic imprinting in plants.

REGULATION OF IMPRINTED GENES IN EMBRYOS

DNA methylation and histone modification are two distinct epigenetic mechanisms involved in the regulation of genomic imprinting in plants. The differential DNA methylation status of parental alleles in the endosperm is due mainly to genome-wide hypomethylation of maternal alleles in the central cell (Gehring et al., 2009, 2011). DNA glycosylase DEMETER *(DME)* with 5-methylcytosine excising activity (Kinoshita et al., 2004; Gehring et al., 2006) and the repression of *MET1* involved in maintaining DNA methylation (Jullien et al., 2008; Hsieh et al., 2011) are responsible for the DNA demethylation at CG sites. However, sometimes DNA methylation alone is not sufficient to establish different imprinting markers of some genes, and the Polycomb repressive complex 2 (PRC2) that catalyzes the trimethylation of histone H3 on lysine 27 (H3K27me3) is required (Baroux et al., 2006; Makarevich et al., 2008; Hennig and Derkacheva, 2009).

To investigate the epigenetic mechanism of genomic imprinting in the embryo, the fertilization-independent endosperm *(fie)* mutant was crossed reciprocally with wild-type plants, and *met1-3* mutants were used to pollinate wild-type plants (Raissig et al., 2013). F1 hybrid embryos were isolated, and mutant embryonic cDNA libraries were created. The detection of the allele-specific expression pattern of 11 embryonic MEGs in *Arabidopsis* demonstrated that imprinted expression of MEGs in embryos are not influenced by the paternal *met1-3* allele. However, disruption of the maternal FIE function changed the monoallelic expression of two MEGs and one PEG. Thus, the function of PRC2 may be comprehensive in regulating imprinted expression in both the embryo and endosperm (Raissig et al., 2013). Furthermore, the role of

asymmetric DNA methylation in the CHG context was negated in the establishment of imprinting in the embryo. Finally, Raissig et al. (2013) indicated that PRC2, but not MET1, played a role in regulating the imprinted genes in the embryo. This is consist with a previous conjecture that DNA methylation is unlikely to be a primary imprinting mark in maize embryos (Gutiérrez-Marcos et al., 2006; Jahnke and Scholten, 2009). Other undiscovered mechanisms may be involved in the establishment of genomic imprinting in the embryo.

PERSPECTIVES

Technological advances in genome-wide sequencing technology and acquisition of gametes, zygotes and early embryos will be important for elucidating the role of parent-of-origin genes during plant early embryogenesis. Such technological progress may lead to the identification of more uniparental transcripts in early embryos, whether gamete-delivered or imprinted gene-derived. Currently, in both gametes and early zygotes, it remains technically difficult to identify the origin of transcripts, which may be gamete-delivered during fertilization or transcribed *de novo* after fertilization. In addition, little is known about the functions of these parent-of-origin genes during fertilization and early embryogenesis. Obviously, no matter how parental transcripts may contribute to the transcriptome of early embryo, the function analysis of the parent-of-origin genes in specific developmental events at specific developmental stage will surely provide imperative knowledge to understand the parental effect in early seed formation.

Recently, the identification of imprinted genes in early embryos in *Arabidopsis* has questioned the concept that imprinted genes are restricted mainly to the endosperm. In light of so many candidate imprinted genes in the embryos of dicots and monocots, optimized methods are required to avoid contamination of maternal tissues and false positives or negatives in data collection and reliable analysis. Determination of the relevant functions of imprinted genes should shed light on our understanding of epigenetic mechanisms in promoting embryogenesis, embryo pattern formation, and cell fate determination during embryogenesis. With further technological advancement, the role of methylation in gene imprinting during embryogenesis might be further elucidated.

ACKNOWLEDGMENTS

We thank Mr. Ce Shi for figure preparation. This project was supported by the "973" Project (2013CB126900; 2014CB943400) and National Natural Science Fund of China (31170297; 31070284).

REFERENCES

Abiko, M., Maeda, H., Tamura, K., Hara-Nishimura, I., and Okamoto, T. (2013). Gene expression profiles in rice gametes and zygotes: identification of gamete-enriched genes and up- or down-regulated genes in zygotes after fertilization. *J. Exp. Bot.* 64, 1927–1940. doi: 10.1093/jxb/ert054

Anderson, S. N., Johnson, C. S., Jones, D. S., Conrad, L. J., Gou, X., Russell, S. D., et al. (2013). Transcriptomes of isolated *Oryza sativa* gametes characterized by deep sequencing: evidence for distinct sex-dependent chromatin and epigenetic states before fertilization. *Plant J.* 76, 729–741. doi: 10.1111/tpj.12336

Angerer, L. M., and Angerer, R. C. (2000). Animal-vegetal axis patterning mechanisms in the early sea urchin embryo. *Dev. Biol.* 218, 1–12. doi: 10.1006/dbio.1999.9553

Autran, D., Baroux, C., Raissig, M. T., Lenormand, T., Wittig, M., Grob, S., et al. (2011). Maternal epigenetic pathways control parental contributions to *Arabidopsis* early embryogenesis. *Cell* 145, 707–719. doi: 10.1016/j.cell.2011.04.014

Babu, Y., Musielak, T., Henschen, A., and Bayer, M. (2013). Suspensor length determines developmental progression of the embryo in *Arabidopsis*. *Plant Physiol.* 162, 1448–1458. doi: 10.1104/pp.113.217166

Baroux, C., Autran, D., Gillmor, C. S., Grimanelli, D., and Grossniklaus, U. (2008). The maternal to zygotic transition in animals and plants. *Cold Spring Harb. Symp. Quant. Biol.* 73, 89–100. doi: 10.1101/sqb.2008.73.053

Baroux, C., Blanvillain, R., and Gallois, P. (2001). Paternally inherited transgenes are down-regulated but retain low activity during early embryogenesis in *Arabidopsis*. *FEBS Lett.* 509, 11–16. doi: 10.1016/S0014-5793(01)03097-6

Baroux, C., Gagliardini, V., Page, D. R., and Grossniklaus, U. (2006). Dynamic regulatory interactions of *Polycomb* group genes: MEDEA autoregulation is required for imprinted gene expression in *Arabidopsis*. *Genes Dev.* 20, 1081–1086. doi: 10.1101/gad.378106

Bateson, P. (1994). The dynamics of parent-offspring relationships in mammals. *Trends Ecol. Evol.* 9, 399–403. doi: 10.1016/0169-5347(94)90066-90063

Bayer, M., Nawy, T., Giglione, C., Galli, M., Meinnel, T., and Lukowitz, W. (2009). Paternal control of embryonic patterning in *Arabidopsis thaliana*. *Science* 323, 1485–1488. doi: 10.1126/science.1167784

Borges, F., Gomes, G., Gardner, R., Moreno, N., McCormick, S., Feijó, J. A., et al. (2008). Comparative transcriptomics of *Arabidopsis* sperm cells. *Plant Physiol.* 148, 1168–1181. doi: 10.1104/pp.108.125229

Chang, F., Wang, Y., Wang, S., and Ma, H. (2011). Molecular control of microsporogenesis in *Arabidopsis*. *Curr. Opin. Plant Biol.* 14, 66–73. doi: 10.1016/j.pbi.2010.11.001

Chaudhuri, S., and Messing, J. (1994). Allele-specific parental imprinting of dzr1, a posttranscriptional regulator of zein accumulation. *Proc. Natl. Acad. Sci. U.S.A.* 91, 4867–4871. doi: 10.1073/pnas.91.11.4867

Constância, M., Kelsey, G., and Reik, W. (2004). Resourceful imprinting. *Nature* 432, 53–57. doi: 10.1038/432053a

Costa, L. M., Yuan, J., Rouster, J., Paul, W., Dickinson, H., and Gutierrez-Marcos, J. F. (2012). Maternal control of nutrient allocation in plant seeds by genomic imprinting. *Curr. Biol.* 22, 160–165. doi: 10.1016/j.cub.2011.11.059

Danilevskaya, O. N., Hermon, P., Hantke, S., Muszynski, M. G., Kollipara, K., and Ananiev, E. V. (2003). Duplicated fie genes in maize: expression pattern and imprinting suggest distinct functions. *Plant Cell* 15, 425–438. doi: 10.1105/tpc.006759

Del Toro-De León, G., García-Aguilar, M., and Gillmor, C. S. (2014). Non-equivalent contributions of maternal and paternal genomes to early plant embryogenesis. *Nature* 514, 624–627. doi: 10.1038/nature13620

Domoki, M., Szűcs, A., Jäger, K., Bottka, S., Barnabás, B., and Fehér, A. (2013). Identification of genes preferentially expressed in wheat egg cells and zygotes. *Plant Cell Rep.* 32, 339–348. doi: 10.1007/s00299-012-1367-0

Engel, M. L., Chaboud, A., Dumas, C., and McCormick, S. (2003). Sperm cells of *Zea mays* have a complex complement of mRNAs. *Plant J.* 34, 697–707. doi: 10.1046/j.1365-313X.2003.01761.x

Feil, R., and Berger, F. (2007). Convergent evolution of genomic imprinting in plants and mammals. *Trends Genet.* 23, 192–199. doi: 10.1016/j.tig.2007.02.004

Fitz Gerald, J. N., Hui, P. S., and Berger, F. (2009). Polycomb group-dependent imprinting of the actin regulator AtFH5 regulates morphogenesis in *Arabidopsis thaliana*. *Development* 136, 3399–3404. doi: 10.1242/dev.036921

Gehring, M., Bubb, K. L., and Henikoff, S. (2009). Extensive demethylation of repetitive elements during seed development underlies gene imprinting. *Science* 324, 1447–1451. doi: 10.1126/science.1171609

Gehring, M., Huh, J. H., Hsieh, T. F., Penterman, J., Choi, Y., Harada, J. J., et al. (2006). DEMETER DNA glycosylase establishes MEDEA polycomb gene self-imprinting by allele-specific demethylation. *Cell* 124, 495–506. doi: 10.1016/j.cell.2005.12.034

Gehring, M., Missirian, V., and Henikoff, S. (2011). Genomic analysis of parent-of-origin allelic expression in *Arabidopsis thaliana* seeds. *PLoS ONE* 6:e23687. doi: 10.1371/journal.pone.0023687

Golden, T. A., Schauer, S. E., Lang, J. D., Pien, S., Mushegian, A. R., Grossniklaus, U., et al. (2002). SHORT INTEGUMENTS1/SUSPENSOR1/CARPEL FACTORY, a Dicer homolog, is a maternal effect gene required for embryo development in *Arabidopsis*. *Plant Physiol.* 130, 808–822. doi: 10.1104/pp.003491

Gou, X. P., Tang, L., Yan, F., and Chen, F. (2001). Representative cDNA library from isolated rice sperm cells. *Acta Bot. Sin.* 43, 1093–1096.

Gou, X., Yuan, T., Wei, X., and Russell, S. D. (2009). Gene expression in the dimorphic sperm cells of *Plumbago zeylanica*: transcript profiling, diversity, and relationship to cell type. *Plant J.* 60, 33–47. doi: 10.1111/j.1365-313X.2009.03934.x

Gregg, C., Zhang, J., Weissbourd, B., Luo, S., Schroth, G. P., Haig, D., et al. (2010). High-resolution analysis of parent-of-origin allelic expression in the mouse brain. *Science* 329, 643–648. doi: 10.1126/science.1190830

Grimanelli, D., Perotti, E., Ramirez, J., and Leblanc, O. (2005). Timing of the maternal-to-zygotic transition during early seed development in maize. *Plant Cell* 17, 1061–1072. doi: 10.1105/tpc.104.029819

Guo, M., Rupe, M. A., Danilevskaya, O. N., Yang, X., and Hu, Z. (2003). Genome-wide mRNA profiling reveals heterochronic allelic variation and a new imprinted gene in hybrid maize endosperm. *Plant J.* 36, 30–44. doi: 10.1046/j.1365-313X.2003.01852.x

Gutiérrez-Marcos, J. F., Costa, L. M., Biderre-Petit, C., Khbaya, B., O'Sullivan, D. M., Wormald, M., et al. (2004). Maternally expressed gene1 is a novel maize endosperm transfer cell-specific gene with a maternal parent-of-origin pattern of expression. *Plant Cell* 16, 1288–1301. doi: 10.1105/tpc.019778

Gutiérrez-Marcos, J. F., Costa, L. M., Dal Prà, M., Scholten, S., Kranz, E., Perez, P., et al. (2006). Epigenetic asymmetry of imprinted genes in plant gametes. *Nat. Genet.* 38, 876–878. doi: 10.1038/ng1828

Gutiérrez-Marcos, J. F., Pennington, P. D., Costa, L. M., and Dickinson, H. G. (2003). Imprinting in the endosperm: a possible role in preventing wide hybridization. *Philos. Trans. R. Soc. Lond. B Biol. Sci.* 358, 1105–1111. doi: 10.1098/rstb.2003.1292

Haun, W. J., Laoueillé-Duprat, S., O'connell, M. J., Spillane, C., Grossniklaus, U., Phillips, A. R., et al. (2007). Genomic imprinting, methylation and molecular evolution of maize Enhancer of zeste (Mez) homologs. *Plant J.* 49, 325–337. doi: 10.1111/j.1365-313X.2006.02965.x

Haun, W. J., and Springer, N. M. (2008). Maternal and paternal alleles exhibit differential histone methylation and acetylation at maize imprinted genes. *Plant J.* 56, 903–912. doi: 10.1111/j.1365-313X.2008.03649.x

Hennig, L., and Derkacheva, M. (2009). Diversity of Polycomb group complexes in plants: same rules, different players? *Trends Genet.* 25, 414–423. doi: 10.1016/j.tig.2009.07.002

Hermon, P., Srilunchang, K. O., Zou, J., Dresselhaus, T., and Danilevskaya, O. N. (2007). Activation of the imprinted Polycomb Group Fie1 gene in maize endosperm requires demethylation of the maternal allele. *Plant Mol. Biol.* 64, 387–395. doi: 10.1007/s11103-007-9160-0

Hsieh, T. F., Shin, J., Uzawa, R., Silva, P., Cohen, S., Bauer, M. J., et al. (2011). Regulation of imprinted gene expression in *Arabidopsis* endosperm. *Proc. Natl. Acad. Sci. U.S.A.* 108, 1755–1762. doi: 10.1073/pnas.1019273108

Ingouff, M., Fitz Gerald, J. N., Guérin, C., Robert, H., Sørensen, M. B., Van Damme, D., et al. (2005). Plant formin AtFH5 is an evolutionarily conserved actin nucleator involved in cytokinesis. *Nat. Cell Biol.* 7, 374–380. doi: 10.1038/ncb1238

Jahnke, S., and Scholten, S. (2009). Epigenetic resetting of a gene imprinted in plant embryos. *Curr. Biol.* 19, 1677–1681. doi: 10.1016/j.cub.2009.08.053

Jeong, S., Bayer, M., and Lukowitz, W. (2011). Taking the very first steps: from polarity to axial domains in the early Arabidopsis embryo. *J. Exp. Bot.* 62, 1687–1697. doi: 10.1093/jxb/erq398

Johnston, S. D. (1995). The intracellular localization of messenger RNAs. *Cell* 81, 161–170. doi: 10.1016/0092-8674(95)90324-0

Jullien, P. E., Kinoshita, T., Ohad, N., and Berger, F. (2006). Maintenance of DNA methylation during the *Arabidopsis* life cycle is essential for parental imprinting. *Plant Cell* 18, 1360–1372. doi: 10.1105/tpc.106.041178

Jullien, P. E., Mosquna, A., Ingouff, M., Sakata, T., Ohad, N., and Berger, F. (2008). Retinoblastoma and its binding partner MSI1 control imprinting in *Arabidopsis*. *PLoS Biol.* 6:e194. doi: 10.1371/journal.pbio.0060194

Kermicle, J. L. (1970). Dependence of the R-mottled aleurone phenotype in maize on mode of sexual transmission. *Genetics* 66, 69–85. doi: 10.3410/f.718486711.793497084

Kinoshita, T., Miura, A., Choi, Y., Kinoshita, Y., Cao, X., Jacobsen, S. E., et al. (2004). One-way control of FWA imprinting in *Arabidopsis* endosperm by DNA methylation. *Science* 303, 521–523. doi: 10.1126/science.1089835

Kinoshita, T., Yadegari, R., Harada, J. J., Goldberg, R. B., and Fischer, R. L. (1999). Imprinting of the MEDEA polycomb gene in the *Arabidopsis* endosperm. *Plant Cell* 11, 1945–1952. doi: 10.1105/tpc.11.10.1945

Köhler, C., Hennig, L., Spillane, C., Pien, S., Gruissem, W., and Grossniklaus, U. (2003). The Polycomb-group protein MEDEA regulates seed development by controlling expression of the MADS-box gene PHERES1. *Genes Dev.* 17, 1540–1553. doi: 10.1101/gad.257403

Köhler, C., Page, D. R., Gagliardini, V., and Grossniklaus, U. (2005). The *Arabidopsis thaliana* MEDEA Polycomb group protein controls expression of PHERES1 by parental imprinting. *Nat. Genet.* 37, 28–30.

Krawetz, S. A. (2005). Paternal contribution: new insights and future challenges. *Nat. Rev. Genet.* 6, 633–642. doi: 10.1038/nrg1654

Le, B. H., Wagmaister, J. A., Kawashima, T., Bui, A. Q., Harada, J. J., and Goldberg, R. B. (2007). Using genomics to study legume seed development. *Plant Physiol.* 144, 562–574. doi: 10.1104/pp.107.100362

Lopes, M. A., and Larkins, B. A. (1993). Endosperm origin, development and function. *Plant Cell* 5, 1383–1399. doi: 10.1105/tpc.5.10.1383

Ludwig, S. R., Habera, L. F., Dellaporta, S. L., and Wessler, S. R. (1989). Lc, a member of the maize R gene family responsible for tissue-specific anthocyanin production, encodes a protein similar to transcriptional activators and contains the myc-homology region. *Proc. Natl. Acad. Sci. U.S.A.* 86, 7092–7096. doi: 10.1073/pnas.86.18.7092

Lund, G., Ciceri, P., and Viotti, A. (1995a). Maternal-specific demethylation and expression of specific alleles of zein genes in the endosperm of *Zea mays* L. *Plant J.* 8, 571–581. doi: 10.1046/j.1365-313X.1995.8040571.x

Lund, G., Messing, J., and Viotti, A. (1995b). Endosperm-specific demethylation and activation of specific alleles of alpha-tubulin genes of *Zea mays* L. *Mol. Gen. Genet.* 246, 716–722. doi: 10.1007/BF00290717

Luo, M., Bilodeau, P., Dennis, E. S., Peacock, W. J., and Chaudhury, A. (2000). Expression and parent-of-origin effects for FIS2, MEA, and FIE in the endosperm and embryo of developing *Arabidopsis* seeds. *Proc. Natl. Acad. Sci. U.S.A.* 97, 10637–10642. doi: 10.1073/pnas.170292997

Luo, M., Taylor, J. M., Spriggs, A., Zhang, H., Wu, X., Russell, S., et al. (2011). A genome-wide survey of imprinted genes in rice seeds reveals imprinting primarily occurs in the endosperm. *PLoS Genet.* 7:e1002125. doi: 10.1371/journal.pgen.1002125

Lukowitz, W., Roeder, A., Parmenter, D., and Somerville, C. (2004). A MAPKK kinase gene regulates extra-embryonic cell fate in *Arabidopsis*. *Cell* 116, 109–119. doi: 10.1016/S0092-8674(03)01067-5

Makarevich, G., Villar, C. B., Erilova, A., and Köhler, C. (2008). Mechanism of PHERES1 imprinting in *Arabidopsis*. *J. Cell Sci.* 121(Pt 6), 906–912. doi: 10.1242/jcs.023077

Martínez, G., and Slotkin, R. K. (2012). Developmental relaxation of transposable element silencing in plants: functional or byproduct? *Curr. Opin. Plant Biol.* 15, 496–502. doi: 10.1016/j.pbi.2012.09.001

Matthys-Rochon, E., Mòl, R., Heizmann, P., and Dumas, C. (1994). Isolation and microinjection of active sperm nuclei into egg cells and central cells of isolated maize embryo sacs. *Zygote* 2, 29–35. doi: 10.1017/S0967199400001738

Meyer, S., and Scholten, S. (2007). Equivalent parental contribution to early plant zygotic development. *Curr. Biol.* 17, 1686–1691. doi: 10.1016/j.cub.2007.08.046

Mohr, S. E., Dillon, S. T., and Boswell, R. E. (2001). The RNA-binding protein Tsunagi interacts with Mago Nashi to establish polarity and localize *oskar* mRNA during *Drosophila* oogenesis. *Genes Dev.* 15, 2886–2899. doi: 10.1101/gad.927001

Ning, J., Peng, X. B., Qu, L. H., Xin, H. P., Yan, T. T., and Sun, M.-X. (2006). Differential gene expression in egg cells and zygotes suggests that the transcriptome is restructed before the first zygotic division in tobacco. *FEBS Lett.* 580, 1747–1752. doi: 10.1016/j.febslet.2006.02.028

Nishida, H. (1997). Cell lineage and timing of fate restriction, determination and gene expression in ascidian embryos. *Semin. Cell Dev. Biol.* 8, 359–365. doi: 10.1006/scdb.1997.0160

Nodine, M. D., and Bartel, D. P. (2012). Maternal and paternal genomes contribute equally to the transcriptome of early plant embryos. *Nature* 482, 94–97. doi: 10.1038/nature10756

Ohnishi, T., Takanashi, H., Mogi, M., Takahashi, H., Kikuchi, S., Yano, K., et al. (2011). Distinct gene expression profiles in egg and synergid cells of rice as revealed by cell type-specific microarrays. *Plant Physiol.* 155, 881–891. doi: 10.1104/pp.110.167502

Ohnishi, Y., Hoshino, R., and Okamoto, T. (2014). Dynamics of male and female chromatin during karyogamy in rice zygotes. *Plant Physiol.* 165, 1533–1543. doi: 10.1104/pp.114.236059

Olsen, O. A. (2004). Nuclear endosperm development in cereals and *Arabidopsis thaliana*. *Plant Cell* 16(Suppl.), S214–S227. doi: 10.1105/tpc.017111

Ostermeier, G. C., Miller, D., Huntriss, J. D., Diamond, M. P., and Krawetz, S. A. (2004). Reproductive biology: delivering spermatozoan RNA to the oocyte. *Nature* 429:154. doi: 10.1038/429154a

Pellettieri, J., and Seydoux, G. (2002). Anterior-posterior polarity in *C. elegans* and *Drosophila*-PARallels and differences. *Science* 298, 1946–1950. doi: 10.1126/science.1072162

Pillot, M., Baroux, C., Vazquez, M. A., Autran, D., Leblanc, O., Vielle-Calzada, J. P., et al. (2010). Embryo and endosperm inherit distinct chromatin and transcriptional states from the female gametes in *Arabidopsis*. *Plant Cell* 22, 307–320. doi: 10.1105/tpc.109.071647

Raissig, M. T., Baroux, C., and Grossniklaus, U. (2011). Regulation and flexibility of genomic imprinting during seed development. *Plant Cell* 23, 16–26. doi: 10.1105/tpc.110.081018

Raissig, M. T., Bemer, M., Baroux, C., and Grossniklaus, U. (2013). Genomic imprinting in the *Arabidopsis* embryo is partly regulated by PRC2. *PLoS Genet.* 9:e1003862. doi: 10.1371/journal.pgen.1003862

Russell, S. D., Gou, X., Wong, C. E., Wang, X., Yuan, T., and Wei, X. (2012). Genomic profiling of rice sperm cell transcripts reveals conserved and distinct elements in the flowering plant male germ lineage. *New Phytol.* 195, 560–573. doi: 10.1111/j.1469-8137.2012.04199.x

Scholten, S., Lörz, H., and Kranz, E. (2002). Paternal mRNA and protein synthesis coincides with male chromatin decondensation in maize zygotes. *Plant J.* 32, 221–231. doi: 10.1046/j.1365-313X.2002.01418.x

Sheldon, C. C., Hills, M. J., Lister, C., Dean, C., Dennis, E. S., and Peacock, W. J. (2008). Resetting of FLOWERING LOCUS C expression after epigenetic repression by vernalization. *Proc. Natl. Acad. Sci. U.S.A.* 105, 2214–2219. doi: 10.1073/pnas.0711453105

Shirzadi, R., Andersen, E. D., Bjerkan, K. N., Gloeckle, B. M., Heese, M., Ungru, A., et al. (2011). Genome-wide transcript profiling of endosperm without paternal contribution identifies parent-of-origin-dependent regulation of AGAMOUS-LIKE36. *PLoS Genet.* 7:e1001303. doi: 10.1371/journal.pgen.1001303

Slotkin, R. K., Vaughn, M., Borges, F., Tanurdzić, M., Becker, J. D., Feijó, J. A., et al. (2009). Epigenetic reprogramming and small RNA silencing of transposable elements in pollen. *Cell* 136, 461–472. doi: 10.1016/j.cell.2008.12.038

Sprunck, S., Baumann, U., Edwards, K., Langridge, P., and Dresselhaus, T. (2005). The transcript composition of egg cells changes significantly following fertilization in wheat (*Triticum aestivum* L.). *Plant J.* 41, 660–672. doi: 10.1111/j.1365-313X.2005.02332.x

Tadros, W., and Lipshitz, H. D. (2009). The maternal-to-zygotic transition: a play in two acts. *Development* 136, 3033–3042. doi: 10.1242/dev.033183

Tiwari, S., Schulz, R., Ikeda, Y., Dytham, L., Bravo, J., Mathers, L., et al. (2008). MATERNALLY EXPRESSED PAB C-TERMINAL, a novel imprinted gene in *Arabidopsis*, encodes the conserved C-terminal domain of polyadenylate binding proteins. *Plant Cell* 20, 2387–2398. doi: 10.1105/tpc.108.061929

Ueda, M., and Laux, T. (2012). The origin of the plant body axis. *Curr. Opin. Plant Biol.* 15, 578–584. doi: 10.1016/j.pbi.2012.08.001

Ueda, M., Zhang, Z., and Laux, T. (2011). Transcriptional activation of *Arabidopsis* axis patterning genes WOX8/9 links zygote polarity to embryo development. *Dev. Cell* 20, 264–270. doi: 10.1016/j.devcel.2011.01.009

Vielle-Calzada, J. P., Baskar, R., and Grossniklaus, U. (2000). Delayed activation of the paternal genome during seed of development. *Nature* 404, 91–94. doi: 10.1038/35003595

Vielle-Calzada, J. P., Thomas, J., Spillane, C., Coluccio, A., Hoeppner, M. A., and Grossniklaus, U. (1999). Maintenance of genomic imprinting at the *Arabidopsis* medea locus requires zygotic DDM1 activity. *Genes Dev.* 13, 2971–2982. doi: 10.1101/gad.13.22.2971

Walbot, V., and Evans, M. M. (2003). Unique features of the plant life cycle and their consequences. *Nat. Rev. Genet.* 4, 369–379. doi: 10.1038/nrg1064

Waters, A. J., Makarevitch, I., Eichten, S. R., Swanson-Wagner, R. A., Yeh, C. T., Xu, W., et al. (2011). Parent-of-origin effects on gene expression and DNA methylation in the maize endosperm. *Plant Cell* 23, 4221–4233. doi: 10.1105/tpc.111.092668

Weijers, D., Geldner, N., Offringa, R., and Jurgens, G. (2001). Seed development: early paternal gene activity in *Arabidopsis*. *Nature* 414, 709–710. doi: 10.1038/414709a

Wolf, J. B., and Hager, R. (2006). A maternal-offspring coadaptation theory for the evolution of genomic imprinting. *PLoS Biol.* 4:e380. doi: 10.1371/journal.pbio.0040380

Wolff, P., Weinhofer, I., Seguin, J., Roszak, P., Beisel, C., Donoghue, M. T., et al. (2011). High-resolution analysis of parent-of-origin allelic expression in the *Arabidopsis* endosperm. *PLoS Genet.* 7:e1002126. doi: 10.1371/journal.pgen.1002126

Wuest, S. E., Vijverberg, K., Schmidt, A., Weiss, M., Gheyselinck, J., Lohr, M., et al. (2010). *Arabidopsis* female gametophyte gene expression map reveals similarities between plant and animal gametes. *Curr. Biol.* 20, 506–512. doi: 10.1016/j.cub.2010.01.051

Wylie, C., Kofron, M., Payne, C., Anderson, R., Hosobuchi, M., Joseph, E., et al. (1996). Maternal beta-catenin establishes a 'dorsal signal' in early *Xenopus* embryos. *Development* 122, 2987–2996.

Xin, H. P., Peng, X. B., Ning, J., Yan, T. T., Ma, L. G., and Sun, M. X. (2011). Expressed sequence-tag analysis of tobacco sperm cells reveals a unique transcriptional profile and selective persistence of paternal transcripts after fertilization. *Sex. Plant Reprod.* 24, 37–46. doi: 10.1007/s00497-010-0151-y

Xin, H. P., Zhao, J., and Sun, M. X. (2012). The maternal-to-zygotic transition in higher plants. *J. Integr. Plant Biol.* 54, 610–615. doi: 10.1111/j.1744-7909.2012.01138.x

Yu, H. S., and Russell, S. D. (1994). Occurrence of mitochondria in the nuclei of tobacco sperm cells. *Plant Cell* 6, 1477–1484. doi: 10.1105/tpc.6.10.1477

Zhang, M., Zhao, H., Xie, S., Chen, J., Xu, Y., Wang, K., et al. (2011). Extensive, clustered parental imprinting of protein-coding and noncoding RNAs in developing maize endosperm. *Proc. Natl. Acad. Sci. U.S.A.* 108, 20042–20047. doi: 10.1073/pnas.1112186108

Zhang, Z., and Laux, T. (2011). The asymmetric division of the *Arabidopsis* zygote: from cell polarity to an embryo axis. *Sex. Plant Reprod.* 24, 161–169. doi: 10.1007/s00497-010-0160-x

Zhao, J., Xin, H., Qu, L., Ning, J., Peng, X., Yan, T., et al. (2011). Dynamic changes of transcript profiles after fertilization are associated with de novo transcription and maternal elimination in tobacco zygote, and mark the onset of the maternal-to-zygotic transition. *Plant J.* 65, 131–145. doi: 10.1111/j.1365-313X.2010.04403.x

19

Arabinogalactan proteins: focus on carbohydrate active enzymes

Eva Knoch , Adiphol Dilokpimol[†] and Naomi Geshi*

Department of Plant and Environmental Sciences, University of Copenhagen, Copenhagen, Denmark

Edited by:
Nausicaä Lannoo, Ghent University, Belgium

Reviewed by:
Ian S. Wallace, University of Nevada, Reno, USA
Peter Ulvskov, Copenhagen University, Denmark

*Correspondence:
Naomi Geshi, Department of Plant and Environmental Sciences, University of Copenhagen, Thorvaldsensvej 40, 1871 Frederiksberg C, Copenhagen, Denmark
e-mail: nge@plen.ku.dk

†Present address:
Adiphol Dilokpimol, Fungal Physiology, CBS-KNAW Fungal Biodiversity Center, Utrecht, Netherlands

Arabinogalactan proteins (AGPs) are a highly diverse class of cell surface proteoglycans that are commonly found in most plant species. AGPs play important roles in many cellular processes during plant development, such as reproduction, cell proliferation, pattern formation and growth, and in plant-microbe interaction. However, little is known about the molecular mechanisms of their function. Numerous studies using monoclonal antibodies that recognize different AGP glycan epitopes have shown the appearance of a slightly altered AGP glycan in a specific stage of development in plant cells. Therefore, it is anticipated that the biosynthesis and degradation of AGP glycan is tightly regulated during development. Until recently, however, little was known about the enzymes involved in the metabolism of AGP glycans. In this review, we summarize recent discoveries of carbohydrate active enzymes (CAZy; http://www.cazy.org/) involved in the biosynthesis and degradation of AGP glycans, and we discuss the biological role of these enzymes in plant development.

Keywords: arabinogalactan proteins, type II arabinogalactan, plant cell wall, carbohydrate active enzymes, glycosyltransferase, glycoside hydrolase

INTRODUCTION

Arabinogalactan proteins (AGPs) are a family of proteoglycans found on the plasma membrane and in the cell walls of diverse species of plants. AGPs are synthesized by several post-translational modifications of proteins in the secretory pathway. The proteins generally contain repetitive dipeptide motifs, e.g., Ala-Pro, Ser-Pro, Thr-Pro, and Val-Pro, which are distinguished from the sequence motifs for extensin type glycosylation [e.g., Ser-(Pro)$_{2-3}$] known as another major class of O-glycosylation in plants (Kieliszewski, 2001). The Pro residues are hydroxylated by prolyl 4-hydroxylases and further O-glycosylated by glycosyltransferases (GTs). Moreover, many AGPs are attached by a glycosylphosphatidylinositol anchor, which attaches AGPs to the plasma membrane, but can be cleaved by phospholipases (Wang, 2001; Schultz et al., 2004). AGPs on the plasma membrane and cell wall may also be processed by proteolytic activities and glycosyl hydrolases or transported by endocytotic multivesicular bodies to the vacuole where they are degraded (Herman and Lamb, 1992).

The glycan moiety of AGPs accounts for more than 90% of their total mass, which has been suggested to play an essential role in the function of AGPs, based on studies using synthetic phenylglycoside dyes (β-Yariv reagents) that specifically binds to the β-1,3-galactan moiety of AGPs (Kitazawa et al., 2013) as well as various monoclonal antibodies that recognize different AGP glycan epitopes (Seifert and Roberts, 2007). However, because of its complexity and heterogeneity, little is known about the structure-function relationship of AGP glycans. In fact, various structures have been reported for AGP glycans depending on samples and analytical methods. The common structural feature is a backbone of β-1,3-galactan, which is often substituted at O6 with side chains

of β-1,6-galactan decorated further with arabinose, and less frequently also with fucose, rhamnose, and (methyl) glucuronic acid (**Figures 1A,B**). Tan et al. (2010) proposed that the backbone is composed of a repeat of a β-1,3-galactotriose unit with or without side chains, which is connected by β-1,6-linkages (kinks). This model is based on the AGPs synthesized onto synthetic peptides expressed in tobacco cells and analyzed by NMR (Tan et al., 2004, 2010). In this model, the side chains are rather short and composed of a single Gal decorated by 1–5 other sugars. However, longer β-1,6-galactan side chains have been reported for AGPs from radish root (Haque et al., 2005), wheat flour (Tryfona et al., 2010) and Arabidopsis leaf (Tryfona et al., 2012) based on the linkage and mass spectroscopy analysis.

Knowledge about each enzyme working on an individual step in the biosynthesis and degradation of AGP glycans is useful to understand the role of a particular sugar moiety of AGPs. This review outlines the recent findings for the carbohydrate active enzymes (CAZy; http://www.cazy.org/, Lombard et al., 2014) identified to be responsible for the biosynthesis and degradation of AGPs. The reader is referred to other excellent reviews for other topics with respect to structure, cell biological functions (Seifert and Roberts, 2007), localization, and commercial interests of AGPs (Nothnagel, 1997; Schultz et al., 1998; Majewska-sawka and Nothnagel, 2000; Gaspar et al., 2001; Showalter, 2001; Ellis et al., 2010; Tan et al., 2012).

GLYCOSYLTRANSFERASES INVOLVED IN AGP BIOSYNTHESIS

A large number of functionally distinct GTs are required for the biosynthesis of complex AGP glycans, e.g.,

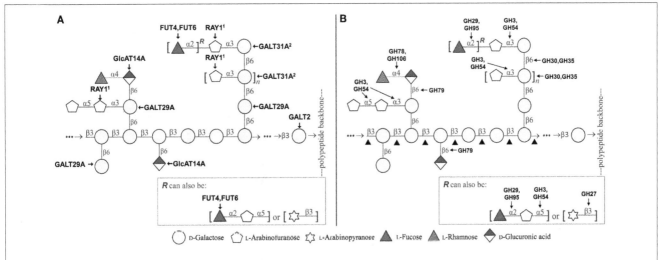

FIGURE 1 | Schematic representation of the type II arabinogalactan and the CAZy enzymes putatively involved in its biosynthesis (A) and degradation (B). [1] Mutant analysis showed reduction of 3-linked Araf, but activity of heterologously expressed protein in *N. benthamiana* showed β-linked product (Gille et al., 2013). [2] Works cooperatively with AtGALT29A by forming a protein complex (Dilokpimol et al., 2014). ▲ includes GH16, GH35, GH43. The arabinogalactan model was modified from Tan et al. (2010) and Tryfona et al. (2012).

β-1,3-galactosyltransferases (GalTs), β-1,6-GalTs, α-1,3- and α-1,5-arabinosyltransferases, fucosyltransferases, rhamnosyltransferases, glucuronosyltransferases, and glucuronic acid methyltransferases. Several GTs identified to date (**Figure 1A** and **Table 1**) are summarized below.

β-GALACTOSYLTRANSFERASES

The first step in the glycosylation of AGPs is the transfer of Gal to hydroxyproline residues present in the peptide backbone. The Arabidopsis enzyme catalyzing this step was identified (At4g21060, AtGALT2, Basu et al., 2013). This enzyme belongs to the CAZy family GT31 and the recombinant protein expressed in *Pichia pastoris* demonstrated GalT activity transferring a Gal to hydroxyproline residues in the synthetic AGP peptides. Arabidopsis T-DNA knockout mutants contained reduced levels of Yariv precipitable AGPs and microsomes purified from mutants exhibited reduced levels of GalT activity compared to wild type. The mutant lines showed no detectable growth phenotype under normal growth conditions. Since the GalT activity was not completely abolished in the mutant microsomes, redundant activities encoded by other genes most likely exist. Nevertheless, based on this study, the quantity of AGP-glycans appears to not be crucial for plant development.

Another Arabidopsis GT from family GT31 encoded by *At1g32930* was also characterized. Recombinant enzyme expressed in *Escherichia coli* and *Nicotiana benthamiana* demonstrated β-1,6-GalT activity elongating β-1,6-galactan side chains of AGP glycans in *in vitro* assays (AtGALT31A; Geshi et al., 2013). *AtGALT31A* is expressed specifically in the suspensor cells of the embryo proper and T-DNA insertion lines showed abnormal cell division in the hypophysis and arrested further development of embryos. Therefore, functional AtGALT31A is essential for normal plant embryogenesis. How *AtGALT31A* that is expressed in suspensor cells influences cell division in hypophysis remains unknown.

AtGALT29A (*At1g08280*) was identified as a gene co-expressed with *AtGALT31A*. Recombinant enzyme expressed in *Nicotiana benthamiana* demonstrated β-1,6-GalT activities elongating β-1,6-galactan and forming 6-Gal branches on β-1,3-galactan of AGP glycans (Dilokpimol et al., 2014). Moreover, Förster resonance energy transfer analysis revealed an interaction between AtGALT29A and AtGALT31A when both proteins are expressed as C-terminal fluorescent fusion proteins in *Nicotiana benthamiana* (Dilokpimol et al., 2014). The protein complex containing heterologously expressed AtGALT29A and AtGALT31A were purified and demonstrated increased levels of β-1,6-GalT activities by the AtGALT29A single enzyme. These results suggest cooperative action between AtGALT31A and AtGALT29A by forming an enzyme complex, which could be an important regulatory mechanism for producing β-1,6-galactan side chains of type II AG during plant development.

β-GLUCURONOSYLTRANSFERASE

An Arabidopsis GT from family GT14 encoded by *At5g39990* was identified as a glucuronosyltransferase involved in the biosynthesis of AGP glycans (AtGlcAT14A; Knoch et al., 2013). The enzyme expressed in *Pichia pastoris* demonstrated β-GlcAT activity by adding GlcA to both β-1,6- and β-1,3-galactan. Arabidopsis possesses 11 proteins in the GT14 family, of which two additional proteins encoded by *At5g15050* and *At2g37585* also demonstrated the same β-GlcAT activities and were named AtGlcAT14B and AtGlcAT14C, respectively (Dilokpimol and Geshi, 2014). The T-DNA insertion lines contained reduced levels of GlcA substitution of β-1,6-galactobiose and β-1,3-galactan in their AGPs compared to wild type. In addition to the altered levels of GlcA, a marked increase of Gal and a decrease of Ara were detected in the mutant AGPs. Mutant lines showed an increased cell elongation rate in dark grown hypocotyls and light grown roots during seedling growth compared to wild type. Since several sugars were altered in the mutant AGPs lines, it is unlikely that the observed phenotype

Table 1 | Characterized GTs and GHs, which process AGP-glycans.

CAZy Family	Activity	Protein name	Origin	No. of genes in Arabidopsis	Evidence[1]	Comments for enzyme activities and genetic manipulations	Selected references
GT14	β-glucuronosyltransferase	AtGlcAT14A	Arabidopsis thaliana	11	HE-P, MA	GlcAT activity to β-1,3 and β-1,6-galactan; [GlcA]↓, [Gal]↑, [Ara]↓ in AG and enhanced cell elongation in seedlings in atglcat14a	Knoch et al., 2013
		AtGlcAT14B AtGlcAT14C	Arabidopsis thaliana Arabidopsis thaliana		HE-P	GlcAT activity to β-1,3 and β-1,6-galactan	Dilokpimol and Geshi, 2014
GT29	β-1,6-galactosyltransferase	AtGALT29A	Arabidopsis thaliana	3	HE-N	β-1,6-GalT activity to β-1,3 and β-1,6-galactan; interaction with AtGALT31A enhances the activity	Dilokpimol et al., 2014
GT31	Hydroxyproline O-galactosyltransferase	AtGALT2	Arabidopsis thaliana	33	HE-P, MA	GalT activity to hydroxylproline; [Yariv-precipitable AG]↓ in galt2; no detectable growth phenotype under normal growth condition	Basu et al., 2013
	β-1,6-galactosyltransferase	AtGALT31A	Arabidopsis thaliana		HE-N, -E; MA	β-1,6-GalT activity elongating β-1,6-galactan; mutant is embryo-lethal	Geshi et al., 2013
GT37	α-1,2-fucosyltransferase	AtFUT4 AtFUT6	Arabidopsis thaliana Arabidopsis thaliana	10	HE-B, MA	FucT activity to AGPs from BY2; [Fuc]↓ in AG in fut4 and fut6; no [Fuc] in AG from fut4/fut6; no detectable growth phenotype under normal growth condition, but reduced root growth under salt stress	Wu et al., 2010; Liang et al., 2013; Tryfona et al., 2014
GT77	Arabinofuranosyltransferase	AtRAY1[2]	Arabidopsis thaliana	19	HE-N, MA	β-ArafT activity to methyl β-Gal; [3-linked Ara]↓ in AG of ray1, slower root growth[2]	Gille et al., 2013
GH3	Exo-α-arabinofuranosidase	RsAraf1	Raphanus sativus L.	16[3]	HE-A	Cleaves α-linked Araf from AGPs, pectic α-1,5-arabinan, arabinoxylan. Overexpression in Arabidopsis resulted in [Ara]↓ in cell walls, but no growth phenotype	Kotake et al., 2006
GH16	Endo-β-1,3-galactanase	FvEn3GAL	Flammulina velutipes	33[4]	PUR, HE-P	Cleaves β-1,3-galactan in endo fashion	Kotake et al., 2011
GH27	Exo-β-arabinopyranosidase	SaArap27A	Streptomyces avermitilis	4	HE-S	Cleaves L-Arap from p-nitrophenyl-β-L-Arap and releases Ara from gum Arabic and larch AG	Fujimoto et al., 2009; Ichinose et al., 2009
GH30	Endo-β-1,6-galactanases	Tv6GAL FoGal1 Sa1,6Gal5A	Trichoderma viride Fusarium oxysporum Streptomyces avermitilis	0	HE-E PUR, HE-E HE-E	Cleaves β-1,6-galactan in endo fashion	Kotake et al., 2004 Sakamoto et al., 2007 Ichinose et al., 2008
		Nc6GAL	Neurospora crassa		HE-P		Takata et al., 2010
GH35	Exo-β-1,3-,1,6-galactosidase	RsBGAL1	Raphanus sativus	18	PUR, HE-P	Exo activity to β-1,3- and β-1,6-Gal, but not to β-1,4-Gal. Cooperative degradation of AG with arabinofuranosidase and glucuronidase	Kotake et al., 2005

(Continued)

Table 1 | Continued

CAZy Family	Activity	Origin	Protein name	No. of genes in Arabidopsis	Evidence[1]	Comments for enzyme activities and genetic manipulations	Selected references
		Arabidopsis thaliana	AtBGAL4		HE-E, HE-I	Preferred cleavage at β-1,3 and β-1,4-linked Gal rather than β-1,6-linked Gal	Ahn et al., 2007
GH43	Exo-β-1,3-galactanase	Phanerochaete chrysosporium	Pc1,3Gal43A	2	HE-P	Cleaves β-1,3-linked Gal regardless the presence or absence of substituted side chains	Ishida et al., 2009
		Clostridium thermocellum	Ct1,3Gal43A		HE-E		Ichinose et al., 2006b
		Streptomyces avermitilis	Sa1,3Gal43A		HE-E		Ichinose et al., 2006a
		Irpex lacteus	Il1,3Gal		HE-P		Kotake et al., 2009
		Streptomyces sp.	SGalase1, 2		HE-E		Ling et al., 2012
GH54	Exo-α-arabinofuranosidase	Neurospora crassa	NcAraf1	0	HE-P	Broad specificity to α-1,3 and α-1,5-Araf, which includes AGPs, pectic arabinan, arabinoxylan	Takata et al., 2010
GH78	Exo-α-rhamnosidase	Streptomyces avermitilis	SaRha78A	0	PUR, HE-E	Releases Rha from gum Arabic	Ichinose et al., 2013
GH79	Exo-β-glucuronidase	Aspergillus niger	AnGlcAase	3	PUR, HE-P	Cleaves both GlcA and methyl GlcA from AG. Methyl GlcA from long β-1,6-galctan is cleaved, but not from short β-1,6-galactan	Haque et al., 2005; Konishi et al., 2008
		Neurospora crassa	NcGlcAase		HE-P	Cleaves GlcA from AG	Konishi et al., 2008
		Arabidopsis thaliana	AtGUS2		PUR, HE-A, MA	Cleaves p-nitrophenyl-β-D-GlcA. Mutant atgus2: [GlcA]↑, [Gal]↑, [Ara]↓, [Xyl]↓ in AG and reduced cell elongation in seedlings; overexpression AtGUS2: no [GlcA]; [Gal]↓, [Ara]↓, [Xyl]↑ in AG and enhanced cell elongation in seedlings	Eudes et al., 2008
GH95	Exo-α-1,2-fucosidase	Bifidobacterium bifidum	AfcA	1[5]	PUR, HE-E	Cleaves α-1,2-linked Fuc (linkage present in AG)	Nagae et al., 2007
GH106	Exo-α-rhamnosidase	Sphingomonas paucimobilis	Rham	0	PUR, HE-E	Broad specificity to α-Rha containing components. Involvement in AG degradation is unclear	Miyata et al., 2005

GTs are only from plants, while GHs are both from plants and microbial origins.

[1]HE, activity demonstrated from heterologously expressed protein in: A, A. thaliana; B, tobacco BY2 cell; E, E. coli; N, N. benthamiana; P, P. pastoris; S, Streptomyces cinnamoneus, I, baculovirus/insect cells; MA, mutant analysis.

[2]Mutant analysis showed reduction of 3-linked Araf in AGPs, but heterologously expressed protein in N. benthamiana showed β-linked Ara to methyl β-Gal product, therefore a role of this protein in AGP glycosylation is not certain since arabinose exists as an α-linked sugar in AGPs (Gille et al., 2013).

[3]Characterized Arabidopsis enzymes demonstrated β-xylosidase activity toward xylan (Goujon et al., 2003; Minic et al., 2004).

[4]Characterized Arabidopsis enzymes demonstrated xyloglucan endo-transferase activity (Rose, 2002).

[5]Characterized Arabidopsis enzymes demonstrated α-1,2-fucosidase activity specifically toward xyloglucan (Léonard et al., 2008; Gunl et al., 2011).

is solely a consequence of the reduced levels of GlcA, but most likely related to the dynamic conformational changes of AGP glycans caused in the mutants.

α-FUCOSYLTRANSFERASE

Two Arabidopsis GTs from family GT37 were identified as α-1,2-fucosyltransferases involved in the biosynthesis of AGP glycans (AtFUT4 and AtFUT6, encoded by *At2g15390* and *At1g14080*, respectively; Wu et al., 2010). AGPs from tobacco BY2 cells contain no fucose, but heterologous expression of AtFUT4 and AtFUT6 in BY2 cells resulted in fucosylated AGPs. The recombinant enzymes purified from BY2 cells demonstrated fucosyltransferase activity to endogenous AGPs. Single Arabidopsis T-DNA insertion lines, *atfut4* and *atfut6*, contained reduced levels of fucose in AGPs, and the double T-DNA insertion line *fut4/fut6* contained no detectable fucose in its AGPs (Liang et al., 2013); however, no obvious phenotype was observed in both types of mutants when they were grown under normal conditions. Differences between wild type and mutants were only seen in seedlings grown under salt stressed condition, where the mutant lines showed reduced root growth compared to wild type (Liang et al., 2013; Tryfona et al., 2014). This was somewhat surprising because the Arabidopsis *mur1* mutant, which is defective in a GDP-mannose-4,6-dehydratase (Bonin et al., 1997), contained a 40% reduction of fucose in root extracts (Reiter et al., 1997) and showed a 50% reduction of cell elongation rate in roots (Van Hengel and Roberts, 2002). The decrease of root cell elongation in *mur1* was previously attributed to the lack of fucose in AGPs (Van Hengel and Roberts, 2002), but the findings by Liang et al. (2013) and Basu et al. (2013) refute that hypothesis. The molecular changes behind the *mur1* phenotype are hard to pinpoint. Mutants affected in *N*-glycan fucosylation show reduced root growth under salt stress conditions (Kaulfürst-Soboll et al., 2011), and *mur2* plants, which lack only xyloglucan fucosylation, have no visible root phenotype (Van Hengel and Roberts, 2002; Vanzin et al., 2002). The *mur1* phenotype might be due to under-fucosylated rhamnogalaturonan II, or to the combination of several cell wall polysaccharides deficient in fucose.

α-ARABINOFURANOSYLTRANSFERASE

An Arabidopsis GT encoded by *At1g70630* (named REDUCED ARABINOSE YARIV1, RAY1; Gille et al., 2013), was characterized as a putative arabinofuranosyltransferase since the mutation caused a reduced level of arabinofuranose (Ara*f*) in its AGPs. This GT belongs to the GT77 family, which also contains *XEG113*, the mutation of which results in the reduction of β-linked arabinose in extensin (Gille et al., 2009). Therefore, Ara*f* transferase activity that makes β-linkages was expected for RAY1, and indeed, microsomes isolated from *Nicotiana benthamiana* after expression of recombinant RAY1 demonstrated β-Ara*f* transferase activity to methyl β-Gal. The T-DNA insertion lines contained reduced levels of 3-linked Ara in its AGP fractions compared to AGPs from wild type, and the mutant plants exhibited slower root growth as well as a reduced rosette size and inflorescence. However, β-1,3-Linked Ara*f* has not been reported in AGPs, therefore the involvement of RAY1 in the biosynthesis of AGP glycans remains unclear.

GLYCOSIDE HYDROLASES

Glycoside hydrolases (GHs) acting on AGPs are potentially very important for the metabolism of these glycoproteins. AGPs from tobacco stylar transmitting tissue are degraded as the pollen tube grows and the released sugars are considered to be used as the carbohydrate resource necessary for the elongation of pollen tubes (Cheung et al., 1995). Similarly, rapid turnover of AGPs are observed in suspension cell culture and millet seedlings (Gibeaut and Carpita, 1991) and a substantial amount of AGPs are considered to be hydrolyzed to free sugars and recycled in the cytosol for the synthesis of new glycans (Gibeaut and Carpita, 1991) or degraded in the vacuole (Herman and Lamb, 1992). The appearance of distinct AGP epitopes in a developmentally regulated manner might be controlled by GHs in the cell walls. Additionally, the occurrence of free AG glycans detached from proteins observed in the cell walls may be a result of GH actions.

For the hydrolysis of AGP glycans, several GHs are required, e.g., β-galactosidases, β-galactanases, α-arabinofuranosidases, β-arabinopyranosidases, β-glucuronidases, α-fucosidases, and α-rhamnosidases. AGP degrading GHs from microbial origin have been relatively well characterized, while only a few plant GHs have been reported to degrade AGP glycans. Below is an overview for those GHs reported to possess hydrolase activity of AGP glycans from both microbial and plant origins (**Figure 1B** and **Table 1**).

β-GALACTOSIDASE AND β-GALACTANASE

A GH16 from the fungus *Flammulina velutipes* was characterized as an endo-β-1,3-galactanase degrading the AGP glycan β-1,3-galactan backbone (Kotake et al., 2011). The enzyme activity is distinct from other GH16 enzymes, which comprise β-1,3- and β-1,3:1,4-glucanases, xyloglucan endo-transglycosylase and β-agarase activities. Arabidopsis contains 33 proteins in the GH16 family and they are characterized as xyloglucan endo-transglycosylases and their homologs (Kaewthai et al., 2013).

Several enzymes of microbial origin in GH30 have been characterized as endo-β-1,6-galactanases that hydrolyze β-1,6-galactan side chains of AGP glycans (Kotake et al., 2004; Sakamoto et al., 2007; Ichinose et al., 2008; Takata et al., 2010). These β-1,6-galactanases were originally categorized as part of the GH5 family, but were moved to the GH30 family after additional bioinformatic analysis by (St John et al., 2010). Tv6GAL from *Trichoderma viride* was the first β-1,6-galactanase cloned and characterized (Kotake et al., 2004). This enzyme specifically recognizes β-1,6-galactan of AGPs and releases galactose and β-1,6-linked galactooligomers with a degree of polymerization from two to five. Efficiency of the hydrolysis of β-1,6-galactan is increased by pretreatment of the AGP substrate with α-L-arabinofuranosidase. The β-1,6-galactanases from *Fusarium oxysporum* (FoGAL1, Sakamoto et al., 2007), *Streptomyces avermitilis* (Sa1,6Gal5A, Ichinose et al., 2008), and *Neurospora crassa* (Nc6GAL, Takata et al., 2010) act in a similar manner, releasing galactose and β-1,6-galactobiose from β-1,6-galactan of AGP. All three enzymes show increased activity on de-arabinosylated AGP, similarly to Tv5GAL. Plants do not have any proteins classified in the GH16 CAZy family.

Several microbial β-1,3-galactosidases from GH43 have been characterized as exo-β-1,3-galactanase that degrades the β-1,3-galactan backbone of AGP glycans (Ichinose et al., 2005, 2006a,b; Kotake et al., 2009). All of these enzymes show similar substrate specificity and degrade β-1,3-linked Gal regardless of the substitution of side chains, which results in free Gal from unsubstituted β-1,3-galactan and side chains attached to β-1,3-linked Gal. Therefore, the GH43 enzymes have been used to release side chains from AGP glycans for structural analysis (Tryfona et al., 2010, 2012; Geshi et al., 2013; Knoch et al., 2013). The GH43 CAZy family contains two uncharacterized Arabidopsis proteins.

In plants, β-1,3-galactosidase has been purified from radish (*Raphanus sativus*) seed extracts (Kotake et al., 2005). Based on the deduced protein sequence, the enzyme RsBGAL1 was classified to the GH35 CAZy family. This enzyme was expressed heterologously in *Pichia pastoris* and demonstrated GH activity by degrading β-1,3- and β-1,6-galactan in an exo manner, but not β-1,4-galactan. The efficiency of degradation of AGP glycans by RsBGAL1 alone was limited, but co-treatment with arabinofuranosidase and glucuronidase resulted in the release of up to 90% of the bound sugars from AGPs, indicating the synergy of those GHs in the degradation of AGP glycans.

α-ARABINOFURANOSIDASE AND β-ARABINOPYRANOSIDASE

An α-arabinofuranosidase from the fungus *Neurospora crassa* with a broad substrate specificity toward AGPs, pectic arabinan, and arabinoxylan was identified and classified to the CAZy GH54 family (NcAraf1, Takata et al., 2010). This enzyme was heterologously expressed in *Pichia pastoris* and demonstrated α-arabinofuranosidase activity on both α-1,3- and α-1,5-linked Ara*f*. NcAraf1 has been used extensively for the structural characterization of AGP glycans together with galactosidases, galactanases, and glucuronidases (Tsumuraya et al., 1990; Okemoto et al., 2003; Kotake et al., 2004, 2009; Konishi et al., 2008; Tryfona et al., 2010, 2012). Arabidopsis does not have proteins in the GH54 family.

Plant α-arabinofuranosidase acting on AGP glycans is classified to family GH3. Kotake et al. (2006) purified an α-arabinofuranosidase from radish seeds and named it RsAraf1. The recombinant enzyme expressed in Arabidopsis demonstrated hydrolytic activity on radish AGPs, pectic α-1,5-arabinan and arabinoxylan. Transgenic Arabidopsis overexpressing RsAraf1 showed decreased levels of Ara in the cell wall, but no obvious growth phenotype was observed compared to wild type plants.

β-Arabinopyranose generally represents only a minor part of Ara in AGP glycans, but has been reported from acacia, larch and wheat flour AGPs (Aspinall et al., 1958; Groman et al., 1994; Odonmazig et al., 1994; Tryfona et al., 2010). β-Arabinopyranosidase has been identified from *Streptomyces avermitilis* (Ichinose et al., 2009). The enzyme, named SaArap27A, belongs to family GH27, and the recombinant enzyme expressed in *Streptomyces* demonstrated the release of L-arabinopyranoside (Ara*p*) from p-nitrophenyl-β-L-arabinopyranoside, as well as the release of L-arabinose from gum Arabic and larch AG. Arabidopsis contains four uncharacterized proteins in the GH27 family.

β-GLUCURONIDASE

Microbial β-glucuronidases are found in family GH79. Two fungal GH79 β-glucuronidases from *Neospora crassa* (NcGlcAase) and *Aspergillus niger* (AnGlcAase) have been cloned and recombinant proteins expressed in *Pichia pastoris* demonstrated β-glucuronidase activity (Konishi et al., 2008). AnGlcAase and NcGlcAase share high homology in their amino acid sequences, but possess slightly different substrate specificity. Both enzymes recognize unsubstituted and 4-methyl substituted β-GlcA on AGPs, but AnGlcAase cleaves both GlcA and 4-methyl GlcA with an equal efficiency, while NcGlcAase preferably cleaves GlcA and only small amounts of 4-methylGlcA. Arabidopsis contains three proteins in the GH79 family.

In plants, β-glucuronidases (GUS) are ubiquitously present and their activity is associated with cell elongation (Sudan et al., 2006). Eudes et al. (2008) partially purified a GUS from Arabidopsis stems, which cleaves p-nitrophenyl-β-D-glucuronide. The corresponding gene was identified and classified to the GH79 family (AtGUS2, At5g07830, Eudes et al., 2008). The T-DNA knockout insertion lines exhibited increased levels of GlcA, whereas plants overexpressing AtGUS2 lacked detectable levels of GlcA in their AGP fractions. The T-DNA insertion mutant of *AtGUS2* showed no clear changes in the elongation rate of plant organs, whereas the overexpression lines exhibited increased elongation of roots and stems. The increase of cell elongation observed in the overexpression lines of *AtGUS2* resembles similar observations of the *atglcat14a* T-DNA insertion lines. Although reduced levels of GlcA is observed in both types of plants, the altered profiles of other sugars present in AGP glycans are inconsistent. Therefore, it is unlikely that the increase of cell elongation is solely caused by the reduction of GlcA levels in AGPs.

α-FUCOSIDASE

α-Fucosidases from various prokaryotic and eukaryotic sources have been characterized, and several of them are commercially available. α-Fucosidases are classified into two GH families: GH29 and GH95 (Lombard et al., 2014). α-Fucosidases from GH29 are capable of hydrolyzing various types of linkages, mainly α-1,3/1,4-linked Fuc, whereas GH95 enzymes are active solely on α-1,2-linked Fuc. Only one GH from Arabidopsis is found in each of these GH families. *At2g28100* (AtFUC1; Zeleny et al., 2006) belongs to GH29 and the recombinant enzyme expressed in *Pichia pastoris* demonstrated α-1,3/1,4-fucosidase activity (Zeleny et al., 2006). *At4g34260* (Fuc95A, AXY8; Léonard et al., 2008; Günl et al., 2011) belongs to GH95 and the enzyme heterologously expressed in *Nicotiana benthamiana* demonstrated α-1,2-fucosidase activity (Léonard et al., 2008). α-1,2-Fuc is present in both xyloglucan and AGP glycans, but Fuc95A (AXY8) acts specifically on α-1,2-Fuc on xyloglucan and not on AGPs (Günl et al., 2011). α-1,2-Fucosidase purified from *Xanthomonas manihotis* apparently cleaves α-1,2-fucose on AGP glycans and has been used for the product analysis of AGP fucosyltransferases (AtFUT4 and AtFUT6, Wu et al., 2010). Wu et al. (2010) also used α-1,3/4-fucosidase from almond meal for the characterization of the Fuc linkage. Both enzymes are commercially available, but are not classified to CAZy GH families.

α-RHAMNOSIDASE

Microbial α-rhamnosidases are classified into three GH families: GH28, GH78, and GH106 (Fujimoto et al., 2013; Lombard et al., 2014). An α-rhamnosidase from *Aspergillus niger* in GH28 was identified as specifically degrading pectic rhamnogalacturonan (RgxB; Martens-Uzunova et al., 2006). Several α-rhamnosidases from GH78 have been characterized and recently an α-rhamnosidase from *Streptomyces avermitilis* expressed in *Escherichia coli* demonstrated an α-rhamnosidase that releases rhamnose (Rha) from gum Arabic AGPs (SaRha78A; Fujimoto et al., 2013; Ichinose et al., 2013). GH106 exclusively contains bacterial α-rhamnosidases, of which only one has been characterized to date. This α-rhamnosidase was purified from *Sphingomonas paucimobilis* FP2001 (Rham; Miyata et al., 2005) and the enzyme expressed in *Escherichia coli* demonstrated α-rhamnosidase activity on a broad range of substrates containing α-Rha. One of those substrates is α-rhamnosyl-1,4-galactose, but whether the enzyme hydrolyzes α-rhamnosyl-1,4-GlcA, which is found as part of the side chains of AGP glycans, remains unknown. Arabidopsis contains 28 proteins in the GH28 family, but no plant proteins are present in GH78 and GH106. Among Arabidopsis GH28s, only pectin polygalacturonase has been characterized (Torki et al., 2000; Markovic and Janecek, 2001). Functions of other plant GHs in GH28 remain unknown.

CONCLUSIONS

Microbial GHs working on the degradation of plant AGPs have been reported in several studies, but little was known about the enzymes working on the biosynthesis and degradation of AGPs in plants. Recent discovery of plant GTs/GHs working on AGPs, together with the technical development of in-depth structural analysis of complex AGP glycans, has broadened our knowledge for AGP metabolism significantly. On the other hand, the attempt to elucidate the biological role of each sugar moiety or a particular part of AGP glycan structure by investigating knockout mutants or overexpressors of those enzymes did not result in straightforward answers. For instance, a mutation in *AtGlcAT14A* did not result in a sole reduction of GlcA but also exhibited an increase of Gal and a reduction of Ara in the AGP glycan as well as enhanced cell elongation in seedlings. Furthermore, the overexpression of *AtGUS2* resulted in a reduction of GlcA, Gal and Ara in AGP glycans and seedlings showed increased cell elongation, similarly to *atglcat14a*. The developmentally regulated appearance of different AGP glycan epitopes is well known, but the results available thus far are inconclusive concerning the molecular role of a particular part of AGP glycans in plant growth and development.

The carbohydrate active enzymes involved in the AGP metabolism have just begun to be identified and characterized. Further investigation of the remaining members in the AGP glycosylation pathway and their role *in vivo* is needed to understand the role of CAZy enzymes in relation to AGP glycans, the cell wall architecture, and in plant growth and development.

REFERENCES

Ahn, Y. O., Zheng, M., Bevan, D. R., Esen, A., Shiu, S.-H., Benson, J., et al. (2007). Functional genomic analysis of *Arabidopsis thaliana* glycoside hydrolase family 35. *Phytochemistry* 68, 1510–1520. doi: 10.1016/j.phytochem.2007.03.021

Aspinall, G. O., Hirst, E. L., Else, R., and Ramsta, E. (1958). The constitution of larch galactan. *J. Chem. Soc.* 593–601. doi: 10.1039/jr9580000593

Basu, D., Liang, Y., Liu, X., Himmeldirk, K., Faik, A., Kieliszewski, M., et al. (2013). Functional identification of a hydroxyproline-o-galactosyltransferase specific for arabinogalactan protein biosynthesis in Arabidopsis. *J. Biol. Chem.* 288, 10132–10143. doi: 10.1074/jbc.M112.432609

Bonin, C. P., Potter, I., Vanzin, G. F., and Reiter, W.-D. (1997). The MUR1 gene of *Arabidopsis thaliana* encodes an isoform of GDP-D-mannose-4,6-dehydratase, catalyzing the first step in the de novo synthesis of GDP-L-fucose. *Proc. Natl. Acad. Sci. U.S.A.* 94, 2085–2090. doi: 10.1073/pnas.94.5.2085

Cheung, A. Y., Wang, H., and Wu, H. M. (1995). A floral transmitting tissue-specific glycoprotein attracts pollen tubes and stimulates their growth. *Cell* 82, 383–393. doi: 10.1016/0092-8674(95)90427-1

Dilokpimol, A., and Geshi, N. (2014). *Arabidopsis thaliana* glucuronosyltransferase in family GT14. *Plant Signal. Behav.* 9:e28891. doi: 10.4161/psb.28891

Dilokpimol, A., Poulsen, C. P., Vereb, G., Kaneko, S., Schulz, A., and Geshi, N. (2014). Galactosyltransferases from *Arabidopsis thaliana* in the biosynthesis of type II arabinogalactan: molecular interaction enhances enzyme activity. *BMC Plant Biol.* 14:90. doi: 10.1186/1471-2229-14-90

Ellis, M., Egelund, J., Schultz, C. J., and Bacic, A. (2010). Arabinogalactan-proteins (AGPs): key regulators at the cell surface? *Plant Physiol.* 153, 403–419. doi: 10.1104/pp.110.156000

Eudes, A., Mouille, G., Thévenin, J., Goyallon, A., Minic, Z., and Jouanin, L. (2008). Purification, cloning and functional characterization of an endogenous beta-glucuronidase in *Arabidopsis thaliana*. *Plant Cell Physiol.* 49, 1331–1341. doi: 10.1093/pcp/pcn108

Fujimoto, Z., Ichinose, H., Harazono, K., Honda, M., Uzura, A., and Kaneko, S. (2009). Crystallization and preliminary crystallographic analysis of beta-L-arabinopyranosidase from *Streptomyces avermitilis* NBRC14893. *Acta Crystallogr. Sect. F Struct. Biol. Cryst. Commun.* 65, 632–634. doi: 10.1107/S1744309109017230

Fujimoto, Z., Jackson, A., Michikawa, M., Maehara, T., Momma, M., Henrissat, B., et al. (2013). The structure of a *Streptomyces avermitilis* α-L-rhamnosidase reveals a novel carbohydrate-binding module CBM67 within the six-domain arrangement. *J. Biol. Chem.* 288, 12376–12385. doi: 10.1074/jbc.M113.460097

Gaspar, Y., Johnson, K. L., McKenna, J. A., Bacic, A., and Schultz, C. J. (2001). The complex structures of arabinogalactan-proteins and the journey towards understanding function. *Plant Mol. Biol.* 47, 161–176. doi: 10.1023/A:1010683432529

Geshi, N., Johansen, J. N., Dilokpimol, A., Rolland, A., Belcram, K., Verger, S., et al. (2013). A galactosyltransferase acting on arabinogalactan protein glycans is essential for embryo development in Arabidopsis. *Plant J.* 76, 128–137. doi: 10.1111/tpj.12281

Gibeaut, D. M., and Carpita, N. C. (1991). Tracing cell wall biogenesis in intact cells and plants: selective turnover and alteration of soluble and cell wall polysaccharides in grasses. *Plant Physiol.* 97, 551–561. doi: 10.1104/pp.97.2.551

Gille, S., Hänsel, U., Ziemann, M., and Pauly, M. (2009). Identification of plant cell wall mutants by means of a forward chemical genetic approach using hydrolases. *Proc. Natl. Acad. Sci. U.S.A.* 106, 14699–14704. doi: 10.1073/pnas.0905434106

Gille, S., Sharma, V., Baidoo, E. E. K., Keasling, J. D., Scheller, H. V., and Pauly, M. (2013). Arabinosylation of a yariv-precipitable cell wall polymer impacts plant growth as exemplified by the arabidopsis glycosyltransferase mutant ray1. *Mol. Plant* 6, 1369–1372. doi: 10.1093/mp/sst029

Goujon, T., Minic, Z., El Amrani, A., Lerouxel, O., Aletti, E., Lapierre, C., et al. (2003). AtBXL1, a novel higher plant (*Arabidopsis thaliana*) putative beta-xylosidase gene, is involved in secondary cell wall metabolism and plant development. *Plant J.* 33, 677–690. doi: 10.1046/j.1365-313X.2003.01654.x

Groman, E. V., Enriquez, P. M., Jung, C., and Josephson, L. (1994). Arabinogalactan for hepatic drug delivery. *Bioconjug. Chem.* 5, 547–556. doi: 10.1021/bc00030a010

Günl, M., Neumetzler, L., Kraemer, F., de Souza, A., Schultink, A., Pena, M., et al. (2011). AXY8 encodes an α-fucosidase, underscoring the importance of apoplastic metabolism on the fine structure of Arabidopsis cell wall polysaccharides. *Plant Cell* 23, 4025–4040. doi: 10.1105/tpc.111.089193

Haque, M. A., Kotake, T., and Tsumuraya, Y. (2005). Mode of action of beta-glucuronidase from *Aspergillus niger* on the sugar chains of arabinogalactan-protein. *Biosci. Biotechnol. Biochem.* 69, 2170–2177. doi: 10.1271/bbb.69.2170

Herman, E. M., and Lamb, C. J. (1992). Arabinogalactan-rich glycoproteins are localized on the cell surface and in intravacuolar multivesicular bodies. *Plant Physiol.* 98, 264–272. doi: 10.1104/pp.98.1.264

Ichinose, H., Fujimoto, Z., Honda, M., Harazono, K., Nishimoto, Y., Uzura, A., et al. (2009). A beta-l-Arabinopyranosidase from *Streptomyces avermitilis* is a novel member of glycoside hydrolase family 27. *J. Biol. Chem.* 284, 25097–25106. doi: 10.1074/jbc.M109.022723

Ichinose, H., Fujimoto, Z., and Kaneko, S. (2013). Characterization of an α-L-Rhamnosidase from *Streptomyces avermitilis*. *Biosci. Biotechnol. Biochem.* 77, 213–216. doi: 10.1271/bbb.120735

Ichinose, H., Kotake, T., Tsumuraya, Y., and Kaneko, S. (2006a). Characterization of an Exo-β-1,3-D-galactanase from *Streptomyces avermitilis* NBRC14893 acting on arabinogalactan-proteins. *Biosci. Biotechnol. Biochem.* 70, 2745–2750. doi: 10.1271/bbb.60365

Ichinose, H., Kotake, T., Tsumuraya, Y., and Kaneko, S. (2008). Characterization of an endo-beta-1,6-Galactanase from *Streptomyces avermitilis* NBRC14893. *Appl. Environ. Microbiol.* 74, 2379–2383. doi: 10.1128/AEM.01733-07

Ichinose, H., Kuno, A., Kotake, T., Yoshida, M., Sakka, K., Hirabayashi, J., et al. (2006b). Characterization of an exo-beta-1,3-galactanase from *Clostridium thermocellum*. *Appl. Environ. Microbiol.* 72, 3515–3523. doi: 10.1128/AEM.72.5.3515-3523.2006

Ichinose, H., Yoshida, M., Kotake, T., Kuno, A., Igarashi, K., Tsumuraya, Y., et al. (2005). An exo-beta-1,3-galactanase having a novel beta-1,3-galactan-binding module from *Phanerochaete chrysosporium*. *J. Biol. Chem.* 280, 25820–25829. doi: 10.1074/jbc.M501024200

Ishida, T., Fujimoto, Z., Ichinose, H., Igarashi, K., Kaneko, S., and Samejima, M. (2009). Crystallization of selenomethionyl exo-beta-1,3-galactanase from the basidiomycete *Phanerochaete chrysosporium*. *Acta Crystallogr. Sect. F Struct. Biol. Cryst. Commun.* 65, 1274–1276. doi: 10.1107/S1744309109043395

Kaewthai, N., Gendre, D., Eklöf, J. M., Ibatullin, F. M., Ezcurra, I., Bhalerao, R. P., et al. (2013). Group III-A XTH genes of Arabidopsis encode predominant xyloglucan endohydrolases that are dispensable for normal growth. *Plant Physiol.* 161, 440–454. doi: 10.1104/pp.112.207308

Kaulfürst-Soboll, H., Rips, S., Koiwa, H., Kajiura, H., Fujiyama, K., and von Schaewen, A. (2011). Reduced immunogenicity of Arabidopsis hgl1 mutant N-glycans caused by altered accessibility of xylose and core fucose epitopes. *J. Biol. Chem.* 286, 22955–22964. doi: 10.1074/jbc.M110.196097

Kieliszewski, M. J. (2001). The latest hype on Hyp-O-glycosylation codes. *Phytochemistry* 57, 319–323. doi: 10.1016/S0031-9422(01)00029-2

Kitazawa, K., Tryfona, T., Yoshimi, Y., Hayashi, K., Kawauchi, S., Antonov, L., et al. (2013). β-galactosyl Yariv reagent binds to the β-1,3-galactan of arabinogalactan proteins. *Plant Physiol.* 161, 1117–1126. doi: 10.1104/pp.112.211722

Knoch, E., Dilokpimol, A., Tryfona, T., Poulsen, C. P., Xiong, G., Harholt, J., et al. (2013). A β-glucuronosyltransferase from *Arabidopsis thaliana* involved in biosynthesis of type II arabinogalactan has a role in cell elongation during seedling growth. *Plant J.* 76, 1016–1029. doi: 10.1111/tpj.12353

Konishi, T., Kotake, T., Soraya, D., Matsuoka, K., Koyama, T., Kaneko, S., et al. (2008). Properties of family 79 beta-glucuronidases that hydrolyze beta-glucuronosyl and 4-O-methyl-beta-glucuronosyl residues of arabinogalactan-protein. *Carbohydr. Res.* 343, 1191–1201. doi: 10.1016/j.carres.2008.03.004

Kotake, T., Dina, S., Konishi, T., Kaneko, S., Igarashi, K., Samejima, M., et al. (2005). Molecular cloning of a b-galactosidase from radish that residues of arabinogalactan protein 1. *Plant Physiol.* 138, 1563–1576. doi: 10.1104/pp.105.062562

Kotake, T., Hirata, N., Degi, Y., Ishiguro, M., Kitazawa, K., Takata, R., et al. (2011). Endo-beta-1,3-galactanase from winter mushroom *Flammulina velutipes*. *J. Biol. Chem.* 286, 27848–27854. doi: 10.1074/jbc.M111.251736

Kotake, T., Kaneko, S., Kubomoto, A., Haque, M. A., Kobayashi, H., and Tsumuraya, Y. (2004). Molecular cloning and expression in *Escherichia coli* of a *Trichoderma viride* endo-beta-(1->6)-galactanase gene. *Biochem. J.* 377, 749–755. doi: 10.1042/BJ20031145

Kotake, T., Kitazawa, K., Takata, R., Okabe, K., Ichinose, H., Kaneko, S., et al. (2009). Molecular cloning and expression in *Pichia pastoris* of a *Irpex lacteus* Exo-β-(1→3)-galactanase Gene. *Biosci. Biotechnol. Biochem.* 73, 2303–2309. doi: 10.1271/bbb.90433

Kotake, T., Tsuchiya, K., Aohara, T., Konishi, T., Kaneko, S., Igarashi, K., et al. (2006). An alpha-L-arabinofuranosidase/beta-D-xylosidase from immature seeds of radish (*Raphanus sativus* L.). *J. Exp. Bot.* 57, 2353–2362. doi: 10.1093/jxb/erj206

Léonard, P., Pabst, M., Bondili, J. S., Chambat, G., Veit, C., Strasser, R., et al. (2008). Identification of an Arabidopsis gene encoding a GH95 alpha1,2-fucosidase active on xyloglucan oligo- and polysaccharides. *Phytochemistry* 69, 1983–1988. doi: 10.1016/j.phytochem.2008.03.024

Liang, Y., Basu, D., Pattathil, S., Xu, W.-L., Venetos, A., Martin, S. L., et al. (2013). Biochemical and physiological characterization of fut4 and fut6 mutants defective in arabinogalactan-protein fucosylation in Arabidopsis. *J. Exp. Bot.* 64, 5537–5551. doi: 10.1093/jxb/ert321

Ling, N. X.-Y., Lee, J., Ellis, M., Liao, M.-L., Mau, S.-L., Guest, D., et al. (2012). An exo-β-(1→3)-d-galactanase from Streptomyces sp. provides insights into type II arabinogalactan structure. *Carbohydr. Res.* 352, 70–81. doi: 10.1016/j.carres.2012.02.033

Lombard, V., Golaconda Ramulu, H., Drula, E., Coutinho, P. M., and Henrissat, B. (2014). The carbohydrate-active enzymes database (CAZy) in 2013. *Nucleic Acids Res.* 42, D490–D495. doi: 10.1093/nar/gkt1178

Majewska-sawka, A., and Nothnagel, E. A. (2000). The multiple roles of arabinogalactan proteins in plant development. *Plant Physiol.* 122, 3–9. doi: 10.1104/pp.122.1.3

Markovic, O., and Janecek, S. (2001). Pectin degrading glycoside hydrolases of family 28: sequence-structural features, specificities and evolution. *Protein Eng.* 14, 615–631. doi: 10.1093/protein/14.9.615

Martens-Uzunova, E. S., Zandleven, J. S., Benen, J. A. E., Awad, H., Kools, H. J., Beldman, G., et al. (2006). A new group of exo-acting family 28 glycoside hydrolases of *Aspergillus niger* that are involved in pectin degradation. *Biochem. J.* 400, 43–52. doi: 10.1042/BJ20060703

Minic, Z., Rihouey, C., Do, C. T., Lerouge, P., and Jouanin, L. (2004). Purification and characterization of enzymes exhibiting beta-D-xylosidase activities in stem tissues of Arabidopsis. *Plant Physiol.* 135, 867–878. doi: 10.1104/pp.104.041269

Miyata, T., Kashige, N., Satho, T., Yamaguchi, T., Aso, Y., and Miake, F. (2005). Cloning, sequence analysis, and expression of the gene encoding *Sphingomonas paucimobilis* FP2001 alpha-L -rhamnosidase. *Curr. Microbiol.* 51, 105–109. doi: 10.1007/s00284-005-4487-8

Nagae, M., Tsuchiya, A., Katayama, T., Yamamoto, K., Wakatsuki, S., and Kato, R. (2007). Structural basis of the catalytic reaction mechanism of novel 1,2-alpha-L-fucosidase from *Bifidobacterium bifidum*. *J. Biol. Chem.* 282, 18497–18509. doi: 10.1074/jbc.M702246200

Nothnagel, E. A. (1997). Proteoglycans and related components in plant cells. *Int. Rev. Cytol.* 174, 195–291. doi: 10.1016/S0074-7696(08)62118-X

Odonmazig, P., Ebringerová, A., Machová, E., and Alföldi, J. (1994). Structural and molecular properties of the arabinogalactan isolated from Mongolian larchwood (*Larix dahurica* L.). *Carbohydr. Res.* 252, 317–324. doi: 10.1016/0008-6215(94)90028-0

Okemoto, K., Uekita, T., Tsumuraya, Y., Hashimoto, Y., and Kasama, T. (2003). Purification and characterization of an endo-beta-(1->6)-galactanase from *Trichoderma viride*. *Carbohydr. Res.* 338, 219–230. doi: 10.1016/S0008-6215(02)00405-6

Reiter, W.-D., Chapple, C., and Somerville, C. R. (1997). Mutants of *Arabidopsis thaliana* with altered cell wall polysaccharide composition. *Plant J.* 12, 335–345. doi: 10.1046/j.1365-313X.1997.12020335.x

Rose, J. K. C. (2002). The XTH family of enzymes involved in xyloglucan endo-transglucosylation and endohydrolysis: current perspectives and a new unifying nomenclature. *Plant Cell Physiol.* 43, 1421–1435. doi: 10.1093/pcp/pcf171

Sakamoto, T., Taniguchi, Y., Suzuki, S., Ihara, H., and Kawasaki, H. (2007). Characterization of *Fusarium oxysporum* beta-1,6-galactanase, an enzyme that hydrolyzes larch wood arabinogalactan. *Appl. Environ. Microbiol.* 73, 3109–3112. doi: 10.1128/AEM.02101-06

Schultz, C., Gilson, P., Oxley, D., Youl, J., and Bacic, A. (1998). GPI-anchors on arabinogalactan- proteins: implications for signalling in plants. *Trends Plant Sci.* 3, 426–431. doi: 10.1016/S1360-1385(98)01328-4

Schultz, C. J., Ferguson, K. L., Lahnstein, J., and Bacic, A. (2004). Post-translational modifications of arabinogalactan-peptides of *Arabidopsis thaliana*. Endoplasmic reticulum and glycosylphosphatidylinositol-anchor signal cleavage sites and hydroxylation of proline. *J. Biol. Chem.* 279, 45503–45511. doi: 10.1074/jbc.M407594200

Seifert, G. J., and Roberts, K. (2007). The biology of arabinogalactan proteins. *Annu. Rev. Plant Biol.* 58, 137–161. doi: 10.1146/annurev.arplant.58.032806.103801

Showalter, A. M. (2001). Arabinogalactan-proteins: structure, expression and function. *Cell. Mol. Life Sci.* 58, 1399–1417. doi: 10.1007/PL00000784

St John, F. J., González, J. M., and Pozharski, E. (2010). Consolidation of glycosyl hydrolase family 30: a dual domain 4/7 hydrolase family consisting of two structurally distinct groups. *FEBS Lett.* 584, 4435–4441. doi: 10.1016/j.febslet.2010.09.051

Sudan, C., Prakash, S., Bhomkar, P., Jain, S., and Bhalla-Sarin, N. (2006). Ubiquitous presence of beta-glucuronidase (GUS) in plants and its regulation in some model plants. *Planta* 224, 853–864. doi: 10.1007/s00425-006-0276-2

Takata, R., Tokita, K., Mori, S., Shimoda, R., Harada, N., Ichinose, H., et al. (2010). Degradation of carbohydrate moieties of arabinogalactan-proteins by glycoside hydrolases from *Neurospora crassa. Carbohydr. Res.* 345, 2516–2522. doi: 10.1016/j.carres.2010.09.006

Tan, L., Qiu, F., Lamport, D. T. A., and Kieliszewski, M. J. (2004). Structure of a hydroxyproline (Hyp)-arabinogalactan polysaccharide from repetitive Ala-Hyp expressed in transgenic *Nicotiana tabacum. J. Biol. Chem.* 279, 13156–13165. doi: 10.1074/jbc.M311864200

Tan, L., Showalter, A. M., Egelund, J., Hernandez-Sanchez, A., Doblin, M. S., and Bacic, A. (2012). Arabinogalactan-proteins and the research challenges for these enigmatic plant cell surface proteoglycans. *Front. Plant Sci.* 3:140. doi: 10.3389/fpls.2012.00140

Tan, L., Varnai, P., Lamport, D. T. A., Yuan, C., Xu, J., Qui, F., et al. (2010). Plant O-hydroxyproline arabinogalactans are composed of repeating trigalactosyl subunits with short bifurcated sidechains. *J. Biol. Chem.* 285, 24575–24583. doi: 10.1074/jbc.M109

Torki, M., Mandaron, P., Mache, R., and Falconet, D. (2000). Characterization of a ubiquitous expressed gene family encoding polygalacturonase in *Arabidopsis thaliana. Gene* 242, 427–436. doi: 10.1016/S0378-1119(99)00497-7

Tryfona, T., Liang, H.-C., Kotake, T., Kaneko, S., Marsh, J., Ichinose, H., et al. (2010). Carbohydrate structural analysis of wheat flour arabinogalactan protein. *Carbohydr. Res.* 345, 2648–2656. doi: 10.1016/j.carres.2010.09.018

Tryfona, T., Liang, H.-C., Kotake, T., Tsumuraya, Y., Stephens, E., and Dupree, P. (2012). Structural characterization of Arabidopsis leaf arabinogalactan polysaccharides. *Plant Physiol.* 160, 653–666. doi: 10.1104/pp.112.202309

Tryfona, T., Theys, T. E., Wagner, T., Stott, K., Keegstra, K., and Dupree, P. (2014). Characterisation of FUT4 and FUT6 α-(1→2)-Fucosyltransferases reveals that absence of root arabinogalactan fucosylation increases arabidopsis root growth salt sensitivity. *PLoS ONE* 9:e93291. doi: 10.1371/journal.pone.0093291

Tsumuraya, Y., Mochizuki, N., Hashimoto, Y., and Kovac, P. (1990). Purification of an exo-beta-(1—3)-D-galactanase of *Irpex lacteus* (*Polyporus tulipiferae*) and its action on arabinogalactan-proteins. *J. Biol. Chem.* 265, 7207–7215.

Van Hengel, A. J., and Roberts, K. (2002). Fucosylated arabinogalactan-proteins are required for full root cell elongation in arabidopsis. *Plant J.* 32, 105–113. doi: 10.1046/j.1365-313X.2002.01406.x

Vanzin, G. F., Madson, M., Carpita, N. C., Raikhel, N. V., Keegstra, K., and Reiter, W.-D. (2002). The mur2 mutant of *Arabidopsis thaliana* lacks fucosylated xyloglucan because of a lesion in fucosyltransferase AtFUT1. *Proc. Natl. Acad. Sci. U.S.A.* 99, 3340–3345. doi: 10.1073/pnas.052450699

Wang, X. (2001). PLANT PHOSPHOLIPASES. *Annu. Rev. Plant Physiol. Plant Mol. Biol.* 52, 211–231. doi: 10.1146/annurev.arplant.52.1.211

Wu, Y., Williams, M., Bernard, S., Driouich, A., Showalter, A. M., and Faik, A. (2010). Functional identification of two nonredundant Arabidopsis alpha(1,2)fucosyltransferases specific to arabinogalactan proteins. *J. Biol. Chem.* 285, 13638–13645. doi: 10.1074/jbc.M110.102715

Zeleny, R., Leonard, R., Dorfner, G., Dalik, T., Kolarich, D., and Altmann, F. (2006). Molecular cloning and characterization of a plant α1,3/4-fucosidase based on sequence tags from almond fucosidase I. *Phytochemistry* 67, 641–648. doi: 10.1016/j.phytochem.2006.01.021

20

On reproduction in red algae: further research needed at the molecular level

Pilar García-Jiménez and Rafael R. Robaina*

Departamento de Biología, Universidad de Las Palmas de Gran Canaria, Las Palmas de Gran Canaria, Spain

Edited by:
Elena M. Kramer, Harvard
University, USA

Reviewed by:
Madelaine E. Bartlett, University of
Massachusetts Amherst, USA
Yoji Nakamura, National Research
Institute of Fisheries Science, Japan

*Correspondence:
Rafael R. Robaina, Departamento
de Biología, Universidad de Las
Palmas de Gran Canaria, 35017 Las
Palmas de Gran Canaria, Canary
Islands, Spain
e-mail: rafael.robaina@ulpgc.es

Multicellular red algae (Rhodophyta) have some of the most complex life cycles known in living organisms. Economically valuable seaweeds, such as phycocolloid producers, have a triphasic (gametophyte, carposporophyte, and tetrasporophyte) life cycle, not to mention the intricate alternation of generations in the edible "sushi-alga" nori. It is a well-known fact that reproductive processes are controlled by one or more abiotic factor(s), including day length, light quality, temperature, and nutrients. Likewise, endogenous chemical factors such as plant growth regulators have been reported to affect reproductive events in some red seaweeds. Still, in the genomic era and given the high throughput techniques at our disposal, our knowledge about the endogenous molecular machinery lags far behind that of higher plants. Any potential effective control of the reproductive process will entail revisiting most of these results and facts to answer basic biological questions as yet unresolved. Recent results have shed light on the involvement of several genes in red alga reproductive events. In addition, a working species characterized by a simple filamentous architecture, easy cultivation, and accessible genomes may also facilitate our task.

Keywords: Rhodophyta, seaweeds, reproduction, light, hormones, photoreceptors, signaling

RHODOPHYTA AND REPRODUCTION. MERGING APPLIED AND FUNDAMENTAL KNOWLEDGE INTEREST

As recently reviewed (Rebours et al., 2014), the seaweed industry produces some 10 billion US$. Among the species exploited, the red seaweeds (Rhodophyta) *Eucheuma/Kappaphycus*, *Porphyra*, and *Gracilaria* occupy a leading position (*Kappaphycus* alone generates 1.3 billion US$ while the nori market is estimated at 1.5 billion US$). Nevertheless, this industry mostly relies on the exploitation of natural populations or primitive aquaculture methods, its expansion being restricted by the lack of technical and knowledge advances. An example of this can be seen in the absence of control of reproductive traits that would allow us to increase production, strain selection and breeding, a major step forward which has been achieved in land plant culture.

Rhodophyta are classified as Archaeplastida, along with glaucophytes and Viridiplantae (land plants and green algae) from which they diverge 1,500 Mya (Yoon et al., 2004). Like other algal groups, red algae comprise a myriad of species with different types of body architecture, ranging from the unicellular and filamentous to the blade or pseudo-parenchymatous as the most complex, particularly in the case of industrially valuable seaweeds (Cole and Sheath, 1995). Their extremely complex life cycles include the transition from unicellularity to complex multicellular bodies, the underlying molecular bases of which are virtually unknown. The diploid "conchocelis" and the meiotic-derived conchospores sustain the industry of "sushi," since the

tiny unicellular meiospore, the conchospore, grows and develops into the haploid leafy thallus, which is the edible phase (Drew, 1949, 1954).

Other economically valuable red algae, such as the producers of the phycocolloids agar or carrageen, have a trigenetic life cycle in which the haploid unicellular meiotic tetraspores germinate to produce both a male and a female multicellular gametophyte thalli (Cole and Sheath, 1995). Fertilization occurs when a spermatium fertilizes a carpogonium on the female gametophyte. The fertilized carpogonium develops into a structure called the cystocarp (diploid) after complex cell differentiation events, leading to the accumulation of mitotic diploid carpospores. Eventually, diploid carpospores are released and develop into tetrasporophytes that produce the meiotic tetraspores (**Figure 1**).

Recent reviews continue to focus on a plethora of external factors that control algal reproduction such as light (intensity, quality, photoperiod), temperature, season, nutrients (be they inorganic or organic), biotic factors (extracellular algal products, bacterial association, animal grazing), osmotic stress, pH of the medium, wave motion and mechanical shock, pollution, and radiations, and the bulk of knowledge accumulated as to the particular conditions on which these external factors exert their control (Dring, 1988; Bornette and Puijalon, 2011; Agrawal, 2012).

Light and temperature are managed effectively to run the intensive cultivation system of *Porphyra*, but there is a general

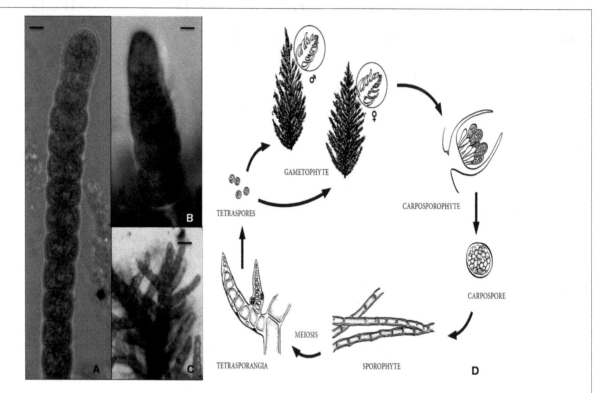

FIGURE 1 | (A) Apical portion of a sporophyte branch of *Bonnemaisonia hamifera*, scale bar = 120 μm. **(B)** Putative gametophyte sporeling, scale bar = 25 μm. **(C)** Apex of a immature gametophytic thalli, scale bar = 100 μm. **(D)** Diagram of putative trigenetic life cycle in the red alga

Bonnemaisonia comprising the gametophytes (haploids), the so-called carposporophyte that develops on the female gametophyte after fertilization, and the sporophyte (diploid) (adapted from *B. geniculata* in Shevlin and Polanshek, 1978. Not a scale).

consensus that increasing our knowledge of the underlying molecular basis of cell growth, development and reproduction in this species, and in economical important seaweeds in general, will improve aquaculture practices (Sahoo et al., 2002; Nakamura et al., 2013).

As we will see below, the same situation occurs when the effect of plant hormones on seaweed reproduction is considered, although some advances at the molecular level have been made on the involvement of certain genes. Almost nothing is known about how the external signals are translated into the molecular mechanisms known to underlie any developmental or reproductive event comprising cell growth and differentiation. Whilst this task was addressed for land plants some time ago, in the genomic era and given the high throughput techniques at our disposal, our knowledge regarding algae in general, and seaweeds in particular, lags far behind that of higher plants and animals. Let us therefore review what is already known about plants and should be revisited in red algae to unveil the secrets of what, no doubt, is also operating at the molecular level to control reproduction.

REPRODUCTION GENES IN RED ALGAE. DISPARATE MODEL SPECIES AND APPROACHES

As seen in the most recent bibliographic reports, and is evident in this special edition, major advances are taking place in the brown alga since the adoption of *Ectocarpus siliculosus* as the model

species and the generalized use of high throughput techniques. This includes key genes in the life cycle transition, developmental pattern, etc. (Peters et al., 2004; Cock et al., 2010; Coelho et al., 2011; Le Bail et al., 2011; Arun et al., 2013). If *E. siliculosus* was chosen mainly because of the particular taxonomic position of brown algae and the evolutionary lineage-related information that could be retrieved from it (Peters et al., 2004), in red algae, no general consensus has been reached as to either the model species or technical approaches and strategies. Consequently, limited advances have been achieved to date, particularly regarding the reproductive event.

Porphyra species (Rhodophyta, Bangiophyceae)—or rather *Porphyra/Pyropia* species, as several species has been reassigned as *Pyropia* (Sutherland et al., 2011)—have been proposed as model species, perhaps because of their economic value. The genome and symbiont-free genome have been sequenced (Chan et al., 2012; Nakamura et al., 2013), and 1% of the genes (10,327 total genes predicted) were annotated as related to reproduction in *P. yezoensis*, using the estimation provided by GO Slim in the Blast2Go software (Nakamura et al., 2013). Therefore, this genomic approach always faces a serious constraint, since these genes are commonly assigned to putative biological processes and functions based on the information available for other organisms, that may simply lack genes and functions related to important life cycle or reproduction events in red algae.

Table 1 | An overview of the complex light sensing–plant hormones interaction, highlighting the key molecular factors implicated.

	Phytochrome	Cryptochrome	Phototropins
Abscisic acid	BBX2, HY5[1] PIF1/PIL5[2]	BBX2, HY5[1] MYC2[?3®] CIB[4,5®]	BBX2, HY5[1] MYC2[?3®]
Ethylene	PIF3, ERF1[6] PIF5, DELLA[?7]	Ethylene synthesis inhibitor[?7]	
Auxins	PAR, HFR1, PIL1[7]		PKS, 14-3-3 I, PIN, PGP, AUX/LAX[8]
Gibberellins	DAG, PIF3, PIF4, PIF1/PIL5 COP1, HY5[2]		
Cytokinins	COP1, HY5[2]		
Jasmonic acid	MYC2[?3]		
Brassinosteroids	PIF4, BZR1[9]		

All of the information comes from higher plants, mostly Arabidopsis thaliana. Prominent role are played by transcription factors (TF) and proteins able to interact with them. PIF are a family of bHLH TF able to bind directly to G-BOX in DNA; HY5 is a nuclear constitutive TF; BZR1 is a TF able to bind to PIF4 during brassinosteroids effect on photomorphogenesis; CIB is a bHLH TF that specifically controls Flowering Time locus; MYC2 is a bHLH TF in Arabidopsis; DELLA are proteins that interacts with PIF; COP1 is a ring finger ubiquitin that promotes HY5 degradation; HFR1 is a protein able to interact with PIF or BRX2, a zinc fingers proteins able to repress or modulate the action of transcription factors. The remaining factors are more specific of the plant hormones or the photoreceptor signaling pathways. ® Denotes participation in reproductive events (i.e., flowering). Details on how they interact can be found in the original references ([1]Xu et al., 2014; [2]Lau and Deng, 2010; [3]Gupta et al., 2012; [4]Liu et al., 2013; [5]Fernando and Coupland, 2012; [6]Zhong et al., 2012; [7]Alabadí and Blázquez, 2009; [8]Hohm et al., 2013; [9]Jaillais and Grégory, 2012). ? Denotes pending confirmation.

The same wide genome approach has recently been used to report the genome and gene annotation of the edible carrageenophyte alga *Chondrus crispus*, the Irish moss (Collén et al., 2013). An important scientific background of knowledge exists for *C. crispus*, to the extent that it has been proposed as the model red alga (Collén et al., 2014). The 9,606 genes annotated for *C. crispus* have produced a very useful bulk of information for the interpretation of the forces driving the evolution of eukaryotic genomes (Collén et al., 2013). Interestingly, *C. crispus* has cryptochromes (Collén et al., 2013), which are important in photosensing and the regulation of the reproduction by light and hormones, as discussed below (**Table 1**). The phases of the life cycle of *C. crispus* are easily accessible and thus frequently used for experiments, but they might be not so easy to handle if the completion of the life cycle is required, as is the case for reproduction studies, for which mutants are required. Moreover, there is an apparent absence or shortage of well-known key elements in the regulatory network (i.e., absence of phytochromes, phototropins, and a rather small amount of transcription-associated proteins). Therefore, the utility of *C. crispus* as a model species for the study of certain aspects of algal growth and development during life cycle completion (i.e., light control of reproduction) remains a matter for debate.

Analysis of the transcriptomes has revealed preferential genes expressed in gametophytes or sporophytes. In *Porphyra purpurea*, an unusual elongation factor (EF-1a) was expressed only in the sporophyte while a second gene, EF, was expressed equally in the sporophyte and the gametophyte (Liu et al., 1996). The PyKPA1 gene, which encoded a sodium pump, was differentially expressed in the gametophyte as compared to the sporophyte, which seems to depend on the presence of specific promoter elements (Uji et al., 2012, 2013). Apart from *Porphyra/Pyropia*, other species also considered to be of economic interest, such as the agarophytic species, have been studied. In this regard, carposporophyte-specific genes were identified in *Gracilariopsis andersonii* (Kamiya et al., 2011). In *Gracilaria lemaneiformis*, a female gametophyte-specific gene, GMF-01, has been reported (Chen et al., 2011) while an ubiquitin gene was also characterized as particularly active during the carposporophyte formation (Ren et al., 2009). In *Griffithsia japonica*, the GjFP-1 gene, encoding a heat-shock protein 90, may be involved in the differentiation of female gametophyte (Lee et al., 1998).

Other approaches have made it possible to reach candidate gene(s) involved in reproduction. This is the case of the GiODC gene in *Grateloupia imbricata*, which encodes the ornithine decarboxylase (ODC, EC. 4.1.1.17). The ODC starts the synthesis of the common polyamines putrescine, spermidine, and spermine by decarboxylating the ornithine to produce putrescine; these substances affect spore maturation and liberation as described below (García-Jiménez et al., 1998; Marián et al., 2000; Guzman-Urióstegui et al., 2002, 2012; Sacramento et al., 2004, 2007). GiODC was cloned using a somewhat laborious approach by means of degenerated primers designed from conserved protein motifs, followed by chromosome walking by iPCR to complete the sequence (García-Jiménez et al., 2009). GiODC expression varied according to cystocarp differentiation with lower levels in the fertile, as compared to the infertile, tissue (García-Jiménez et al., 2009).

All of these findings are clearly contributing to our knowledge about reproduction in red seaweeds, whether achieved through a wide genome strategy using high throughput methods as done in *Porphyra/Pyropia* or *Chondrus*, or using a candidate gene approach as in the case of ODC in *Grateloupia imbricata*. Nevertheless, the weakest point still remains on how this—perhaps species-specific—information can be translated into data that is relevant to most red seaweeds; how to construct a reliable "red seaweed conceptual framework" of knowledge on reproduction from these disparate approaches and species strategies. From our point of view this could only be started to achieve using a species that is easy to handle, with relatively short generation times, and fulfills the criteria needed to undergo genetic transformation, which constitutes the current bottleneck in the molecular biology of seaweeds (see Mikami, 2014).

LIGHT AND PLANT HORMONES SIGNALING AND INTERACTION. THE WAY TO UNVEIL THE MOLECULAR SECRETS OF RED ALGAL REPRODUCTION?

In photosynthetic eukaryotes, like algae, light is the driving force for growth and development; it is the source of energy, but also the signal triggering both vegetative and reproductive developmental events. Light is perceived through families of photoreceptors: phytochromes (red/far red radiation), UVR8 (UV-B), and membrane associated phototropins, cryptochromes, and the members of the ZTL/FKF1/LKP2 family which absorb UV-A/blue light (Hohm et al., 2013 and references therein). In turn, the existence and the type of photoreceptors in aquatic organisms have attracted scientific attention, due to the peculiar characteristics of the interaction of light in the aquatic environment. Thus, the presence of diverse genuine photoreceptors, such as phototropins, aurochromes (blue absorbing), neochromes—a kind of chimeric phytochrome, cryptochromes, and phytochromes in marine algae has been reported and reviews have been produced, which include future applied dimensions (Kianianmomeni and Hallmann, 2014). Interestingly, as far as signal transduction is concerned, the phototropin mechanism seems to be conserved between algae and higher plant (Huang et al., 2002; Onodera et al., 2005; Prochnik et al., 2010).

In seaweeds, plant hormones have been reported to affect growth and development (Chan et al., 2006; Baweja et al., 2009). Concerning reproduction events, in *Grateloupia imbricata* (as *G. doryphora*), the levels of the polyamines putrescine, spermidine, and spermine changed as the cystocarps maturate. Subsequently, it was observed that these polyamines, particularly spermine, favored the maturation, liberation, and growth of carpospores in *Grateloupia imbricata* and *Hydropuntia cornea* (as *Gracilaria cornea*; García-Jiménez et al., 1998; Marián et al., 2000; Guzman-Urióstegui et al., 2002, 2012; Sacramento et al., 2004, 2007). In addition, ethylene has been reported to accelerate the maturation of tetrasporangia in *Pterocladiella capillacea* (García-Jiménez and Robaina, 2012)

In recent years extensive knowledge has been accumulated about light, photoreceptors and plant hormone interaction, and crosstalk at the molecular level in higher plants, particularly during the events occurring at two physiological scenarios: seedling photomorphogenesis and shade avoidance (Gyula et al., 2003; Lau and Deng, 2010). Transcription factors of several families, protein–protein interaction, as well as post-translational protein modification are involved (Alabadí and Blázquez, 2009; Lau and Deng, 2010; Fernando and Coupland, 2012; Gupta et al., 2012; Jaillais and Grégory, 2012; Zhong et al., 2012; Liu et al., 2013). **Table 1** provides an overview of the complex system operating in the light–photoreceptors–plant hormones integrated network. All of this important information has so far proved to be relevant to higher plants, particularly for the model species *Arabidopsis thaliana*, and it is completely unknown whether the key elements highlighted also operate in seaweeds, despite the fact that the influence of photoreceptors and plant hormones on reproduction has been reported, as previously mentioned.

Transcription factors controlled by photoreceptors, such as the PIF family (Alabadí and Blázquez, 2009; Lau and Deng, 2010; Jaillais and Grégory, 2012; Zhong et al., 2012), along with others under the control of plant hormones, such as HY5 (Lau and Deng, 2010; Xu et al., 2014), are very important players in the integrated network (**Table 1**). Other factors affecting plant growth and development, like circadian clock sensors that control endogenous levels of plant hormones (i.e., auxin), or temperature also seem to act by modulating the activity of PIF transcription factors, thus connecting important abiotic factors and development (Leivar and Quail, 2011).

Finally, by way of future perspective, in our laboratory we have recently adopted *Bonnemaisonia hamifera* (Bonnemaisoniaceae) as a working species. Cultures of the sporophyte (*Trailliella*) phase have been established so far, and work is progressing toward the induction of the differentiation of gametophytes, using temperature and photoperiod, and plant hormones (**Figures 1A–C**). Should the completion of the life cycle may be accomplished in the next future as a basic requirement for a working species, it remains to find an appropriate genomic structure (a compact or simple genome, a small number of genes, important functional transcriptomic information, etc.), but there is little information on the *B. hamifera* genome so far. Nevertheless, even in the long run, with *B. hamifera* or any other similar and more adequate species, it is time to revisit this scenario in seaweeds but focusing on the molecular standpoint; why not start identifying within the rising genomic/transcriptomic data for all or any of these regulating factors shown in **Table 1**?

ACKNOWLEDGMENT

The authors want to thank the support obtained from the Spanish Ministerio de Economía y Competitividad (Plan Nacional, grants # BFU2003-01244; 2006-06918; 2010-17248).

REFERENCES

Agrawal, S. C. (2012). Factors controlling induction of reproduction in algae. *Folia Microbiol.* 57, 387–407. doi: 10.1007/s12223-012-0147-0

Alabadí, D., and Blázquez, M. A. (2009). Molecular interaction between light and hormone signalling to control plant growth. *Plant Mol. Biol.* 69, 409–417. doi: 10.1007/s11103-008-9400-y

Arun, A., Peters, N. T., Scornet, D., Peters, A. F., Mark Cock, J., and Coelho, S. M. (2013). Non-cell autonomous regulation of life cycle transitions in the model brown alga *Ectocarpus*. *New Phytol.* 197, 503–510. doi: 10.1111/nph.12007

Baweja, P., Sahoo, D., García-Jiménez, P., and Robaina, R. R. (2009). Seaweed tissue culture as applied to biotechnology: problems, achievements and prospects. *Phycol. Res.* 57, 45–58. doi: 10.1111/j.1440-1835.2008.00520.x

Bornette, G., and Puijalon, S. (2011). Response of aquatic plants to abiotic factors: a review. *Aquat. Sci.* 73, 1–14. doi: 10.1007/s00027-010-0162-7

Chan, C. X., Blouin, N. A., Zhuang, Y., Zäuner, S., Prochnik, S. E., Lindquist, E., et al. (2012). *Porphyra* (Bangiophyceae) transcriptomes provide insights into red algal development and metabolism. *J. Phycol.* 48, 1328–1342. doi: 10.1111/j.1529-8817.2012.01229.x

Chan, C. Y., Ho, C., and Phang, S. M. (2006). Trends in seaweed research. *Trends Plant Sci.* 11, 165–166. doi: 10.1016/j.tplants.2006.02.003

Chen, P., Shao, H., and Xu, D. (2011). Cloning and characterization of a female gametophyte-specific gene in *Gracilaria lemaneiformis* (Gracilariales, Rhodophyte). *Afr. J. Microbiol. Res.* 5, 2590–2595. doi: 10.5897/AJMR11.133

Cock, J. M., Sterck, L., Rouzé, P., Scornet, D., Allen, A. E., Amoutzias, G., et al. (2010). The *Ectocarpus* genome and the independent evolution of multicellularity in brown algae. *Nature* 465, 617–621. doi: 10.1038/nature09016

Coelho, S. M., Godfroy, O., Arun, A., Le Corguillé, G., Peters, A. F., and Cock, J. M. (2011). OUROBOROS is a master regulator of the gametophyte to sporophyte life cycle transition in the brown alga *Ectocarpus*. *Proc. Natl. Acad. Sci. U.S.A.* 108, 11518–11523. doi: 10.1073/pnas.1102274108

Cole, K. M., and Sheath, R. G. (1995). *Biology of the Red Seaweeds*. New York: Cambridge University Press.

Collén, J., Cornish, M. L., Craigie, J., Ficko-Bean, E., Hervé, C., Krueger-Hadfield, S. A., et al. (2014). "*Chondrus crispus*. A present and historical model organism for red seaweeds," in *Advances in Botanical Research*, Vol. 71, *Sea Plants*, ed. N. Bourgougnon (London: Elsevier), 53–89.

Collén, J., Porcel, B., Carré, W., Ball, S. G., Chaparro, C., Tonon, T., et al. (2013). Genome structure and metabolic features in the red seaweed *Chondrus crispus* shed light on evolution of the Archaeplastida. *Proc. Natl. Acad. Sci. U.S.A.* 26, 5247–5252. doi: 10.1073/pnas.1221259110

Drew, K. M. (1949). Conchocelis-phase in the life-history of *Porphyra umbilicalis* (L.) Kutz. *Nature* 164, 748–749. doi: 10.1038/164748a0

Drew, K. M. (1954). Life-history of *Porphyra*. *Nature* 173, 1243–1244. doi: 10.1038/1731243a0

Dring, M. J. (1988). Photo control of development in algae. *Annu. Rev. Plant Physiol. Plant Mol. Biol.* 39, 157–174. doi: 10.1146/annurev.pp.39.060188.001105

Fernando, A., and Coupland, G. (2012). The genetic basis of flowering responses to seasonal cues. *Nat. Rev. Genet.* 9, 627–639. doi: 10.1038/nrg3291

García-Jiménez, P., García-Maroto, F., Garrido-Cárdenas, J. A., Ferrandiz, C., and Robaina, R. R. (2009). Differential expression of the ornithine decarboxylase gene during carposporogenesis in the thallus of the red seaweed *Grateloupia imbricata* (Halymeniaceae). *J. Plant Physiol.* 166, 1745–1754. doi: 10.1016/j.jplph.2009.04.018

García-Jiménez, P., and Robaina, R. R. (2012). Effects of ethylene on tetrasporogenesis in *Pterocladiella capillacea* (Rhodophyta). *J. Phycol.* 48, 710–715. doi: 10.1111/j.1529-8817.2012.01156.x

García-Jiménez, P., Rodrigo, M., and Robaina, R. R. (1998). Influence of plant growth regulators, polyamines and glycerol interaction on growth and morphogenesis in carposporeling of *Grateloupia doryphora* cultured in vitro. *J. Appl. Phycol.* 10, 95–100. doi: 10.1023/A:1008063532233

Gupta, N., Prassad, V. B. R., and Chattopadhyay, S. (2012). LeMYC2 acts as a negative regulator of blue light mediated photomorphogenic growth, and promotes the growth of adult tomato plants. *BMC Plant Biol.* 14:38. doi: 10.1186/1471-2229-14-38

Guzman-Urióstegui, A., García-Jiménez, P., Marián, F., Robledo, D., and Robaina, R. (2002). Polyamines influence maturation in reproductive structures of *Gracilaria cornea* (Gracilariales, Rhodophyta). *J. Phycol.* 38, 1169–1175. doi: 10.1046/j.1529-8817.2002.01202.x

Guzman-Urióstegui, A., Robaina, R. R., Freile-Pelegri, Y., and Robledo, D. (2012). Polyamines increase carpospore output and growth during in vitro cultivation of *Hydropuntia cornea*. *Biotechnol. Lett.* 34, 755–761. doi: 10.1007/s10529-011-0823-1

Gyula, P., Schäfer, E., and Nagy, F. (2003). Light perception and signalling in higher plants. *Curr. Opin. Plant Biol.* 6, 446–452. doi: 10.1016/S1369-5266(03) 00082-7

Hohm, T., Preuten, T., and Fankhauser, C. H. (2013). Phototropism: translating light into directional growth. *Am. J. Bot.* 100, 47–59. doi: 10.3732/ajb.1200299

Huang, K. Y., Merkle, T., and Beck, C. F. (2002). Isolation and characterization of a *Chlamydomonas* gene that encodes a putative blue-light photoreceptor of the phototropin family. *Physiol. Plant.* 115, 613–622. doi: 10.1034/j.1399-3054.2002.1150416.x

Jaillais, Y., and Grégory, V. (2012). Brassinosteroids, gibberellins and light-mediated signalling are the three-way controls of plant sprouting. *Nat. Cell Biol.* 8, 788–790. doi: 10.1038/ncb2551

Kamiya, M., Kawai, H., Moon, D., and Goff, L. J. (2011). Isolation and characterization of phase-specific cDNAs from carposporophytes of *Gracilariopsis andersonii* (Gracilariales, Rhodophyta). *Eur. J. Phycol.* 46, 27–35. doi: 10.1080/09670262.2010.548101

Kianianmomeni, A., and Hallmann, A. (2014). Algal photoreceptors: in vivo functions and potential applications. *Planta* 293, 1–26. doi: 10.1007/s00425-013-1962-5

Lau, O. S., and Deng, X. W. (2010). Plant hormone signalling lightens up: integrators of light and hormones. *Curr. Opin. Plan Biol.* 13, 571–577. doi: 10.1016/j.pbi.2010.07.001

Le Bail, A., Billoud, B., Le Panse, S., Chenivesse, S., and Charrier, B. (2011). ETOILE regulates developmental patterning in the filamentous brown alga *Ectocarpus siliculosus*. *Plant Cell* 23, 1666–1678. doi: 10.1105/tpc.110.081919

Lee, Y. K., Kim, S. H., Hong, C. B., Chah, O.-K., Kim, G. H., and Lee, I. K. (1998). Heat-shock protein 90 may be involved in differentiation of the female gametophytes in *Griffithsia japonica* (Ceramiales, Rhodophyta). *J. Phycol.* 34, 1017–1023. doi: 10.1046/j.1529-8817.1998.341017.x

Leivar, P., and Quail, P. H. (2011). PIFs: pivotal component in a cellular signalling hub. *Trends Plant Sci.* 16, 19–28. doi: 10.1016/j.tplants.2010.08.003

Liu, Q. Y., Baldauf, S. L., and Reith, M. E. (1996). Elongation factor 1 alpha genes of the red alga *Porphyra purpurea* include a novel, developmentally specialized variant. *Plant Mol. Biol.* 31, 77–85. doi: 10.1007/BF00020608

Liu, Y., Li, X., Li, K., Liu, H., and Lin, C. (2013). Multiple bHLH proteins form heterodimers to mediate CRY2-dependent regulation of flowering-time in *Arabidopsis*. *PLoS Genet.* 9:e1003861. doi: 10.1371/journal.pgen.1003861

Marián, F. D., García-Jiménez, P., and Robaina, R. R. (2000). Polyamine in marine macroalgae: natural levels of putrescine, spermidine and spermine in the thalli and change of their concentration during glycerol-induced cell growth in vitro. *Physiol. Plant.* 110, 530–534. doi: 10.1111/j.1399-3054.2000.1100416.x

Mikami, K. (2014). A technical breakthrough close at hand: feasible approaches toward establishing a gene-targeting genetic transformation system in seaweeds. *Front. Plant Sci.* 5:498. doi: 10.3389/fpls.2014.00498

Nakamura, Y., Sasaki, N., Kobayashi, M., Ojima, N., Yasuike, M., Shigenobu, Y., et al. (2013). The first symbiont-free genome sequence of marine red alga, Susabi-nori (*Pyropia yezoensis*). *PLoS ONE* 8:e57122. doi: 10.1371/journal.pone.0057122

Onodera, A., Kong, S. G., Doi, M., Shimazaki, K., Christie, J., Mochizuki, N., et al. (2005). Phototropin from *Chlamydomonas reinhardtii* is functional in *Arabidopsis thaliana*. *Plant Cell Physiol.* 46, 367–374. doi: 10.1093/pcp/pci037

Peters, A. F., Marie, D., Scornet, D., Kloareg, B., and Cock, J. M. (2004). Proposal of *Ectocarpus siliculosus* (Ectocarpales, Phaeophyceae) as a model organism for brown algal genetics and genomics. *J. Phycol.* 40, 1079–1088. doi: 10.1111/j.1529-8817.2004.04058.x

Prochnik, S. E., Umen, J., Nedelcu, A. M., Hallmann, A., Miller, S. M., Nishii I., et al. (2010). Genomic analysis of organismal complexity in the multicellular green alga *Volvox carteri*. *Science* 329, 223–226. doi: 10.1126/science.1188800

Rebours, C., Marinho-Soriano, E., Zertuche-González, J. A., Hayashi, L., Vásquez, J. A., Kradolfer, P., et al. (2014). Seaweeds: an opportunity for wealth and sustainable livelihood for coastal communities. *J. Appl. Phycol.* 26, 1939–1951. doi: 10.1007/s10811-014-0304-8

Ren, X., Sui, Z., Mao, Y., Zang, X., Xu, D., and Zhang, X. (2009). Cloning and characterization of two types of ubiquitin genes from *Gracilariopsis lemaneiformis* (Gracilariales, Rhodophyta). *J. Appl. Phycol.* 21, 273–278. doi: 10.1007/s10811-008-9361-1

Sacramento, A. T., García-Jiménez, P., Alcázar, R., Tiburcio, A. F., and Robaina, R. R. (2004). Influence of polyamines on the sporulation of *Grateloupia* (Halymeniaceae, Rhodophyta). *J. Phycol.* 40, 887–894. doi: 10.1111/j.1529-8817.2004.03183.x

Sacramento, A. T., García-Jiménez, P., and Robaina, R. R. (2007). Spermine induces cystocarp development in marine alga. *Plant Growth Regul.* 53, 147–154. doi: 10.1007/s10725-007-9212-0

Sahoo, D., Tang, X., and Yarish, C. (2002). *Porphyra*—the economic seaweed as a new experimental system. *Curr. Sci.* 83, 1313–1316.

Shevlin, D. E., and Polanshek, A. R. (1978). Life history of *Bonnemaisonia geniculata* (Rhodophyta): a laboratory and field study. *J. Phycol.* 14, 282–289. doi: 10.1111/j.1529-8817.1978.tb00300.x

Sutherland, E. J., Lindstrom, S. C., Nelson, W. A., Brodie, J., and Lynch, D. J. M. (2011). A new look at an ancient order: generic revision of the Bangiales (Rhodophyta). *J. Phycol.* 47, 1131–1151. doi: 10.1111/j.1529-8817.2011.01052.x

Uji, T., Hirata, R., Mikami, K., Mizuta, H., and Saga, N. (2012). Molecular characterization and expression analysis of sodium pump genes in the marine red alga *Porphyra yezoensis*. *Mol. Biol. Rep.* 39, 7973–7980. doi: 10.1007/s11033-012-1643-7

Uji, T., Mizuta, H., and Saga, N. (2013). Characterization of the sporophyte-preferential gene promoter from the red alga *Porphyra yezoensis* using transient gene expression. *Mar. Biotechnol.* 15, 188–196. doi: 10.1007/s10126-012-9475-y

Xu, D., Li, J., Gangappa, S. N., Hettiarachchi, C., Lin, F., Andersson, M. X., et al. (2014). Convergence of light and ABA signaling on the ABI5 promoter. *PLoS Genet.* 10:e1004197. doi: 10.1371/journal.pgen.1004197

Yoon, H. S., Hackett, J. D., Ciniglia, C., Pinto, G., and Bhattacharya, D. (2004). A molecular timeline for the origin of photosynthetic eukaryotes. *Mol. Biol. Evol.* 21, 809–818. doi: 10.1093/molbev/msh075

Zhong, S. H., Shi, H., Xue, C., Wang, L., Xi, Y., Li, J., et al. (2012). A molecular framework of light-controlled phytohormone action in *Arabidopsis*. *Curr. Biol.* 22, 1530–1535. doi: 10.1016/j.cub.2012.06.039

21

Plant responses to *Agrobacterium tumefaciens* and crown gall development

Jochen Gohlke[1] and Rosalia Deeken[2] *

[1] *School of Plant Sciences, University of Arizona, Tucson, AZ, USA*
[2] *Department of Molecular Plant Physiology and Biophysics, Julius-von-Sachs-Institute, University of Wuerzburg, Wuerzburg, Germany*

Edited by:
Stanton B. Gelvin, Purdue University, USA

Reviewed by:
Erh-Min Lai, Academia Sinica, Taiwan
Ze-Chun Yuan, University of Western Ontario, Canada

*Correspondence:
Rosalia Deeken, Department of Molecular Plant Physiology and Biophysics, Julius-von-Sachs-Institute, University of Wuerzburg, Julius-von-Sachs-Platz 2, 97082 Wuerzburg, Germany
e-mail: deeken@ botanik.uni-wuerzburg.de

Agrobacterium tumefaciens causes crown gall disease on various plant species by introducing its T-DNA into the genome. Therefore, *Agrobacterium* has been extensively studied both as a pathogen and an important biotechnological tool. The infection process involves the transfer of T-DNA and virulence proteins into the plant cell. At that time the gene expression patterns of host plants differ depending on the *Agrobacterium* strain, plant species and cell-type used. Later on, integration of the T-DNA into the plant host genome, expression of the encoded oncogenes, and increase in phytohormone levels induce a fundamental reprogramming of the transformed cells. This results in their proliferation and finally formation of plant tumors. The process of reprogramming is accompanied by altered gene expression, morphology and metabolism. In addition to changes in the transcriptome and metabolome, further genome-wide ("omic") approaches have recently deepened our understanding of the genetic and epigenetic basis of crown gall tumor formation. This review summarizes the current knowledge about plant responses in the course of tumor development. Special emphasis is placed on the connection between epigenetic, transcriptomic, metabolomic, and morphological changes in the developing tumor. These changes not only result in abnormally proliferating host cells with a heterotrophic and transport-dependent metabolism, but also cause differentiation and serve as mechanisms to balance pathogen defense and adapt to abiotic stress conditions, thereby allowing the coexistence of the crown gall and host plant.

Keywords: plant defenses, phytohormones, morphological adaptions, metabolomic changes, epigenetics

INTRODUCTION

Agrobacterium tumefaciens causes crown gall disease on a wide range of host species by transferring and integrating a part of its own DNA, the T-DNA, into the plant genome (Chilton et al., 1977). This unique mode of action has also made the bacterium an important tool in plant breeding. After attachment of *Agrobacterium* to plant cells and expression of multiple virulence (vir) genes, several effector proteins, together with T-DNA, are transported into the plant cell by a type-IV-secretion system (Thompson et al., 1988; Ward et al., 1988, 2002; Kuldau et al., 1990; Shirasu et al., 1990; Beijersbergen et al., 1994). Plant factors assist with T-DNA integration into the plant genome (Gelvin, 2000; Mysore et al., 2000; Tzfira et al., 2004; Magori and Citovsky, 2012). After integration, expression of the T-DNA-encoded oncogenes iaaH, iaaM, and ipt induces biosynthesis of auxin and cytokinin (Morris, 1986; Binns and Costantino, 1998). Increased levels of these phytohormones result in enhanced proliferation and formation of crown galls. Despite the transfer of bacterial proteins into the plant cell, most *Agrobacterium* strains do not elicit a hypersensitive response (HR), which is associated with rapid and localized death of cells (Staskawicz et al., 1995). Such a response often occurs when plants are challenged by bacterial pathogens and serves to restrict the growth and spread of pathogens to other parts of the plant. Accordingly, no systemic, broad-spectrum resistance response throughout the plant (systemic acquired resistance,

SAR) is induced. Within the first several hours of co-cultivation, pathogen defense response pathways are activated more or less strongly depending on the plant system and *Agrobacterium* genotype used for infection (Ditt et al., 2001, 2006; Veena et al., 2003; Lee et al., 2009). Defense responses become stronger during crown gall development. Furthermore, the physiological behavior of the transformed cells changes drastically. In contrast to the articles which focus on the molecular mechanism utilized by the bacterium to transform the plant cell, here we review the latest findings on the responses of the host plant and in the crown gall to *Agrobacterium* infection. Special attention is paid to the role of gene expression regulation, phytohormones, and metabolism.

HOST RESPONSES TO *Agrobacterium tumefaciens* BEFORE T-DNA TRANSFER
PATHOGEN DEFENSE

The recognition of microbial pathogens plays a central role in the induction of active defense responses in plants. The conserved flagellin peptide flg22 is recognized by the receptor kinase FLS2 and induces the expression of numerous defense-related genes to trigger resistance to pathogenic bacteria (Gómez-Gómez et al., 1999, 2001; Zipfel et al., 2004; Chinchilla et al., 2006). However, the genus *Agrobacterium* fails to induce this type of rapid and general defense response because of an exceptional divergence in the N-terminal conserved domain of flagellin (Felix et al., 1999). When comparing early gene expression changes after infection with the virulent *Agrobacterium* strain C58 with application of the bacterial peptide elf26 (after 1 and 3 h, respectively), dampening of host responses becomes apparent with *Agrobacterium* treatment. The elf26 peptide, a highly conserved motif of one of the most abundant proteins in microbes recognized by the receptor kinase EFR, is a fragment of the elongation factor Tu (EF-Tu). EF-Tu triggers innate immunity responses associated with disease resistance in *Arabidopsis* (Kunze et al., 2004). While treatment with pure elf26 induces gene expression changes of 948 *Arabidopsis* genes (Zipfel et al., 2006), only 35 genes are induced after infection with the virulent *Agrobacterium* strain C58, suggesting that the bacterium somehow neutralizes the response to elf26 by the host plant (Lee et al., 2009). It should be mentioned that the *Arabidopsis* ecotype and age (seedling vs. adult stalk) used in the studies may also account for some of the differences in defense response.

Concerning the transcriptional activation of genes involved in early plant defense responses, several studies have come to different conclusions. *Ageratum conyzoides* cell cultures showed differential expression of defense genes as early as 24 h post infection with a non-oncogenic hypervirulent *Agrobacterium* strain (Ditt et al., 2001). In tobacco suspension cultures infected with different *Agrobacterium* strains, transcription of defense genes increased within 3–6 h, but started to decrease with the onset of T-DNA-transfer (Veena et al., 2003). A study using suspension-cultured cells of *Arabidopsis* did not show changes in transcript levels within 4 to 24 h but activation of defense genes 48 h after infection (Ditt et al., 2006). When agrobacteria are inoculated at the base of wounded *Arabidopsis* stems just very few defense genes are activated 3 h post infection compared to uninfected wounded stems (Lee et al., 2009). In contrast to cell cultures, the latter experimental setup does neither require phytohormone pre-treatment nor virulence gene induction prior to infection. Phytohormone pre-treatment of the cell culture systems of the earlier studies may alter host cell defense responses. Thus, discrepancies between these studies probably result from the different plant inoculation systems used. Nevertheless, agrobacteria can abuse host defense responses for T-DNA delivery. The mitogen-activated protein kinase MPK3 phosphorylates the *Arabidopsis* VIP1 protein, inducing VIP1 relocalization from the cytoplasm to the nucleus. Nuclear localization of VIP1 increases T-DNA transfer and transformation efficiency (Djamei et al., 2007).

PHYTOHORMONES

Agrobacteria produce auxin and cytokinin themselves in order to modulate plant responses (**Figure 1A**). These phytohormones have been determined in the cells as well as cultivation medium (Morris, 1986). It was postulated that biosynthesis of the phytohormones is catalyzed by enzymes of the T-DNA encoded oncogenes, as transcripts and proteins of these genes were detected in agrobacterial cells (Schröder et al., 1983; Janssens et al., 1984). Pronounced amounts of auxin have been determined in the virulent *Agrobacterium* strain C58 and at lower levels also in plasmidless and T-DNA depleted strains (Liu and

Kado, 1979; Kutáĉek and Rovenská, 1991). More recent data have confirmed the latter results (Lee et al., 2009). The finding that a strain without a Ti-plasmid still can make auxin implies localization of genes also outside of the Ti-plasmid. However, this assumption is not supported by sequencing data for strain C58 (Wood et al., 2001). Genes known to be involved in auxin biosynthesis seem to be encoded only by the T-DNA of the Ti-plasmid. Recently, these authors determined the presence of iaaH and iaaM transcripts by PCR in *Agrobacterium* cells of strain C58 and confirmed the earlier findings. It remains to be proven whether these genes are responsible for auxin production or if auxin is synthesized by a different mechanism in *Agrobacterium* cells. The mechanism for cytokinin biosynthesis by agrobacteria is far better understood. In nopaline utilizing *Agrobacterium* strains cytokinin is produced in high amounts by the Ti-plasmid encoded *trans*-zeatin synthesizing (tzs) enzyme of which the gene is located in the vir regulon (Akiyoshi et al., 1985, 1987; Hwang et al., 2010). A substantial smaller source for cytokinin production is isopentenylated transfer RNA (tRNA) catalyzed by the chromosomal-encoded enzyme tRNA:isopentenyltransferase (MiaA) present in all *Agrobacterium* strains (Gray et al., 1996).

Earlier studies have shown that pre-treatment of explants with either auxin alone or both auxin and cytokinin increase T-DNA transfer efficiency and stable transformation (Krens et al., 1996; Chateau et al., 2000) as well as crown gall growth (Gafni et al., 1995). In this respect, *Agrobacterium* produced phytohormones play a role at very early time points of infection (**Figure 1A**), before T-DNA-encoded enzymes catalyze synthesis of cytokinin and auxin in the transformed host cell. Concerning the mechanism causing an increase in susceptibility it was speculated that phytohormones induce plant cell division and that the cell cycle phase influences agrobacterial attachment and stable transformation. It seems likely that phytohormone-mediated modification of the physiological state of the cell increases competence for T-DNA transformation and integration. More recent investigations addressed the question about the molecular mechanism and the signaling pathways by which these phytohormones influence host cell susceptibility. Transcriptome microarray data from 3 h after inoculation of *Agrobacterium* strain C58 into *Arabidopsis* stems revealed that the genes known to be involved in phytohormone-dependent signaling are not induced in host cells at this very early time point of infection before transfer of the T-DNA (Lee et al., 2009). It has been shown that indole-3-acetic acid (IAA) has an impact on agrobacterial virulence by inhibiting vir gene induction and growth of agrobacteria (Liu and Nester, 2006). However, this effect was observed with relatively high concentrations of auxin (25–250 µM). In *Agrobacterium* cells the total (free and conjugated) IAA content is 0.3 ± 0.1 µM and in *Arabidopsis* stems 3 h after inoculation with strain C58 it is 2.1 ± 1 µM, whereas in *Arabidopsis* crown galls the content is ca. 10 times higher (17.3 ± 8.8 µM) due to the expression of the T-DNA encoded iaaH and iaaM genes and their enzyme activity (own data and Thomashow et al., 1986). Application of 1 µM IAA, a concentration found in wounded and uninfected *Arabidopsis* stems (0.8 ± 0.2 µM), stimulated growth of *Agrobacterium* cells, whereas growth stimulation vanished at

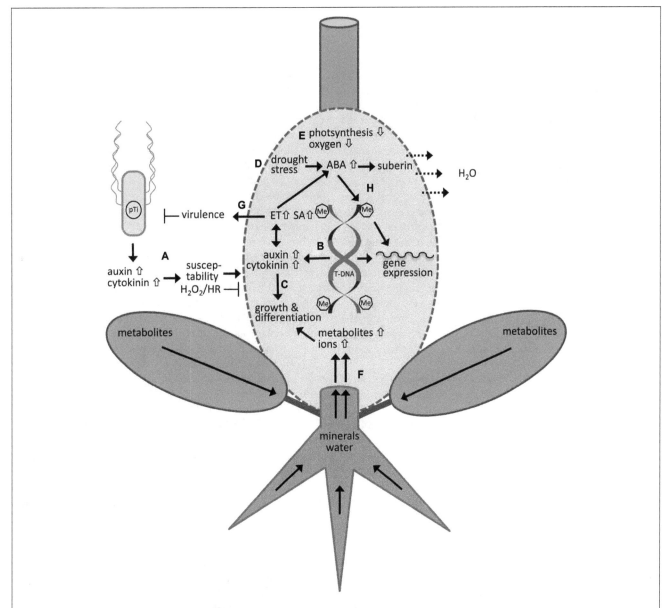

FIGURE 1 | Responses of the model plant *Arabidopsis thaliana* to *Agrobacterium tumefaciens* and crown gall development. (A) Virulent (pTi) agrobacteria cells themselves produce and release cytokinin and auxin, which increase host susceptibility and inhibit hydrogen peroxide production (H_2O_2) and hypersensitive response (HR) at initiation of infection. **(B)** After integration of the bacterial T-DNA into the plant genome, cytokinin and auxin is synthesized by T-DNA encoded enzymes and accumulate inside the tumor. **(C)** This causes massive changes in the gene expression pattern, resulting in metabolomic and morphological adaptations that are necessary for tumor growth and differentiation. **(D)** Loss of water is minimized by drought stress protecting mechanism, which causes an increase in the levels of the stress hormone ABA, and ABA-dependent suberization of cells to prevent water loss. Evaporation of water (H_2O) from the disrupted crown gall surface drives the flow of water and minerals into crown galls. **(E)** Because photosynthesis is down-regulate the oxygen levels are low, the tumor produces C and N compounds heterotrophically and gains energy mainly anaerobically by alcoholic fermentation. **(F)** Consequently the developing tumor becomes a metabolic sink for the host plant, which accumulates metabolites produced by source leaves and minerals taken up by the roots. **(G)** Auxin and cytokinin also cause an increase in ethylen (ET) which together with salicylic acid (SA) inhibits agrobacterial virulence. **(H)** ABA also induces DNA methylation of the plant genome, thereby regulating gene expression of drought-stress responsive genes. Overall, the crown gall genome becomes hypermethylated (Me) after *Agrobacterium* infection and possibly contributes to the strong changes in gene expression during tumor growth. The oncogenes of the T-DNA remain unaffected by methylation of the plant genome.

10 μM and higher IAA concentrations (personal communication, J. Ludwig-Mueller, Technical University Dresden, Germany). It is known that the effect of auxin is strongly dose dependent with a growth promoting effect at low concentrations and an inhibitory effect at high concentrations, which slightly varies dependent on the plant and tissue type. One may speculate that at initiation of infection, the relatively low auxin levels of agrobacterial cells and/or of wounded plant tissue stimulate growth of agrobacteria, whereas the higher concentrations produced in the crown gall inhibit virulence as well as growth of *Agrobacterium*. Such an

antagonistic auxin effect would promote transformation of the host cell at the beginning of the infection process and inhibit agrobacterial virulence and growth to prevent further transformation events in developing crown galls. In contrast to auxin, the role of cytokinin signaling in plant susceptibility is well known. Recently, it has been shown that cytokinin secreted by *Agrobacterium* controls virulence via bacterial cell growth and vir gene expression at early stages of the infection process (Hwang et al., 2010). Some, but not all plant species showed a cytokinin-dependent increase in transformation efficiency (Hwang et al., 2013). *Agrobacterium*-derived cytokinin not only acts on bacterial physiology but also influences host gene expression via the classical cytokinin-dependent signaling pathway including cytokinin receptors and the phosphotransfer cascade (Sardesai et al., 2013). Activation of this signaling cascade through agrobacterial-derived cytokinin results in inhibition of gene expression of the *Arabidopsis* MYB family transcription factor, MTF1 (Sardesai et al., 2013). MTF1 turned out to be a negative regulator of transformation susceptibility by blocking expression of the integrin-like protein At14a, a plant membrane receptor. At14A serves as anchor points for bacterial attachment at the host cell surface. Thus, at early stages of infection agrobacterial auxin and cytokinin manipulates plant phytohormone signaling pathways to prepare the host cell for transformation.

In addition to auxin and cytokinin, plant defense signaling involves a network of interconnected pathways in which salicylic acid (SA) and jasmonic acid (JA) together with ethylene (ET) function as essential signaling molecules (Kunkel and Brooks, 2002). Exogenous application of the plant defense molecule SA to *Agrobacterium* cells inhibited expression of vir genes including tzs, bacterial growth, bacterial attachment to plant cells and virulence (Yuan et al., 2007; Anand et al., 2008). However, at initiation of infection (3 h post infection) neither SA nor JA levels nor the genes of these signaling pathways are elevated in *Agrobacterium*-infected *Arabidopsis* tissues (Lee et al., 2009). At this time point only the level of 1-amino-cyclopropane-1-carboxylic acid (ACC), an ET precursor, is increased in the presence of both virulent and disarmed *Agrobacterium* strains, but not expression of marker genes of the ET-dependent defense-signaling pathway. Inoculation of melon (*Cucumis melo*) explants with *Agrobacterium* also increases ET production (Ezura et al., 2000). ET is known to trigger plant auxin production due to increased expression of plant genes involved in auxin biosynthesis (Stepanova et al., 2005). Auxin enhances host susceptibility whereas plant ET production has a negative effect on agrobacterial virulence. Application of ACC reduces *Agrobacterium*-mediated gene transfer to melon explants whereas addition of aminoethoxyvinylglycin, an inhibitor of ACC synthase, increased it (Ezura et al., 2000). A reduction in transformation efficiency results from suppression of vir gene expression, but not *Agrobacterium* growth (Nonaka et al., 2008). The promoting effect of low auxin concentrations on agrobacterial growth and the inhibiting effect of ET on virulence illustrates that both, *Agrobacterium* and the host plant control host cell transformation. Taken together, at early stages of the infection process, cytokinin and auxin produced by *Agrobacterium* cells have a promoting effect on transformation efficiency, which is in part counteracted by the inhibitory effect of host plant-derived ET and SA on agrobacterial virulence. Thus, the correct phytohormone balance decides on the success of infection.

HYPERSENSITIVE RESPONSE

Examination of early events in pathogenesis has demonstrated that virulent *Agrobacterium* does not induce HR in *Arabidopsis* (**Figure 1A**; Lee et al., 2009). Moreover, *Agrobacterium* is able to suppress HR induced by *Pseudomonas syringae pv. phaseolus* in plants (Robinette and Matthysse, 1990). This suppression is dependent on the activity of the *iaaH* and *iaaM* oncogenes which encode enzymes for auxin synthesis, since several *Agrobacterium* transposon mutants in the *iaa* genes failed to inhibit a HR. Likewise, transcription of several genes involved in oxidative stress signaling are only induced by the oncogenic, but not the T-DNA-depleted *Agrobacterium* strain (Lee et al., 2009). Production of H_2O_2 precedes HR, which is degraded via a chromosomally encoded catalase of *Agrobacterium* (Xu and Pan, 2000). H_2O_2 acts both as a local trigger for the programmed cell death and as a diffusible signal for the induction of cellular protectant genes in surrounding cells (Levine et al., 1994). Apart from its signaling functions, H_2O_2 is also involved in toughening of cell walls in the initial stages of plant defense by cross-linking of cell wall structural proteins (Bradley et al., 1992). Accumulation of H_2O_2 is prevented only at the early stages of agrobacterial infection, but proceeds in the course of tumor development (Lee et al., 2009).

HOST RESPONSES TO CROWN GALL DEVELOPMENT
MORPHOLOGICAL ADAPTATIONS

Development of crown galls is accompanied by profound changes in the gene expression profile, metabolism, and morphology. The uncontrolled synthesis of auxin and cytokinin by cells transformed with a T-DNA of tumorigenic Ti-plasmids drives tumor development, while the auxin to cytokinin ratios determine the crown gall morphology (**Figure 1B**). In the early days of studies about the molecular basis of crown gall development it was observed that mutations in the *tmr* locus encoding *ipt* cause rooty crown galls and those in the *tms* loci coding for *iaaH* and *iaaM* induce shooty phenotypes (Garfinkel et al., 1981; Akiyoshi et al., 1983; Barry et al., 1984; Buchmann et al., 1985; Black et al., 1994). A recent study on the T-DNA locus Atu6002 of strain C58 indicated that when the encoding protein C is expressed, it increases host cell sensitivity to auxin (Lacroix et al., 2013). In addition to the T-DNA-encoded genes, the expression of several host genes involved in auxin and cytokinin metabolism and signaling are expressed in crown galls (Lee et al., 2009). Cytokinin and auxin together with ET are known to be essential for growth of crown gall tumors and differentiation of cell types with different morphology and function (**Figure 1C**). Particularly, ET has been shown to be essential for the formation of vascular tissue and crown gall tumor development (Aloni et al., 1998; Wächter et al., 1999, 2003; Ullrich and Aloni, 2000). Application of the ET synthesis inhibitor aminoethoxyvinyl-glycine prevents vascularization in castor bean (*Ricinus communis*) stems and inhibits tumor growth completely (Wächter et al., 2003). When the ET-insensitive tomato (*Lycospersicon esculentum*) mutant, *never ripe*, is infected with virulent *Agrobacterium* cells it does not develop tumors despite integration and expression of the T-DNA encoded oncogenes for auxin and cytokinin biosynthesis (Aloni et al., 1998). Thus, neovascularization is a prerequisite for crown gall development.

Growth and expansion of crown gall tumors cause disruption of the epidermal cell layer and thereby loss of guard cells and an intact cuticle. Accordingly, expression of genes involved in cutin biosynthesis is downregulated (Deeken et al., 2006). As a disrupted surface area provides access for pathogens and leads to uncontrolled loss of water for the host plant, the crown gall surface has to be sealed. This is achieved by differentiating a periderm-like surface layer (Efetova et al., 2007). The polymerization of suberin monomers involves peroxidases for which H_2O_2 is the electron donor. Thus, H_2O_2 produced in crown galls functions in strengthening of cell walls rather than in induction of a HR. The stimulus for inducing suberization is drought stress-mediated ABA signaling (**Figure 1D**). Drought stress signaling seems to play a central role in crown gall development. ABA accumulates in crown galls in high amounts and transcription of a set of drought and/or ABA-inducible genes is elevated (Mistrik et al., 2000; Veselov et al., 2003; Efetova et al., 2007). ABA synthesis is triggered by ET as demonstrated by the application of various inhibitors of ET or ABA biosynthesis and the use of ET-insensitive or ABA-deficient tomato mutants (Hansen and Grossmann, 2000). Among the genes which play a role in drought stress protection of crown gall tumors is *FAD3*, encoding a fatty acid desaturase. The *fad3-2* mutant with impaired biosynthesis in α-linolenic acid (C18:3) develops much smaller crown gall tumors particularly in low but not high relative humidity (Klinkenberg et al., 2014). Elevated levels of C18:3 were found in the phospholipid fraction of *Arabidopsis* crown gall tumors and maintain membrane integrity under drought stress conditions. In addition to gene expression changes, crown galls accumulate high amounts of osmoprotectants, such as proline (Pro), gamma aminobutyric acid (GABA), and alpha-aminoadipinic acid. The retarded tumor growth in *abi* and *aba* mutant plants underlines the importance of an ABA-mediated drought stress-signaling pathway in crown gall development (Efetova et al., 2007).

NUTRIENT TRANSLOCATION AND METABOLISM

Expression profiles of genes involved in energy metabolism, such as photosynthesis, mitochondrial electron transport, and fermentation together with physiological data revealed that *Arabidopsis* tumors produce C and N compounds heterotrophically and gain energy mainly anaerobically by alcoholic fermentation (**Figure 1E**; Deeken et al., 2006). The change from autotrophy to heterotrophy reduces the oxygen level in crown gall tumors thereby inducing expression of hypoxia-sensitive genes, such as *SAD6*. This gene encodes a stearoyl-acyl carrier protein desaturase, which belongs to a class of enzymes known to catalyze the first step in fatty acid desaturation, an oxygen-dependent process. Despite limited oxygen availability in crown galls, SAD6 provides the monounsaturated fatty acid, oleic acid, for membrane phospholipids (Klinkenberg et al., 2014). Thus, expression of SAD6 maintains fatty acid desaturation under hypoxic conditions.

Crown gall tumors primarily use organic carbon and nitrogen for growth and are therefore a strong sink for the host plant.

Metabolites and minerals have to be provided by the host plant and translocated into the crown gall tumor (**Figure 1F**). The mechanisms of nutrient translocation and their accumulation have been studied on crown gall tumors by applying cytological staining, eletrophysiological, and $^{14}CO_2$ tracer techniques as well as a viral movement protein (Marz and Ullrich-Eberius, 1988; Malsy et al., 1992; Pradel et al., 1999). Solutes enter the crown gall tumor via vascular tissue, which is connected to that of the host plant and consists of phloem for the transport of assimilates and xylem for water and minerals (Aloni et al., 1995; Deeken et al., 2003). Assimilates are produced by source leaves and are apoplastically and symplastically unloaded from the phloem in crown gall tumors. High apoplastic invertase activity indicated that sucrose is unloaded apoplastically (Malsy et al., 1992). After cleavage of sucrose by sucrose-degrading enzymes, hexoses can be taken up via hexose transporters into tumor cells. *Arabidopsis* crown galls show elevated expression of several genes encoding sucrose degrading enzymes and a monosaccharide transporter (Deeken et al., 2006). In addition, a high-affinity hexose transporter has been isolated from meristematic tobacco cells transformed with a tumor inducing T-DNA and was characterized as energy independent hexose uptake transporter (Verstappen et al., 1991). Application of the membrane impermeable fluorescent probe, carboxyfluorescein (CF) to source leaves and transient expression of the GFP-labeled potato virus X (PVX) coat protein (CP), exclusively exploiting plasmodesmata for distribution, demonstrated the existence of a symplastic transport pathway between the phloem and tumor cells (Pradel et al., 1999). Both reporters show extensive cell-to-cell movement in the parenchyma of crown gall tumors but not in uninfected stem tissues of different plant species ranging from symplastic (*Curcubita maxima*) to apoplastic loaders (*R. communis, Nicotiana benthamiana*). The disrupted and enlarged surface of the crown gall tumor drives water and mineral translocation into crown gall tumors since the evaporation rate of crown galls exceeds that of leaves and non-infected stems (Schurr et al., 1996; Wächter et al., 2003). The periderm-like layer of suberized cells that covers the crown gall surface provides a considerable diffusion resistance against water vapor, but it is not an impermeable barrier for water (**Figure 1D**; Kolattukudy and Dean, 1974; Vogt et al., 1983; Schreiber et al., 2005). Cations and anions are taken up into the tumor cells through the function of membrane-localized channels and transporters expressed in the crown gall (Deeken et al., 2003). Potassium channel mutants with impaired crown gall growth underline the importance of optimal nutrient supply for growth.

DEFENSE RESPONSES

Gamma aminobutyric acid and Pro not only serve as osmoprotectants in drought-stress related processes of the host plant, but have also an impact on *Agrobacterium* virulence (Haudecoeur et al., 2009a,b). GABA produced in crown gall tumors can be taken up by *Agrobacterium* cells and causes a delay in accumulation of 3-oxo-octanoylhomoserine lactone (OC8HSL) and Ti plasmid conjugation. GABA activates the AttKLM operon of which the AttM lactonase degrades the quorum sensing signal, OC8HSL, thereby turning on quorum quenching to protect the host plant against infections with bacterial pathogens (Yuan et al., 2008). However, Pro interferes with the import of GABA and thereby prevents GABA-induced degradation of the bacterial quorum sensing signal OC8HSL. Thus, Pro antagonizes the GABA-induced degradation of OC8HSL and therefore may be used by the pathogen to by-pass the GABA-based host plant defense.

In addition to growth and developmental processes regulated by auxin and cytokinin, crown gall biology also involves pathogen defense signaling pathways. Hormones such as SA, JA, and ET are the primary signals inducing defense responses (López et al., 2008). In *Arabidopsis* crown galls the levels of SA and ET, but not JA, are elevated (**Figure 1G**). JA has no obvious impact on crown gall tumor development, as the development on *Arabidopsis* JA-insensitive mutants is wildtype-like (Lee et al., 2009). SA and ET contents together with the expression of pathogen-related marker genes of the SA- and ET-dependent signaling pathways increase with accumulation of the T-DNA-encoded iaa and ipt transcripts. Thus, auxin and/or cytokinin seem to be important for defense signaling in crown gall tumors, since the non-tumorigenic *Agrobacterium* strain which contains a disarmed pTiC58 does not induce expression of marker genes of the SA- and ET-dependent signaling pathways (Lee et al., 2009). It is known that high levels of auxin and cytokinin stimulate ET synthesis and its accumulation in crown galls (Goodman et al., 1986; Aloni et al., 1998; Johnson and Ecker, 1998; Vogel et al., 1998; Wächter et al., 1999). In contrast to ET, the classical marker genes of the SA-dependent signaling pathways are not induced most likely as a result of the high auxin content, which has been shown to inhibit SA responses to avoid the induction of SAR (Robert-Seilaniantz et al., 2011). Despite the lack of induction of SA-dependent defense signaling, *Arabidopsis* mutant plants with high SA levels strongly reduce while those with low SA levels promote tumor growth (Lee et al., 2009). Instead of inducing host defense pathways, high SA levels act directly on oncogenic agrobacteria by inhibiting vir gene expression and thereby reducing agrobacterial virulence (Yuan et al., 2007; Anand et al., 2008). Besides SA-mediated inhibition of *Agrobacterium* virulence, SA activates the AttKLM operon, just like GABA does, to down regulate quorum sensing in *Agrobacterium* (Yuan et al., 2008). Thus, activation of quorum quenching by auxin, SA, and GABA, is part of the plant defense program against *Agrobacterium* in the developing crown gall. In addition to SA, ET and IAA also inhibit the vir regulon and T-DNA transfer into plant cells (**Figure 1G**; Ezura et al., 2000; Nonaka et al., 2008). Thus, the interaction between the host plant and *Agrobacterium* is very much based on phytohormone cross talk which provides a balance between pathogen-defense by the host and crown gall development promoted by *Agrobacterium*.

EPIGENETIC PROCESSES IN DNA INTEGRATION, ONCOGENE EXPRESSION, AND CROWN GALL DEVELOPMENT
EPIGENETIC CHANGES ASSOCIATED WITH T-DNA INTEGRATION AND ONCOGENE EXPRESSION

Epigenetic changes that affect chromatin structure play an important role in regulating a wide range of cellular processes. Histones for example are subject to post-translational modification including acetylation, phosphorylation, methylation, and

ubiquitination. These modifications may influence crown gall development on different levels, either by affecting chromatin structure and DNA integration or by influencing gene expression in the host tissue. Up-regulation of several members from the core histone gene families after *Agrobacterium* infection indicates that they are important for the transformation process (Veena et al., 2003). For example, *Arabidopsis* mutants lacking histone H2A are defective in T-DNA integration (Mysore et al., 2000). In addition, a truncated version of VIP1, an *Arabidopsis* protein proposed to interact with the T-DNA-protein-complex (T-complex), which is not able to interact with histone H2A, strongly decreases *Agrobacterium* tumorigenicity (Li et al., 2005). As this decrease is most likely due to a reduced T-DNA integration efficiency, this suggests that association of the VIP1 with the host chromatin is critical for integration of the T-DNA. One hypothesis of how epigenetic information affects DNA integration is that chromatin modifications surrounding double-strand breaks (DSBs) of the DNA can be recognized by the T-complex. The resulting chromatin-T-complex may then bring T-DNA into close proximity to DSBs and facilitate its integration by the DSB repair pathway (Magori and Citovsky, 2011). Alternatively, histones may also enhance transformation by protecting incoming DNA from nuclease digestion during the initial stages of transformation. Indeed, overexpression of several histone genes in *Arabidopsis* results in higher amounts of transferred DNA and increased transient transgene expression in transformed cells (Tenea et al., 2009). Other epigenetic modifications like DNA methylation do not correlate with the T-DNA integration pattern, suggesting that T-DNA integration occurs without regard to this type of modification (Kim et al., 2007). Concerning post-translational modifications of histones, RNA-mediated knockdown of two histone deacetylases (HDT1 and HDT2) decreases *Agrobacterium*-mediated transformation efficiency of *Arabidopsis* root segments (Crane and Gelvin, 2007). Histone deacetylation functions in chromatin compaction and transcriptional repression (Strahl and Allis, 2000). Therefore, the observed effect on transformation may either be a result of effects on chromatin structure or gene expression of plant factors involved in the integration process. Histone deacetylation may also influence DNA integration by affecting DSB repair, as several histone deacetylases are critical for the DNA repair process in yeast (Munoz-Galvan et al., 2013).

After T-DNA is integrated into the plant genome, the host plant often silences transgenes. Gene silencing can occur by two different mechanisms. Transcriptional gene silencing (TGS) is a result of promoter inactivation while post-TGS (PTGS) occurs when the promoter is active but the mRNA fails to accumulate. DNA methylation of promoter sequences is frequently associated with inactivation of transgenes (Linne et al., 1990; Matzke and Matzke, 1991; Kilby et al., 1992). Screening of a large collection of transgenic *Arabidopsis* lines with single T-DNA copies including a pNOS-*NPTII* reporter gene has shown that promoter methylation is required but not sufficient for transcriptional inactivation (Fischer et al., 2008). Silencing only occurs when the plants, challenged by the silencer transgene, also provide an RNA signal. Concerning local features of the host genome affecting gene silencing, repeats flanking the site of integration seem to promote inactivation whereas flanking genes rather attenuate it. RNA silencing is triggered only if the transcript level of a transgene surpasses a gene-specific threshold, suggesting that the inactivation is part of plant defense mechanism corresponding to excessively transcribed genes (Schubert et al., 2004).

Apart from the down-regulation of transgenes that are integrated into the plant genome along with the T-DNA, the T-DNA itself may also be subject to modification by the plant silencing machinery. The first comprehensive analysis of T-DNA methylation revealed that methylation can occur in different plant tumor lines induced by *Agrobacterium*. At least one T-DNA copy in each tumor genome remained unmethylated, thereby allowing oncogene expression and crown gall proliferation (Gelvin et al., 1983). Experiments using the demethylating agent 5-azacytidine indicates that methylation negatively correlates with gene expression in plant tumors (Hepburn et al., 1983). A more recent study on T-DNA methylation in crown gall tumors induced on *Arabidopsis* stems demonstrates that the oncogene sequences are only methylated to a very low degree (Gohlke et al., 2013). The two intergenic regions, which serve as promoters for expression of the oncogenes *iaaH*, *iaaM*, and *ipt*, are completely unmethylated in *Arabidopsis* crown galls. As the gene products of these oncogenes are essential for an increase in levels of cytokinin and auxin, they are always actively transcribed in crown gall tumors of *Arabidopsis* stems (Deeken et al., 2006). The low degree of T-DNA methylation in crown galls suggests that this is a prerequisite to maintain the expression levels of oncogenes required for tumor formation. Indeed, induction of DNA oncogene methylation by production of double-stranded RNAs is sufficient to repress oncogene transcription and prevent tumor development (Gohlke et al., 2013).

EPIGENETIC MODIFICATIONS IN THE CROWN GALL GENOME

Analysis of *Agrobacterium*-infected inflorescence stalks allowed monitoring of gene expression in the crown gall tumor at later developmental stages and revealed massive changes in its transcriptome (Deeken et al., 2006). A large part of the *Arabidopsis* genome (about 22% of genes) was found to be expressed differentially between crown galls and mock-infected stems. Of these genes, a slightly higher percentage was found to be downregulated in crown galls (12%) compared to up-regulated genes (10%). Distinct expression changes occur at genes pivotal for energy metabolism, such as those involved in photosynthesis, mitochondrial electron transport, and fermentation. This reflects the induced host cell changes from an auxotrophic, aerobic metabolism to a heterotrophic, transport-dependent, sugar-dependent anaerobic metabolism (see Nutrient Translocation and Metabolism).

Considering that a high percentage of the *Arabidopsis* genome is differentially regulated in crown gall tumors, transcriptional reprogramming probably occurs on several levels. For example, the transcript levels of several transcription factor families (MYB, bHLH, bZIP, AP domain) change after *Agrobacterium* infection (Ditt et al., 2006; Sardesai et al., 2013), thereby inducing a tumor-specific gene expression pattern. Gene expression may also be regulated by epigenetic mechanisms like chromatin modification or DNA methylation. Apart from modifications which play a role during T-DNA integration and silencing of oncogenes (see Epigenetic Processes in DNA Integration, Oncogene Expression

and Crown Gall Development), DNA methylation of plant genes can also influence tumor growth (**Figure 1H**). Indeed, 8% of protein-coding genes are differentially methylated in crown galls compared to mock-infected stems, with an overall tendency toward being hypermethylated (Gohlke et al., 2013). Depending on the position of DNA methylation, different effects on the gene expression levels are observed. In agreement with trends observed for DNA methylation changes in *Arabidopsis* (Zhang et al., 2006), increased methylation at transcription start and end sites has a negative impact on gene expression, while the two processes are positively correlated in the transcribed region. Mapping of DNA methylation in tumors revealed hypomethylation in the upstream regions of genes as well as hypermethylation in transcribed regions. Both of these may, in turn, influence gene expression and contribute to the tumor-specific expression pattern. Not surprisingly, pathways that are associated with tumor development like genes associated with cell division, biotic stress, and redox regulation are differentially methylated. Changes in the methylation pattern also have an impact on tumor growth, as *Arabidopsis* mutants in *de novo* methylation pathways promote crown gall development. Intriguingly, callus induction, which like crown gall development is also associated with dedifferentiation of plant cells, is increased in the methyltransferase mutant *cmt3* (Berdasco et al., 2008). In addition, treatment with the methyltransferase inhibitor 5-acacytidine results in increased callus formation. Recently, the DNA methylation pattern has been extensively studied in calli from *Populus trichocarpa* and *Oryza sativa*. In *Oryza sativa* calli, hypermethylation was detected compared to wild-type plants (Stroud et al., 2013). Gene bodies are hypermethylated in *Populus trichocarpa* calli compared to explants, while promoter methylation is reduced (Vining et al., 2013). Consistent with the methylation pattern in crown galls, DNA hypermethylation seems to be a general feature of a dedifferentiated status.

An attempt to identify internal plant signals which may influence DNA methylation suggests that high levels of ABA induce DNA methylation of promoter sequences (**Figure 1H**; Gohlke et al., 2013). Therefore, this phytohormone may at least partly be responsible for the methylation pattern found in crown galls. It is tempting to speculate that ABA induces DNA methylation as a response to abiotic stresses such as drought stress acclimation due to the increased water loss in crown gall tumors (Schurr et al., 1996). Possibly, ABA signaling pathways are interconnected with methylation processes in crown galls, as has been suggested for *Physcomitrella patens* (Khraiwesh et al., 2010). In the future, it would be interesting to analyze ABA knockout mutants concerning their methylation pattern in order to map ABA-induced methylation changes in a comprehensive manner and thereby improve our understanding of the connection between the different pathways. In addition, other phytohormones would also be interesting to study regarding their influence on the DNA methylation pattern in crown galls, as they display not only increased levels of ABA, but also of cytokinin, auxin, ET, and JA (Veselov et al., 2003; Lee et al., 2009).

SUMMARY AND OUTLOOK

At the beginning of infection, sensing of *Agrobacterium* does not induce a strong defense response of the host plant. *Agrobacterium* rather exploits defense responses to increase host susceptibility for transformation and host signaling pathways to promote bacterial growth. In crown galls, however, pathogen defense pathways are considerably activated and inhibit *Agrobacterium* virulence. Accordingly, the host plant is able to limit the number of further T-DNA transformation events and to control the growth dimension of crown galls, which represent a strong metabolic sink for the host plant. Metabolic and morphological adaptations accompany the development of crown galls and generate an import-oriented tissue. The heterotrophic metabolism together with anaerobically gain of energy requires translocation of metabolites, water and minerals from the plant into the proliferating crown gall tissue. As a basis for nutrient translocation the vascular tissue needs to differentiate and the disrupted and suberized crown gall surface provides the driving force for nutrient flow. In fact, the suberized surface minimizes water loss, but still allows enough evaporation of water. Membrane integrity is maintained under the low oxygen and elevated ROS levels in crown galls by adaptation of lipid metabolism. The transcriptional changes underlying the physiological changes are partially caused by differential DNA methylation of the crown gall genome. In conclusion, both *Agrobacterium* infection and crown gall growth are highly regulated processes, which are accompanied by pathogen defense of the host and counter-defense launched by *Agrobacterium*. This regulation takes place on different levels including epigenetic control of gene expression, changes in phytohormone content as well as metabolic and morphological adaptions.

Despite the fact that the *Agrobacterium*-plant-interaction has been studied since more than 100 years and is most likely one of the best-known pathogen-host-relationships, there are still some questions left, which one may aim to answer. In addition to the one raised about the role of phytohormones other than ABA on DNA methylation in crown gall development, another one would be about the molecular mechanisms of how *Agrobacterium* cells produce auxin and how auxin increases host susceptibility for transformation. Furthermore, the status and type of plant cell susceptible for T-DNA integration is as yet unknown. The knowledge about the cellular identity sensitive for transformation will improve our understanding of transformation recalcitrant plant species. Moreover, differentiation processes in crown galls do not follow the usual patterning, unlike the situation in plant organs where developmental patterning underlies a precise spatiotemporal expression of signals and their cognate receptors. Since the original/typical developmental program seems to be overruled, crown gall tumors provide a unique opportunity for studying the molecular and biochemical mechanisms underlying cellular de-differentiation as well as differentiation processes. Not all of the questions raised may be easy to address, as some require sophisticated techniques, which at first have to be developed and established. However, invention of new techniques will benefit the entire scientific community as they have done before when *Agrobacterium* became the biotechnological tool for generation of genetically modified plants.

ACKNOWLEDGMENTS

Studies were financially supported by the Deutsche Forschungsgemeinschaft (project B5, SFB 567, and project A5 of the Research Training Group 1342 – Lipid Signaling). We thank Stanton Gelvin for critical reading of the manuscript.

REFERENCES

Akiyoshi, D. E., Morris, R., Hinz, R., Mischke, B., Kosuge, T., Garfinkel, D., et al. (1983). Cytokinin/auxin balance in crown gall tumors is regulated by specific loci in the T-DNA. *Proc. Natl. Acad. Sci. U.S.A.* 80, 407–411. doi: 10.1073/pnas.80.2.407

Akiyoshi, D. E., Regier, D. A., and Gordon, M. P. (1987). Cytokinin production by *Agrobacterium* and *Pseudomonas* spp. *J. Bacteriol.* 169, 4242–4248.

Akiyoshi, D. E., Regier, D. A., Jen, G., and Gordon, M. P. (1985). Cloning and nucleotide sequence of the tzs gene from *Agrobacterium tumefaciens* strain T37. *Nucleic Acids Res.* 13, 2773–2788. doi: 10.1093/nar/13.8.2773

Aloni, R., Pradel, K. S., and Ullrich, C. I. (1995). The three-dimensional structure of vascular tissues in *Agrobacterium tumefaciens*-induced crown galls and in the host stems of *Ricinus communis* L. *Planta* 196, 597–605. doi: 10.1007/BF00203661

Aloni, R., Wolf, A., Feigenbaum, P., Avni, A., and Klee, H. J. (1998). The never ripe mutant provides evidence that tumor-induced ethylene controls the morphogenesis of *Agrobacterium tumefaciens*-induced crown galls on tomato stems. *Plant Physiol.* 117, 841–849. doi: 10.1104/pp.117.3.841

Anand, A., Uppalapati, S. R., Ryu, C.-M., Allen, S. N., Kang, L., Tang, Y., et al. (2008). Salicylic acid and systemic acquired resistance play a role in attenuating crown gall disease caused by *Agrobacterium tumefaciens*. *Plant Physiol.* 146, 703–715. doi: 10.1104/pp.107.111302

Barry, G. F., Rogers, S. G., Fraley, R. T., and Brand, L. (1984). Identification of a cloned cytokinin biosynthetic gene. *Proc. Natl. Acad. Sci. U.S.A.* 81, 4776–4780. doi: 10.1073/pnas.81.15.4776

Beijersbergen, A., Smith, S. J., and Hooykaas, P. J. (1994). Localization and topology of VirB proteins of *Agrobacterium tumefaciens*. *Plasmid* 32, 212–218. doi: 10.1006/plas.1994.1057

Berdasco, M., Alcázar, R., García-Ortiz, M. V., Ballestar, E., Fernández, A. F., Roldán-Arjona, T., et al. (2008). Promoter DNA hypermethylation and gene repression in undifferentiated *Arabidopsis* cells. *PLoS ONE* 3:e3306. doi: 10.1371/journal.pone.0003306

Binns, A. N., and Costantino, P. (1998). "The *Agrobacterium* oncogenes," in *The Rhizobiaceae*, eds H. P. Spaink, A. Kondorosi, and P. J. J. Hooykaas (Netherlands: Springer), 251–266.

Black, R. C., Binns, A. N., Chang, C. F., and Lynn, D. G. (1994). Cell-autonomous cytokinin-independent growth of tobacco cells transformed by *Agrobacterium tumefaciens* strains lacking the cytokinin biosynthesis gene. *Plant Physiol.* 105, 989–998. doi: 10.1104/pp.105.3.989

Bradley, D. J., Kjellbom, P., and Lamb, C. J. (1992). Elicitor- and wound-induced oxidative cross-linking of a proline-rich plant cell wall protein: a novel, rapid defense response. *Cell* 70, 21–30. doi: 10.1016/0092-8674(92)90530-P

Buchmann, I., Marner, F. J., Schroder, G., Waffenschmidt, S., and Schroder, J. (1985). Tumour genes in plants: T-DNA encoded cytokinin biosynthesis. *EMBO J.* 4, 853–859.

Chateau, S., Sangwan, R. S., and Sangwan-Norreel, B. S. (2000). Competence of *Arabidopsis thaliana* genotypes and mutants for *Agrobacterium tumefaciens*-mediated gene transfer: role of phytohormones. *J. Exp. Bot.* 51, 1961–1968. doi: 10.1093/jexbot/51.353.1961

Chilton, M. D., Drummond, M. H., Merio, D. J., Sciaky, D., Montoya, A. L., Gordon, M. P., et al. (1977). Stable incorporation of plasmid DNA into higher plant cells: the molecular basis of crown gall tumorigenesis. *Cell* 11, 263–271. doi: 10.1016/0092-8674(77)90043-5

Chinchilla, D., Bauer, Z., Regenass, M., Boller, T., and Felix, G. (2006). The *Arabidopsis* receptor kinase FLS2 binds flg22 and determines the specificity of flagellin perception. *Plant Cell* 18, 465–476. doi: 10.1105/tpc.105.036574

Crane, Y. M., and Gelvin, S. (2007). RNAi-mediated gene silencing reveals involvement of *Arabidopsis* chromatin-related genes in *Agrobacterium*-mediated root transformation. *Proc. Natl. Acad. Sci. U.S.A.* 104. doi: 10.1073/pnas.0706986104

Deeken, R., Engelmann, J. C., Efetova, M., Czirjak, T., Müller, T., Kaiser, W. M., et al. (2006). An integrated view of gene expression and solute profiles of *Arabidopsis* tumors: a genome-wide approach. *Plant Cell* 18, 3617–3634. doi: 10.1105/tpc.106.044743

Deeken, R., Ivashikina, N., Czirjak, T., Philippar, K., Becker, D., Ache, P., et al. (2003). Tumour development in *Arabidopsis thaliana* involves the Shaker-like

K+ channels AKT1 and AKT2/3. *Plant J.* 34, 778–787. doi: 10.1046/j.1365-313X.2003.01766.x

Ditt, R. F., Kerr, K. F., de Figueiredo, P., Delrow, J., Comai, L., and Nester, E. W. (2006). The *Arabidopsis thaliana* transcriptome in response to *Agrobacterium tumefaciens*. *Mol. Plant Microbe Interact.* 19, 665–681. doi: 10.1094/MPMI-19-0665

Ditt, R. F., Nester, E. W., and Comai, L. (2001). Plant gene expression response to *Agrobacterium tumefaciens*. *Proc. Natl. Acad. Sci. U.S.A.* 98, 10954–10959. doi: 10.1073/pnas.191383498

Djamei, A., Pitzschke, A., Nakagami, H., Rajh, I., and Hirt, H. (2007). Trojan horse strategy in *Agrobacterium* transformation: abusing MAPK defense signaling. *Science* 318, 453–456. doi: 10.1126/science.1148110

Efetova, M., Zeier, J., Riederer, M., Lee, C.-W., Stingl, N., Mueller, M., et al. (2007). A central Role of abscisic acid in drought stress protection of *Agrobacterium*-induced tumors on *Arabidopsis*. *Plant Physiol.* 145, 853–862. doi: 10.1104/pp.107.104851

Ezura, R. F., Yuhashi, K.-I., Yasuta, T., and Minamisawa, K. (2000). Effect of ethylene on *Agrobacterium tumefaciens*-mediated gene transfer to melon. *Plant Breed.* 119, 75–79. doi: 10.1046/j.1439-0523.2000.00438.x

Felix, G., Duran, J. D., Volko, S., and Boller, T. (1999). Plants have a sensitive perception system for the most conserved domain of bacterial flagellin. *Plant J.* 18, 265–276. doi: 10.1046/j.1365-313X.1999.00265.x

Fischer, U., Kuhlmann, M., Pecinka, A., Schmidt, R., and Mette, M. F. (2008). Local DNA features affect RNA-directed transcriptional gene silencing and DNA methylation. *Plant J.* 53, 1–10. doi: 10.1111/j.1365-313X.2007.03311.x

Gafni, Y., Icht, M., and Rubinfeld, B.-Z. (1995). Stimulation of *Agrobacterium tumefaciens* virulence with indole-3-acetic acid. *Lett. Appl. Microbiol.* 20, 98–101. doi: 10.1111/j.1472-765X.1995.tb01295.x

Garfinkel, D. J., Simpson, R. B., Ream, L. W., White, F. F., Gordon, M. P., and Nester, E. W. (1981). Genetic analysis of crown gall: fine structure map of the T-DNA by site-directed mutagenesis. *Cell* 27, 143–153. doi: 10.1016/0092-8674(81)90368-8

Gelvin, S. B. (2000). *Agrobacterium* and plant genes involved in T-DNA transfer and integration. *Annu. Rev. Plant Physiol. Plant Mol. Biol.* 51, 223–256. doi: 10.1146/annurev.arplant.51.1.223

Gelvin, S., Karcher, S. J., and DiRita, V. J. (1983). Methylation of the T-DNA in *Agrobacterium tumefaciens* and in several crown gall tumors. *Nucleic Acids Res.* 11, 159–174. doi: 10.1093/nar/11.1.159

Gohlke, J., Scholz, C.-J., Kneitz, S., Weber, D., Fuchs, J., Hedrich, R., et al. (2013). DNA methylation mediated control of gene expression is critical for development of crown gall tumors. *PLoS Genet.* 9:e1003267. doi: 10.1371/journal.pgen.1003267

Gómez-Gómez, L., Bauer, Z., and Boller, T. (2001). Both the extracellular leucine-rich repeat domain and the kinase activity of FSL2 are required for flagellin binding and signaling in *Arabidopsis*. *Plant Cell* 13, 1155–1163.

Gómez-Gómez, L., Felix, G., and Boller, T. (1999). A single locus determines sensitivity to bacterial flagellin in *Arabidopsis thaliana*. *Plant J.* 18, 277–284. doi: 10.1046/j.1365-313X.1999.00451.x

Goodman, T. C., Montoya, A. L., Williams, S., and Chilton, M. D. (1986). Sustained ethylene production in *Agrobacterium*-transformed carrot disks caused by expression of the T-DNA tms gene products. *J. Bacteriol.* 167, 387–388.

Gray, J., Gelvin, S. B., Meilan, R., and Morris, R. O. (1996). Transfer RNA is the source of extracellular isopentenyladenine in a Ti-plasmidless strain of *Agrobacterium tumefaciens*. *Plant Physiol.* 110, 431–438. doi: 10.1104/pp.110.2.431

Hansen, H., and Grossmann, K. (2000). Auxin-induced ethylene triggers abscisic acid biosynthesis and growth inhibition. *Plant Physiol.* 124, 1437–1448. doi: 10.1104/pp.124.3.1437

Haudecoeur, E., Planamente, S., Cirou, A., Tannières, M., Shelp, B. J., Moréra, S., et al. (2009a). Proline antagonizes GABA-induced quenching of quorum-sensing in *Agrobacterium tumefaciens*. *Proc. Natl. Acad. Sci.* 106, 14587–14592. doi: 10.1073/pnas.0808005106

Haudecoeur, E., Tannières, M., Cirou, A., Raffoux, A., Dessaux, Y., and Faure, D. (2009b). Different regulation and roles of lactonases AiiB and AttM in *Agrobacterium tumefaciens* C58. *Mol. Plant Microbe Interact.* 22, 529–537. doi: 10.1094/MPMI-22-5-0529

Hepburn, A. G., Clarke, L. E., Pearson, L., and White, J. (1983). The role of cytosine methylation in the control of nopaline synthase gene expression in a plant tumor. *J. Mol. Appl. Genet.* 2, 315–329.

Hwang, H.-H., Wang, M. H., Lee, Y.-I., Tsai, Y.-L., Li, Y.-H., Yang, F.-J., et al. (2010). *Agrobacterium*-produced and exogenous cytokinin-modulated

Agrobacterium-mediated plant transformation. *Mol. Plant Pathol.* 11, 677–690. doi: 10.1111/j.1364-3703.2010.00637.x

Hwang, H.-H., Yang, F.-J., Cheng, T.-F., Chen, Y.-C., Lee, Y.-L., Tsai, Y.-L., et al. (2013). The Tzs protein and exogenous cytokinin affect virulence gene expression and bacterial growth of *Agrobacterium tumefaciens*. *Phytopathology* 103, 888–899. doi: 10.1094/PHYTO-01-13-0020-R

Janssens, A., Engler, G., Zambryski, P., and Montagu, M. V. (1984). The nopaline C58 T-DNA region is transcribed in *Agrobacterium tumefaciens*. *Mol. Gen. Genet.* 195, 341–350. doi: 10.1007/BF00332769

Johnson, P. R., and Ecker, J. R. (1998). The ethylene gas signal transduction pathway: a molecular perspective. *Annu. Rev. Genet.* 32, 227–254. doi: 10.1146/annurev.genet.32.1.227

Khraiwesh, B., Arif, M. A., Seumel, G. I., Ossowski, S., Weigel, D., Reski, R., et al. (2010). Transcriptional control of gene expression by microRNAs. *Cell* 140, 111–122. doi: 10.1016/j.cell.2009.12.023

Kilby, N. J., Leyser, H. M., and Furner, I. J. (1992). Promoter methylation and progressive transgene inactivation in *Arabidopsis*. *Plant Mol. Biol.* 20, 103–112. doi: 10.1007/BF00029153

Kim, S.-I., Veena, and Gelvin, S. B. (2007). Genome-wide analysis of *Agrobacterium* T-DNA integration sites in the *Arabidopsis* genome generated under non-selective conditions. *Plant J.* 51, 779–791. doi: 10.1111/j.1365-313X.2007.03183.x

Klinkenberg, J., Faist, H., Saupe, S., Lamberts, S., Krischke, M., Stingl, N., et al. (2014). Two fatty acid desaturases, SAD6 and FAD3, are involved in drought and hypoxia stress signaling in *Arabidopsis* crown galls. *Plant Physiol.* 164, 570–583. doi: 10.1104/pp.113.230326

Kolattukudy, P. E., and Dean, B. B. (1974). Structure, gas chromatographic measurement, and function of suberin synthesized by potato tuber tissue slices. *Plant Physiol.* 54, 116–121. doi: 10.1104/pp.54.1.116

Krens, F. A., Trifonova, A., Paul Keizer, L. C., and Hall, R. D. (1996). The effect of exogenously-applied phytohormones on gene transfer efficiency in sugarbeet (*Beta vulgaris* L.). *Plant Sci.* 116, 97–106. doi: 10.1016/0168-9452(96)04357-9

Kuldau, G. A., De Vos, G., Owen, J., McCaffrey, G., and Zambryski, P. (1990). The virB operon of *Agrobacterium tumefaciens* pTiC58 encodes 11 open reading frames. *Mol. Gen. Genet. MGG* 221, 256–266. doi: 10.1007/BF00261729

Kunkel, B. N., and Brooks, D. M. (2002). Cross talk between signaling pathways in pathogen defense. *Curr. Opin. Plant Biol.* 5, 325–331. doi: 10.1016/S1369-5266(02)00275-3

Kunze, G., Zipfel, C., Robatzek, S., Niehaus, K., Boller, T., and Felix, G. (2004). The N terminus of bacterial elongation factor Tu elicits innate immunity in *Arabidopsis* plants. *Plant Cell* 16, 3496–3507. doi: 10.1105/tpc.104.026765

Kutáček, M., and Rovenská, J. (1991). Auxin synthesis in *Agrobacterium tumefaciens* and *A. tumefaciens-transformed plant tissue*. *Plant Growth Regul.* 10, 313–327. doi: 10.1007/BF00024591

Lacroix, B., Gizatullina, D. I., Babst, B. A., Gifford, A. N., and Citovsky, V. (2013). *Agrobacterium* T-DNA-encoded protein Atu6002 interferes with the host auxin response. *Mol. Plant Pathol.* 15, 275–283. doi: 10.1111/mpp.12088

Lee, C.-W., Efetova, M., Engelmann, J. C., Kramell, R., Wasternack, C., Ludwig-Muller, J., et al. (2009). *Agrobacterium tumefaciens* promotes tumor induction by modulating pathogen defense in *Arabidopsis thaliana*. *Plant Cell* 21, 2948–2962. doi: 10.1105/tpc.108.064576

Levine, A., Tenhaken, R., Dixon, R., and Lamb, C. (1994). H_2O_2 from the oxidative burst orchestrates the plant hypersensitive disease resistance response. *Cell* 79, 583–593. doi: 10.1016/0092-8674(94)90544-4

Li, J., Krichevsky, A., Vaidya, M., Tzfira, T., and Citovsky, V. (2005). Uncoupling of the functions of the *Arabidopsis* VIP1 protein in transient and stable plant genetic transformation by *Agrobacterium*. *Proc. Natl. Acad. Sci. U.S.A.* 102, 5733–5738. doi: 10.1073/pnas.0404118102

Linne, F., Heidmann, I., Saedler, H., and Meyer, P. (1990). Epigenetic changes in the expression of the maize A1 gene inPetunia hybrida: role of numbers of integrated gene copies and state of methylation. *Mol. Gen. Genet. MGG* 222, 329–336. doi: 10.1007/BF00633837

Liu, P., and Nester, E. W. (2006). Indoleacetic acid, a product of transferred DNA, inhibits vir gene expression and growth of *Agrobacterium tumefaciens* C58. *Proc. Natl. Acad. Sci. U.S.A.* 103, 4658–4662. doi: 10.1073/pnas.0600366103

Liu, S. T., and Kado, C. I. (1979). Indoleacetic acid production: a plasmid function of *Agrobacterium tumefaciens* C58. *Biochem. Biophys. Res. Commun.* 90, 171–178. doi: 10.1016/0006-291X(79)91605-X

López, M. A., Bannenberg, G., and Castresana, C. (2008). Controlling hormone signaling is a plant and pathogen challenge for growth and survival. *Curr. Opin. Plant Biol.* 11, 420–427. doi: 10.1016/j.pbi.2008.05.002

Magori, S., and Citovsky, V. (2011). Epigenetic control of *Agrobacterium* T-DNA integration. *Biochim. Biophys. Acta* 1809, 388–394. doi: 10.1016/j.bbagrm.2011.01.007

Magori, S., and Citovsky, V. (2012). The role of the ubiquitin-proteasome system in *Agrobacterium tumefaciens*-mediated genetic transformation of plants. *Plant Physiol.* 160, 65–71. doi: 10.1104/pp.112.200949

Malsy, S., Bel, V., Kluge, M., Hartung, W., and Ullrich, C. I. (1992). Induction of crown galls by *Agrobacterium tumefaciens* (strain C58) reverses assimilate translocation and accumulation in *Kalanchoë daigremontiana*. *Plant Cell Environ.* 15, 519–529. doi: 10.1111/j.1365-3040.1992.tb01485.x

Marz, S., and Ullrich-Eberius, C. I. (1988). Solute accumulation and electrical membrane potential in *Agrobacterium tumefaciens*-induced crown galls in *Kalanchoë daigremontiana* leaves. *Plant Sci.* 57, 27–36. doi: 10.1016/0168-9452(88)90138-0

Matzke, M. A., and Matzke, A. J. M. (1991). Differential inactivation and methylation of a transgene in plants by two suppressor loci containing homologous sequences. *Plant Mol. Biol.* 16, 821–830. doi: 10.1007/BF00015074

Mistrik, I., Pavlovkin, J., Wächter, R., Pradel, K. S., Schwalm, K., Hartung, W., et al. (2000). Impact of *Agrobacterium tumefaciens*-induced stem tumors on NO_3^- uptake in *Ricinus communis*. *Plant Soil* 226, 87–98. doi: 10.1023/A:1026465606865

Morris, R. O. (1986). Genes specifying auxin and cytokinin biosynthesis in phytopathogens. *Annu. Rev. Plant Physiol.* 37, 509–538. doi: 10.1146/annurev.pp.37.060186.002453

Munoz-Galvan, S., Jimeno, S., Rothstein, R., and Aguilera, A. (2013). Histone H3K56 acetylation, Rad52, and Non-DNA repair factors control double-strand break repair choice with the sister chromatid. *PLoS Genet.* 9:e1003237. doi: 10.1371/journal.pgen.1003237

Mysore, K. S., Nam, J., and Gelvin, S. B. (2000). An *Arabidopsis* histone H2A mutant is deficient in *Agrobacterium* T-DNA integration. *Proc. Natl. Acad. Sci. U.S.A.* 97, 948–953. doi: 10.1073/pnas.97.2.948

Nonaka, S., Yuhashi, K.-I., Takada, K., Sugaware, M., Minamisawa, K., and Ezura, H. (2008). Ethylene production in plants during transformation suppresses vir gene expression in *Agrobacterium tumefaciens*. *New Phytol.* 178, 647–656. doi: 10.1111/j.1469-8137.2008.02400.x

Pradel, K. S., Ullrich, C. I., Cruz, S. S., and Oparka, K. J. (1999). Symplastic continuity in *Agrobacterium tumefaciens*-induced tumours. *J. Exp. Bot.* 50, 183–192. doi: 10.1093/jxb/50.331.183

Robert-Seilaniantz, A., Grant, M., and Jones, J. D. G. (2011). Hormone crosstalk in plant disease and defense: more than just jasmonate-salicylate antagonism. *Annu. Rev. Phytopathol.* 49, 317–343. doi: 10.1146/annurev-phyto-073009-114447

Robinette, D., and Matthysse, A. G. (1990). Inhibition by *Agrobacterium tumefaciens* and *Pseudomonas savastanoi* of development of the hypersensitive response elicited by *Pseudomonas syringae* pv. *phaseolicola*. *J. Bacteriol.* 172, 5742–5749.

Sardesai, N., Lee, L.-Y., Chen, H., Yi, H., Olbricht, G. R., Stirnberg, A., et al. (2013). Cytokinins secreted by *Agrobacterium* promote transformation by repressing a plant Myb transcription factor. *Sci. Signal.* 6, ra100. doi: 10.1126/scisignal.2004518

Schreiber, L., Franke, R., and Hartmann, K. (2005). Wax and suberin development of native and wound periderm of potato (*Solanum tuberosum* L.) and its relation to periderm transpiration. *Planta* 220, 520–530. doi: 10.1007/s00425-004-1364-9

Schröder, G., Klipp, W., Hillebrand, A., Ehring, R., Koncz, C., and Schroder, J. (1983). The conserved part of the T-region in Ti-plasmids expresses four proteins in bacteria. *EMBO J.* 2, 403–409.

Schubert, D., Lechtenberg, B., Forsbach, A., Gils, M., Bahadur, S., and Schmidt, R. (2004). Silencing in *Arabidopsis* T-DNA transformants – the predominant role of a gene-specific RNA sensing mechanism versus position effects. *Plant Cell* 104.024547. doi: 10.1105/tpc.104.024547

Schurr, U., Schuberth, B., Aloni, R., Pradel, K. S., Schmundt, D., Jaehne, B., et al. (1996). Structural and functional evidence for xylem-mediated water transport and high transpiration in *Agrobacterium tumefaciens*-induced tumors of *Ricinus communis*. *Bot. Acta* 109, 405–411. doi: 10.1111/j.1438-8677.1996.tb00590.x

Shirasu, K., Morel, P., and Kado, C. I. (1990). Characterization of the virB operon of an *Agrobacterium tumefaciens* Ti plasmid: nucleotide sequence and protein analysis. *Mol. Microbiol.* 4, 1153–1163. doi: 10.1111/j.1365-2958.1990.tb00690.x

Staskawicz, B. J., Ausubel, F. M., Baker, B. J., Ellis, J. G., and Jones, J. D. (1995). Molecular genetics of plant disease resistance. *Science* 268, 661–667. doi: 10.1126/science.7732374

Stepanova, A. N., Hoyt, J. M., Hamilton, A. A., and Alonso, J. M. (2005). A link between ethylene and auxin uncovered by the characterization of two root-specific ethylene-insensitive mutants in *Arabidopsis*. *Plant Cell* 17, 2230–2242. doi: 10.1105/tpc.105.033365

Strahl, B. D., and Allis, C. D. (2000). The language of covalent histone modifications. *Nature* 403, 41–45. doi: 10.1038/47412

Stroud, H., Ding, B., Simon, S. A., Feng, S., Bellizzi, M., Pellegrini, M., et al. (2013). Plants regenerated from tissue culture contain stable epigenome changes in rice. *eLife* 2:e00354. doi: 10.7554/eLife.00354

Tenea, G. N., Spantzel, J., Lee, L.-Y., Zhu, Y., Lin, K., Johnson, S. J., et al. (2009). Over-expression of several *Arabidopsis* histone genes increases *Agrobacterium*-mediated transformation and transgene expression in plants. *Plant Cell* 21, 3350–3367. doi: 10.1105/tpc.109.070607

Thomashow, M. F., Hugly, S., Buchholz, W. G., and Thomashow, L. S. (1986). Molecular basis for the auxin-independent phenotype of crown gall tumor tissues. *Science* 231, 616–618. doi: 10.1126/science.3511528

Thompson, D. V., Melchers, L. S., Idler, K. B., Schilperoort, R. A., and Hooykaas, P. J. (1988). Analysis of the complete nucleotide sequence of the *Agrobacterium tumefaciens* virB operon. *Nucleic Acids Res.* 16, 4621–4636. doi: 10.1093/nar/16.10.4621

Tzfira, T., Li, J., Lacroix, B., and Citovsky, V. (2004). *Agrobacterium* T-DNA integration: molecules and models. *Trends Genet.* 20, 375–383. doi: 10.1016/j.tig.2004.06.004

Ullrich, C. I., and Aloni, R. (2000). Vascularization is a general requirement for growth of plant and animal tumours. *J. Exp. Bot.* 51, 1951–1960. doi: 10.1093/jexbot/51.353.1951

Veena, Jiang, H., Doerge, R. W., and Gelvin, S. B. (2003). Transfer of T-DNA and Vir proteins to plant cells by *Agrobacterium tumefaciens* induces expression of host genes involved in mediating transformation and suppresses host defense gene expression. *Plant J.* 35, 219–236. doi: 10.1046/j.1365-313X.2003.01796.x

Verstappen, R., Ranostaj, S., and Rausch, T. (1991). The hexose transporters at the plasma membrane and the tonoplast of transformed plant cells: kinetic characterization of two distinct carriers. *Biochim. Biophys. Acta* 1073, 366–373. doi: 10.1016/0304-4165(91)90144-6

Veselov, D., Langhans, M., Hartung, W., Aloni, R., Feussner, I., Götz, C., et al. (2003). Development of *Agrobacterium tumefaciens* C58-induced plant tumors and impact on host shoots are controlled by a cascade of jasmonic acid, auxin, cytokinin, ethylene and abscisic acid. *Planta* 216, 512–522. doi: 10.1007/s00425-002-0883-5

Vining, K., Pomraning, K. R., Wilhelm, L. J., Ma, C., Pellegrini, M., Di, Y., et al. (2013). Methylome reorganization during in vitro dedifferentiation and regeneration of *Populus trichocarpa*. *BMC Plant Biol.* 13:92. doi: 10.1186/1471-2229-13-92

Vogel, J. P., Schuerman, P., Woeste, K., Brandstatter, I., and Kieber, J. J. (1998). Isolation and characterization of *Arabidopsis* mutants defective in the induction of ethylene biosynthesis by cytokinin. *Genetics* 149, 417–427.

Vogt, E., Schönherr, J., and Schmidt, H. W. (1983). Water permeability of periderm membranes isolated enzymatically from potato tubers (*Solanum tuberosum* L.). *Planta* 158, 294–301. doi: 10.1007/BF00397330

Wächter, R., Fischer, K., Gäbler, R., Kühnemann, F., Urban, W., Bögemann, G. M., et al. (1999). Ethylene production and ACC-accumulation in *Agrobacterium tumefaciens*-induced plant tumours and their impact on tumour and host stem structure and function. *Plant Cell Environ.* 22, 1263–1273. doi: 10.1046/j.1365-3040.1999.00488.x

Wächter, R., Langhans, M., Aloni, R., Götz, S., Weilmünster, A., Koops, A., et al. (2003). Vascularization, high-volume solution flow, and localized roles for enzymes of sucrose metabolism during tumorigenesis by *Agrobacterium tumefaciens*. *Plant Physiol.* 133, 1024–1037. doi: 10.1104/pp.103.028142

Ward, D. V., Draper, O., Zupan, J. R., and Zambryski, P. C. (2002). Peptide linkage mapping of the *Agrobacterium tumefaciens* vir-encoded type IV secretion system reveals protein subassemblies. *Proc. Natl. Acad. Sci. U.S.A.* 99, 11493–11500. doi: 10.1073/pnas.172390299

Ward, J. E., Akiyoshi, D. E., Regier, D., Datta, A., Gordon, M. P., and Nester, E. W. (1988). Characterization of the virB operon from an *Agrobacterium tumefaciens* Ti plasmid. *J. Biol. Chem.* 263, 5804–5814.

Wood, D. W., Setubal, J. C., Kaul, R., Monks, D. E., Kitajima, J. P., Okura, V. K., et al. (2001). The genome of the natural genetic engineer *Agrobacterium tumefaciens* C58. *Science* 294, 2317–2323. doi: 10.1126/science.1066804

Xu, X. Q., and Pan, S. Q. (2000). An *Agrobacterium* catalase is a virulence factor involved in tumorigenesis. *Mol. Microbiol.* 35, 407–414. doi: 10.1046/j.1365-2958.2000.01709.x

Yuan, Z.-C., Edlind, M. P., Liu, P., Saenkham, P., Banta, L. M., Wise, A. A., et al. (2007). The plant signal salicylic acid shuts down expression of the vir regulon and activates quormone-quenching genes in *Agrobacterium*. *Proc. Natl. Acad. Sci. U.S.A.* 104, 11790–11795. doi: 10.1073/pnas.0704866104

Yuan, Z. C., Haudecoeur, E., Faure, D., Kerr, K. F., and Nester, E. W. (2008). Comparative transcriptome analysis of *Agrobacterium tumefaciens* in response to plant signal salicylic acid, indole-3-acetic acid and gamma-amino butyric acid reveals signalling cross-talk and *Agrobacterium*-plant co-evolution. *Cell Microbiol.* 10, 2339–2354. doi: 10.1111/j. 1462-5822.2008.01215

Zhang, X., Yazaki, J., Sundaresan, A., Cokus, S., Chan, S. W.-L., Chen, H., et al. (2006). Genome-wide high-resolution mapping and functional analysis of DNA methylation in *Arabidopsis*. *Cell* 126, 1189–1201. doi: 10.1016/j.cell.2006.08.003

Zipfel, C., Kunze, G., Chinchilla, D., Caniard, A., Jones, J. D. G., Boller, T., et al. (2006). Perception of the bacterial PAMP EF-Tu by the receptor EFR restricts *Agrobacterium*-mediated transformation. *Cell* 125, 749–760. doi: 10.1016/j.cell.2006.03.037

Zipfel, C., Robatzek, S., Navarro, L., Oakeley, E. J., Jones, J. D. G., Felix, G., et al. (2004). Bacterial disease resistance in *Arabidopsis* through flagellin perception. *Nature* 428, 764–767. doi: 10.1038/nature02485

22

Horizontal gene transfer from *Agrobacterium* to plants

Tatiana V. Matveeva * and Ludmila A. Lutova

Department of Genetics and Biotechnology, St. Petersburg State University, St. Petersburg, Russia

Edited by:
Stanton B. Gelvin, Purdue
University, USA

Reviewed by:
Leon Otten, Université de
Strasbourg, France
Konstantin Skryabin, Centre
Bioengineering of the Russian
Academy of Sciences, Russia

*Correspondence:
Tatiana V. Matveeva, Department of
Genetics and Biotechnology,
St. Petersburg State University,
University emb., 7/9, St. Petersburg
199034, Russia
e-mail: radishlet@gmail.com

Most genetic engineering of plants uses *Agrobacterium* mediated transformation to introduce novel gene content. In nature, insertion of T-DNA in the plant genome and its subsequent transfer via sexual reproduction has been shown in several species in the genera *Nicotiana* and *Linaria*. In these natural examples of horizontal gene transfer from *Agrobacterium* to plants, the T-DNA donor is assumed to be a mikimopine strain of *A. rhizogenes*. A sequence homologous to the T-DNA of the Ri plasmid of *Agrobacterium rhizogenes* was found in the genome of untransformed *Nicotiana glauca* about 30 years ago, and was named "cellular T-DNA" (cT-DNA). It represents an imperfect inverted repeat and contains homologs of several T-DNA oncogenes (Ng*rolB*, Ng*rolC*, NgORF13, NgORF14) and an opine synthesis gene (Ng*mis*). A similar cT-DNA has also been found in other species of the genus *Nicotiana*. These presumably ancient homologs of T-DNA genes are still expressed, indicating that they may play a role in the evolution of these plants. Recently T-DNA has been detected and characterized in *Linaria vulgaris* and *L. dalmatica*. In *Linaria vulgaris* the cT-DNA is present in two copies and organized as a tandem imperfect direct repeat, containing Lv*ORF2*, Lv*ORF3*, Lv*ORF8*, Lv*rolA*, Lv*rolB*, Lv*rolC*, Lv*ORF13*, Lv*ORF14*, and the Lv*mis* genes. All *L. vulgaris* and *L. dalmatica* plants screened contained the same T-DNA oncogenes and the *mis* gene. Evidence suggests that there were several independent T-DNA integration events into the genomes of these plant genera. We speculate that ancient plants transformed by *A. rhizogenes* might have acquired a selective advantage in competition with the parental species. Thus, the events of T-DNA insertion in the plant genome might have affected their evolution, resulting in the creation of new plant species. In this review we focus on the structure and functions of cT-DNA in *Linaria* and *Nicotiana* and discuss their possible evolutionary role.

Keywords: *Agrobacterium*, T-DNA, horizontal gene transfer, *Nicotiana*, *Linaria*

INTRODUCTION

Horizontal gene transfer (HGT) takes place widely in prokaryotes, where its ecological and evolutionary effects are well-studied (Koonin et al., 2001). Comparative and phylogenetic analyses of eukaryotic genomes show that considerable numbers of genes have been acquired by HGT. Gene acquisition by HGT is therefore a potential creative force in both eukaryotic and prokaryotic genome evolution. However, mechanisms of HGT are poorly understood in the Eukaryota in comparison to gene transfer among the Procaryotae. The persistence of horizontally transferred genes in some organisms may confer selective advantages (Koonin et al., 2001; Richardson and Palmer, 2007). Most examples of HGT in higher plants involve the transfer of chloroplast or mitochondrial DNA and have been the subject of numerous reviews (Dong et al., 1998; Richardson and Palmer, 2007). There are few descriptions of horizontal transfer of nuclear genes between species. One example is transfer of the gene that codes for the cytosolic enzyme phosphoglucose isomerase predicted to have occurred between *Festuca ovina* and some species from the genus *Poa* (Ghatnekar et al., 2006; Vallenback et al., 2008, 2010). Evidence of gene transfer from bacteria to the nuclei of multi-cellular eukaryotes is rare (Richards

et al., 2006; Acuna et al., 2012). HGT from bacteria to plants has been restricted to *Agrobacterium rhizogenes* and representatives of genera *Nicotiana* and *Linaria*, and represents some of the most recent transfers in evolution (White et al., 1983; Intrieri and Buiatti, 2001; Matveeva et al., 2012; Pavlova et al., 2013).

A. rhizogenes, and the related bacterium *A. tumefaciens*, transform a wide variety of host plants by transferring a segment of the large tumor-inducing plasmid, called T-DNA, into host cells (White et al., 1982; Otten et al., 1992; Veena et al., 2003; Tzfira and Citovsky, 2006; Vain, 2007). The T-DNA is integrated through non-homologous recombination into the host cell genome where it is expressed. Expression of T-DNA genes results in the formation of hairy roots or crown galls, that are transgenic tissues, formed on a non-transgenic plant. This phenomenon is called "genetic colonization," one of the examples of the host-parasite relationship (Tzfira and Citovsky, 2006). It is unclear whether or not colonized plants have received benefits from such colonization, however, we could expect that it is beneficial in some cases since there are footprints of HGT from *Agrobacterium* to plants in the genomes of several present day plant species.

T-DNA IN *NICOTIANA GLAUCA*

In early investigations of *Agrobacterium* mediated transformation of plants, most researchers assumed that there was no significant homology to the T-DNA in untransformed plant genomes. White et al. (1982) attempted to detect pRiA4b T-DNA sequences in the genome of *Nicotiana glauca*, transformed in laboratory conditions by *Agrobacterium rhizogenes* strain A4. Southern analysis detected a fragment of pRiA4 in the transgenic tissue. Surprisingly, a hybridization signal was also detected in uninfected tissues of *N. glauca*. Further analysis confirmed the presence of DNA homologous to T-DNA in the *N. glauca* genome. This homologous DNA was referred to as "cellular T-DNA" (cT-DNA) (White et al., 1983).

Furner et al. (1986) investigated *Nicotiana glauca* plants, collected in geographically separated territories. Southern analyses showed the presence of cT-DNA in all studied varieties of *N. glauca*. Sequencing of the *N. glauca* cT-DNA demonstrated that it was organized as an imperfect inverted repeat. The left arm of cT-DNA, containing *rolB* and *rolC* homologs (Ng*rolB* and Ng*rolC*L) was more extended than the right arm, which contained only the *rolC* homolog (Ng*rolC*R). The coding sequences of Ng*rolB* and Ng*rolC*R were found to contain early stop codons.

Subsequent analysis of the nucleotide sequence of this cT-DNA identified open reading frame 13 (ORF13) and ORF14 homologs in both the left and right arms, called NgORF13L, NgORF14L, NgORF13R, and NgORF14R, respectively (Aoki et al., 1994).

In 2001 Suzuki et al. characterized *A. rhizogenes* strain MAFF301724 and described a new opine synthase gene (mikimopine synthase gene *mis*). A part of the *mis* gene displayed strong homology to distal fragments of *N. glauca* cT-DNA, called Ng*mis*L and Ng*mis*R, respectively (Suzuki et al., 2002). Suzuki et al. (2002) suggested that the complete cT-DNA region of *N.glauca* is comprised of the 7968 bp left arm and 5778 bp right arm that were derived from the T-DNA of a mikimopine Ri plasmid similar to pRi1724. The level of nucleotide sequence similarity between the left and right arms is greater than 96% and the gene order is conserved suggesting a duplication event. The structure of the *N.glauca* cT-DNA is summarized in **Figure 1**. Since cT-DNA has been identified in all studied varieties of *N. glauca* (Furner et al., 1986), it is reasonable to suggest that

the transformation event occurred before the formation of this species. This suggests that other related species may contain cT-DNA.

T-DNA IN OTHER *NICOTIANA SPECIES*

The genus *Nicotiana* is one of the largest genera in the Solanaceae and contains 75 species that are characterized by a wide range of variations among their floral and vegetative morphology (Clarkson et al., 2004). The different *Nicotiana* species evidence interspecific crosses which complicates *Nicotiana* phylogeny. Goodspeed hypothesized that there are two distinct lineages in *Nicotiana* which arose from two ancestral pre-petunioid and pre-cestroid lineages. He supposed that the base chromosome number of the genus was 12 and stressed the role of doubling and hybridization in *Nicotiana* evolution. Goodspeed divided *Nicotiana* into three sub-genera *Rustica, Tabacum*, and *Petunioides* and 14 sections (Goodspeed, 1954). Since then, the number of subgenera of *Nicotiana* has remained constant, while the number and composition of the sections has been revised (Clarkson et al., 2004). 75% of tobacco species originate from the Americas and 25% of species are from Australia (Goodspeed, 1954; Clarkson et al., 2004).

Identification of T-DNA in *N. glauca* raises two questions: what other *Nicotiana* species contain cT-DNA, and what was the pattern of dissemination within the group?

To answer the first question Furner et al. (1986) examined the genomes of 17 species of the genus *Nicotiana*. Using Southern analyses he showed that only six species from the subgenera *Rustica* and *Tabacum* contained sequences homologous to the *rol* genes of *Agrobacterium rhizogenes*. These species are *N. glauca, N. otophora, N. tomentosiformis, N. tomentosa, N. benavidesii, N. tabacum*. Examination of T-DNA- like sequences in *N. tabacum* has shown that it contains a *rolC* homolog and two ORF13 homologs (t*rolC*, tORF13-1 *and* tORF13-2, respectively) (Meyer et al., 1995; Frundt et al., 1998). Intrieri and Buiatti studied the distribution and evolution of *Agrobacterium rhizogenes* genes in the genus *Nicotiana*. Forty two species representing all *Nicotiana* sections were examined for the presence of *rolB*, *rolC*, ORF13, and ORF14 homologs in their genomes. T-DNA-like sequences detected were compared with each other and with contemporary sequences of *Agrobacterium*. The results demonstrated the presence of at least one T-DNA gene in each of 15 *Nicotiana* species representing all three subgenera. All currently available data on the distribution of cT-DNA among *Nicotiana* species are summarized in **Table 1**.

It is important to note, that there are some inconsistencies among the data by Furner et al. (1986) and Intrieri and Buiatti (2001). For example, Intrieri and Buiatti (2001) showed that T-DNA is present in *N. debneyi* and *N. cordifolia*. Furner et al. (1986) found no T-DNA in these species. This contradiction requires additional studies.

Thus, to date, T-DNA was found in every *Nicotiana* subgenus which include species, native to America and Australia (Goodspeed, 1954).

Phylogenetic analyses were performed by Intrieri and Buiatti (2001) to compare nucleotide sequences of cT-DNA in several *Nicotiana* species with the T-DNA of *Agrobacterium*.

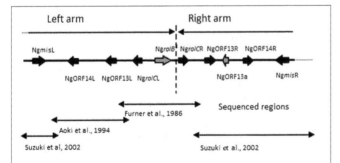

FIGURE 1 | Structure of cT-DNA in the *Nicotiana glauca* genome (based on Suzuki et al., 2002). The cT-DNA and its flanking regions are indicated. Lines with a single arrowhead indicate the imperfect inverted repeat. Lines with arrowheads at both ends indicate regions sequenced by each of three groups.

Table 1 | Distribution of T-DNA-like sequences among *Nicotiana* species.

	Section*	Species	T-DNA genes	Sequence Acc#	References
Rustica	*Paniculatae*	*N.glauca* *	+(rolB-mis)	X03432.1; D16559.1 AB071334.1; AB071335.1	1, 2, 3, 4**
		N.paniculata	–		2, 5
		N.knightiana	–		2, 5
		N.solanifolia	–		5
		N.benavidesii	+(rolC)	n/a***	2, 5
		N.cordifolia	+(rolB-ORF14)	AF281252.1 AF281248.1 AF281244.1	2, 5
		N.raimondi	–		5
	Rusticae	*N.rustica*	–		2, 5
Tabacum	*Tomentosae*	*N.tomentosa*	+(ORF13-mis)	n/a	2, 4
		N.tomentosiformis	+(rolC-mis)	AF281249.1 AF281245.1 AF281241.1	2, 4,5
		N.otophora	+(rolC-ORF14)	AF281250.1 AF281247.1 AF281243.1	2, 5
		N.setchelli	+(rolC)	n/a	2
	Nicotiana	*N.tabacum*	+(rolC-mis)	AF281246.1 AF281242.1	2,4,5
Petunioides	*Undulatae*	*N.glutinosa*	–		2, 5
		N.undulata	–		5
		N.arentsii	+(rolC)	n/a	5
	Trigonophyllae	*N.trigonophylla*	–		5
	Sylvestris	*N.sylvestris*	–		5
	Alatae	*N.langsdorffi*	–		2, 5
		N.alata	–		5
		N.longiflora	–		5
		N.forgetiana	–		5
		N.sanderae	–		5
		N.plumbaginifolia	–		5
	Repandae	*N.nesophila*	–		5
		N.stocktonii	–		5
		N.repanda	–		5
		N.nudicaulis	–		5
	Noctiflorae	*N.noctiflora*	–		5
		N.petunioides	–		5
	Petunioides	*N.acuminata*	(rolC)	n/a	5
		N.pauciflora	–		5
		N.attenuata	–		5
		N.miersii	+(rolB)	n/a	5
	Bigelovianae	*N.bigelovi*	+(rolB)	n/a	5
	Polydiclae	*N.clevelandi*	–		5
	Suaveolentes	*N.umbratica*	–		5
		N.debneyi	+(rolC)	AF281251.1	5
		N.gossei	+(rolC)	n/a	5
		N.rotundifolia	–		5
		N.suaveolens	+(rolC)	n/a	5
		N.exigua	+(rolC)	n/a	5
		N.goodspeedii	–		5

*Nicotiana sections from Knapp et al. (2004) and N.glauca section is from Goodspeed (1954); ** 1, White et al., 1983; 2, Furner et al., 1986; 3, Aoki et al., 1994; 4, Suzuki et al., 2002; 5, Intrieri and Buiatti, 2001; ***n/a, not available.

The following species were used in the analyses and represented all three subgenera: *N. cordifolia* (subgenus *Rustica* sec. *Paniculatae*); *N. tomentosiformis* and *N. otophora* (subgenus *Tabacum*), *N. tomentosiformis* (participated in *N. tabacum* speciation together with *N. sylvestris*); *N. glauca* used to be included in the subgenus *Rustica* sec. *Paniculatae* (Goodspeed, 1954), but later it was moved to the sec. *Noctiflorae* of the subgenus *Petunioides* (Knapp et al., 2004); and *N. debneyi*

[*Suaveolentes*, an Australian section of *Nicotiana*, and a polyploid species of the subgenus *Petunioides* (Knapp et al., 2004)].

Analysis of nucleotide sequences revealed that *N. cordifolia* and *N. glauca rolB, rolC*, ORF13, and ORF14 genes show a high level of sequence similarity (93.5–98.5%). These data indicate that *N. cordifolia* and *N. glauca* are related species and are consistent with the proposal of Goodspeed (1954) that both species should be included in subgenus *Rustica* sec. *Paniculatae*. Similar clustering was found between the representatives of the subgenus *Tabacum*. Sequence similarities were lower between *Rustica* and *Tabacum* species, ranging from 66.3 to 68.6% for *rolC* and from 70.2 to 82.9% for ORF13 but was higher for ORF14 (94–97%). Surprisingly, the petunioid *N. debneyi rolC* gene demonstrates high sequence similarity (93.4%) with the *N. glauca rolC* gene, but lower similarity (around 67%) with those found in species belonging to the subgenus *Tabacum*. It was speculated that the polyploid species *N. debneyi* got cT-DNA from an ancestor of sec. *Paniculatae*. The homologies suggest that the *Nicotiana rol* genes parallel *Nicotiana spp.* evolution, being divided into two clusters, one that includes *N. glauca*, *N. cordifolia*, and *N. debneyi*, the second comprising species from the subgenus *Tabacum* (Intrieri and Buiatti, 2001).

Present day cT-DNA genes clustered with each other, making it difficult to predict which Ri-plasmid would be the source of the cT-DNA in each *Nicotiana* species. Since the pace of evolution differs between bacteria and plants, and since the pRi T-DNAs may have undergone rearrangements with each other (Moriguchi et al., 2001), it is difficult to define which ancient T-DNA was the origin of cT-DNA in different *Nicotiana* species using such phylogenetic analysis (Tanaka, 2008).

Another option for exploring the origin of cT-DNA is opine typing which was performed by Suzuki et al. (2002). They identified opine gene homologs in *N. glauca* (Ng*misL* and Ng*misR*, respectively) and screened 12 *Nicotiana* species for *mis* homologs using Southern blot hybridization. The analyses included five species from the subgenus *Rustica* (*N. glauca, N. benavidesii, N. paniculata, N. knightiana, N. rustica*), five species from the subgenus *Tabacum* (*N. tomentosa, N. tomentosiformis, N. otophora, N. tabacum* and *N. glutinosa*), and two species from the subgenus *Petunioides* (*N. langsdorfii* and *N. sylvestris*). Homologs of gene *mis* were detected in the genomes of *N. glauca, N. tomentosa, N. tomentosiformis* and *N. tabacum*, however, the size of the hybridized fragments was different between *N. glauca* and species in the subgenus *Tabacum* and the hybridization pattern in *N. tomentosa* was different from that of the two species in the subgenus *Tabacum* (*N. tabacum* and *N. tomentosiformis*). Since T-DNA fragments of *N. tabacum* were identical to those of *N.tomentosiformis* and were not detected in the genome of *N. sylvestris*, the *mis* gene of *N. tabacum* likely came from *N. tomentosiformis*.

Suzuki et al. (2002) sequenced DNA in *N. glauca* adjacent to the cT-DNA. To investigate regions adjacent to the cT-DNA in other species, Southern hybridization was carried out using either a DNA fragment outside the left or right arms of the cT-DNA of *N. glauca* as a probe. All examined *Nicotiana* genomes showed the presence of sequences homologous to both sides of the cT-DNA suggesting that these are original sequences existing

in the genomes of *Nicotiana* plants. Similar size fragments were found in most species of the subgenus *Rustica* and *Tabacum*, which likely represent subgenus-specific restriction fragments. Interestingly, the signals using NgL and NgR as probes fell into the same fragment in the genomes of *N. tomentosa, N. tomentosiformis, N. tabacum*. Although the same DNA sequences bordering the cT-DNA in *N. glauca* were found in the genome of these three species, the sequences were not contiguous to the cT-DNA therefore the location of the cT-DNA in *N. glauca* is different from that in the species of the subgenus *Tabacum*.

Thus, the phylogenetic analysis undertaken by Intrieri and Buiatti (2001), and the study of opine genes and T-DNA integration sites performed Suzuki et al. (2002), suggest that there have been no less than two acts of *Agrobacterium* mediated transformation in the evolution of *Nicotiana* species (**Figure 2**). While the comparison of DNA sequences and the detection of *mis* homologs clearly demonstrates that the origin of the cT-DNA in *N. glauca, N. tabacum, N. tomentosa*, and *N. tomentosiformis* is derived from a mikimopine-type Ri plasmid similar to pRi1724, the origins of the cT-DNA in other species are still unknown.

EXPRESSION OF *NICOTIANA* cT-DNA GENES

pRi transgenic plants exhibit a specific phenotype (dwarfing, loss of apical dominance, increased root mass, and decreased rate of fertilization) (Tepfer, 1984). However, *Nicotiana* species that contain cT-DNA in their genomes show no such phenotype. Are these genes expressed and functional, or they are pseudogenes?

Early experiments by Taylor et al. (1985) did not detect transcripts of the cT-DNA genes in *Nicotiana glauca*. Other researchers were able to detect transcripts of Ng*rolB*, Ng*rolC*, NgORF13, NgORF14 in callus tissues of *N. glauca* (Aoki and Syono, 1999a) as well as in genetic tumors of F1 *N. glauca* × *N. langsdorffii* hybrids, but not in leaf tissues of the same hybrid (Ichikawa et al., 1990; Aoki et al., 1994). Northern analyses of *N. tabacum* oncogenes showed that a t*rolC* transcript accumulated in shoot tips and young leaves. The expression pattern of tORF13 was similar to t*rolC*, however, tORF13 expression was detected in flowers (Meyer et al., 1995; Frundt et al., 1998).

Intrieri and Buiatti studied transcription of the cT-DNA genes using RT-PCR in a number of species, using *N. langsdorffii* as a negative control. For their analyses authors used leaves from young *in vitro*-grown plantlets and hormone autotrophic (habituated) callus tissues grown on a hormone-free cultural medium as previously described by Bogani et al. (1985).

In this study *rolB* transcripts were found to be present in all cases in habituated callus tissues, but not in leaves; *rolC* was expressed in calli as well as in leaves of *Rustica* species (*N. glauca* and *N. cordifolia*), and only in calli in species representing the subgenus *Petunioides*; no expression was demonstrated to occur in species from the subgenus *Tabacum*. ORF13 and 14 mRNA was always detected in calli and ORF13 transcripts were found in leaf tissues of *N. tabacum* and *N. tomentosiformis*.

Aoki and Syono (1999b, 2000) analyzed the function of Ng*rol* genes by transforming leaf explants of *N. tabacum* and *N. debneyi* with *A. tumefaciens* that harbored either a *rolB-rolC*-ORF13-ORF14 fragment from pRi or cT-DNA of *N. glauca*. Nearly all of the leaf segments inoculated with pRi fragment developed

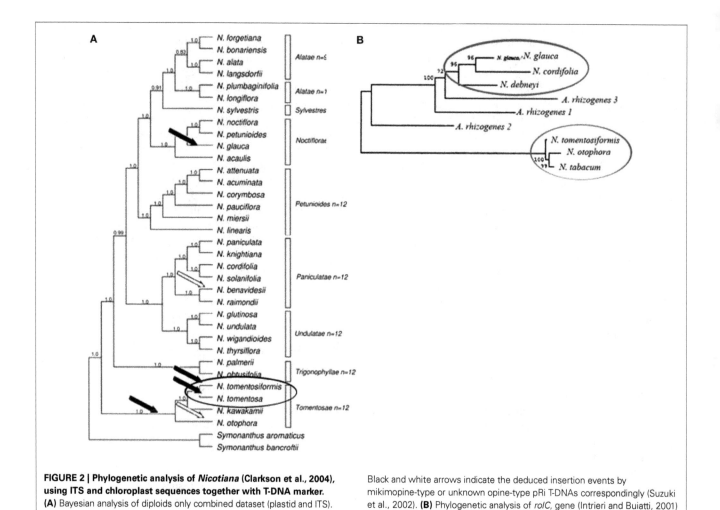

FIGURE 2 | Phylogenetic analysis of *Nicotiana* (Clarkson et al., 2004), using ITS and chloroplast sequences together with T-DNA marker.
(A) Bayesian analysis of diploids only combined dataset (plastid and ITS). Consensus of 40,001 trees with posterior probabilities shown above branches. Bars indicate *Nicotiana* sections according to Knapp et al. (2004),

Black and white arrows indicate the deduced insertion events by mikimopine-type or unknown opine-type pRi T-DNAs correspondingly (Suzuki et al., 2002). **(B)** Phylogenetic analysis of *rolC*, gene (Intrieri and Buiatti, 2001) by neighbor-joining method. Ovals show results of possible independent transformation events.

hairy roots. No significant root growth, however, appeared on the explants treated with *A. tumefaciens* that harbored Ng cT-DNA.

A comparison of the nucleotide sequences of Ng*rolB* and Ri*rolB* indicates that these oncogenes have different length coding regions. Each ORF starts at the same position, but Ng*rolB* ends at early termination codon 633 bp from the initiation site. The authors suggested that *N. glauca* plants do not exhibit the hairy-root phenotype due to the truncation of the Ng*rolB* reading frame.

A comparison of the DNA sequences of Ng*rolC* and Ri*rolC* indicates that the reading frame of Ng*rolC* begins and terminates at the same positions as Ri*rolC*. *Nicotiana tabacum* leaf disks were transformed with the P35S-Ng*rolC* chimeric construction, yielded transformants that expressed a dramatically dwarfed phenotype, probably because of the reduced length of internodes. The leaves of these P35S-Ng*rolC* transgenic plants were lanceolated and pale green, with floral organs that were thin and small. These characteristics were identical to the phenotype of the P35S-Ri*rolC* transgenic plants, described earlier (Schmülling et al., 1988). Transgene expression was detected only in transformants that demonstrated these characteristic morphological shifts, no transcripts were detected in leaf tissues from a

comparable T0 plant demonstrating a normal phenotype (Aoki and Syono, 1999b).

To compare the expression patterns between the Ng*rol* genes of N. glauca and the Ri*rol* genes of Agrobacterium rhizogenes, Nagata et al. (1996) carried out fluorometric and histochemical analyses of the tissues from transgenic genetic tumors, growing on the hybrid of Nicotiana glauca × N. langsdorffii (F1) that contained a beta-glucuronidase (GUS) reporter gene fused to the promoter of (Ng*rolB*, Ng*rolC*, Ri*rolB*, or Ri*rolC*. In all constructs they studied, significantly higher GUS activity was found in tumors than in the other organs (roots, stems, and leaves) of transgenic plants. The tendency toward higher GUS activities in tumors than in normal tissues seen with the Ri*rolB* and Ri*rolC* promoters was also seen with the Ng*rolB* and Ng*rolC* promoters. GUS activities from the rolB promoter expressed in normal F1 plants were, however, different from those seen from the rolC promoter. The expression of the Ri*rolB* and Ng*rolB* promoters in stems, roots, and leaves were 10–100 fold lower than in genetic tumors. Almost no activity was detected in leaves. By contrast, expression from the Ri*rolC* and Ng*rolC* promoters was only 1.5–10 fold lower than in genetic tumors and a significant activity was detected in leaves.

Histochemical analysis of transgenic normal Fl plant tissues showed that NgrolB-GUS and NgrolC-GUS, as well as RirolB-GUS and RirolC-GUS, had common tissue-specific expression patterns. NgrolB-GUS normal Fl transgenic plant tissues displayed high GUS activity in the meristematic zones of roots and in the apexes of shoots. A similar pattern of staining was ob-served in the RirolB-GUS transgenic plants. In the case of NgrolC-GUS and RirolC-GUS normal Fl transgenic plant tissues, GUS activity was observed primarily in the apices, vascular bundles of leaves, stems, and roots.

Expression of the *mis* gene homologs in *N. glauca* was detected by RT-PCR (Suzuki et al., 2002). It was shown that both homologs of the *mis* gene were amplified by RT-PCR using separate ortholog specific primers. These data support the hypothesis that the *mis* homologs are not pseudogenes. Transgenic plants transformed by the T-DNA of the wild type plasmid Ri1724 or by the *mis* gene alone, synthesize mikimopine in different organs (Suzuki et al., 2001), although no mikimopine accumulation was detected in wild-type *N. glauca* by paper electrophoresis. It is therefore likely that Ng*mis* homologs are transcribed at a very low level.

A full-length Ng*mis*R homolog was isolated and integrated into an expression vector in *Escherichia coli*. The purified Mis protein was able to catalyze synthesis of mikimopine from L-histidine and α-ketoglutaric acid in a reaction buffer supplemented with NADH as a co-factor (Suzuki et al., 2002).

Thus, the oncogenes of *Nicotiana* cT-DNA are expressed in different tissues of present day tobacco plants at a low level and are therefore not pseudogenes.

cT-DNA AND GENETIC TUMORS IN *NICOTIANA*

Genetic tumors appear in certain genotypes spontaneously without being induced by any detectable environmental factor, and the tumor state is hereditary. Spontaneous genetic tumors in *Nicotiana* were first reported by Tanaka (2008). They have been detected throughout the plant and in whole progeny populations of certain crosses of *Nicotiana* species (Kehr and Smith, 1954). In some hybrids, genetic tumors have been reported to be formed irregularly in some of the offspring or limited to certain organs of the plants (Smith, 1958). It has been proposed that certain genes appropriately combined in a hybrid promote the development of these genetic tumors (Naf, 1958; Ahuja, 1968). Naf (1958) divided *Nicotiana* species into two groups, so called "plus" and "minus" groups. The "plus" group consists mainly of the species of the section *Alatae* whereas the minus group contains species from several sections. Crosses between the species within "plus" or "minus" groups do not produce tumorous progeny, while crosses between species from "plus" and "minus" groups do. Ahuja (1968) hypothesized that the species belonging to the "plus" group have a gene or a locus defined as initiator (I) and the species belonging to the "minus" group have a number of genes or loci (ee) for tumor enhancement and expression. For tumor formation both I and ee loci must be present.

Fujita (1994) expect that most species belonging to the minus group contain cT-DNA and that its genes could somehow be associated with the formation of genetic tumors on the *Nicotiana* hybrids. However, since there are no reports showing a connection between the ee genes and cT-DNA genes so far, this promising hypothesis has not yet been validated (Tanaka, 2008). As already mentioned, NgrolB, NgrolC, NgORF13 and NgORF14 genes are transcribed in genetic tumors on *N. glauca* × *N. langsdorffii* F1 hybrids (Ichikawa et al., 1990; Aoki et al., 1994). Some of these genes function in several organs of non-tumorous hybrid plants, like their counterparts in pRi T-DNA (Nagata et al., 1995, 1996; Udagawa et al., 2004). As soon as tumorigenesis is initiated by aging or stress, these genes are active in the developing outgrowth in a regulated manner. This means that a high level of expression of Ngrol genes is correlated with tumor formation on an F1 hybrid. However, it has not been determined if the formation of tumors is caused by the expression of Ngrol genes (Tanaka, 2008). Moreover, the stem and leaf tissues of *Nicotiana* species accumulate transcripts of the Ngrol genes (Meyer et al., 1995; Frundt et al., 1998). These observations suggest that the expression of these Ngrol genes might be unrelated to the induction of tumors. Overexpression of NgrolC, NgORF13, or tORF13 cause the proliferation of cells on carrot disks (Frundt et al., 1998), and morphological alterations of tobacco explants, similar to hairy root syndrome on transgenic plants (Aoki and Syono, 1999a,b). Therefore, cT-DNA oncogenes may be responsible for enhancing the development of genetic tumors (Tanaka, 2008).

It is widely discussed that phytohormones contribute to genetic tumor formation. The role of auxin and cytokinins in genetic tumor formation in *Nicotiana*, however, has been disputed. On the one hand, in the light-grown tissues of genetic tumors, indole acetic acid (IAA) was found to be the predominant auxin and its level increased during tumor initiation (Bayer, 1967; Ichikawa et al., 1989). On the other hand, in dark conditions endogeneous IAA remained at a constant, low level throughout the tumorigenetic process (Fujita et al., 1991). A higher cytokinin level was associated with tumorigenesis in tumor-prone hybrid tissues. While analyzing the role of cytokinins in *Nicotiana* genetic tumor formation, Feng et al. (1990) have shown that tumor formation of X-ray-induced non-tumorous mutants of *N. glauca* × *N. langsdorffii* was restored either by the insertion of the *A. tumefaciens ipt* gene, which encodes the key enzyme of cytokinin biosynthesis, or by the addition of cytokinin. Nandi et al. (1990) determined the profile of endogenous cytokinins in genetic tumors of *N. glauca* (Grah.) × *N. langsdorffii* (Weinm.) hybrids. They showed that while zeatin is predicted to be the predominant endogenous cytokinin in tissues of all ages, the genetic tumor tissue derived from this hybrid does not contain notably high endogenous cytokinin levels.

Since tumor growth may be caused not only by high concentrations of hormones, but also by enhanced sensitivity to them, this may explain the contradictory data on the content of hormones in tumors. Even if hormone levels are increased in tumors, it is not necessarily caused by the expression of the cT-DNA genes: *rol* genes can be regulated by hormones. For example, it was shown that expression of NgrolB was induced by auxin, as was RirolB, probably through the presence of the auxin-responsive *cis*-element ACTTTA found in the promoters of NgrolB and RirolB which is acted upon by the trans-factor NtBBF1 (Tanaka, 2008).

It is clear that further work will be needed to establish the relationship between cT-DNA oncogenes and genetic tumorigenesis in *Nicotiana*.

SEARCH FOR T-DNA-LIKE SEQUENCES IN OTHER SOLANACEAE SPECIES

The Solanaceae is a large angiosperm family containing many economically important crops. *A. rhizogenes* is known to infect species belonging to different Solanaceae genera. Intrieri and Buiatti (2001) attempted to identify T-DNA-like sequences in species belonging to genera *Cestrum*, *Petunia*, and *Solanum* (*C. parqui*, *C. foetidus*, *P. hybrida*, *S. tuberosa*, *S. melongena*, *C. annuum*, and *S. lycopersicon*) using the same screening procedure, as they did for *Nicotiana*. None of the species screened showed amplification by PCR and no hybridization was obtained using *A. rhizogenes* and *N. glauca* probes. Kulaeva et al. (2013) extended the analysis to species from genus *Solanum* looking for T-DNA-like sequences. The authors used TaqMan real-time PCR with degenerate primers and probes for *rolB*, *rolC*, ORF13, ORF14 to analyze the following species: *S. chmielewskii*, *S. esculentum var. cerasiforme*, *S. glabratum*, *S. habrochaites*, *S. peruvianum*, *S. pimpinellifolium*, *S. cheesmanii*, *S. parviflorum*, *S. chilense*, *S. acaule*, *S. ajanhuiri*, *S. albicans*, *S. andigenum*, *S. berthaultii*, *S. boyacense*, *S. boyacense*, *S. canarense*, *S. canarense*, *S. cardiophyllum*, *S. chacoense*, *S. chaucha*, *S. chocclo*, *S. curtilobum S. demissum*, *S. demissum*, *S. doddsii*, *S. dulcamara*, *S. fendleri*, *S. goniocalyx*, *S. hjertingii*, *S. hondelmanii*, *S. hougassi*, *S. jamesii*, *S. juzepczukii*, *S. kurtzianum*, *S. mamilliferum*, *S. phureja*, *S. pinnatisectum*, *S. pinnatisectum*, *S. polytrichon*, *S. riobambense*, *S. rybinii*, *S. sparsipilum*, *S. spegazzinii*, *S. stenotomum*, *S. stoloniferum*, *S. tarijense*, *S. tenuifilamentum*, *S. tuberosum*, *S. vernei*, *S. verrucosum*, *S. oplocense*. They used *N. tabacum* DNA as positive control and *N. langsdorffii* as negative control. Amplification of specific sequences was not detected in any of the tested species.

This data shows that the presence of cT-DNA is not a feature of whole Solanaceae family, but has only been described to date for members of the genus *Nicotiana*.

T-DNA IN OTHER DICOTYLEDONOUS FAMILIES

Given the documented occurrence of cT-DNA, it is reasonable to hypothesize that other plant species, outside of the family Solanaceae, would have been transformed by *Agrobacterium* and contain at least remnants of cT-DNA.

The existence of cT-DNAs in species outside of the family Solanaceae has been reported by several groups. Using Southern analyses, sequences similar to pRi T-DNA were found in the genomic DNA of normal carrot (*Daucus carota*) (Spano et al., 1982), field bindweed (*Convolvulus arvensis*) (Tepfer, 1982); and carpet bugleweed (*Ajuga reptans*) (Tanaka, 2008). These studies did not involve DNA sequencing so they were not able to confirm whether there is T-DNA in the analyzed species.

Prior work in our group (Matveeva et al., 2012) attempted to clarify whether fixed T-DNA is present in other species outside the genus *Nicotiana*, and to evaluate the evolutionary relevance of natural T-DNA transfer. We sought to quickly screen a large number of plant genomes for the presence of T-DNA

from both *A. tumefaciens* and *A. rhizogenes* using a modification of TaqMan-based real-time PCR (Livak et al., 1995) that combines the positive features of PCR and DNA blot hybridization in a single reaction (Matveeva et al., 2006). The search was limited to dicotyledonous plants native to temperate zones (mild winter, warm summer, and sufficient rainfall) due to the common occurrence of *Agrobacterium* in soils under these climate conditions.

This work analyzed 127 dicotyledonous plant species belonging to 38 different families. Species included carrot (*Daucus carota*) and field bindweed (*Convolvulus arvensis*), mentioned above. Each of the plants was screened for DNA sequences homologous to two different sets of T-DNA oncogenes: The first set included sequences homologous to *A. rhizogenes* oncogenes (*rolB*, *rolC*, ORF13, and ORF14), and the second contained sequences homologous to *A. tumefaciens* oncogenes (*tms1* and *tmr*). Plant DNA samples from 126 species did not display detectable fluorescent signals for the T-DNA genes from either *A. rhizogenes* or *A. tumefaciens*. However, DNA samples isolated from several plants of *L. vulgaris* gave a positive result. In contrast, amplification was not observed using primers for the *tms1* and *tmr* genes of *A. tumefaciens* (Matveeva et al., 2012).

The absence of T-DNA homologs in most of the plant species investigated leads us to the conclusion that HGT from *Agrobacterium* is a rare event in the plants. However, finding cT-DNA sequences in the genomes of species other plants than *Nicotiana* indicates that HGT from *Agrobacterium* to plants occurs outside of this genus.

It is interesting to note that the only examples of HGT demonstrated thus far occur in plants transformed by *A. rhizogenes*, but not *A. tumefaciens*. This may suggest that infection induced by *A. rhizogenes* is more efficient than that induced by *A. tumefaciens* (Tepfer, 1984).

cT-DNA IN THE GENOMES OF THE GENUS *LINARIA*

Sequences homologous to T-DNA oncogenes in *Linaria vulgaris*, were identified by real time PCR and named LvrolB, LvrolC, LvORF13, LvORF14. BLAST analyses demonstrated the highest level of sequence identity (93%) between LvrolC and RirolC from the pRiA4 of *A. rhizogenes*. LvORF14 had the lowest similarity to the corresponding *Agrobacterium* oncogene (85%). A homolog of a gene for mikimopine synthase (*mis*) was also identified. To define the full extent of the cT-DNA integrated into *Linaria vulgaris* genome, a chromosome walking approach was performed to identify the upstream fragment of the Lv cT-DNA. This work indicated that the *L. vulgaris* genome contains two copies of cT-DNA which are organized as an imperfect direct repeat. Analysis of the cT-DNA copies demonstrated that both of them contain sequences similar to the following genes: ORF2, ORF3, ORF8, *rolA*, *rolB*, ORF 13, ORF1, and *mis*. The left side of the repeat contains additional sequence, homologous to part of the agrocinopine synthase (*acs*) gene. Analysis of the flanking regions of the Lv T-DNA was performed by real-time thermal asymmetrical interlaced (TAIL)-PCR with primers and probes to the Lv*mis* gene. The flanking plant DNA identified in this analysis was found to be similar to the Ty3/gypsy-like retrotransposon (Matveeva et al., 2012).

Samples of *L. vulgaris* were collected in the European part of Russia in Moscow, Voronezh and Krasnodar regions, and in the Asian part of Russia in the Novosibirsk, Tumen, and Chelyabinsk regions. The distance between the most western to the most eastern points was about 4000 km. The distance from the most northern to the most southern points was about 2000 km. Two to three plants were analyzed from each of these collection points. All of the samples contained T-DNA-like sequences, however, there was polymorphism among their nucleotide sequences (Matveeva et al., 2012).

Analysis of the sequences for both *L. vulgaris* homologs of *rolC* demonstrated that they contained intact open reading frames. The fragments corresponding to the coding regions of genes L*vrolB*, L*v*ORF13, L*v*ORF14, and L*vmis* contain several stop codons or frameshifts that alter the ORFs. An analysis of the expression of these genes was carried out in tissues from the internodes, leaves, and roots of 1 month old, *in vitro* aseptically grown plants using RT real-time PCR. No mRNA corresponding to L*vrolB*, L*vrolC*, L*v*ORF13, L*v*ORF14 and L*vmis* genes was amplifiable from these samples, therefore, the *rol* genes do not appear to be transcribed in *L. vulgaris* (Matveeva et al., 2012).

Toadflaxes (*Linaria* Mill.) form the largest genus within the tribe *Antirrhineae*. *Linaria* includes about 150 species that are widely spread in the Palearctic region, but the representatives of the genus are the most variable in the Mediterranean basin. The origin of the genus has been placed in the Miocene era (Fernandez-Mazuecos and Vargas, 2011) predating the Messinian Salinity Crisis (Hsu et al., 1977). The latest classification of the genus *Linaria* accepts seven sections (*Linaria, Speciosae, Diffusae, Supinae, Pelisserianae, Versicolores, and Macrocentrum*) (Sutton, 1988).

Studies indicate that cT-DNA exists in a number of *Linaria* species belonging to the sections *Linaria* and *Speciosae* (Matveeva and Kosachev, 2013; Pavlova et al., 2013). No cT-DNA was detected in *Linaria* species outside these sections. It appears that *rolC* is concerved among studied *Linaria* species based on the sequencing analysis of *rolC* homologs in *L. genistifolia* subsp. *dalmatica* (sec. *Speciosae*) and *L. acutiloba* (sec. *Linaria*) (Matveeva and Kosachev, 2013; Pavlova et al., 2013).

Thus, HGT of T-DNA from *Agrobacterium* to plants is not limited to *Nicotiana* spp, it has also occurred in the genus *Linaria*. The *rolC* homolog is the most conserved gene among the cT-DNA genes in *Linaria* and *Nicotiana* spp. In both genera plants were transformed by a mikimopine strain of *A. rhizogenes*.

POSSIBLE FUNCTION OF T-DNA IN PLANT GENOMES

The existence of several independent acts of *Agrobacterium* mediated transformation of plants and the maintenance of the cT-DNA in plant genomes during the process of evolution propose, that T-DNA-like sequences may give some selective advantages to the transformed plants (Ichikawa et al., 1990; Matveeva et al., 2012).

Suzuki et al. (2002) mentioned two possible functions of cT-DNA: increasing root mass leading to tolerance to drought, and changing the biological environment, particularly the soil microbiome represented by root-associated bacterial populations.

Increasing root mass would seem beneficial for tolerance to dry conditions. Hence, ancient transformed plants with increased root mass might have demonstrated increased tolerance to dry environments surviving in arid conditions (Tanaka, 2008). However, no phenotype of the hairy root disease is observed in *Nicotiana* and *L. vulgaris* plants. In contrast, *L. vulgaris* explants show *in vitro* a shooty phenotype and in representatives of both genera *rolB* is mutated. Among the oncogenes of pRi T-DNA, *rolB* gene function seems to be the most important for hairy root induction because transformation of plants by the R*irolB* gene alone can induce hairy root formation. In contrast to the pR*irolB*, the Ng*rolB* gene alone or in combination with other *N. glauca* homologs of *A. rhizogenes* oncogenes did not induce adventitious roots (Aoki and Syono, 1999a,b).

Aoki and Syono (1999b) performed base substitutions at two nucleotide positions, using site-directed mutagenesis, with the aim of producing a full-length form of Ng*rolB* capable of stimulating adventitious root induction. Transgenic plants overexpressing this altered Ng*rolB* demonstrated typical morphogenetic abnormalities. This experiment shows the possibility that a functional *rolB* gene may have operated during early steps of the evolution of transgenic *Nicotiana*.

Identification and sequencing of the *mis* homologs in *Nicotiana* and *Linaria* suggests that the origin of their cT-DNA is probably the mikimopine Ri plasmid. The presence of this gene may be related to plant–microbe interactions. Oger et al. (1997, 2000) reported that producing opines in genetically modified plants alters their ecological environment, in particular, changing the soil microbiome and root-associated microbe populations. If the synthesis of opines were beneficial for a plant species (even at a low level, in a specific tissue, or at a specific stage of oncogenesis), it may impact the appearance of advantageous plant–bacterium interactions. Plants maintaining cT-DNA in the genome could potentially maintain certain species of microorganisms in their rhizosphere via the secretion of opines in the root zone. Such potentially beneficial bacteria in the rhizosphere may in turn influence the root microbiome and convey nutritional and/or defensive features.

Early flowering or a shift from biennial to annual lifecycle without vernalization can take place on pRi transgenic *Cichorium intybus* and *Daucus carota* plants (Limami et al., 1998). These flowering features are beneficial when propagating such transgenic plants over the untransformed parentals. When considering the adaptational potential of natural transformation, the authors focused on the occurrence of flowering in the absence of a cold treatment. Given the mobility of seeds by wind, animals, and water, it is likely that biennial varieties or ecotypes may be transported to the southern latitudes where annualism would be beneficial. However, *Nicotiana* and *Linaria* species are not biennials (Goodspeed, 1947; Sutton, 1988; Blanco-Pastor et al., 2012). In addition, cT-DNA containing *Linaria* species from the sections *Linaria* and *Speciosae* are perennial, while other sections contain annual species. It is interesting to note that the cT-DNA containing sections are found worldwide while other sections are in the Mediterranean region and the Pyrenees (**Table 2**) (Sutton, 1988). It is unclear if this observation is due to the rarity by which plants acquire permanent cT-DNA, or if its foundation is related

Table 2 | Major features of infrageneric taxa of the genus *Linaria* (according to Sutton, 1988).

Section	Habit	Distribution
Linaria	Perennial	Eurasia
Speciosae	Perennial	Europe
Diffusae	Annual or perennial	Mediterranean
Supinae	Annual or perennial	Mediterranean
Pelisserinae	Annual or perennial	Mediterranean
Versicolores	Annual or perennial	Mediterranean, Iberian Peninsula
Macrocentrum	Annual	Mediterranean

to some fitness benefit conferred by the cT-DNA. It can be speculated, however, that the ecological plasticity of species within the sections *Linaria* and *Speciosae* is somehow associated with the presence of cT-DNA in their genomes.

It would appear, therefore, that annualism is not related to natural transformation in *Nicotiana* and *Linaria*.

It is interesting to note that *rolC* is the most conserved gene among the cT-DNA oncogenes found in *Nicotiana* and *Linaria* (Intrieri and Buiatti, 2001; Mohajjel-Shoja et al., 2011). In some representatives of the *Nicotiana*, only *rolC* is able to encode a functional product (Mohajjel-Shoja et al., 2011). The same trend was observed for *Linaria* T-DNA-like sequences (Matveeva et al., 2012; Matveeva and Kosachev, 2013). The function of *rolC*, however, is poorly understood. It has been speculated that the product of *rolC* releases cytokinins from conjugates (Estruch et al., 1991). Other researchers demonstrated that the RolC protein participates in the processes of sucrose metabolism and/or transport (Nilsson and Olsson, 1997; Mohajjel-Shoja et al., 2011). RolC has also been proposed to promote somatic embryogenesis in plants (Gorpenchenko et al., 2006). Such data are consistent with a cytokinin function of the gene. Constitutive expression of *rolC* in cultured plant tissues activates secondary metabolism: the *rolC* gene alone increases production of tropane alkaloids, pyridine alkaloids, ginsenosides, and anthraquinones among others (Bulgakov et al., 1998, 2002; Palazon et al., 1998a; Bonhomme et al., 2000a,b; Bulgakov, 2008) and stimulates the expression of pathogenesis-related proteins (Kiselev et al., 2007). It is unclear how the *rolC* gene product mediates such pleiotropic effects, further biochemical characterization of RolC is required. This is complicated by the fact that *rolC* has no significant homology with any other genes (of prokaryotic or eukaryotic organisms) whose function is known (Bulgakov, 2008).

Activation of secondary metabolism in transformed cells may be due to the action of other *rol* genes (Chandra, 2012). Shkryl et al. (2008) studied the influence of *rol* genes products on secondary metabolism of *Rubia cordifolia*. They investigated *rol* genes individually and studied their combined action. They found that individual *rolA*, *rolB*, and *rolC* genes were able to stimulate biosynthesis of anthraquinones in transformed calli. The strongest anthraquinone—stimulating activity was detected for an *R. cordifolia* culture overproducing RolB where they saw a 15-fold increase of anthraquinone accumulation as compared to untransformed calli. The *rolA*- and *rolC*-expressing calli produced 2.8- and 4.3-fold higher amounts of anthraquinones, correspondingly. Palazon et al.

(1998b) reported that the *rolA* gene stimulated production of nicotine.

Thus, increasing the amount of secondary metabolites is a characteristic of tissues where *rol* genes are expressed. This property can be useful for plants, because secondary metabolites may contribute to the resistance of plants to pests. It seems likely that a possible function cT-DNA is to mediate how plants interact with their environment by secreting opines and/or by changing the amounts of secondary metabolites. It will be essential to confirm such hypotheses through additional experimentation that might include silencing or excision experiments that are now possible using CRISPR technology (Qi et al., 2013).

The study of the long term impacts of HGT by *Agrobacterium* in plant lineages is in the early stages. However, we can note some trends:

– HGT of T-DNA from *Agrobacterium* to plants occurred in the evolution of several genera, at least *Nicotiana* (family Solanaceae) and *Linaria* (family Plantaginaceae);

– in both genera plants were transformed by a mikimopine strain of *A. rhizogenes*;

– a *rolC* homologe is the most conserved gene among the T-DNA genes in *Linaria* and *Nicotiana* spp;

– In *Linaria vulgaris* and *Nicotiana glauca* there are more than one copy of T-DNA per genome.

Continued studies of the genetic and biochemical effects of cT-DNA integration in naturally transgenic plants are important and will continue to provide insights into the impact of such rare acquisitions on plant evolution.

ACKNOWLEDGMENTS

The authors thank Derek Wood (Seattle Pacific University) for useful discussion and critical reading of the manuscript. This paper was prepared within the framework of the thematic plan of St. Petersburg State University ## 0.37.526.2013; 1.39.315.2014 and supported by a grant to Tatiana V. Matveeva from the Russian Foundation for Basic Research #14-04-01480.

REFERENCES

Acuna, R., Padilla, B. E., Florez-Ramos, C. P., Rubio, J. D., Herrera, J. C., Benavides, P., et al. (2012). Adaptive horizontal transfer of a bacterial gene to an invasive insect pest of coffee. *Proc. Natl. Acad. Sci. U.S.A.* 109, 4197–4202. doi: 10.1073/pnas.1121190109

Ahuja, M. A. (1968). An hypothesis and evidence concerning the genetic components controlling tumor formation in *Nicotiana*. *Mol. Gen. Genet.* 103, 176–184. doi: 10.1007/BF00427144

Aoki, S., Kawaoka, A., Sekine, M., Ichikawa, T., Fujita, T., Shinmyo, A., et al. (1994). Sequence of the cellular T-DNA in the untransformed genome of *Nicotiana glauca* that is homologous to ORFs 13 and 14 of the Ri plasmid and analysis of its expression in genetic tumors of *N. glauca* × *N. langsdorffii*. *Mol. Gen. Genet.* 243, 706–710.

Aoki, S., and Syono, K. (1999a). Function of Ngrol genes in the evolution of *Nicotiana glauca*: conservation of the function of NgORF13 and NgORF14 after ancient infection by an *Agrobacterium rhizogenes*-like ancestor. *Plant Cell Physiol.* 40, 222–230.

Aoki, S., and Syono, K. (1999b). Horizontal gene transfer and mutation: Ngrol genes in the genome of *Nicotiana glauca*. *Proc. Natl. Acad. Sci. U.S.A.* 96, 13229–13234.

Aoki, S., and Syono, K. (2000). The roles of Rirol and Ngrol genes in hairy root induction in *Nicotiana debneyi*. *Plant Sci.* 159, 183–189. doi: 10.1016/S0168-9452(00)00333-2

Bayer, M. H. (1967). Thin-layer chromatography of auxin and inhibitors in *Nicotiana glauca, N. langsdorffii* and three of their tumor-forming hybrids. *Planta*. 72, 329–337.

Blanco-Pastor, J. L., Vargas, P., and Pfeilm, B. E. (2012). Coalescent Simulations Reveal Hybridization and Incomplete Lineage Sorting in Mediterranean *Linaria*. *PLoS ONE* 7:e39089. doi: 10.1371/journal.pone.0039089

Bogani, P., Buiatti, M., Tegli, S., Pellegrini, M. G., Bettini, P., and Scala, A. (1985). Interspecific differences in differentiation and dedifferentiation patterns in the genus *Nicotiana*. *Plant Syst. Evol.* 151, 19–29. doi: 10.1007/BF02418016

Bonhomme, V., Laurain, M. D., and Fliniaux, M. A. (2000a). Effects of the *rolC* gene on hairy root: induction development and tropane alkaloid production by *Atropa belladonna. J. Nat. Prod.* 63, 1249–1252. doi: 10.1021/np990614l

Bonhomme, V., Laurain-Mattar, D., Lacoux, J., Fliniaux, M., and Jacquin-Dubreuil, A. (2000b). Tropane alkaloid production by hairy roots of *Atropa belladonna* obtained after transformation with *Agrobacterium rhizogenes* 15834 and *Agrobacterium tumefaciens* containing rol A, B, C genes only. *J. Biotechnol.* 81, 151–158. doi: 10.1016/S0168-1656(00)00287-X

Bulgakov, V. P. (2008). Functions of *rol* genes in plant secondary metabolism. *Biotechnol. Adv.* 26, 318–324. doi: 10.1016/j.biotechadv.2008.03.001

Bulgakov, V. P., Khodakovskaya, M. V., Labetskaya, N. V., Chernoded, G. K., and Zhuravlev, Y. N. (1998). The impact of plant *rolC* oncogene on ginsenoside production by ginseng hairy root cultures. *Phytochemistry* 49, 1929–1934. doi: 10.1016/S0031-9422(98)00351-3

Bulgakov, V. P., Tchernoded, G. K., Mischenko, N. P., Khodakovskay, M. V., Glazunov, V. P., Zvereva, E. V., et al. (2002). Effects of salicylic acid, methyl jasmonate, etephone and cantharidin on anthraquinone production by *Rubia cordifolia* callus cultures transformed with *rolB* and *rolC* genes. *J. Biotechnol.* 97, 213–221. doi: 10.1016/S0168-1656(02)00067-6

Chandra, S. (2012). Natural plant genetic engineer *Agrobacterium rhizogenes*: role of T-DNA in plant secondary metabolism. *Biotechnol. Lett.* 34, 407–415. doi: 10.1007/s10529-011-0785-3

Clarkson, J. J., Knapp, S., Garcia, V. F., Olmstead, R. G., Leitch, A. R., and Chase, M. W. (2004). Phylogenetic relationships in *Nicotiana* (Solanaceae) inferred from multiple plastid DNA regions. *Mol. Phylogenet. Evol.* 33, 75–90. doi: 10.1016/j.ympev.2004.05.002

Dong, F., Wilson, K. G., and Makaroff, C. A. (1998). Analysis of the four *cox2* genes found in turnip (*Brassica campestris*, Brassicaceae) mitochondria. *Am. J. Bot.* 85, 153–161. doi: 10.2307/2446303

Estruch, J. J., Chriqui, D., Grossmann, K., Schell, J., and Spena, A. (1991). The plant oncogene *rolC* is responsible for the release of cytokinins from glucoside conjugates. *EMBO J.* 10, 2889–2895.

Feng, X. H., Dube, S. K., Bottino, P. J., and Kung, S. D. (1990). Restoration of shooty morphology of a nontumorous mutant of *Nicotiana glauca × N. langsdorffii* by cytokinin and the isopentenyltransferase gene. *Plant Mol. Biol.* 15, 407–420.

Fernandez-Mazuecos, M., and Vargas, P. (2011). Historical Isolation versus recent long-distance connections between Europe and Africa in Bifid Toadflaxes (*Linaria* sect. *Versicolores*). *PLoS ONE* 6:e22234 doi: 10.1371/journal.pone.0022234

Frundt, C., Meyer, A. D., Ichikawa, T., and Meins, F. Jr. (1998). A tobacco homologue of the Ri-plasmid *orf13* gene causes cell proliferation in carrot root discs. *Mol. Gen. Genet.* 259, 559–568. doi: 10.1007/s004380050849

Fujita, T. (1994). Screening of genes related to tumor formation in tobacco genetic tumors. *Plant Tiss. Cult. Lett.* 11, 171–177. doi: 10.5511/plantbiotechnology1984.11.171

Fujita, T., Ichikawa, T., and Syono, K. (1991). Changes in morphology, levels of endogenous IAA and protein composition in relation to the development of tobacco genetic tumor induced in the dark. *Plant Cell Physiol.* 32, 169–177.

Furner, I. J., Huffman, G. A., Amasino, R. M., Garfinkel, D. J., Gordon, M. P., and Nester, E. W. (1986). An *Agrobacterium* transformation in the evolution of the genus *Nicotiana. Nature* 329, 422–427. doi: 10.1038/319422a0

Ghatnekar, L., Jaarola, M., and Bengtsson, B. O. (2006). The introgression of a functional nuclear gene from *Poa* to *Festuca ovina. Proc. Biol. Sci.* 273, 395–399. doi: 10.1098/rspb.2005.3355

Goodspeed, T. H. (1947). On the evolution of the genus *Nicotiana. Proc. Natl. Acad. Sci. U.S.A.* 33, 158–171. doi: 10.1073/pnas.33.6.158

Goodspeed, T. H. (1954). *The Genus Nicotiana*. (Waltham, MA: Chronica Botanica).

Gorpenchenko, T. Y., Kiselev, K. V., Bulgakov, V. P., Tchernoded, G. K., Bragina, E. A., Khodakovskaya, M. V., et al. (2006). The *Agrobacterium rhizogenes*

rolC-gene induced somatic embryogenesis and shoot organogenesis in *Panax ginseng* transformed calluses. *Planta* 223, 457–467. doi: 10.1007/s00425-005-0102-2

Hsu, K. J., Montadert, L., Bernoulli, D., Cita, M. B., and Erickson, A. (1977). History of the Mediterranean salinity crisis. *Nature* 267, 399–403. doi: 10.1038/267399a0

Ichikawa, T., Kobayashi, M., Nakagawa, S., Sakurai, A., and Syono, K. (1989). Morphological observations and qualitative and quantitative studies of auxins after induction of tobacco genetic tumor. *Plant Cell. Physiol.* 30, 57–63.

Ichikawa, T., Ozeki, Y., and Syono, K. (1990). Evidence for the expression of the *rol* genes of *Nicotiana glauca* in genetic tumors of *N. glauca × N. langsdorffii. Mol. Gen. Genet.* 220, 177–180.

Intrieri, M. C., and Buiatti, M. (2001). The horizontal transfer of *Agrobacterium rhizogenes* genes and evolution of the genus *Nicotiana. Mol. Phylogenet. Evol.* 20, 100–110. doi: 10.1006/mpev.2001.0927

Kehr, A. E., and Smith, H. H. (1954). Genetic tumors in *Nicotiana* hybrids. *Brookhaven Symposia Biol.* 6, 55–78.

Kiselev, K. V., Dubrovina, A. S., Veselova, M. V., Bulgakov, V. P., Fedoreyev, S. A., and Zhuravlev, Y. N. (2007). The *rolB* gene induced overproduction of resveratrol in *Vitis amurensis* transformed cells. *J. Biotechnol.* 128, 681–692. doi: 10.1016/j.jbiotec.2006.11.008

Knapp, S., Chase, M. W., and Clarkson, J. J. (2004). Nomenclatural changes and a new sectional classification in *Nicotiana* (Solanaceae). *Taxon* 53, 73–82. doi: 10.2307/4135490

Koonin, E. V., Makarova, K. S., and Aravind, L. (2001). Horizontal gene transfer in prokaryotes: quantification and classification. *Annu. Rev. Microbiol.* 55, 709–742. doi: 10.1146/annurev.micro.55.1.709

Kulaeva, O. A., Matveeva, T. V., Lutova, L. A., Kulaeva, O. A., Matveeva, T. V., and Lutova, L. A. (2013). Study of the possibility of horizontal gene transfer from Agrobacterium to some representatives of family Solanaceae. *Ecol. Genet.* 11, 3–9. Available online at: http://ecolgenet.ru/ru/node/1625

Limami, M. A., Sun, L. Y., Douatm, C., Helgeson, J., and Tepfer, D. (1998). Natural genetic transformation by *Agrobacterium rhizogenes*. Annual flowering in two biennials, belgian endive and carrot. *Plant Physiol.* 118, 543–550.

Livak, K. J., Flood, S. J., Marmaro, J., Giusti, W., and Deetz, K. (1995). Oligonucleotides with fluorescent dyes at opposite ends provide a quenched probe system useful for detecting PCR product and nucleic acid hybridization. *PCR Methods Appl.* 4, 357–362. doi: 10.1101/gr.4.6.357

Matveeva, T. V., Bogomaz, D. I., Pavlova, O. A., Nester, E. W., and Lutova, L. A. (2012). Horizontal Gene Transfer from Genus *Agrobacterium* to the Plant *Linaria* in Nature. *Mol. Plant Microbe Interact.* 25, 1542–1551. doi: 10.1094/MPMI-07-12-0169-R

Matveeva, T. V., and Kosachev, P. A. (2013). "Sequences homologous to *Agrobacterium rhizogenes* rolC in the genome of Linaria acutiloba," in *International Conference on Frontiers of Environment, Energy and Bioscience (ICFEEB 2013)*. (Lancaster, PA: DEStech Publications, Inc.), 541–546.

Matveeva, T. V., Lutova, L. A., and Bogomaz, D. I. (2006). "Search for TDNA- like sequences in plant genomes, using real-time PCR with degenerate primers and probe," in *Biotechnology in the Agriculture and Food Industry*, ed G. E. Zaikov (New York, NY: Nova Science Publishers), 101–104.

Meyer, A. D., Ichikawa, T., and Meins, F. (1995). Horizontal gene transfer: regulated expression of a tobacco homologue of the *Agrobacterium rhizogenes* rolC gene. *Mol. Gen. Genet.* 249, 265–273. doi: 10.1007/BF00290526

Mohajjel-Shoja, H., Clément, B., Perot, J., Alioua, M., and Otten, L. (2011). Biological activity of the *Agrobacterium rhizogenes*-derived *trolC* gene of *Nicotiana tabacum* and its functional relation to other plast genes. *Mol. Plant Microbe Interact.* 24, 44–53. doi: 10.1094/MPMI-06-10-0139

Moriguchi, K., Maeda, Y., Satou, M., Hardayani, N. S., Kataoka, M., Tanaka, N., et al. (2001). The complete nucleotide sequence of a plant root-inducing (Ri) plasmid indicates its chimeric structure and evolutionary relationship between tumor-inducing (Ti) and symbiotic (Sym) plasmids in Rhizobiaceae. *J. Mol. Biol.* 307, 771–784. doi: 10.1006/jmbi.2001.4488

Naf, U. (1958). Studies on tumor formation in *Nicotiana* hybrids. I. The classification of the parents into two etiologically significant groups. *Growth* 22, 167–180.

Nagata, N., Kosono, S., Sekine, M., Shinmyo, A., and Syono, K. (1995). The regulatory functions of the *rolB* and *rolC* genes of *Agrobacterium rhizogenes* are conserved in the homologous genes (*Ng rol*) of *Nicotiana glauca* in tobacco genetic tumors. *Plant Cell. Physiol.* 36, 1003–1012.

Nagata, N., Kosono, S., Sekine, M., Shinmyo, A., and Syono, K. (1996). Different expression patterns of the promoter of the NgrolB and NgrolC genes during the development of tobacco genetic tumors. *Plant Cell. Physiol.* 37, 489–493. doi: 10.1093/oxfordjournals.pcp.a028971

Nandi, S. K., Palni, L. S. M., and Parker, C. W. (1990). Dynamic of endogenous cytokinin during the growth cycle of a hormone-autotrophic genetic tumor line of tobacco. *Plant Physiol.* 94, 1084–1089. doi: 10.1104/pp.94.3.1084

Nilsson, O., and Olsson, O. (1997). Getting to the root: The role of the *Agrobacterium rhizogenes rol* genes in the formation of hairy roots. *Physiol. Plant.* 100, 463–473. doi: 10.1111/j.1399-3054.1997.tb03050.x

Oger, P., Mansouri, H., and Dessaux, Y. (2000). Effect of crop rotation and soil cover on alteration of the soil microflora generated by the culture of transgenic plants producing opines. *Mol. Ecol.* 9, 881–890. doi: 10.1046/j.1365-294x.2000.00940.x

Oger, P., Petit, A., and Dessaux, Y. (1997). Genetically engineered plants producing opines alter their biological environment. *Nat. Biotechnol.* 15, 369–372. doi: 10.1038/nbt0497-369

Otten, L., Canaday, J., Gerard, J. C., Fournier, P., Crouzet, P., and Paulus, F. (1992). Evolution of agrobacteria and their Ti plasmids. *Mol. Plant Microbe Interact.* 5, 279–287. doi: 10.1094/MPMI-5-279

Palazon, J., Cusido, R. M., Gonzalo, J., Bonfill, M., Morales, S., and Pinol, M. T. (1998a). Relation between the amount the *rolC* gene product and indole alkaloid accumulation in *Catharanthus roseus* transformed root cultures. *J. Plant Physiol.* 153, 712–718.

Palazon, J., Cusido, R. M., Roig, C., and Pino, M. T. (1998b). Expression of the *rolC* gene and nicotine production in transgenic roots and their regenerated plants. *Plant Cell Rep.* 17, 384–390.

Pavlova, O. A., Matveeva, T. V., and Lutova, L. A. (2013). *Linaria dalmatica* genome contains a homologue of *rolC* gene of *Agrobacterium rhizogenes*. *Ecol. Genet.* 11, 10–15. Available online at: http://ecolgenet.ru/ru/node/1629

Qi, L. S., Larson, M. H., Gilbert, L. A., Doudna, J. A., Weissman, J. S., Arkin, A. P., et al. (2013). Repurposing CRISPR as an RNA-Guided platform for sequence-specific control of gene expression. *Cell* 152, 1173 doi: 10.1016/j.cell.2013.02.022

Richards, T. A., Dacks, J. B., Campbell, S. A., Blanchard, J. L., Foster, P. G., McLeod, R., et al. (2006). Evolutionary origins of the eukaryotic shikimate pathway: gene fusions, horizontal gene transfer, and endosymbiotic replacements. *Eukaryot. Cell* 5, 1517–1531. doi: 10.1128/EC.00106-06

Richardson, A. O., and Palmer, J. D. (2007). Horizontal gene transfer in plants. *J. Exp. Bot.* 58, 1–9. doi: 10.1093/jxb/erl148

Schmülling, T., Schell, J., and Spena, A. (1988). Single genes from *Agrobacterium rhizogenes* influence plant development. *EMBO J.* 7, 2621–2629.

Shkryl, Y. N., Veremeichik, G. N., and Bulgakov, V. P. (2008). Individual and combined effects of the rolA, B, and C genes on anthraquinone production in *Rubia cordifolia* transformed calli. *Biotechnol. Bioeng.* 100, 118–125. doi: 10.1002/bit.21727

Smith, H. H. (1958). Genetic plant tumors in Nicotiana. *Ann. N.Y Acad. Sci.* 71, 1163–1177. doi: 10.1111/j.1749-6632.1958.tb46832.x

Spano, L., Pomponi, M., Costantino, P., van Slogteren, G. M. S., and Tempé, J. (1982). Identification of T-DNA in the root-inducing plasmid of the agropine type *Agrobacterium rhizogenes* 1855. *Plant. Mol. Biol.* 1, 291–304. doi: 10.1007/BF00027560

Sutton, D. A. (1988). *A Revision of the Tribe Antirrhineae*. London: Oxford University Press.

Suzuki, K., Tanaka, N., Kamada, H., and Yamashita, I. (2001). Mikimopine synthase (*mis*) gene on pRi1724. *Gene* 263, 49–58. doi: 10.1016/S0378-1119(00)00578-3

Suzuki, K., Yamashita, I., and Tanaka, N. (2002). Tobacco plants were transformed by *Agrobacterium rhizogenes* infection during their evolution. *Plant J.* 32, 775–787. doi: 10.1046/j.1365-313X.2002.01468.x

Tanaka, N. (2008). Horizontal gene transfer in *Agrobacterium: from Biology to Biotechnology*, eds T. Tzfira and V. Citovsky (New York, NY: Springer), 623–647.

Taylor, B. H., White, F. F., Nester, E. W., and Gordon, M. P. (1985). Transcription of *Agrobacterium rhizogenes* A4 T-DNA. *Mol. Gen. Genet.* 201, 546–553. doi: 10.1007/BF00331354

Tepfer, D. (1982). "La transformation genetique de plantes superieures par *Agrobacterium rhizogenes*," in *2e Colloque sur les Recherches Fruitieres* (Bordeaux: Cent Tech Interprofessionnel des Fruits et Legumes), 47–59.

Tepfer, D. (1984). Transformation of several species of higher plants by *Agrobacterium rhizogenes*: sexual transmission of the transformed genotype and phenotype. *Cell* 37, 959–967. doi: 10.1016/0092-8674(84)90430-6

Tzfira, T., and Citovsky, V. (2006). *Agrobacterium*-mediated genetic transformation of plants: biology and biotechnology. *Curr. Opin. Biotechnol.* 17, 147–154. doi: 10.1016/j.copbio.2006.01.009

Udagawa, M., Aoki, S., and Syono, K. (2004). Expression analysis of the NgORF13 promoter during the development of tobacco genetic tumors. *Plant Cell Physiol.* 45, 1023–1031. doi: 10.1093/pcp/pch123

Vain, P. (2007). Thirty years of plant transformation technology development. *Plant Biotechnol. J.* 5, 221–229. doi: 10.1111/j.1467-7652.2006.00225.x

Vallenback, P., Ghatnekar, L., and Bengtsson, B. O. (2010). Structure of the natural transgene PgiC2 in the common grass *Festuca ovina*. *PLoS ONE* 5:e13529. doi: 10.1371/journal.pone.0013529

Vallenback, P., Jaarola, M., Ghatnekar, L., and Bengtsson, B. O. (2008). Origin and timing of the horizontal transfer of a PgiC gene from *Poa* to *Festuca ovina*. *Mol. Phylogenet. Evol.* 46, 890–896. doi: 10.1016/j.ympev.2007.11.031

Veena, H.-J., Doerge, R. W., and Gelvin, S. B. (2003). Transfer of T-DNA and Vir proteins to plant cells by *Agrobacterium tumefaciens* induces expression of host genes involved in mediating transformation and suppresses host defense gene expression. *Plant J.* 5, 219–236. doi: 10.1046/j.1365-313X.2003.01796.x

White, F. F., Garfinkel, D. J., Huffman, G. A., Gordon, M. P., and Nester, E. W. (1983). Sequence homologous to *Agrobacterium rhizogenes* TDNA in the genomes of uninfected plants. *Nature* 301, 348–350. doi: 10.1038/301348a0

White, F. F., Ghidossi, G., Gordon, M. P., and Nester, E. W. (1982). Tumor induction by *Agrobacterium rhizogenes* involves the transfer of plasmid DNA to the plant genome. *Proc. Natl. Acad. Sci. U.S.A.* 79, 3193–3319. doi: 10.1073/pnas.79.10.3193

23

Agrobacterium infection and plant defense—transformation success hangs by a thread

Andrea Pitzschke *

Department of Applied Genetics and Cell Biology, University of Natural Resources and Applied Life Sciences, Vienna, Austria

Edited by:
Stanton B. Gelvin, Purdue
University, USA

Reviewed by:
Lois Banta, Williams College, USA
Saikat Bhattacharjee, Regional
Centre for Biotechnology, India

*Correspondence:
Andrea Pitzschke, Department of
Applied Genetics and Cell Biology,
University of Natural Resources and
Applied Life Sciences, Muthgasse
18, Vienna A-1190, Austria
e-mail: andrea.pitzschke@boku.ac.at

The value of *Agrobacterium tumefaciens* for plant molecular biologists cannot be appreciated enough. This soil-borne pathogen has the unique capability to transfer DNA (T-DNA) into plant systems. Gene transfer involves both bacterial and host factors, and it is the orchestration of these factors that determines the success of transformation. Some plant species readily accept integration of foreign DNA, while others are recalcitrant. The timing and intensity of the microbially activated host defense repertoire sets the switch to "yes" or "no." This repertoire is comprised of the specific induction of mitogen-activated protein kinases (MAPKs), defense gene expression, production of reactive oxygen species (ROS) and hormonal adjustments. *Agrobacterium tumefaciens* abuses components of the host immunity system it mimics plant protein functions and manipulates hormone levels to bypass or override plant defenses. A better understanding of the ongoing molecular battle between agrobacteria and attacked hosts paves the way toward developing transformation protocols for recalcitrant plant species. This review highlights recent findings in agrobacterial transformation research conducted in diverse plant species. Efficiency-limiting factors, both of plant and bacterial origin, are summarized and discussed in a thought-provoking manner.

Keywords: *Agrobacterium tumefaciens*, transformation, plant defense, reactive oxygen species, VIP1

INTRODUCTION

In their natural habitats, plants live in close contact with a myriad microorganisms. Plant-microbe associations can be mutually beneficial, such as the root nodule symbiosis with nitrogen-fixing bacteria or the more wide-spread association of plant roots with arbuscular mycorrhizal fungi (reviewed in Parniske, 2008; Markmann and Parniske, 2009). In contrast, pathogenic fungi or bacteria impair plant development and cause various disease symptoms in their hosts. The gram-negative *Agrobacterium tumefaciens* of the family Rhizobeaceae is a "special case." It is a biotroph pathogen, which markedly alters the physiology and morphology of infected host plants. What makes *Agrobacterium* so special is its capability for interkingdom gene transfer. In nature, wild type *A. tumefaciens* (as well as *A. rhizogenes* and *A. vitis*) causes "crown gall disease," characterized by the growth of tumor-like structures (calli) on host species. The genetic information for this anatomical reprogramming is encoded on the tumor-inducing (Ti) plasmid. The transfer DNA (T-DNA) derived from the Ti plasmid is imported into the host cell's cytoplasm and subsequently into the nucleus (Gelvin, 2003, 2005; Dafny-Yelin et al., 2008; Pitzschke and Hirt, 2010b). T-DNA transport is mediated by agrobacterial virulence factors, and—involuntarily—supported by proteins of the attacked host. Over the last decade, microbiologists and plant scientists have disclosed an impressive portfolio of agrobacterial infection strategies, some of which resemble those in other pathogen-host interactions. Plant defense mechanisms counteracting these strategies are equally diverse and impressive.

PRINCIPAL STEPS

The principal steps and factors involved in *Agrobacterium*-mediated plant transformation are comparatively well-understood, and reviews can be found in e.g., (Gelvin, 2009, 2010a,b; Pitzschke and Hirt, 2010b). Briefly, agrobacteria sense phenolic substances that are secreted by wounded plant tissue. Reception of these signals drives the expression of bacterial virulence (*vir*) genes. Subsequently, Vir proteins are produced, and single-stranded T-DNA molecules are synthesized from the Ti plasmid. The T-complex, i.e., T-DNA associated with certain Vir proteins, is injected into the host cytoplasm. A sophisticated network of bacterial and plant factors mediates translocation of the T-DNA to its final destination, the host cell's nucleus.

Agrobacterium inserts substrates (T-DNA and virulence proteins including VirD2, VirE2, VirE3, VirD5, and VirF) into the host cell by a type IV secretion system (Cascales and Christie, 2003). This strategy is also employed for the delivery of microbial factors by other plant pathogens, including *Xanthomonas campestris* (Thieme et al., 2005) and *Burkholderia* (Engledow et al., 2004). Likewise, mammalian pathogens including *Bordetella pertussis*, *Legionella pneumophila*, *Brucella* spp., and *Helicobacter pylori*, use type IV machineries to export effector proteins to the extracellular milieu or the cell cytosol (Christie and Vogel, 2000). Remarkably, under laboratory conditions, agrobacteria can genetically transform virtually any type of eukaryote, ranging from yeast (Bundock et al., 1995) to human cells (Kunik et al., 2001) (reviewed in Michielse et al., 2005; Lacroix et al., 2006). The T-complex, consisting of T-DNA, bacterial virulence proteins (VirE2, VirD2) and the host factor

VIP1 (VirE2-interacting protein 1) is imported into the nucleus. Subsequently, the proteinaceous components are stripped off, releasing the T-DNA from the T-complex. This step relies on degradation of VirE2, VirD2, and VIP1 by the plant SCF proteasomal machinery (see below). The bacterial F-box protein VirF, which is contained in and confers substrate specificity to the SCF complex, participates in this degradation. If the T-complex disintegrates *before* it is in contact with the host's chromatin, the delivered transgenes are expressed for only a few days. The loss of transgene activity at later stages likely results from the T-DNA being degraded by host nucleases (Gelvin, 2003). In contrast, if the T-DNA is shielded *until* the T-complex is in contact with chromatin, stable transformants can be obtained. Due to its affinity for histones, VIP1 most probably guides the T-DNA to its target destination, the chromatin (Lacroix et al., 2008).

Since the discovery of the gene transfer mechanism (Schell and Van Montagu, 1977; Holsters et al., 1978), *Agrobacterium* strains have been converted ("disarmed") into efficient delivery systems for the genetic manipulation of plants. While transient expression approaches can provide rapid answers on e.g., subcellular localization, protein-protein interaction and promoter/effector relationships (Andrews and Curtis, 2005; Li et al., 2009; Pitzschke, 2013b), genetic engineering requires the transgene(s) to be stably integrated in the host genome.

The so-called disarmed/non-oncogenic *A. tumefaciens* strains employed are deprived of their Ti properties, and the T-DNA region is used as a vehicle for the introduction of tailor-made DNA sequences. Any DNA sequence placed between T-DNA "border sequences" (Ti-plasmid-derived 25-bp direct repeats) can be transferred (Gelvin, 2012). Disarmed strains, therefore, facilitate transformation, but do not provoke callus growth or other abnormalities caused by oncogenic strains. Consequently, phenotypic abnormalities that may be exhibited by transformed plants are primarily due to the particular transgene being expressed. Furthermore, by using armed and disarmed strains side-by-side, host responses that are independent of or dependent on Ti sequences can be distinguished.

TRANSCRIPTIONAL RE-PROGRAMMING OF HOST CELLS

The advent of full genome sequencing and microarray technologies has created the opportunity to draw a complete picture on *Agrobacterium*-induced changes at the transcript level. Gene expression profiling data have been generated for various plant species, and comprehensive databases (e.g., http://www.plexdb. org) and bioinformatics resources even allow comparison of transcriptional responses across multiple plant species (Dash et al., 2012). One major finding from diverse microarray studies was that agrobacteria largely modify host gene expression, particularly that of defense-related genes.

This fact had already been recognized in the "pre-microarray era." cDNA-AFLP analysis of *Ageratum conyzoides* plant cell cultures enabled the identification of (non-oncogenic) *Agrobacterium*-induced transcripts, many of which encoded putative defense factors (Ditt et al., 2001). In a subsequent study the same research group observed an anti-correlation between *Agrobacterium*-mediated transformation efficiency and defense gene expression levels (Ditt et al., 2005). By the

approach of suppression subtractive hybridization and DNA macroarrays, Veena Jiang et al. (2003) provided the first insight into the molecular kinetics of *Agrobacterium*-plant interactions. Transcriptional responses of tobacco BY-2 cell cultures to a subset of agrobacterial strains, impaired in T-DNA and/or Vir protein transfer, were monitored over a 36-h-period. All strains elicited a general defense response during early stages of infection. However, expression of defense-related genes was repressed at later stages—exclusively by the transfer-competent strains. More detailed expression profiling of selected genes furthermore disclosed the "unintentional" participation of the host cellular machinery in the transformation process (Veena Jiang et al., 2003).

MICROBIAL ATTACK AND PLANT DEFENSE

Microbes attempting to invade their hosts betray themselves by the presence of so-called microbe- or pathogen-associated molecular patterns (MAMPs or PAMPs). These molecules, which are recognized as "non-self" initiate the first line of defense, known as PAMP-triggered immunity (PTI) (Nurnberger et al., 2004; Sanabria et al., 2008; Boller and He, 2009) (see below). Pathogens, in turn, aim to overcome PTI activation by injecting certain effector proteins into the host cytoplasm. Perception of these pathogen-encoded effectors by cognate intracellular plant proteins raises the second line of defense, effector-triggered immunity (ETI) (Bonardi and Dangl, 2012; Gassmann and Bhattacharjee, 2012). This response is characterized by the induction of localized apoptosis (hypersensitive response, HR) and systemic defense signaling. Plants capable of activating ETI can thus not only restrict pathogen spread, but they can also fortify themselves against subsequent attacks (Shah and Zeier, 2013).

MAMPs AND THEIR PERCEPTION

MAMPs are best described as molecular "signatures" typical of whole classes of microbes (Boller and Felix, 2009). MAMP perception through specific cell-surface-located proteins ("pattern recognition receptors") is a conserved strategy of eukaryotic innate immune systems. Because MAMPs initiate defense responses in many plant species, they are also referred to as "general elicitors" (Nurnberger et al., 2004). Prominent examples of MAMPs include oligopeptide elicitors such as those derived from EF-Tu (elongation factor thermo unstable), flagellin, and cryptogein (a fungal sterol-scavenging protein), as well as glycol-conjugates, including bacterial lipopolysaccharides and peptido-glycan, and the fungal MAMPs beta-glucan, chitin and chitosan oligosaccharides (reviewed in Silipo et al., 2010).

The two undoubtedly best-characterized MAMP receptors in plants, FLS2 and EFR, recognize the oligopeptides flagellin and EF-Tu, respectively. Owing to their composite structure, these membrane-located leucine-rich repeat-receptor-like kinases (LRR-RLK) convert and transmit perceived "attack signals" into the interior of cells to initiate appropriate defense responses. On the contrary, the primary "aims" of pathogens are to claim nutrients from and multiply to high levels in their hosts. To avoid or block defense responses during early stages of infection, pathogens have two options: (1) evade recognition and "sneak in" or (2) "step in self-consciously" and counteract the elicited warfare attack. Biotrophs, such as *Pseudomonas syringae*, *A. tumefaciens*, *Xanthomonas campestris,* and *Botrytis cinerea,* have developed sophisticated strategies to block defense signaling in their hosts at several steps (Pitzschke et al., 2009c).

A total of 292 and 165 *LRR-RLK* genes were retrieved from the rice and *Arabidopsis* genomes, respectively (Hwang et al., 2011). These large numbers provide an idea of the versatility of LRR-RLK applications. Specific roles have been ascribed to individual family members. Studies in individual LRR-RLK mutants have contributed to our understanding of pathogen perception in general. They also demonstrate the similarity of early plant responses to agrobacteria and other microbial pathogens.

For instance, *fls2* mutants fail to recognize flagellin and are more susceptible to infection by the pathogen *Pseudomonas syringae* (Zipfel et al., 2004). Similarly, mutants deficient in EFR, the receptor for the agrobacterial MAMP EF-Tu, are hypersensitive to *Agrobacterium*-mediated transformation (Zipfel et al., 2006). These examples demonstrate that "ignoring" the invader is not advisable. Instead, perception is the first and mandatory step to restrict bacterial invasion. *FLS2* gene induction upon pathogen exposure or flagellin treatment (Boutrot et al., 2010), as well as *EFR1* induction by EF-Tu-derived peptides (Zipfel et al., 2006) reflect additional host mechanisms to better target the suspected invaders.

MAPK SIGNALING

One of the early intracellular events following pathogen perception is signal transduction and amplification through mitogen-activated protein kinases (MAPKs) (Nakagami et al., 2005; Pitzschke et al., 2009c; Huang et al., 2012; Rasmussen et al., 2012). MAPK cascades are conserved eukaryotic signaling modules. Their minimal components, a MAPK kinase kinase (MAPKKK), a MAPKK and a MAPK, represent multigene families. Exogenous or developmental signals are perceived by a receptor which subsequently (directly or indirectly) initiates the MAPK cascade. Once activated, a MAPKKK phosphorylates its downstream MAPKK which in turn phosphorylates and thereby activates its downstream MAPK (Nakagami et al., 2005). MAPK-mediated phosphorylation of target proteins can alter their properties, such as subcellular location, DNA-binding specificity, enzymatic activity or stability. There is ample evidence for disturbed MAPK signaling markedly affecting biotic and abiotic stress tolerance (Rohila and Yang, 2007; Pitzschke and Hirt, 2009; Pitzschke et al., 2009a; Rodriguez et al., 2010; Sinha et al., 2011; Persak and Pitzschke, 2013; Zhang et al., 2013b). It is very likely that such a scenario will hold true in many plant species.

MAPK SIGNALING AND THE MULTIFUNCTIONAL PROTEIN VIP1

In the context of agrobacteria and pathogen defense, one member of the *Arabidopsis* MAPK family has merited special attention: MPK3. This protein is activated within few minutes upon treatment with pathogens or bacterial elicitor-derived peptides such as flg22 and elf18 (Djamei et al., 2007; Lu et al., 2009). MPK3 is an important positive regulator in defense signaling (Nakagami et al., 2005; Pitzschke et al., 2009c). From a pathogen's point of view, activation of MPK3 should be avoided to circumvent repelling. Accordingly, agrobacteria have evolved strategies to

co-opt induction of this kinase. MPK3 phosphorylates the host protein VIP1 and thereby triggers cyto-nuclear translocation of this bZIP transcription factor (Djamei et al., 2007). VIP1, which enters the nucleus *via* interaction with importin alpha (Citovsky et al., 2004) subsequently induces expression of defense genes such as *PR1* (pathogenesis-related protein 1) (Djamei et al., 2007; Pitzschke et al., 2009b; Pitzschke and Hirt, 2010a). Agrobacteria, on the other hand, hijack VIP1 as a shuttle for nuclear import of the T-complex (Citovsky et al., 2004). A number of plant species lack putative VIP1 homologs; yet these species *are* transformable. This apparent paradox was solved by the discovery and characterization of virulence factor VirE3. VirE3 functionally replaces the "shuttle" function of VIP1, thus ensuring nuclear import of the T-DNA (Lacroix et al., 2005). In contrast to VIP1, VirE3 is not a transcription factor and is therefore unlikely to (directly) induce defense gene expression. VirE3 may thus be an attractive target for biotechnological approaches.

VIP1 as transcriptional regulator

A random-DNA-selection-assay (RDSA) enabled the identification of putative VIP1 target sequences. The DNA consensus motif recognized by VIP1 (VRE—VIP1 response element) was found to be enriched in promoters of stress-responsive genes (Pitzschke et al., 2009b). Notably, this motif does not resemble known regulatory DNA elements. *In vivo*, VIP1 directly binds to VRE sites in the promoter of *MYB44* (Pitzschke et al., 2009b), a stress-related transcription factor (Jung et al., 2008; Persak and Pitzschke, 2013). Importantly, this binding occurs in a stress-dependent manner that correlated with the MPK3 activation profile (Pitzschke et al., 2009b). Through binding to VRE sites, VIP1 might directly regulate expression of another stress-responsive gene, *thioredoxin Trxh8*. In protoplast cotransfection experiments, VIP1 triggered the expression of the pathogen-responsive PR1 gene (Djamei et al., 2007). However, this *PR1* induction is likely an indirect effect. The *PR1* promoter is devoid of VRE sites; and *PR1* is known as a late stress-responsive gene, in contrast to the early and transient nature of MPK3 activation and VIP1 cyto-nuclear translocation. A very recent report (Lacroix and Citovsky, 2013) provides a deeper insight into the VRE-VIP1 mechanism. In agreement with the original study (Pitzschke et al., 2009b), VIP1 bound VRE *in vitro*, and VIP1-VRE binding strongly correlated with transcriptional activation levels *in vivo*. Presence of the agrobacterial F-box protein VirF did not affect VIP1-VRE binding *in vitro*. In contrast, coexpression of *virF* markedly decreased VIP1 transcriptional activation ability *in vivo*. The most likely explanation for this effect is that *in vivo*, VirF prevents VRE induction by triggering proteasomal degradation of VIP1 (Lacroix and Citovsky, 2013). In fact, agrobacteria have learned to control VIP1 abundance by abusing the host proteasome machinery (see below). Being aware of the ongoing host-pathogen arms race, it is tempting to speculate that VIP1 may not only turn on expression of host defense genes. Instead, agrobacteria *may* benefit from one or more VIP1-induced gene products involuntarily provided by the plant. Discovering the VIP1-targetome seems a highly rewarding undertaking. Screening of the *Arabidopsis* genome for promoters enriched in VRE and related motifs isolated by RDSA (Pitzschke et al., 2009b) could be a first step in that direction (Pitzschke, unpublished).

Overexpression studies in tobacco have shown that VIP1 also promotes transformation efficiency in heterologous systems (Tzfira et al., 2002). The cross-species functionality of VIP1 as transcription factor was further documented in a rather non-conventional expression system: protoplasts from red leaves of poinsettia (*Euphorbia pulcherrima*). Polyethylenglycol-mediated cotransfection experiments showed that VIP1 efficiently induces VRE-mediated gene expression (Pitzschke and Persak, 2012). For this transactivation to occur neither a tissue context, chloroplasts nor external stimuli are required.

In its unquestionable key role in *Agrobacterium*-mediated transformation, VIP1 presents an attractive target for manipulation. It appears feasible to uncouple the T-complex-vehicle from the defense-gene-inducer function. Experiments with a C-terminally truncated VIP1 variant have shown that full-length VIP1 is required for stable, but not for transient transformation (Li et al., 2005a). The transgenesis-enhancing effect most likely derives from VIP1 acting as mediator between host nucleosomes and T-DNA/VirE2 complexes. Therefore, replacing critical residues rather than deleting certain domains/peptides seems a more purposeful approach. Indeed, mutation of Lys212, located in the bZIP domain, rendered VIP1 fully incapable of transactivating the *PR1* promoter or a synthetic VRE promoter (Pitzschke et al., 2009b).

THE SCF PROTEASOMAL MACHINERY, VirF AND VBF

Many biological processes, including host-pathogen interactions, are controlled by SCF (Skp1-Cul1-F-box protein) ubiquitin ligase complexes. These complexes mediate the proteasomal degradation of specific target proteins. The F-box protein contained in SCF complexes confers substrate specificity (Lechner et al., 2006).

Although prokaryotes lack SCF complexes, F-box-encoding genes are found in some pathogenic bacteria. The translocation of F-box effectors appears to be a wide-spread "infection strategy." Pathogens secrete F-box proteins into their hosts to abuse the SCF machinery, resulting in high infection rates. However, F-box effectors are intrinsically unstable proteins which are rapidly degraded by the host proteasome pathway (Magori and Citovsky, 2011b). The Citovsky laboratory uncovered yet another level of agrobacterial cleverness and callousness: Destabilization of the agrobacterial F-box protein VirF is counteracted by the bacterial effector, VirD5 (Magori and Citovsky, 2011a). As if this was not enough, agrobacteria also exploit additional host factors to maximize infection: Diverse pathogens, including *Agrobacterium*,

induce expression of VBF (VIP1-binding factor), a host-encoded F-box protein. VBF can functionally replace the agrobacterial VirF in regulating VIP1 and VirE2 protein levels (Zaltsman et al., 2010b). Analogous to VirF, VBF destabilizes VirE2 and VIP1, most likely *via* SCF-mediated proteasomal degradation (Zaltsman et al., 2010a). A very recent study extends on this finding and highlights the importance of VBF at the final stage of T-DNA pre-integration (Zaltsman et al., 2013). As reported earlier, T-complexes can be reconstituted from ssDNA and VirE2 *in vitro* (Zupan et al., 1996). Its tight packaging by VirE2 molecules shields the ssDNA from the outside and makes it inaccessible to degradation by exogenously added DNAse. In the presence of extracts from wild type, but not from *VBF* antisense plants, this "shielding effect" was found to be rapidly lost. Thus, VBF-mediated un*coating* of the T-complex indeed results in *unmasking* of the T-DNA (Zaltsman et al., 2013).

Micro-bombardment studies in *N. benthamiana* leaves have disclosed a cytoplasmic-nuclear distribution of VBF. In contrast, VBF/VIP1 complexes occur exclusively in the nucleus. Based on these observations, VBF may have additional functions in the cytoplasm, besides acting in T-complex disassembly in the nucleus, (Zaltsman et al., 2010b). Alternatively, VBF may re-locate upon pathogen attack (similar to VIP1). If this—currently hypothetic—scenario was true, a straight-forward question arises. Is VBF distribution phosphorylation-dependent; is it controlled by MAPKs? At least *in silico*, such scenario appears possible (Pitzschke, unpublished). MAPKs phosphorylate their targets at serine or threonine residues adjacent to a proline. A kinase interaction motif [KIM; R/K-x2-6-I/Lx(/L)], known to be recognized by mammalian MAPKs (Tanoue and Nishida, 2003), assists MAPK binding also in substrate proteins of plant MAPKs (Schweighofer et al., 2007). The VBF protein sequence contains one Ser-Pro dipeptide motif as well as one KIM (position 164-171) (**Figure 1**). Pathogen-activated MAPK(s), such as MPK3, *may* phosphorylate residue Ser17 and thereby initiate VBF nuclear translocation.

THE ROLE OF PLANT HORMONES IN TRANSFORMATION AND TUMOR FORMATION

A plethora of developmental and stimulus-triggered responses are signaled *via* phytohormones. Auxin is involved in essentially all aspects of plant growth and development (Benjamins and Scheres, 2008; Ljung, 2013). Ethylene controls fruit ripening and plant senescence. It also mediates biotic stress and numerous other environmental responses (Merchante et al., 2013). Abscisic acid controls seed germination, stomatal movement

---

VBF AT1G56250

MMMLPEACIANILAFT█ADAFSSSEVSSVFRLAGDSDFVWEKFLPSDYKSLISQSTDHHWN
ISSKKEIYRCLCDSLLIDNARKLFKINKFSGKISYVLSARDISITHSDHASYWSWSNVSDSR
FSESAELIITDRLEIEGKIQTRVLSANTRYGAYLIVKVT**K**GAYG**L**D**L**VPAETSIKSKNGQIS
KSATYLCCLDEKKQQMKRLFYGNREERMAMTVEAVGGDGKRREPKCRDDGWMEIELGEFETR
EGEDDEVNMTLTEVKGYQLKGGILIDGIEVRPKT

FIGURE 1 | Arabidopsis VBF protein sequence. A peptide matching the consensus motif for MAPK interaction [R/K-x2-6-I/Lx(/L)], and a putative MAPK phosphorylation site are highlighted.

and is tightly connected with diverse abiotic and biotic stress responses (Nakashima and Yamaguchi-Shinozaki, 2013). Salicylic acid (SA), jasmonate and ethylene primarily act in biotic stress protection. There is ample evidence for the existence of substantial crosstalk between plant hormone defense pathways (De Torres Zabala et al., 2009; Robert-Seilaniantz et al., 2011a; Boatwright and Pajerowska-Mukhtar, 2013). These reports highlighted the importance of the plant's need to dynamically balance absolute and relative levels of phytohormones. A complex and comprehensive review on plant hormones and pathogen response was published very recently (Denance et al., 2013).

Agrobacteria largely shift the "hormone balance" in their infected hosts. This effect on endogenous growth regulators will ultimately lead to agrobacterium-induced tumor formation. An elaborate study provided an insight into *Agrobacterium*-induced phytohormonal changes, and it allowed the researchers to separate tumor-dependent and -independent host responses. Lee et al. (2009) examined the physiological changes and adaptations during tumor development provoked by an oncogenic strain (C58) or a disarmed derivate (GV3101), which only lacks the T-DNA but not the Vir factors (VirD2, VirE2, VirE3, VirF) (Holsters et al., 1980). The oncogenic strain was found to cause much stronger host responses than the disarmed strain. The authors monitored the kinetics of *Agrobacterium*-induced concentration changes of plant hormones, including SA, ethylene, jasmonic acid and indole-3-acetic acid (IAA, the most important auxin). In parallel, they assessed transcriptional changes, with a focus on hormone biosynthesis genes. At the early stage of infection, IAA and ethylene started to accumulate, while later, after T-DNA integration, primarily SA levels increased.

In the subsequent sections particular attention is given to the roles of auxin and SA in the agrobacterium/plant interaction.

AUXIN

Auxin-controlled processes are tightly linked to the intracellular auxin gradient. As reviewed recently (Korbei and Luschnig, 2011), this asymmetric hormone distribution arises from polar deployment and intracellular trafficking of auxin carriers. The stability and activity of these auxin transport proteins, in turn, is controlled by a number of post-translational modifications (Lofke et al., 2013; Rahman, 2013).

Upon its perception by a small number of F-box proteins, auxin rapidly induces the expression of two types of transcriptional regulators, encoded by the aux/IAA and ARF (auxin response factor) gene families. In fact, each physiological response might result from the combinatorial interaction between individual members of these two families (Kim et al., 1997). ARFs directly induce or repress the transcription of their target genes that contain auxin responsive elements in the promoter. By binding to their partner ARFs, aux/IAA proteins keep ARFs in an inactive state. In the presence of auxin, this inhibition is released by degradation of the aux/IAA protein. Recent comprehensive reviews on these principles of auxin responses can e.g., be found in (Korbei and Luschnig, 2011; Lofke et al., 2013; Rahman, 2013).

Several plant pathogens interfere with auxin signaling. This interference can occur at several levels. For instance, *Pseudomonas syringae* was shown to alter *Arabidopsis* auxin physiology *via*

its type III effector protein AvrRpt2 (Cui et al., 2013). In this scenario, AvrRpt2 promotes auxin response by stimulating the turnover of aux/IAA proteins, the key negative transcriptional regulators in auxin signaling. Furthermore, some *P. syringae* strains were found to produce auxin themselves (Glickmann et al., 1998).

miR393 as regulator of auxin signaling and bactericide synthesis

Agrobacteria employ an impressive strategic repertoire to manipulate host auxin levels and signal transduction. First, auxin is one of the T-DNA products introduced by oncogenic *A. tumefaciens* (Weiler and Schroder, 1987). Because auxin stimulates cell growth and gall formation, T-DNA-based auxin biosynthesis serves the pathogen directly in remodeling its host. Attacked host plants, on the other hand, try to evade or at least restrict this remodeling. They employ a gene silencing-based mechanism involving production of a particular micro RNA. *miR393* targets three major auxin receptors (F-box proteins TIR1, AFB2, AFB3) and contributes to antibacterial resistance (Navarro et al., 2006). Increased levels of *miR393* were found in C58-infiltrated zones, but not in areas infiltrated with the disarmed control (Pruss et al., 2008). *miR393* appears to be a versatile instrument to keep pathogen invasion in check. *miR393* expression is induced by the PAMP-derived peptide flg22 (Robert-Seilaniantz et al., 2011b). Notably, flagellin sequences from *Agrobacterium* (as well as *Rhizobium*) are exceptionally divergent from this PTI-triggering conserved 22-amino-acid motif (Felix et al., 1999). *Arabidopsis* plants overexpressing *miR393* have a higher resistance to biotrophic pathogens (Robert-Seilaniantz et al., 2011b). The authors showed that miR393/auxin-related resistance is due to interference with another hormone pathway, SA. Generally, auxin and SA act as negative and positive regulators of plant defense, respectively (Denance et al., 2013). These opposing effects are largely due to the repressive effect of auxin on SA levels and signaling, although auxin also represses defense in an SA-pathway-independent manner (Kazan and Manners, 2009; Mutka et al., 2013). As proposed by (Robert-Seilaniantz et al., 2011b), *miR393* represses auxin signaling and thereby prevents auxin from antagonizing SA signaling. Infection studies with auxin signaling mutants furthermore indicated that the auxin-regulated transcription factor ARF9 induces accumulation of camalexin, but represses accumulation of glucosinolate (Robert-Seilaniantz et al., 2011b). Compared to camalexin, glucosinolates are considered more effective protectants against biotrophic invaders. Therefore, *miR393*-related stabilization of ARF9 in inactive complexes may present a means to shift camalexin toward glucosinolate production. Whether *miR393* synthesis upon agrobacterial attack "only" serves to repress auxin-related callus growth or whether it has additional functions in the defense remains to be established. As noticed recently, naturally high contents of glucosinolates *per se* are no obstacle to transformation. *Tropaeolum majus*, a glucosinolate-rich plant of the order Brassicales, is transformed by agro-infiltration of leaves (GV3101, disarmed strain) to high efficiency (Pitzschke, 2013b).

Besides camalexin and glucosinolates, plants produce various other secondary metabolites to defend themselves against biotrophic pathogens. Agrobacteria can defy at least one major group of bactericides. Several phenolic compounds are enzymatically converted by the agrobacterial protein VirH; and a *virH2* mutant was found to be more susceptible to growth inhibition by these substances (Brencic et al., 2004).

One member of the bactericidal polyamines deserves special attention, putrescine. A recent study (Kim et al., 2013) documented that putrescine accumulation is controlled by MAPK signaling involving MPK3 and MPK6. In *Arabidopsis*, *ADC* genes, encoding key enzymes for putrescine biosynthesis, are induced by infection with *P. syringae*. *adc*-deficient mutants are impaired in *P. syringae*-induced *PR1* expression. Disease susceptibility in these mutants can be recovered by exogenous putrescine. *ADC* transcript and putrescine levels are elevated in transgenic *Arabidopsis* plants expressing a constitutively active MAPK3/6 regulatory kinase in the wild-type background. In the *mpk3* or *mpk6* mutant background, however, this effect is largely reduced. An earlier study in tobacco had shown that plants accumulate putrescine derivatives also to combat agrobacterial infection. Auxin likely is involved in this response (Galis et al., 2004). It remains elusive whether *P. syringae*- and *A. tumefaciens*-induced putrescine synthesis are mediated by a common MPK3/MPK6 signaling pathway.

SALICYLIC ACID

Plants produce SA in response to pathogen attack or microbial elicitors. Mutants with constitutively elevated SA levels are generally more resistant toward biotrophic pathogens (Boatwright and Pajerowska-Mukhtar, 2013). Previously, SA was shown to attenuate *A. tumefaciens*-induced tumors (Yuan et al., 2007; Anand et al., 2008). Additional experimental data documented that the antagonism of auxin to SA responses (see above) is reciprocal. SA represses expression of several auxin-related genes. Moreover, by stabilizing Aux/IAA proteins, SA inhibits auxin responses (Wang et al., 2007). Elevated SA levels were observed in *Arabidopsis* stalks during later stages (>6 dpi) of agrobacterial infection, indicating defense activation. This response was provoked by both the oncogenic (C58) and the disarmed strain (GV3101) (Lee et al., 2009). However, *Arabidopsis* stems infected with C58 contained higher levels of SA, which further increased in 35-day-old tumors. The authors (Lee et al., 2009) also found that high SA levels in mutant plants (*npr1*, *cpr5*) prevented tumor development, while low levels promoted it (*nahG*, *eds1*, *pad4*). One specific role of SA in the *Agrobacterium*-plant interaction is its inhibitory effect on *vir* gene expression, which is accomplished by shut-down of the *vir* regulon (Yuan et al., 2007). What is more, SA indirectly interferes with pathogen multiplication by activating the expression of quormone-degrading enzymes (Yuan et al., 2007). In summary, SA appears to counteract agrobacterial invasion at several levels. It represses *vir* regulon genes (Yuan et al., 2007; Anand et al., 2008) and induces quormone-quenching genes (Yuan et al., 2007). Furthermore, SA antagonises auxin responses (Wang et al., 2007) and acts as antimicrobial agent (Gershon and Parmegiani, 1962). Interestingly, SA accumulation in *Agrobacterium*-infected *Arabidopsis* stalks was not accompanied by the induction of SA-responsive pathogenesis-related genes (3 h, 6 d, 35 dpi tested) (Lee et al., 2009). This effect is different from what is known from other plant-pathogen interactions and from pharmacological

studies. Generally, in pathogen-infected plants, elevated SA synthesis triggers PR gene expression. Likewise, *PR* genes are induced by exogenous application of SA or its analog BTH (Lawton et al., 1996). Despite the lack of *PR* gene induction, SA does play a role in agrobacterial infection, as evidenced by the altered tumor size in SA-deficient/accumulating mutants (Yuan et al., 2007; Lee et al., 2009). Apparently, *A. tumefaciens* cannot prevent SA accumulation, but it can suppress some SA-related defense responses. As suggested by (Lee et al., 2009), abnormally high SA levels in the host may have overextended the agrobacterial control machinery.

A recent comprehensive survey of *Arabidopsis* transcriptome profiling data (including diverse stress treatments and biotic stress signaling mutants *sid2*, *npr1*, *coi1*, *ein2*) provided a deeper insight into the SA/PR gene relation (Gruner et al., 2013). In *P syringae*-treated *Arabidopsis*, *PR1* expression fully depends on (isochorismate-synthase1) ICS1-mediated SA biosynthesis and on (non-expressor of PR1) NPR1-mediated downstream signaling. *PR1* is not induced by exogenous hydrogen peroxide, abscisic acid or flg22, and it is independent of jasmonic acid and ethylene signaling (Gruner et al., 2013).

The small set of genes induced by *Agrobacterium* (strain C58: 35genes; strain GV3101: 28 genes) (Lee et al., 2009) is in striking contrast to the high number (948) of elicitor-responsive (EF-Tu-derived peptide elf26) transcripts. Agrobacteria clearly dampen host responses (Lee et al., 2009). This dampening is not restricted to the transcriptional level. Histological analysis (using diaminobenzidine) revealed that agrobacteria efficiently repressed H_2O_2 accumulation in wounded stalks over several days post-infection. The agrobacterial interference with the host's redox-regulatory machinery is also mirrored by the differential expression of several oxidative-stress-related genes (Ditt et al., 2001; Veena Jiang et al., 2003; Lee et al., 2009). By repressing H_2O_2 production agrobacteria may also avoid activation of ROS-dependent defense genes. Given the known sensitivity of any living cell to reactive oxygen species (ROS), the blocking of accumulation appears an agrobacterial strategy to protect both itself and its living food source, i.e., the host.

PLANT ATTEMPTS TO REPRESS ONCOGENE EXPRESSION

Plants exhibit an admirable perseverance in their battle against microbial manipulation. Even *after* unsuccessful attempts to escape *Agrobacterium*-induced genetic re-programming, the host cell does not surrender. Instead, transformed cells employ gene silencing mechanisms to limit the levels of T-DNA-derived transcripts. Evidence for the involvement of post-transcriptional gene silencing had been provided in a pioneering work by Dunoyer et al. (2006). Small interfering RNAs (siRNAs) directed against T-DNA oncogenes (*tryptophan 2-monooxygenase* and *agropine synthase*) were detected in *Nicotiana benthamiana* leaves 3 days after infiltration with virulent agrobacteria. Additional experiments in *Arabidopsis* further stressed the importance of gene silencing as a disease-limiting strategy. RNA interference-deficient mutant plants (*rdr6*, lacking a RNA-dependent RNA polymerase) were found to be hypersusceptible to agrobacterial infection, as evidenced by extensive tumor formation (Dunoyer et al., 2006). The researchers also conducted infection studies in leaves and stems of *Nicotiana bethamiana* carrying a post-transcriptionally-silenced reporter gene (green fluorescent protein, GFP). This approach enabled them to show that the siRNA protection strategy against T-DNA genes is efficient only at early stages of infection: Strong green fluorescence, high *GFP* mRNA concentrations and low siRNA concentrations were detected specifically in young tumors. Later in the infection process, the pathogen takes command. By specifically inhibiting siRNA synthesis, agrobacteria induce an anti-silencing state—thereby ensuring oncogene expression and tumor maturation (Dunoyer et al., 2006).

A more recent study furthermore documented that DNA methylation also plays a critical role in the regulation of T-DNA transcript levels (Gohlke et al., 2013). The authors compared the methylation pattern of mock- and *Agrobacterium*-inoculated *Arabidopsis* inflorescence stalks on a genome-wide level. Four-week-old tumors, arising from inoculation with the oncogenic *A. tumefaciens* strain C58 contained a globally hypermethylated genome. Intriguingly, a specifically low degree of methylation was observed in T-DNA-derived oncogenes (*Ipt IaaH, IaaM*). Data obtained from experiments with DNA methylation mutants lead to the conclusion that crown gall formation and oncogene expression correlate with the unmethylated state and, consequently, that hypermethylation is a strategy to inhibit plant tumor growth.

RECALCITRANCE TO *AGROBACTERIUM*-MEDIATED TRANSFORMATION

Agrobacterium naturally has a wide host range in plants, primarily dicot species. Driven by the demand for higher yields and improved stress tolerance the accessibility to transformation has become a prime issue in crop science. Despite intensive research it is still poorly understood why some plant species can be transformed easily, while others are recalcitrant to *Agrobacterium*-mediated transformation. Transformation methods of model plants and important crop species are frequently updated, documenting the striving for simpler, more robust and more efficient protocols (reviewed in e.g., Pitzschke, 2013a). These protocols primarily focus on optimizing the conditions of *Agrobacterium*—explant co-incubation. Here, duration, light conditions and the concentration of supplemented acetosyringone and plant hormones are key parameters.

One central message emerges from enumerable transformation studies. The outcome of co-cultivation is primarily determined by the timing and intensity at which host defense responses are activated. Understanding the molecular language of the plant—*Agrobacterium* dialogue is therefore of substantial interest both to basic research and agricultural science.

Studies that compare different cultivars of the same species are particularly informative, and one such study shall be mentioned here. Transformation efficiencies between rice cultivars differ greatly. The indica variety lags far behind the japonica cultivars. A comparative study of the two cultivars in transient and stable transformation assays revealed that the lower transformation efficiency in indica rice was mainly due to less-efficient T-DNA integration into the host genome (Tie et al., 2012). Microarray analyses (1, 6, 12, and 24 h post-infection) revealed major differences in the *Agrobacterium*-induced changes in transcriptome profiles of the two cultivars. These differences were most pronounced at the early stages of infection (within the first

6 h). The authors observed an overall stronger response in the indica cultivar (Zs), with several genes being repressed, and they postulated that some of these genes may be required for the transformation process. From this study, one may conclude that (1) although T-DNA integration represents a late step in the transformation process, the "decision" that leads to failure or success is made early. This decision is made in a narrow time window, since many Zs-specific transcripts are repressed only transiently (at the 1 OR 6 h time-point only). (2) Agrobacteria manage to actively prevent repression of integration-assisting genes in the susceptible cultivar. Among others, gene ontology (GO) annotations "stress-responsive" and "lipid transport" are overrepresented in the group of indica-specific transcripts. The lower T-DNA integration efficiency in the indica cultivar may also be attributable to the specific repression of genes related to DNA damage repair. This assumption is in good agreement with the importance of the host DNA repair machinery in T-DNA integration reported earlier (Li et al., 2005b; Citovsky et al., 2007).

THE ROLE OF REACTIVE OXYGEN SPECIES IN RECALCITRANCE

A promising approach for converting hitherto non-transformable plant species is to determine the basis of this recalcitrance. Poor transformation rates can have entirely different reasons. As outlined above, bacterial and host factors contribute and need to be well-balanced. In pro- and eukaryotic organisms alike, ROS play important roles in the transmission of information. ROS- and MAPK signaling in plants is strongly inter-connected (Pitzschke and Hirt, 2009; Meng and Zhang, 2013). Because high ROS levels trigger cell death, their targeted stress-dependent production serves host organisms to restrict pathogen spread. Inappropriate ROS concentration or distribution can therefore be a barrier to successful transformation. For instance, recalcitrance in *Hypericum perforatum* (St. John's wart; medicinal herb), cell cultures was found to be due to an early oxidative burst, which killed 99% of the co-cultivated agrobacteria within 12 h of infection. Interestingly, the oxidative burst only affected agrobacterial viability but did not trigger plant apopotosis (Franklin et al., 2008). Antimicrobial factors likely also have a negative effect on transformation efficiency and agrobacterial viability in *H. perforatum*. A 12-fold increase in xanthone levels was observed in *H. perforatum* cells 1 day after infection. Increased xanthone levels correlated with an elevated antimicrobial and antioxidative competence. On the basis of these observations one may conclude that the plant can divert its antioxidant capacity to prevent itself, but not the invader, from oxidative damage.

One known agrobacterial factor determining oxidative resistance levels is the ferric uptake regulator Fur. A *fur*-deficient mutant was found to be hypersensitive to H_2O_2 and to have reduced catalase activity (a H_2O_2-detoxifying enzyme). Agrobacterial *fur* mutants were also compromized in tumorigenesis on tobacco leaves (Kitphati et al., 2007). Similarly, *A. tumefaciens* mutants in the *RirA* gene (*rhizobial iron regulator*; repressor of iron uptake) exhibited a peroxide-sensitive phenotype and were impaired in tumor formation on tobacco. In addition, induction of the virulence genes *virB* and *virE* was reduced in *rirA* mutants (Ngok-Ngam et al., 2009). Furthermore, *A. tumefaciens* mutants affected in oxidative stress tolerance

Agrobacterial strategies

Turn host defense steps into advantage
Benefit from stress-induced MAPK pathway to trigger activation of the VIP1 vehicle

Abuse host molecules
Hijack VIP1 for nuclear import of T-complex
Make use of VIP1 properties (chromatin affinity, histone interaction) to target the T-DNA to the host's chromatin

Repress defense gene expression

Protect transferred factors from destruction
Mask T-DNA with VirE2 proteins to avoid nucleolytic cleavage
VirD5 stabilises (intrinsically unstable) VirF

Ensure oncogene expression to drive tumor formation
Inhibit synthesis of oncogene-targeting siRNAs
Repress methylation of T-DNA-encoded oncogenes

Counteract host-derived killer substances
VirH detoxifies (bacteriostatic) phenolics
Inhibit ROS production

Force plants to co-operate
Introduce oncogenes and trigger auxin production
→ to promote callus growth
→ as defense-repressing agent

Functional mimicry of host proteins
VirE3 mimics VIP1 in the T-complex nuclear import function
VirF mimics VBF as part of the SCF proteaasome machinery
→ to facilitate release of T-DNA for integration into the plant genome
→ to prevent VIP1-induced defense gene expression

Plant strategies

Sense microbial elicitors as "non-self" to raise the alarm
Elevate expression of receptor proteins

MAPK cascade signalling to transduce AND amplify the stress information
→ to activate and translocate stress-responsive transcription factors, incl. VIP1
→ to trigger putrescine synthesis

Express defense-related genes

Degrade bacterial factors
Nucleolytic cleavage of T-DNA
SCF-mediated proteasomal degradation of VirD, VirE2

Prevent or repress oncogene expression
Express *miR393* to shut down auxin receptors
Produce siRNAs for post-transcriptional silencing of oncogenes
DNA methylation to repress gene expression in tumors

Kill the invader
Produce antimicrobial compounds
Produce reactive oxygen species

Salicylic acid as as multi-purpose weapon
→ SA attenuates tumorigenesis
→ SA shuts down expression of *vir* regulon
→ SA activates quormone-quenching genes
→ SA has antimicrobial activity *in vitro*
→ SA blocks auxin signaling by stabilising aux/IAA proteins

Biotechnological strategies to improve transformation efficiencies

1. Reported strategies
Use ornithine lipid-deficient agrobacterial strain (potato)
VIP1 overexpression (tobacco; *Arabidopsis*)
Myo-inositol-free medium and cold pre-treatment (perennial ryegrass)
Expression of *P. syringae* effector *AvrPto* prior to agroinfiltration (*Arabidopsis*, transient)
Add L-glutamine to protect agrobacteria from bactericidal polyphenols (tea)

2. Promising strategies (hypothetical)

Manipulate the microbe
Use alternatives to acetosyringone as pre-infection stimulant
Use *A. tumefaciens* strain with higher ROS-tolerance to enhance bacterial viability in infected plant tissue
Develop agrobacterial strains with hyperactive VirE3 (esp. for plant species lacking a VIP1 ortholog)

Manipulate the host
Pre-expose plants to a mild stress ("priming") to enhance agrobacterium-induced MAPK activation and thus VIP1 nuclear translocation
→ Chance: more efficient nuclear import of the T-complex / → Risk: stronger defense response

Transient repression of MAPK activation (pharmacologically) to minimize defense responses

Co-transform an optimized variant of VIP1, with the following characteristics:
→ cyto-nuclear translocation - (like wild-type VIP1)
→ full-length (needed for stable transformation) – (like wild-type VIP1)
→ manipulated DNA binding site (e.g. Lys212Thr replacement) to avoid stress gene induction

Manipulate VBF activity/kinetics to unmask the T-DNA *after* nuclear T-complex import

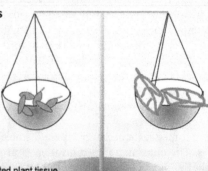

FIGURE 2 | The molecular arms race between host and microbe in Agrobacterium-mediated plant transformation. The activities of both partners need to be well-balanced for successful transformation. Numbers in brackets refer to the corresponding sections in the manuscript.

have been characterized, e.g., *mbfA* (membrane-bound ferritin) (Ruangkiattikul et al., 2012).

The above examples document the vital importance of ROS balancing for both invader and invaded cell. It is tempting to speculate that, the reduced tumor formation in the *fur*/tobacco and *rirA*/tobacco interaction is caused by the poor viability of agrobacteria in a ROS-rich environment of infected host cells. Such a scenario would be in analogy to the situation in *H. perforatum* (Franklin et al., 2008), At this point, concerted efforts of microbiologists and plant biologists are needed to systematically define the proportion and identity of ROS-related agrobacterial factors playing a limiting role in plant transformation.

Another recalcitrant species of agricultural importance that has attracted attention is grapevine (*Vitis vinifera*). Proteomic profiling in grapevine calli grown in the absence or presence of agrobacteria allowed identification of 38 differentially expressed proteins (Zhao et al., 2011). ROS scavenging enyzmes were down-regulated in co-cultivated cells (ascorbate peroxidase, tocopherol cyclase). The authors concluded that low transformation rates and extensive necrosis in *A. tumefaciens*-treated grapevine derive from an impaired ROS scavenging system and an over-activation of apoptotic/hypersensitive response pathways.

APPROACHES TO OVERCOME RECALCITRANCE

Because strong and prolonged host defense responses generally correlate with reduced transformation success (**Figure 2**), external attenuation of these responses may be a means to improve transformation efficiencies. The experimental approaches that can be taken to manipulate host defenses are as manifold as the defense strategies themselves. The problem can be tackled from different sides: (1) by using modified agrobacterial strains that elicit a weaker defense, as e.g., shown in a study on potato (Vences-Guzman et al., 2013); (2) by modifying the composition of plant media and/or growth conditions to keep defense levels low, e.g., Zhang et al. (2013a); (3) by transient and targeted manipulation of the plants non-self-recognition machinery (see below); (4) by counteracting the effect of antimicrobial substances. This strategy proved successful in tea, where L-glutamine was found to overcome the bactericidity of polyphenols (Sandal et al., 2007).

In an innovative study Tsuda and colleagues demonstrated how detailed knowledge on plant-microbe interactions can be employed for successful transformation. *AvrPto* encodes an effector protein from the bacterial plant pathogen *Pseudomonas syringae*. The protein suppresses plant immunity by interfering with plant immune receptors. The *AvrPto* gene was placed under the control of a dexamethasone-inducible promoter. In transgenic *Arabidopsis* plants carrying the inducible construct, dexamethasone pre-treatment largely improved transformation in agro-infiltrated leaves (Tsuda et al., 2012).

An entirely different "pre-treatment strategy" proved successful in perennial ryegrass (*Lolium perenne* L.) (Zhang et al., 2013a). Stable transformants were obtained at an impressively high rate (84%), and 60% of the transgenic calli were regenerated into green plantlets. This was achieved by combining two strategies, while either treatment alone had little effect (10–20% transformation efficiency): (1) Myo-inositol, a component of many standard

media, was removed from the callus culture medium. (2) A cold shock pre-treatment was applied prior to agrobacterial infection.

Myo-inositol levels in plants are primarily controlled by a specific oxygenase, which catalyses the first step in the conversion of this sugar into plant cell wall polysaccharides (Endres and Tenhaken, 2009). The basis of the effect observed by Zhang and colleagues is still largely elusive. It appears that myo-inositol acts in different ways and at multiple levels: omission of myo-inositol promoted *Agrobacterium* binding to the cell surface. It also repressed H_2O_2 production in infected tissue. One indirect consequence of ROS production, callus browning, could furthermore be suppressed when including the cold pre-treatment (Zhang et al., 2013a). Worthwhile questions are: Does growth of cold-pre-treated calli on myo-inositol-free medium alter cell wall composition to support agrobacterial attraction, invasion and/or survival in *L. perenne* cells? If so, what is the critical difference? Can such favorable cell wall characteristics be imitated to facilitate agrobacterial transformation of other recalcitrant species?

CONCLUSIONS

The molecular battle between agrobacteria and plants is impressive, instructive and challenging (**Figure 2**). Impressive, because the arms race takes so many forms. Instructive, because discoveries from *Agrobacterium*-plant interaction studies may drive progress in other fields of microbe-host association research. Challenging, because the external conditions that permit or prohibit transformation including transgene expression are diverse, and the balance needs to be determined empirically. The current state of research provides substantial breeding ground for plant scientists to search for this balance in their favorite species in a more targeted manner.

ACKNOWLEDGMENTS

I gratefully acknowledge the Austrian Science Foundation for financial support (FWF Elise-Richter project V167-B09). I thank Helene Persak, Jürgen Kleine-Vehn and Barbara Korbei for critical reading of the manuscript. Constructive and valuable comments of two anonymous reviewers helped to improve the manuscript.

REFERENCES

Anand, A., Uppalapati, S. R., Ryu, C. M., Allen, S. N., Kang, L., Tang, Y., et al. (2008). Salicylic acid and systemic acquired resistance play a role in attenuating crown gall disease caused by *Agrobacterium tumefaciens*. *Plant Physiol.* 146, 703–715. doi: 10.1104/pp.107.111302

Andrews, L. B., and Curtis, W. R. (2005). Comparison of transient protein expression in tobacco leaves and plant suspension culture. *Biotechnol. Prog.* 21, 946–952. doi: 10.1021/bp049569k

Benjamins, R., and Scheres, B. (2008). Auxin: the looping star in plant development. *Annu. Rev. Plant Biol.* 59, 443–465. doi: 10.1146/annurev.arplant.58.032806.103805

Boatwright, J. L., and Pajerowska-Mukhtar, K. (2013). Salicylic acid: an old hormone up to new tricks. *Mol. Plant Pathol.* 14, 623–634. doi: 10.1111/mpp.12035

Boller, T., and Felix, G. (2009). A renaissance of elicitors: perception of microbe-associated molecular patterns and danger signals by pattern-recognition receptors. *Annu. Rev. Plant Biol.* 60, 379–406. doi: 10.1146/annurev.arplant.57.032905.105346

Boller, T., and He, S. Y. (2009). Innate immunity in plants: an arms race between pattern recognition receptors in plants and effectors in microbial pathogens. *Science* 324, 742–744. doi: 10.1126/science.1171647

Bonardi, V., and Dangl, J. L. (2012). How complex are intracellular immune receptor signaling complexes? *Front. Plant Sci.* 3:237. doi: 10.3389/fpls.2012.00237

Boutrot, F., Segonzac, C., Chang, K. N., Qiao, H., Ecker, J. R., Zipfel, C., et al. (2010). Direct transcriptional control of the *Arabidopsis* immune receptor FLS2 by the ethylene-dependent transcription factors EIN3 and EIL1. *Proc. Natl. Acad. Sci. U.S.A.* 107, 14502–14507. doi: 10.1073/pnas.1003347107

Brencic, A., Eberhard, A., and Winans, S. C. (2004). Signal quenching, detoxification and mineralization of vir gene-inducing phenolics by the VirH2 protein of *Agrobacterium tumefaciens*. *Mol. Microbiol.* 51, 1103–1115. doi: 10.1046/j.1365-2958.2003.03887.x

Bundock, P., Den Dulk-Ras, A., Beijersbergen, A., and Hooykaas, P. J. (1995). Transkingdom T-DNA transfer from *Agrobacterium tumefaciens* to saccharomyces cerevisiae. *EMBO J.* 14, 3206–3214.

Cascales, E., and Christie, P. J. (2003). The versatile bacterial type IV secretion systems. *Nat. Rev. Microbiol.* 1, 137–149. doi: 10.1038/nrmicro753

Christie, P. J., and Vogel, J. P. (2000). Bacterial type IV secretion: conjugation systems adapted to deliver effector molecules to host cells. *Trends Microbiol.* 8, 354–360. doi: 10.1016/S0966-842X(00)01792-3

Citovsky, V., Kapelnikov, A., Oliel, S., Zakai, N., Rojas, M. R., Gilbertson, R. L., et al. (2004). Protein interactions involved in nuclear import of the *Agrobacterium* VirE2 protein *in vivo* and *in vitro*. *J. Biol. Chem.* 279, 29528–29533. doi: 10.1074/jbc.M403159200

Citovsky, V., Kozlovsky, S. V., Lacroix, B., Zaltsman, A., Dafny-Yelin, M., Vyas, S., et al. (2007). Biological systems of the host cell involved in *Agrobacterium* infection. *Cell. Microbiol.* 9, 9–20. doi: 10.1111/j.1462-5822.2006.00830.x

Cui, F., Wu, S., Sun, W., Coaker, G., Kunkel, B., He, P., et al. (2013). The Pseudomonas syringae type III effector AvrRpt2 promotes pathogen virulence via stimulating *Arabidopsis* auxin/indole acetic acid protein turnover. *Plant Physiol.* 162, 1018–1029. doi: 10.1104/pp.113.219659

Dafny-Yelin, M., Levy, A., and Tzfira, T. (2008). The ongoing saga of *Agrobacterium*-host interactions. *Trends Plant Sci.* 13, 102–105. doi: 10.1016/j.tplants.2008.01.001

Dash, S., Van Hemert, J., Hong, L., Wise, R. P., and Dickerson, J. A. (2012). PLEXdb: gene expression resources for plants and plant pathogens. *Nucleic Acids Res.* 40, D1194–D1201. doi: 10.1093/nar/gkr938

Denance, N., Sanchez-Vallet, A., Goffner, D., and Molina, A. (2013). Disease resistance or growth: the role of plant hormones in balancing immune responses and fitness costs. *Front. Plant Sci.* 4:155. doi: 10.3389/fpls.2013.00155

De Torres Zabala, M., Bennett, M. H., Truman, W. H., and Grant, M. R. (2009). Antagonism between salicylic and abscisic acid reflects early host-pathogen conflict and moulds plant defence responses. *Plant J.* 59, 375–386. doi: 10.1111/j.1365-313X.2009.03875.x

Ditt, R. F., Nester, E., and Comai, L. (2005). The plant cell defense and *Agrobacterium tumefaciens*. *FEMS Microbiol. Lett.* 247, 207–213. doi: 10.1016/j.femsle.2005.05.010

Ditt, R. F., Nester, E. W., and Comai, L. (2001). Plant gene expression response to *Agrobacterium tumefaciens*. *Proc. Natl. Acad. Sci. U.S.A.* 98, 10954–10959. doi: 10.1073/pnas.191383498

Djamei, A., Pitzschke, A., Nakagami, H., Rajh, I., and Hirt, H. (2007). Trojan horse strategy in *Agrobacterium* transformation: abusing MAPK defense signaling. *Science* 318, 453–456. doi: 10.1126/science.1148110

Dunoyer, P., Himber, C., and Voinnet, O. (2006). Induction, suppression and requirement of RNA silencing pathways in virulent *Agrobacterium tumefaciens* infections. *Nat. Genet.* 38, 258–263. doi: 10.1038/ng1722

Endres, S., and Tenhaken, R. (2009). Myoinositol oxygenase controls the level of myoinositol in *Arabidopsis*, but does not increase ascorbic acid. *Plant Physiol.* 149, 1042–1049. doi: 10.1104/pp.108.130948

Engledow, A. S., Medrano, E. G., Mahenthiralingam, E., Lipuma, J. J., and Gonzalez, C. F. (2004). Involvement of a plasmid-encoded type IV secretion system in the plant tissue watersoaking phenotype of *Burkholderia cenocepacia*. *J. Bacteriol.* 186, 6015–6024. doi: 10.1128/JB.186.18.6015-6024.2004

Felix, G., Duran, J. D., Volko, S., and Boller, T. (1999). Plants have a sensitive perception system for the most conserved domain of bacterial flagellin. *Plant J.* 18, 265–276. doi: 10.1046/j.1365-313X.1999.00265.x

Franklin, G., Conceicao, L. F., Kombrink, E., and Dias, A. C. (2008). *Hypericum perforatum* plant cells reduce *Agrobacterium* viability during co-cultivation. *Planta* 227, 1401–1408. doi: 10.1007/s00425-008-0691-7

Galis, I., Kakiuchi, Y., Simek, P., and Wabiko, H. (2004). *Agrobacterium tumefaciens* AK-6b gene modulates phenolic compound metabolism in tobacco. *Phytochemistry* 65, 169–179. doi: 10.1016/j.phytochem.2003.10.015

Gassmann, W., and Bhattacharjee, S. (2012). Effector-triggered immunity signaling: from gene-for-gene pathways to protein-protein interaction networks. *Mol. Plant Microbe Interact.* 25, 862–868. doi: 10.1094/MPMI-01-12-0024-IA

Gelvin, S. B. (2003). *Agrobacterium*-mediated plant transformation: the biology behind the "gene-jockeying" tool. *Microbiol. Mol. Biol. Rev.* 67, 16–37. doi: 10.1128/MMBR.67.1.16-37.2003

Gelvin, S. B. (2005). Agricultural biotechnology: gene exchange by design. *Nature* 433, 583–584. doi: 10.1038/433583a

Gelvin, S. B. (2009). *Agrobacterium* in the genomics age. *Plant Physiol.* 150, 1665–1676. doi: 10.1104/pp.109.139873

Gelvin, S. B. (2010a). Finding a way to the nucleus. *Curr. Opin. Microbiol.* 13, 53–58. doi: 10.1016/j.mib.2009.11.003

Gelvin, S. B. (2010b). Plant proteins involved in *Agrobacterium*-mediated genetic transformation. *Annu. Rev. Phytopathol.* 48, 45–68. doi: 10.1146/annurev-phyto-080508-081852

Gelvin, S. B. (2012). Traversing the cell: *Agrobacterium* T-DNA's journey to the host genome. *Front. Plant Sci.* 3:52. doi: 10.3389/fpls.2012.00052

Gershon, H., and Parmegiani, R. (1962). Antimicrobial activity of 8-quinolinols, salicylic acids, hydroxynaphthoic acids, and salts of selected quinolinols with selected hydroxy-acids. *Appl. Microbiol.* 10, 348–353.

Glickmann, E., Gardan, L., Jacquet, S., Hussain, S., Elasri, M., Petit, A., et al. (1998). Auxin production is a common feature of most pathovars of *Pseudomonas syringae*. *Mol. Plant Microbe Interact.* 11, 156–162. doi: 10.1094/MPMI.1998.11.2.156

Gohlke, J., Scholz, C. J., Kneitz, S., Weber, D., Fuchs, J., Hedrich, R., et al. (2013). DNA methylation mediated control of gene expression is critical for development of crown gall tumors. *PLoS ONE Genet.* 9:e1003267. doi: 10.1371/journal.pgen.1003267

Gruner, K., Griebel, T., Navarova, H., Attaran, E., and Zeier, J. (2013). Reprogramming of plants during systemic acquired resistance. *Front. Plant Sci.* 4:252. doi: 10.3389/fpls.2013.00252

Holsters, M., Silva, B., Van Vliet, F., Genetello, C., De Block, M., Dhaese, P., et al. (1980). The functional organization of the nopaline a tumefaciens plasmid pTiC58. *Plasmid* 3, 212–230. doi: 10.1016/0147-619X(80)90110-9

Holsters, M., Silva, B., Van Vliet, F., Hernalsteens, J. P., Genetello, C., Van Montagu, M., et al. (1978). In vivo transfer of the ti-plasmid of *Agrobacterium tumefaciens* to *Escherichia coli*. *Mol. Gen. Genet.* 163, 335–338. doi: 10.1007/BF00271963

Huang, G. T., Ma, S. L., Bai, L. P., Zhang, L., Ma, H., Jia, P., et al. (2012). Signal transduction during cold, salt, and drought stresses in plants. *Mol. Biol. Rep.* 39, 969–987. doi: 10.1007/s11033-011-0823-1

Hwang, S. G., Kim, D. S., and Jang, C. S. (2011). Comparative analysis of evolutionary dynamics of genes encoding leucine-rich repeat receptor-like kinase between rice and *Arabidopsis*. *Genetica* 139, 1023–1032. doi: 10.1007/s10709-011-9604-y

Jung, C., Seo, J. S., Han, S. W., Koo, Y. J., Kim, C. H., Song, S. I., et al. (2008). Overexpression of AtMYB44 enhances stomatal closure to confer abiotic stress tolerance in transgenic *Arabidopsis*. *Plant Physiol.* 146, 623–635. doi: 10.1104/pp.107.110981

Kazan, K., and Manners, J. M. (2009). Linking development to defense: auxin in plant-pathogen interactions. *Trends Plant Sci.* 14, 373–382. doi: 10.1016/j.tplants.2009.04.005

Kim, J., Harter, K., and Theologis, A. (1997). Protein-protein interactions among the Aux/IAA proteins. *Proc. Natl. Acad. Sci. U.S.A.* 94, 11786–11791. doi: 10.1073/pnas.94.22.11786

Kim, S. H., Yoo, S. J., Min, K. H., Nam, S. H., Cho, B. H., and Yang, K. Y. (2013). Putrescine regulating by stress-responsive MAPK cascade contributes to bacterial pathogen defense in *Arabidopsis*. *Biochem. Biophys. Res. Commun.* 437, 502–508. doi: 10.1016/j.bbrc.2013.06.080

Kitphati, W., Ngok-Ngam, P., Suwanmaneerat, S., Sukchawalit, R., and Mongkolsuk, S. (2007). *Agrobacterium tumefaciens* fur has important physiological roles in iron and manganese homeostasis, the oxidative stress response, and full virulence. *Appl. Environ. Microbiol.* 73, 4760–4768. doi: 10.1128/AEM.00531-07

Korbei, B., and Luschnig, C. (2011). Cell polarity: PIN it down! *Curr. Biol.* 21, R197–R199. doi: 10.1016/j.cub.2011.01.062

Kunik, T., Tzfira, T., Kapulnik, Y., Gafni, Y., Dingwall, C., and Citovsky, V. (2001). Genetic transformation of HeLa cells by *Agrobacterium*. *Proc. Natl. Acad. Sci. U.S.A.* 98, 1871–1876. doi: 10.1073/pnas.98.4.1871

Lacroix, B., and Citovsky, V. (2013). Characterization of VIP1 activity as a transcriptional regulator *in vitro* and in planta. *Sci. Rep.* 3:2440. doi: 10.1038/srep02440

Lacroix, B., Li, J., Tzfira, T., and Citovsky, V. (2006). Will you let me use your nucleus? how *Agrobacterium* gets its T-DNA expressed in the host plant cell. *Can. J. Physiol. Pharmacol.* 84, 333–345. doi: 10.1139/y05-108

Lacroix, B., Loyter, A., and Citovsky, V. (2008). Association of the *Agrobacterium* T-DNA-protein complex with plant nucleosomes. *Proc. Natl. Acad. Sci. U.S.A.* 105, 15429–15434. doi: 10.1073/pnas.0805641105

Lacroix, B., Vaidya, M., Tzfira, T., and Citovsky, V. (2005). The VirE3 protein of *Agrobacterium* mimics a host cell function required for plant genetic transformation. *EMBO J.* 24, 428–437. doi: 10.1038/sj.emboj.7600524

Lawton, K. A., Friedrich, L., Hunt, M., Weymann, K., Delaney, T., Kessmann, H., et al. (1996). Benzothiadiazole induces disease resistance in *Arabidopsis* by activation of the systemic acquired resistance signal transduction pathway. *Plant J.* 10, 71–82. doi: 10.1046/j.1365-313X.1996.10010071.x

Lechner, E., Achard, P., Vansiri, A., Potuschak, T., and Genschik, P. (2006). F-box proteins everywhere. *Curr. Opin. Plant Biol.* 9, 631–638. doi: 10.1016/j.pbi.2006.09.003

Lee, C. W., Efetova, M., Engelmann, J. C., Kramell, R., Wasternack, C., Ludwig-Muller, J., et al. (2009). *Agrobacterium tumefaciens* promotes tumor induction by modulating pathogen defense in *Arabidopsis* thaliana. *Plant Cell* 21, 2948–2962. doi: 10.1105/tpc.108.064576

Li, J. F., Park, E., Von Arnim, A. G., and Nebenfuhr, A. (2009). The FAST technique: a simplified *Agrobacterium*-based transformation method for transient gene expression analysis in seedlings of *Arabidopsis* and other plant species. *Plant Methods* 5:6. doi: 10.1186/1746-4811-5-6

Li, J., Krichevsky, A., Vaidya, M., Tzfira, T., and Citovsky, V. (2005a). Uncoupling of the functions of the *Arabidopsis* VIP1 protein in transient and stable plant genetic transformation by *Agrobacterium*. *Proc. Natl. Acad. Sci. U.S.A.* 102, 5733–5738. doi: 10.1073/pnas.0404118102

Li, J., Vaidya, M., White, C., Vainstein, A., Citovsky, V., and Tzfira, T. (2005b). Involvement of KU80 in T-DNA integration in plant cells. *Proc. Natl. Acad. Sci. U.S.A.* 102, 19231–19236. doi: 10.1073/pnas.0506437103

Ljung, K. (2013). Auxin metabolism and homeostasis during plant development. *Development* 140, 943–950. doi: 10.1242/dev.086363

Lofke, C., Luschnig, C., and Kleine-Vehn, J. (2013). Posttranslational modification and trafficking of PIN auxin efflux carriers. *Mech. Dev.* 130, 82–94. doi: 10.1016/j.mod.2012.02.003

Lu, X., Tintor, N., Mentzel, T., Kombrink, E., Boller, T., Robatzek, S., et al. (2009). Uncoupling of sustained MAMP receptor signaling from early outputs in an *Arabidopsis* endoplasmic reticulum glucosidase II allele. *Proc. Natl. Acad. Sci. U.S.A.* 106, 22522–22527. doi: 10.1073/pnas.0907711106

Magori, S., and Citovsky, V. (2011a). *Agrobacterium* counteracts host-induced degradation of its effector F-box protein. *Sci. Signal.* 4, ra69. doi: 10.1126/scisignal.2002124

Magori, S., and Citovsky, V. (2011b). Hijacking of the host SCF ubiquitin ligase machinery by plant pathogens. *Front. Plant Sci.* 2:87. doi: 10.3389/fpls.2011.00087

Markmann, K., and Parniske, M. (2009). Evolution of root endosymbiosis with bacteria: How novel are nodules? *Trends Plant Sci.* 14, 77–86. doi: 10.1016/j.tplants.2008.11.009

Meng, X., and Zhang, S. (2013). MAPK cascades in plant disease resistance signaling. *Annu. Rev. Phytopathol.* 51, 245–266. doi: 10.1146/annurev-phyto-082712-102314

Merchante, C., Alonso, J. M., and Stepanova, A. N. (2013). Ethylene signaling: simple ligand, complex regulation. *Curr. Opin. Plant Biol.* 16, 554–560. doi: 10.1016/j.pbi.2013.08.001

Michielse, C. B., Hooykaas, P. J., Van Den Hondel, C. A., and Ram, A. F. (2005). *Agrobacterium*-mediated transformation as a tool for functional genomics in fungi. *Curr. Genet.* 48, 1–17. doi: 10.1007/s00294-005-0578-0

Mutka, A. M., Fawley, S., Tsao, T., and Kunkel, B. N. (2013). Auxin promotes susceptibility to Pseudomonas syringae via a mechanism independent of suppression of salicylic acid-mediated defenses. *Plant J.* 74, 746–754. doi: 10.1111/tpj.12157

Nakagami, H., Pitzschke, A., and Hirt, H. (2005). Emerging MAP kinase pathways in plant stress signalling. *Trends Plant Sci.* 10, 339–346. doi: 10.1016/j.tplants.2005.05.009

Nakashima, K., and Yamaguchi-Shinozaki, K. (2013). ABA signaling in stress-response and seed development. *Plant Cell Rep.* 32, 959–970. doi: 10.1007/s00299-013-1418-1

Navarro, L., Dunoyer, P., Jay, F., Arnold, B., Dharmasiri, N., Estelle, M., et al. (2006). A plant miRNA contributes to antibacterial resistance by repressing auxin signaling. *Science* 312, 436–439. doi: 10.1126/science.1126088

Ngok-Ngam, P., Ruangkiattikul, N., Mahavihakanont, A., Virgem, S. S., Sukchawalit, R., and Mongkolsuk, S. (2009). Roles of *Agrobacterium tumefaciens* RirA in iron regulation, oxidative stress response, and virulence. *J. Bacteriol.* 191, 2083–2090. doi: 10.1128/JB.01380-08

Nurnberger, T., Brunner, F., Kemmerling, B., and Piater, L. (2004). Innate immunity in plants and animals: striking similarities and obvious differences. *Immunol. Rev.* 198, 249–266. doi: 10.1111/j.0105-2896.2004.0119.x

Parniske, M. (2008). Arbuscular mycorrhiza: the mother of plant root endosymbioses. *Nat. Rev. Microbiol.* 6, 763–775. doi: 10.1038/nrmicro1987

Persak, H., and Pitzschke, A. (2013). Tight interconnection and multi-level control of *Arabidopsis* MYB44 in MAPK cascade signalling. *PLoS ONE* 8:e57547. doi: 10.1371/journal.pone.0057547

Pitzschke, A. (2013a). From bench to barn: plant model research and its applications in agriculture. *Adv. Genet. Eng.* 2, 1–9. doi: 10.4172/2169-0111.1000110

Pitzschke, A. (2013b). Tropaeolum tops tobacco - simple and efficient transgene expression in the order brassicales. *PLoS ONE* 8:e73355. doi: 10.1371/journal.pone.0073355

Pitzschke, A., and Hirt, H. (2009). Disentangling the complexity of mitogen-activated protein kinases and reactive oxygen species signaling. *Plant Physiol.* 149, 606–615. doi: 10.1104/pp.108.131557

Pitzschke, A., and Hirt, H. (2010a). Mechanism of MAPK-targeted gene expression unraveled in plants. *Cell Cycle* 9, 18–19. doi: 10.4161/cc.9.1.10329

Pitzschke, A., and Hirt, H. (2010b). New insights into an old story: *Agrobacterium*-induced tumour formation in plants by plant transformation. *EMBO J.* 29, 1021–1032. doi: 10.1038/emboj.2010.8

Pitzschke, A., and Persak, H. (2012). Poinsettia protoplasts - a simple, robust and efficient system for transient gene expression studies. *Plant Methods* 8, 14. doi: 10.1186/1746-4811-8-14

Pitzschke, A., Djamei, A., Bitton, F., and Hirt, H. (2009a). A major role of the MEKK1-MKK1/2-MPK4 pathway in ROS signalling. *Mol. Plant* 2, 120–137. doi: 10.1093/mp/ssn079

Pitzschke, A., Djamei, A., Teige, M., and Hirt, H. (2009b). VIP1 response elements mediate mitogen-activated protein kinase 3-induced stress gene expression. *Proc. Natl. Acad. Sci. U.S.A.* 106, 18414–18419. doi: 10.1073/pnas.0905599106

Pitzschke, A., Schikora, A., and Hirt, H. (2009c). MAPK cascade signalling networks in plant defence. *Curr. Opin. Plant Biol.* 12, 421–426. doi: 10.1016/j.pbi.2009.06.008

Pruss, G. J., Nester, E. W., and Vance, V. (2008). Infiltration with *Agrobacterium tumefaciens* induces host defense and development-dependent responses in the infiltrated zone. *Mol. Plant Microbe Interact.* 21, 1528–1538. doi: 10.1094/MPMI-21-12-1528

Rahman, A. (2013). Auxin: a regulator of cold stress response. *Physiol. Plant.* 147, 28–35. doi: 10.1111/j.1399-3054.2012.01617.x

Rasmussen, M. W., Roux, M., Petersen, M., and Mundy, J. (2012). MAP kinase cascades in *Arabidopsis* innate immunity. *Front. Plant Sci.* 3:169. doi: 10.3389/fpls.2012.00169

Robert-Seilaniantz, A., Grant, M., and Jones, J. D. (2011a). Hormone crosstalk in plant disease and defense: more than just jasmonate-salicylate antagonism. *Annu. Rev. Phytopathol.* 49, 317–343. doi: 10.1146/annurev-phyto-073009-114447

Robert-Seilaniantz, A., Maclean, D., Jikumaru, Y., Hill, L., Yamaguchi, S., Kamiya, Y., et al. (2011b). The microRNA miR393 re-directs secondary metabolite biosynthesis away from camalexin and towards glucosinolates. *Plant J.* 67, 218–231. doi: 10.1111/j.1365-313X.2011.04591.x

Rodriguez, M. C., Petersen, M., and Mundy, J. (2010). Mitogen-activated protein kinase signaling in plants. *Annu. Rev. Plant Biol.* 61, 621–649. doi: 10.1146/annurev-arplant-042809-112252

Rohila, J. S., and Yang, Y. (2007). Rice mitogen-activated protein kinase gene family and its role in biotic and abiotic stress response. *J. Integr. Plant Biol.* 49, 751–759. doi: 10.1111/j.1744-7909.2007.00501.x

Ruangkiattikul, N., Bhubhanil, S., Chamsing, J., Niamyim, P., Sukchawalit, R., and Mongkolsuk, S. (2012). *Agrobacterium tumefaciens* membrane-bound ferritin plays a role in protection against hydrogen peroxide toxicity and is negatively regulated by the iron response regulator. *FEMS Microbiol. Lett.* 329, 87–92. doi: 10.1111/j.1574-6968.2012.02509.x

Sanabria, N., Goring, D., Nurnberger, T., and Dubery, I. (2008). Self/nonself perception and recognition mechanisms in plants: a comparison of self-incompatibility and innate immunity. *New Phytol.* 178, 503–514. doi: 10.1111/j.1469-8137.2008.02403.x

Sandal, I., Saini, U., Lacroix, B., Bhattacharya, A., Ahuja, P. S., and Citovsky, V. (2007). *Agrobacterium*-mediated genetic transformation of tea leaf explants: effects of counteracting bactericidity of leaf polyphenols without loss of bacterial virulence. *Plant Cell Rep.* 26, 169–176. doi: 10.1007/s00299-006-0211-9

Schell, J., and Van Montagu, M. (1977). Transfer, maintenance, and expression of bacterial Ti-plasmid DNA in plant cells transformed with *A. tumefaciens*. *Brookhaven Symp. Biol.* 36–49.

Schweighofer, A., Kazanaviciute, V., Scheikl, E., Teige, M., Doczi, R., Hirt, H., et al. (2007). The PP2C-type phosphatase AP2C1, which negatively regulates MPK4 and MPK6, modulates innate immunity, jasmonic acid, and ethylene levels in *Arabidopsis*. *Plant Cell* 19, 2213–2224. doi: 10.1105/tpc.106.049585

Shah, J., and Zeier, J. (2013). Long-distance communication and signal amplification in systemic acquired resistance. *Front. Plant Sci.* 4:30. doi: 10.3389/fpls.2013.00030

Silipo, A., Erbs, G., Shinya, T., Dow, J. M., Parrilli, M., Lanzetta, R., et al. (2010). Glyco-conjugates as elicitors or suppressors of plant innate immunity. *Glycobiology* 20, 406–419. doi: 10.1093/glycob/cwp201

Sinha, A. K., Jaggi, M., Raghuram, B., and Tuteja, N. (2011). Mitogen-activated protein kinase signaling in plants under abiotic stress. *Plant Signal. Behav.* 6, 196–203. doi: 10.4161/psb.6.2.14701

Tanoue, T., and Nishida, E. (2003). Molecular recognitions in the MAP kinase cascades. *Cell. Signal.* 15, 455–462. doi: 10.1016/S0898-6568(02)00112-2

Thieme, F., Koebnik, R., Bekel, T., Berger, C., Boch, J., Buttner, D., et al. (2005). Insights into genome plasticity and pathogenicity of the plant pathogenic bacterium *Xanthomonas campestris* pv. vesicatoria revealed by the complete genome sequence. *J. Bacteriol.* 187, 7254–7266. doi: 10.1128/JB.187.21.7254-7266.2005

Tie, W., Zhou, F., Wang, L., Xie, W., Chen, H., Li, X., et al. (2012). Reasons for lower transformation efficiency in indica rice using *Agrobacterium tumefaciens*-mediated transformation: lessons from transformation assays and genome-wide expression profiling. *Plant Mol. Biol.* 78, 1–18. doi: 10.1007/s11103-011-9842-5

Tsuda, K., Qi, Y., Nguyen, L. V., Bethke, G., Tsuda, Y., Glazebrook, J., et al. (2012). An efficient *Agrobacterium*-mediated transient transformation of *Arabidopsis*. *Plant J.* 69, 713–719. doi: 10.1111/j.1365-313X.2011.04819.x

Tzfira, T., Vaidya, M., and Citovsky, V. (2002). Increasing plant susceptibility to *Agrobacterium* infection by overexpression of the *Arabidopsis* nuclear protein VIP1. *Proc. Natl. Acad. Sci. U.S.A.* 99, 10435–10440. doi: 10.1073/pnas.162304099

Veena Jiang, H., Doerge, R. W., and Gelvin, S. B. (2003). Transfer of T-DNA and Vir proteins to plant cells by *Agrobacterium tumefaciens* induces expression of host genes involved in mediating transformation and suppresses host defense gene expression. *Plant J.* 35, 219–236. doi: 10.1046/j.1365-313X.2003.01796.x

Vences-Guzman, M. A., Guan, Z., Bermudez-Barrientos, J. R., Geiger, O., and Sohlenkamp, C. (2013). Agrobacteria lacking ornithine lipids induce more rapid tumour formation. *Environ. Microbiol.* 15, 895–906. doi: 10.1111/j.1462-2920.2012.02867.x

Wang, D., Pajerowska-Mukhtar, K., Culler, A. H., and Dong, X. (2007). Salicylic acid inhibits pathogen growth in plants through repression of the auxin signaling pathway. *Curr. Biol.* 17, 1784–1790. doi: 10.1016/j.cub.2007.09.025

Weiler, E. W., and Schroder, J. (1987). Hormone genes and crown gall disease. *Trends Biochem. Sci.* 12, 271–275. doi: 10.1016/0968-0004(87)90133-2

Yuan, Z. C., Edlind, M. P., Liu, P., Saenkham, P., Banta, L. M., Wise, A. A., et al. (2007). The plant signal salicylic acid shuts down expression of the vir regulon and activates quormone-quenching genes in *Agrobacterium*. *Proc. Natl. Acad. Sci. U.S.A.* 104, 11790–11795. doi: 10.1073/pnas.0704866104

Zaltsman, A., Krichevsky, A., Kozlovsky, S. V., Yasmin, F., and Citovsky, V. (2010a). Plant defense pathways subverted by *Agrobacterium* for genetic transformation. *Plant Signal. Behav.* 5, 1245–1248. doi: 10.4161/psb.5.10.12947

Zaltsman, A., Krichevsky, A., Loyter, A., and Citovsky, V. (2010b). *Agrobacterium* induces expression of a host F-box protein required for tumorigenicity. *Cell Host Microbe.* 7, 197–209. doi: 10.1016/j.chom.2010.02.009

Zaltsman, A., Lacroix, B., Gafni, Y., and Citovsky, V. (2013). Disassembly of synthetic *Agrobacterium* T-DNA-protein complexes via the host SCF(VBF) ubiquitin-ligase complex pathway. *Proc. Natl. Acad. Sci. U.S.A.* 110, 169–174. doi: 10.1073/pnas.1210921110

Zhang, W. J., Dewey, R. E., Boss, W., Phillippy, B. Q., and Qu, R. (2013a). Enhanced *Agrobacterium*-mediated transformation efficiencies in monocot cells is associated with attenuated defense responses. *Plant Mol. Biol.* 81, 273–286. doi: 10.1007/s11103-012-9997-8

Zhang, X., Cheng, T., Wang, G., Yan, Y., and Xia, Q. (2013b). Cloning and evolutionary analysis of tobacco MAPK gene family. *Mol. Biol. Rep.* 40, 1407–1415. doi: 10.1007/s11033-012-2184-9

Zhao, F., Chen, L., Perl, A., Chen, S., and Ma, H. (2011). Proteomic changes in grape embryogenic callus in response to *Agrobacterium tumefaciens*-mediated transformation. *Plant Sci.* 181, 485–495. doi: 10.1016/j.plantsci.2011.07.016

Zipfel, C., Kunze, G., Chinchilla, D., Caniard, A., Jones, J. D., Boller, T., et al. (2006). Perception of the bacterial PAMP EF-Tu by the receptor EFR restricts *Agrobacterium*-mediated transformation. *Cell* 125, 749–760. doi: 10.1016/j.cell.2006.03.037

Zipfel, C., Robatzek, S., Navarro, L., Oakeley, E. J., Jones, J. D., Felix, G., et al. (2004). Bacterial disease resistance in *Arabidopsis* through flagellin perception. *Nature* 428, 764–767. doi: 10.1038/nature02485

Zupan, J. R., Citovsky, V., and Zambryski, P. (1996). *Agrobacterium* VirE2 protein mediates nuclear uptake of single-stranded DNA in plant cells. *Proc. Natl. Acad. Sci. U.S.A.* 93, 2392–2397. doi: 10.1073/pnas.93.6.2392

24

Chromatin dynamics during plant sexual reproduction

Wenjing She and Célia Baroux*

Institute of Plant Biology – Zürich-Basel Plant Science Center, University of Zürich, Zürich, Switzerland

Edited by:
Olga Pontes, University of New Mexico, USA

Reviewed by:
Paula Casati, Centro de Estudios Fotosinteticos – Consejo Nacional de Investigaciones Científicas y Técnicas, Argentina
Karen McGinnis, Florida State University, USA

*Correspondence:
Célia Baroux, Institute of Plant Biology – Zürich-Basel Plant Science Center, University of Zürich, Zollikerstrasse 107, 8008 Zürich, Switzerland
e-mail: cbaroux@botinst.uzh.ch

Plants have the remarkable ability to establish new cell fates throughout their life cycle, in contrast to most animals that define all cell lineages during embryogenesis. This ability is exemplified during sexual reproduction in flowering plants where novel cell types are generated in floral tissues of the adult plant during sporogenesis, gametogenesis, and embryogenesis. While the molecular and genetic basis of cell specification during sexual reproduction is being studied for a long time, recent works disclosed an unsuspected role of global chromatin organization and its dynamics. In this review, we describe the events of chromatin dynamics during the different phases of sexual reproduction and discuss their possible significance particularly in cell fate establishment.

Keywords: chromatin dynamics, epigenetic reprogramming, histone modification, DNA methylation, nucleosome remodeling, small RNA, histone variants, plant sexual reproduction

INTRODUCTION

Flowering plants have a life cycle alternating between a dominant, diploid sporophytic phase and a short haploid gametophytic phase. Sexual reproduction can be divided into three phases: sporogenesis, gametogenesis, embryo- and endosperm-genesis (**Figure 1**). Unlike animals, plants do not set aside a germline lineage during embryogenesis. Instead, the reproductive lineage is established late in development. Cells that will share a meiotic fate and hence initiate a "reproductive lineage" differentiate from and within a somatic tissue in dedicated floral organs of adult plants. Sporogenesis is initiated by the differentiation of spore mother cells (SMCs) that engage somatic cells into a meiotic fate entailing the development of haploid, multicellular gametophytes. The female SMC, also called megaspore mother cell (MMC) differentiates in a subepidermal position in an ovule primordium – composed of the L1-outer layer of cells and the nucellus –; the male SMC differentiates from a mitotic division of the archesporial cell within the sporangium of the anther locule (**Figure 1**, and see section Chromatin Dynamics During Sporogenesis). Gametogenesis is the process by which the gametes are formed within the gametophytes. The male and female gametophytes develop from one haploid spore through a limited number of mitosis and cellularization events that will give rise to highly distinct cell types. A vast majority of flowering plants share the seven-celled type of female gametophyte comprising two gametes – the egg cell and the central cell – and five accessory cells – two synergids and three antipodals. All cells are haploid except for the central cell that inherits two polar nuclei, which following fusion generate a di-haploid maternal genome in the central cell. In contrast, the mature male gametophyte contained in the pollen grain is highly reduced and is composed of one vegetative – accessory – cell and two gametes, the sperm cells (Maheshwari, 1950; **Figure 1**). During double fertilization, the egg cell fuses with one sperm to give rise to the diploid zygote, while the central cell is fertilized by the

second sperm cell – from the same pollen – to produce the triploid endosperm (**Figure 1**). Strikingly, although genetically identical the two fertilization products share distinct developmental fates. The totipotent zygote engages into embryogenesis that establishes the basic body plan and the symmetries (axial and radial) of the future seedling; in contrast, the primary endosperm cell engages in a syncytial phase of proliferation, before cellularization, to form an extra-embryonic, nurturing tissue (Maheshwari, 1950).

Genetic analyses uncovered several molecular factors responsible for cell fate establishment during plant sporogenesis, gametogenesis, and embryogenesis that shed light on the principles of cell specification during these developmental processes, underlying both commonalities and differences with cell specification in the animal reproductive lineage. Several putative intercellular signaling components, non-cell autonomous epigenetic regulators, environmental cues fueled the idea that SMC specification results from a cross-talk within the cells of the founder niche involving molecular, epigenetic, and physiological cues (reviewed in Feng et al., 2013). In contrast, cell specification in the multicellular female gametophytes involves position cues, nuclear migration, and spatially controlled cellularization (reviewed in Drews and Koltunow, 2011; Sprunck and Gross-Hardt, 2011; Rabiger and Drews, 2013). In the male gametophyte, germ cell fate commitment is contributed by factors that influence asymmetric division, cytokinesis and cell cycle (Berger and Twell, 2011), in addition to a cross-talk between the gametophytes and its surrounding tissue (reviewed in Feng et al., 2013). During embryogenesis, cell fate establishment is contributed by embryo-specific transcription factors, signaling components, and local auxin gradients overriding geometric rules of morphogenesis (reviewed in Wendrich and Weijers, 2013; Yoshida et al., 2014), but also by peptides acting non-cell autonomously (Costa et al., 2014). While still incomplete, our understanding of cell specification during plant reproduction

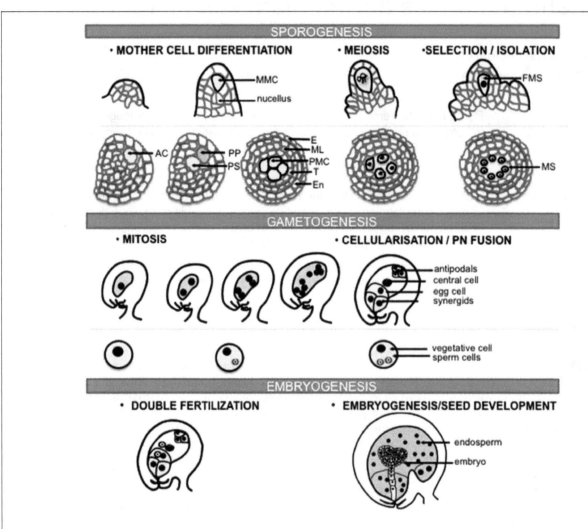

FIGURE 1 | Sexual reproduction in flowering plants. The process of sexual reproduction begins with sporogenesis where spore mother cells (SMCs) differentiate in the floral organs of adult plants. The female SMC, also called megaspore mother cell (MMC) differentiates from a subepidermal nucellar cell within the ovule primordium, the MMC then undergoes meiosis to produce four haploid spores while only one survives to form the functional megaspore (FM). In the stamen primordium, one subepidermal cell enlarges to from the archesporial cell (AC). The archesporial cell then divides to form one primary sporogenous cell (PS) on the inner side and one primary parietal cell (PP) toward the outside. The primary parietal cell divides periclinally and anticlinally to generate the anther wall that is composed of epidermis (E), endothecium (En), the middle layer (ML), and the tapetum (T), while the primary sporogenous cell divides to give rise to the male SMCs, also called the pollen mother cells (PMCs). Each PMC then undergoes meiosis to form four haploid microspores (MS). During gametogenesis, the FM undergoes three rounds of mitosis and cellularization to generate the female gametophyte that harbors two gametes: the egg cell and the central cell, accompanied with three antipodals and two synergids. While for the male side, each microspore undergoes an asymmetric division to give rise to a larger vegetative cell and a smaller generative cell within the bicellular pollen grain. The generative cell divides further to produce the gametes: two sperm cells. During double fertilization, the egg cell is fertilized by one sperm to form the zygote that will give rise to the embryo, while the central cell fuses with the other sperm to generate the triploid endosperm. Original drawings were made after microscopy pictures (female sporogenesis) or inspired from Zhang et al. (2011) (male sporogenesis).

at the genetic, molecular, physiological, and biomechanical levels improved tremendously. Yet, the current models omit a deeper level of possible control over those processes conveyed by nuclear organization. This level is, yet, still difficult to comprehend as it remains at its infancy of formulation, particularly in the field of research in plants. Nuclear organization is a collective term that describes structural and functional arrangements of the chromatin and chromatin-associated structures or factors, at the global, nuclear level, that influences *in fine* genome expression, hence the cellular phenotype; we will focus the discussion in this review onto chromatin dynamics events underlying, and perhaps partly driving, cellular fate transitions during sporogenesis, gametogenesis, and embryogenesis.

In multicellular organisms, cellular identities are the output of distinctive transcriptional programs, which in turn reflect differential, epigenetic instructions encoded beyond the genetic sequence information. Genome expression is modulated in part by the chromatin structure which influences the accessibility and processivity of the transcription machinery

(Jenuwein and Allis, 2001). Two manifestations of chromatin can be discerned: an open, transcriptionally permissive state, and a compact, transcriptionally repressive state. Large-scale manifestations of these two chromatin states are microscopically visible in the nucleus as euchromatin and heterochromatin, respectively. At the cytogenetic level, while heterochromatin is typically enriched in DNA methylation, H3K9me1/2, H3K27me1/2, and H4K20me1, euchromatin is characterized by bivalent instructions such as those associated with a transcriptionally repressive (e.g., H3K27me3), and transcriptionally permissive state (H3K4me2/3, H3K9me3, H3K36me3, H3K56Ac, and H2Bub; Fransz et al., 2006; Roudier et al., 2011). The distribution of histone and DNA methylation marks along the genome is described by chromatin profiling methods. These approaches revealed that, in somatic tissues, their differential combination within promoter or core gene regions indexes distinct chromatin states (Roudier et al., 2011). Moreover, DNA methylation is observed in three sequence contexts that are enriched with gene bodies (CG) or repeat regions (CHG, CHH, respectively, Chan et al., 2005). Histone modifications and DNA methylation are set and maintained by a cohort of enzymes, with complex interplay between themselves and chromatin remodelers but also with small RNAs acting as *trans* signals that reinforce heterochromatic states (reviewed in Tariq and Paszkowski, 2004).

Heterochromatin domains cytologically defined as chromocenters contain rDNA, transposons, centromeric, and pericentromeric repeats, while euchromatin domains are composed of the distal chromosome arms deployed as rosette loops around chromocenters at interphase (Fransz and de Jong, 2002). Although chromosome territories are arranged in randomly in somatic *Arabidopsis* cells (Pecinka et al., 2004), the regular spacing of chromocenters indicates spatial constraints among chromosomes (Andrey et al., 2010). Chromosome capture-based interaction mapping revealed multiple sites that may associate with regions sharing similar chromatin states among distal chromosomal regions (Grob et al., 2013). Whether those interactions causally influence gene expression remains to be determined.

Chromatin dynamics are referred to as the processes that modify the organization of eu- and hetero-chromatin domains in the nucleus, the distribution of genomic sequences within these domains, the arrangement of chromosome territories, and the distribution of functional chromatin proteins and histone modifications. How chromatin dynamics underlie genome expression, or vice versa, particularly during cellular differentiation remains largely unknown. The aim of this review is to discuss the emerging concept that chromatin dynamics contributes to the establishment of new cell fates during sexual reproduction, and probably to the resetting of the epigenome to a ground-state toward pluripotency in the gametophyte and totipotency in the zygote.

CHROMATIN DYNAMICS DURING SPOROGENESIS

Sporogenesis initiates with the differentiation of SMCs. The female SMC, also called MMC corresponds in *Arabidopsis* to a single sub-epidermal cell at the distal end of each ovule

primordium (**Figure 1**, Maheshwari, 1950). In some species, the archesporial cell undergoes division to give rise to several MMCs (Maheshwari, 1950). The MMC undergoes meiosis to produce four haploid spores, while only one survives to form the functional megaspore (**Figure 1**). Male SMCs, also called pollen mother cells (PMCs), or microspore mother cells, differentiate within the sporangium formed in the anther locule. In *Arabidopsis*, the hypodermal cell in the sporangium enlarges to form the archesporial cell that then divides to generate the primary sporogenous cell toward the inside and the primary parietal cell in the outside. The sporogenous cell undergoes mitosis to give rise to PMCs, while the primary parietal cell divides to form the anther wall comprising the epidermis, the endothecium, the middle layer, and the tapetum (**Figure 1**, Maheshwari, 1950). Male sporogenesis is completed after meiosis resulting in four viable haploid microspores.

CHROMATIN DYNAMICS DURING SMC DIFFERENTIATION

Here, we would like to review more particularly epigenetic events occurring and contributing locally to the somatic-to-reproductive transition taking place during sporogenesis. Specific chromatin dynamics related to meiotic execution will be described elsewhere in this issue (Plant Meiosis – Global Approaches).

The first visible signs of SMC differentiation are cellular and nuclear enlargement in the sporogenous tissue. Visible changes in nuclear morphology during MMC differentiation were reported on early drawings or micrographs with clear nuclear and nucleolar enlargement compared to the surrounding nucellar cells (Cooper, 1937; Schulz and Jensen, 1981; Armstrong and Jones, 2003; Sniezko, 2006). In light of our current understanding, these observations suggest large-scale chromatin reorganization. Nuclear swelling and chromatin decondensation in differentiating MMC was recently confirmed and quantified (**Figure 2A**, She et al., 2013). Interestingly, it correlates with the depletion of canonical linker histones and the concomitant, yet progressive reduction in heterochromatin content (She et al., 2013). This H1 depletion is the earliest event of MMC differentiation at a stage where cellular differentiation is barely visible strongly suggests a causal link between chromatin dynamics and the somatic-to-reproductive fate transition in this cell. Following this event, the MMC chromatin undergoes further nucleosome remodeling and biphasic changes in histone modifications (**Figure 2C**). Nucleosome remodeling is illustrated by a presumably dynamic turnover of the centromeric-specific H3 variant (CENH3). This was incidentally detected in the MMC by the depletion of a C-terminally tagged CENH3 variant that failed to be reloaded, in contrast to its N-terminally tagged counterpart (She et al., 2013), in agreement with the model established in male SMCs (Ravi et al., 2011; Schubert et al., 2014). Moreover, the incorporation of a specific H3.3 variant (HTR8) in the MMC suggests global changes in nucleosome composition. Further chromatin dynamics events affecting histone modifications occur along a long meiotic S-phase and seem to establish a transcriptionally permissive state (She et al., 2013). This is suggested by a quantitative increase in the permissive-associated mark H3K4me3, and the reduction of repressive-related marks including H3K27me1, H3K27me3, and H3K9me1 in MMCs, compared to that in surrounding

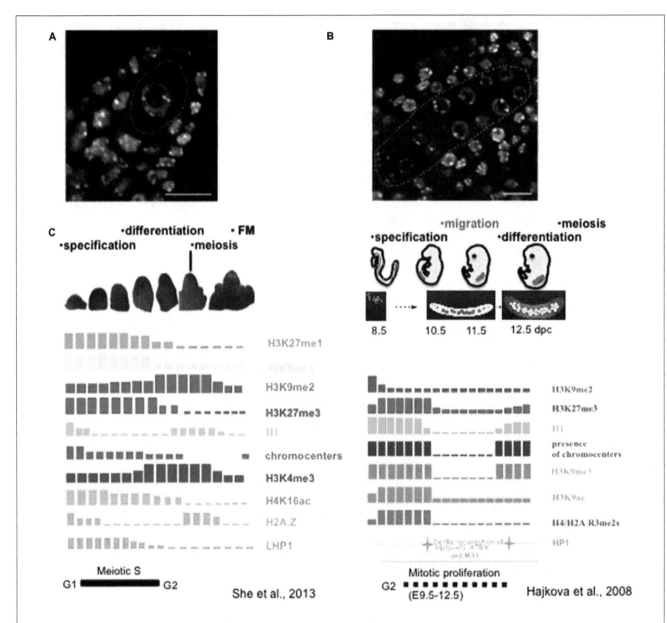

FIGURE 2 | Chromatin dynamics in plant MMCs shows similarities to that in animal PGCs. (A) The MMC (red contour) originates from a subepidermal somatic cell in the ovule primordium, it is distinct from the surrounding nucellar cells by its enlarged nuclear size, as shown by whole-mount DNA staining using propidium iodide of the early ovule primordium as described (She et al., 2013). Scale bar: 10 μm. **(B)** Specification of PMCs (red contour) in the anther, which are marked by the enlarged nuclear and nucleolar size compared to the surrounding somatic cells. The anther was stained by propidium iodide in whole-mount as described for ovule primordia

(She et al., 2013). Scale bar: 10 μm. **(C)** Likewise in animal PGCs, plant MMCs undergo drastic changes in chromatin modification patterns. The schemes summarize studies from Hajkova et al. (2008) and She et al. (2013). However and in contrast, events are asynchronous in plant MMCs and are characterized by both gain and depletion of marks, while animal PGCs at stage 10.5 show a marked depletion of all marks analyzed (Reprinted by permission from Macmillan Publishers Ltd: Nature, Hajkova et al., 2008),© 2008 and Prof. Azim Surani (The Gurdon Institute, University of Cambridge). The schematic images for PGCs development were modified after Ohno et al. (2013).

nucellar cells (She et al., 2013). However, decreasing levels of Ser2-phosphorylated RNA PolII and H4Kac16 indicated a moderate transcriptional competence.

The events described in the MMC are reminiscent of those observed in mouse primordial germ cells (PGCs) that can be seen as functional equivalent of plant SMCs: mouse PGCs undergo large-scale chromatin reprogramming characterized by chromatin decondensation, DNA demethylation, depletion of linker

histone, histone replacement, and extensive erasure of the histone marks such as H3K9me2, H3K9ac, H3K9me3 and H3K27me3 (**Figure 2C**, Hajkova et al., 2002, 2008).

Whether pre-meiotic reprogramming of the DNA methylation landscape occurs, in the MMC, remains a fundamental question to address. At least, genetic evidence showed that DNA methylation landscape influences meiotic recombination in *Arabidopsis* (Mirouze et al., 2012). Post-meiotic reprogramming has been suggested largely based on the expression dynamics of DNA methyltransferases in the female gametophyte (see Chromatin Dynamics During Female Gametogenesis). However, the specific impact on the actual gametic epigenome remains unknown. Possibly, given their mechanistic link with DNA methylation, H1 and H2A.Z depletion in the MMC may enable profound remodeling of the methylome already in the MMC (Wierzbicki and Jerzmanowski, 2005; Kumar and Wigge, 2010; Zemach et al., 2013). Resolving the genomic loci targeted by those epigenetic reprogramming events, at the DNA or histone modification level, is the next challenge to address. However, the techniques that would enable MMC-specific chromatin profiling are not yet established.

The mechanisms controlling chromatin reprogramming in the MMC are likely to be diverse, including both active and passive processes. For instance, proteasome-mediated degradation controls histone variants eviction such as H1 (She et al., 2013) and possibly H2A.Z too. Yet, upstream modifications such as phosphorylation, ubiquitinylation, or citrullination may contribute to destabilize these variants (Contreras et al., 2003; Christophorou et al., 2014). Furthermore, some changes in histone modifications may be coupled with replication occurring during meiotic S phase: the reduction in H3K27me3 levels (relative to the increasing DNA content) may be caused by incorporation of new, non-modified nucleosomes during DNA replication. This, however, does not hold true for marks such as H3K4me3 and H3K9me2 that do show a relative increase during MMC differentiation and are likely involving the activity of chromatin-modifying enzymes. Yet, the process may still be mechanistically coupled: it is noteworthy that H3K9me2 increases at chromocenters at stages where DNA replication is mostly detected in these domains while H3K4me3 increases in euchromatin at later stages where DNA replication is mostly detected in this nuclear compartment (She et al., 2013). Finally, we may speculate that part of the chromatin dynamics may be mediated in trans as suggested by the large representation of small-RNA silencing effectors in the MMC transcriptome (Schmidt et al., 2011).

In contrast, chromatin dynamics events underlying PMC differentiation in the anther are barely known. Yet, similar to MMCs, PMC nuclei enlarge in the male sporangium compared to the surrounding tapetum in different species (Maheshwari, 1950, **Figure 2B**). The finding that transposable elements become expressed in PMCs may further suggest decondensation at heterochromatin loci (Yang et al., 2011) like in MMCs. In addition, remodeling of the nucleosome composition is very likely to occur in PMC likewise in MMCs, as suggested by the dynamic turnover of the centromeric-specific H3 variant (CENH3) detected in both rye and *Arabidopsis* (Ravi et al., 2011; Schubert et al., 2014). H1 linker histones are dynamically phosphorylated – hence potentially destabilized – during the meiotic S-phase of wheat meiocytes (Greer et al., 2012), consistent with the observation of reduced levels in *Arabidopsis* PMCs (Célia Baroux, unpublished). It would be interesting to determine whether the PMC chromatin undergoes a selective replacement of histone H1 with a male-specific variant, possibly resembling that of mouse testis (Sasaki et al., 1990). Collectively, these observations suggest that large-scale, chromatin

dynamics may operate PMC fate establishment similar to that in MMCs, but detailed investigations remain necessary to confirm this proposal.

FUNCTIONS FOR CHROMATIN DYNAMICS IN THE SMCs
Preparation for meiosis
The differentiation of SMCs is followed by meiotic prophase I, with homologous chromosome pairing, synapsis, and recombination. In mice, H3K9me2 deposition is critical for synapsis and in yeast, H3K4me3 marks meiotic recombination initiation sites and regulates double-stranded DNA breaks (Tachibana et al., 2007; Borde et al., 2009; Kniewel and Keeney, 2009). H3K9me2 and H3K4me3 enrichment in the chromatin of plant MMCs during the meiotic S-phase but also during prophase I (She et al., 2013) may suggest a similar role for these marks in synapsis and recombination initiation. Furthermore, the role of DNA methylation in determining the recombination landscape in *Arabidopsis* meiocytes (Melamed-Bessudo and Levy, 2012; Mirouze et al., 2012) may be contributed by H1 and H2A.Z dynamics in the MMC, two histone variants shown to influence DNA methylation patterns in *Arabidopsis* (Wierzbicki and Jerzmanowski, 2005; Zemach et al., 2013). But whether these epigenetic marks directly instruct the meiotic machinery is not known. Alternatively, an intuitive interpretation of chromatin dynamics in the MMC is to enable the expression of meiotic genes and the repression of the mitotic pathway. For instance, it was recently proposed that the female meiotic gene *DMC1* (*DISRUPTED MEIOTIC cDNA1*) is repressed in somatic cells by ACTIN RELATED PROTEIN6 (ARP6), thought to belong to chromatin modulating complexes, possibly via H2A.Z deposition (Qin et al., 2014). This model and the reported expression of *DMC1* in MMC of ovule primordia at stage 2-II is consistent with the eviction of H2A.Z from the MMC chromatin that thus likely enables meiotic gene derepression (She et al., 2013). Similarly in yeast, H1 depletion is a prerequisite to activate meiotic effectors and in mouse oocytes, H3K27 demethylation at key developmental genes in mouse is also essential to meiotic progression (Agger et al., 2007; Bryant et al., 2012). Thus, global remodeling of the meiocyte chromatin likely favors meiotic gene expression. However, it may not be the sole function, since ameiotic *ago9* MMCs resume similar chromatin dynamics than meiotic MMCs (She et al., 2013).

Repression of the somatic program
The SMC fate is not inherited, but it is established locally within a niche of somatic cells in floral sex organs. Intuitively, SMC specification may thus require to exit the somatic program. It was formerly proposed that a globally, epigenetic repressive landscape is established in the nucellus that may favor this transition (reviewed in Baroux et al., 2011; Feng et al., 2013). Several lines of evidence suggest that small-RNA-mediated silencing mechanisms may contribute to this process. ARGONAUTE proteins are central players in microRNAs (miRNAs) and small-interfering RNAs (siRNAs) directed post-transcriptional gene silencing (PTGS) and RNA directed DNA methylation (Vaucheret, 2008). In rice, *MEL1* encodes an AGO protein specifically expressed in SMCs before meiosis. Most SMCs cannot complete sporogenesis and arrest at early meiosis in the loss-of-function mutant, suggesting that MEL1 is important for switching from a mitotic to a meiotic

program, a prerequisite for the somatic-to-reproductive cell fate transition. Possibly as well, *MEL1* may contribute to repress other somatic features as *mel1* mutant PMCs harbor somatic type of mitochondria (Nonomura et al., 2007). In maize, AGO104 specifically accumulates in the nucellar cells of ovule primordium during sporogenesis. MMCs lacking *ago104* activity fail to undergo meiosis, resulting in unreduced (diploid) embryo sacs. Transcriptional profiling of the *ago104* mutant suggests that it represses somatic gene expression in a non-cell autonomous way (Singh et al., 2011). Collectively, the above studies allow to propose a small-RNA-mediated repression of the somatic cell fate during SMC specification. Interestingly, this situation is reminiscent of the animal germline which differentiation requires the inhibition of the somatic transcriptional program, partially relying on piwiRNA-mediated silencing (Nakamura et al., 2010). A non-coding RNA transcribed by the gene *polar granule component* (*pgc*) represses somatic gene expression in *Drosophila* germ cells (pole cells; Martinho et al., 2004).

TE silencing during sporogenesis?

Transmitting the genetic information to the next generation without accumulated mutations is a considerable challenge for sexually reproducing organisms. Transposable elements (TE) are potentially mobile sequences within the genome that pose a threat to genome integrity. Epigenetic reprogramming during germline formation in animals, during sporogenesis in plants, is a potential risky window for TE to escape silencing. Both plants and animals have evolved different strategies to restrict TE activity, particularly in the germline (reviewed in Bao and Yan, 2012, see Companion Cell-Dependent TE Silencing in the Gametes to Preserve Genome Integrity). Chromatin decondensation, loss of heterochromatin, and genome-wide remodeling of the epigenetic landscape during MMC, and likely PMC, specification in plants create a favorable environment for TE escape, thus control mechanisms are likely in place for restricting TE activity in these cells. In somatic plant cells, TEs are kept silenced via an RNA-dependent DNA Methylation (RdDM) pathway, with 24 nt long siRNA targeting DNA and H3K9 methylation at TE loci (Xu et al., 2013). In the MMC, despite a very low heterochromatin content (10.51% compared to 32.3% of somatic cells), the remaining chromocenters are highly enriched in H3K9me2 (She et al., 2013), whereby the immunostaining signals largely overcome the chromocenter foci. This suggests the possibility that TE silencing is reinforced although heterochromatin domains are not maintained. Furthermore, TE silencing could be mediated in trans by siRNAs produced by the surrounding, somatic cells of the nucellus (Olmedo-Monfil et al., 2010). Plants deficient in RdDM-mediated silencing are unable to exert a control on TE proliferation when the parental plant was subjected to heat stress and transmit novel TE copies to their progeny. Genetic analyses suggested that this control normally takes place in the floral tissue and not during gametogenesis (Ito et al., 2011). This heat-activated TEs proliferate during chromatin reprogramming in the MMC of RdDM-deficient nucellus respectively, is the most plausible explanation. Consistent with this, the transcriptionally activated retrotransposon, EVADE, was shown to be actively, maternally suppressed via an siRNA-mediated heterochromatin pathway before meiosis (Reinders et al., 2013) suggesting further a siRNA-based mechanism to doom TE activity during chromatin reprogramming in the MMC.

Epigenetic reprogramming toward pluripotency establishment

Sporogenesis achieves the formation of a haploid, pluripotent spore, which will generate several distinct cell types upon gametophyte development. It has been proposed that chromatin reprogramming in the MMC contributes to establish competence to the gametophytic, pluripotent development of the spore. This proposal is based on the analysis of mutants forming ectopic, ameiotic gametophytes in the ovule (*ago9*, Olmedo-Monfil et al., 2010) and the *sdg2* mutant that lost female gametophytic competence (Berr et al., 2010); in those mutants with altered gametophytic competence, chromatin dynamics was either ectopically expressed (H1 eviction, H3.3 incorporation, H3K27me1 and H3K27me3 reduction) or with altered H3K4me3 levels, respectively.

Although a systematic functional dissection and a challenging, single-cell epigenome profiling remain to be done to confirm this hypothesis, large-scale chromatin dynamics in the MMC likely enables reprogramming the epigenetic landscape to prime a gametophytic developmental program. This situation is also highly reminiscent of that in mice where epigenetic reprogramming in PGCs establishes a ground-state epigenome and alleviates barriers against pluripotency in the germline (Yamaji et al., 2008; Hajkova, 2011; Hackett et al., 2012). Specifically, it would be interesting to test whether H3K27 demethylation in the MMC underlies transcriptional derepression of gametophytic genes, similar to the derepression of pluripotency genes in mice and humans, mediated by the H3K27 demethylase Utx (Mansour et al., 2012). The only H3K27 demethylase characterized so far in *Arabidopsis*, REF6 (Lu et al., 2011) does not seem to be involved in this process (She et al., 2013); thus determining the possible role of H3K27me3 on gametophytic gene expression awaits the elucidation of the mechanisms by which the MMC chromatin is depleted of H3K27me3.

CHROMATIN DYNAMICS DURING GAMETOGENESIS

In plants, gametogenesis is the last step of gametophyte development. The gametes are differentiated, together with accessory cells, within the multicellular male and female gametophytes. In both cases, the establishment of distinct cell fates from genetically identical haploid cells is underlined by distinct chromatin organization.

CHROMATIN DYNAMICS DURING MALE GAMETOGENESIS

Microgametogenesis begins with an asymmetric and atypical mitosis in the microspore, resulting in the formation of a large vegetative cell engulfing a smaller generative cell in *Arabidopsis*. The vegetative cell arrests at G1-phase, while the generative cell undergoes another mitosis to produce two sperm cells (Berger and Twell, 2011). The vegetative cell serves the function of delivering the male gametes toward the ovule during fertilization. The structurally and functionally different cell types are also marked by their dimorphic chromatin states (**Figure 3**).

The chromatin of the vegetative cell is largely decondensed compared to that of the somatic cells, with, notably, low levels

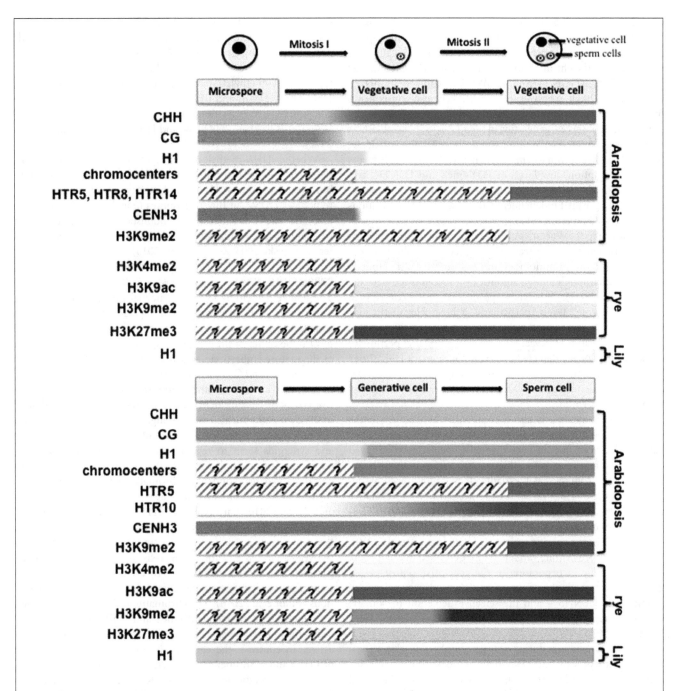

FIGURE 3 | Chromatin dynamics during male gametogenesis. This scheme summarizes cytogenetic and molecular profiling data suggesting large-scale chromatin dynamics events during male gametophyte development. Although disparate in the level of investigation and plant species analyzed it provides a conceptual framework, yet to be completed, for apprehending the extent and potential significance of chromatin dynamics during this developmental stage. In *Arabidopsis*, the microspore harbors low levels of CHH methylation at retrotransposon loci, but retains CG methylation. After the first mitosis, the vegetative nucleus restores CHH methylation, but undergoes CG demethylation at a subset of TE loci (Calarco et al., 2012). The chromatin of the vegetative cell is highly decondensed, mostly deprived of linker H1 (Wenjing She and Célia Baroux, unpublished) and H3K9me2 (Schoft et al., 2009). Additionally, the somatic patterns of histone H3 variants are erased, and only a few H3 variants are retained

including HTR5, HTR8, and HTR14 (Ingouff et al., 2010). Compared to that in somatic nuclei, the chromatin of vegetative cell in rye lost H3K4me2, H3K9ac and H3K9me2, but retains H3K27me3, which can be traced back to the bicellular stage (Houben et al., 2011). In contrast, the sperm chromatin inherits the pattern of DNA methylation from the microspore nucleus, with low levels of CHH methylation, and enrichment of methylated CG (Calarco et al., 2012). It accumulates linker histone H1.1 (Wenjing She and Célia Baroux, unpublished) and H3K9me2 (Schoft et al., 2009). Dynamic changes in the histone H3 repertoire are also observed, with erasure of the somatic variants, but enrichment in HTR5, HTR10 in the sperm nucleus (Ingouff et al., 2010). In rye, it was shown that the sperm chromatin is enriched in H3K4me2, H3K9ac and H3K9me2 modifications, but depleted of H3K27me3, a state that can be traced back to the generative cell at the bicellular stage (Houben et al., 2011).

of H3K9me2 in both eudicots and monocot species (Schoft et al., 2009; Houben et al., 2011). In *Arabidopsis*, the observed disperse of 180-bp centromeric repeats (180CEN) is possibly caused by the absence of the chromatin remodeler DDM1 (DECREASE IN DNA METHYLATION 1) from the SWI/SNF-family of in this cell (Probst et al., 2003; Schoft et al., 2009). Likely as a consequence of this chromatin state, massive transcription of transposable elements (TE) is observed, generating in turn TE-specific small-RNAs (Slotkin et al., 2009). While chromatin decondensation and depletion of repressive chromatin marks such as H3K9me2 likely favors active transcription, low levels of H3K4me2 and H3K9ac, two permissive marks, at least in rye, suggests that transcriptional competence is established independently of these usual modifications (**Figure 3**, Houben et al., 2011).

In contrast to the vegetative cell, the chromatin of the sperm cell is highly condensed. There, transcriptional activity is almost undetectable, based on immunolocalization of Ser2-P-PolII (Houben et al., 2011), although a large amount of transcripts are detected (Borges et al., 2008). This landscape may be partly contributed by high H3K9me2 levels, particularly at heterochromatin loci. However, and paradoxically, the sperm chromatin is enriched in H3K4me2 and H3K9ac, two transcriptionally permissive marks, while globally depleted in the repressive mark H3K27me3 (**Figure 3**, Houben et al., 2011). Collectively, these observations could suggest that the sperm chromatin acquires a poised state as in the animal germline.

Male gametogenesis is also accompanied by changes in the histone H3 variant repertoire, with distinct patterns established between the sperm and the vegetative cells, which can be observed early at the bicellular stage (**Figure 3**). While both cells are devoid of the somatic H3.1 variants, they contain each a specific repertoire of H3.3 variants: the chromatin of the vegetative cell includes a few canonical H3.3 variants (HTR5 and HTR8) and the variant HTR14, while the sperm chromatin contains HTR5 and a sperm-specific variant (HTR10; Ingouff et al., 2010). Dynamics of core histone variants is also described in Lily pollen, with the specific incorporation in the generative cell of gH2A, gH2B, gH3 – which shares common structural properties with Arabidospsis CENH3 – and the selective depletion of somatic H1 in the vegetative cells (Tanaka et al., 1998; Xu et al., 1999; Ueda et al., 2000).

Chromatin dynamics during male gametophyte development is also reflected by the distinct DNA methylation patterns established between the vegetative cell and the gametes, which can be traced back to the microspore stage before mitosis I (**Figure 3**). Comparatively to somatic cells, the microspore chromatin is devoid of CHH methylation mostly from retrotransposon loci. Gametogenesis entails antagonist changes in the sperm and vegetative cells: while the sperm cells inherit the CHH DNA methylation patterns from the microspore, with more pronounced depletion, the vegetative cells restore CHH methylation at TE loci. In contrast, CG methylation is globally retained, in the sperm cells, but depleted from a subset of TE loci and intergenic regions in the vegetative cell. While compared to that in the sperm cells, CHG methylation is generally higher in the vegetative cell, albeit depleted from the same demethylated CG TE loci (Calarco et al., 2012; Ibarra et al., 2012). This profound, dimorphic remodeling of DNA methylomes

during microgametogenesis is likely a consequence of differential activity of key factors in the gametes and vegetative cell: the de novo DNA methyltransferase DRM2 and the 24nt siRNA-based machinery, that normally act together in establishing and maintaining CHH methylation, respectively, and the DNA glycosylases DEMETER (DME) and REPRESSOR OF SILENCING 1 (ROS1) enabling CG demethylation via a base-pair excision-repair process (Morales-Ruiz et al., 2006; Law and Jacobsen, 2010; Calarco et al., 2012).

Whether DNA methylome reprogramming is a cause or consequence of large-scale chromatin dynamics is unclear. Possibly, however, depletion of H1 linker histones and of the chromatin remodeler DDM1 in the microspores (Tanaka et al., 1998; Wenjing She and Célia Baroux, unpublished) may underscore a mechanistic link with DNA methylation changes (Wierzbicki and Jerzmanowski, 2005; Zemach et al., 2013).

CHROMATIN DYNAMICS DURING FEMALE GAMETOGENESIS

The female gametophyte has a syncytial mode of development until the eight-nuclear stage. The bipolar organization of the gametophyte is short lived and migration of two polar nuclei toward the center of the syncytium quickly sets the future pattern of the mature embryo sac, which is definitively set at cellularization (Sprunck and Gross-Hardt, 2011). A microscopic observation of the nuclear size and chromatin appearance at the consecutive stages of development suggests a rather decondensed state of the chromatin but also rapid changes entailed by cellularization (Célia Baroux, unpublished). Particularly, while the antipodals and synergids seem to regain a chromatin organization similar to that of sporophytic cells, the egg and the central cells reveal globally less condensed chromatin state, with fewer heterochromatin foci compared to that of the somatic cells (Jullien and Berger, 2010; Baroux et al., 2011). Yet, the gametes appear clearly dimorphic with a more pronounced decondensation in the central cell and this dimorphism, similar to that between the vegetative cell and the sperm cells, respectively, in the male gametophyte, is further illustrated by the distinct epigenetic and transcriptional landscapes detected using cytogenetic investigations (Pillot et al., 2010). The chromatin in the central cell shows a dramatic reduction of H3K9me2 and LHP1 induced at/after cellularization of the gametophyte, while being transcriptionally active. In contrast, the egg cell chromatin harbors high levels of LHP1 and H3K9me2 at conspicuous foci, coincidentally with low-to-undetectable levels of active RNA PolII, reflecting a relatively transcriptional quiescent state (Pillot et al., 2010; **Figure 4**). Concomitantly, unequal expression of DNA methyltransferases in the central cell and egg cell – with notably undetectable level of these enzymes in the central cell contrasting with the presence of *de novo* DNA methyltransferases DRM1/2 in the egg – may contribute to reinforce the epigenetic dimorphism (Jullien et al., 2012).

The dimorphic epigenetic state between the egg cell and the central cell is also reflected by the establishment of distinct core histone variant patterns (**Figure 4**). Similar to that in the male gametes, both of the female gametes are devoid of most of the canonical, somatic H3 variants. The mature egg cell only harbors the H3.3 variant HTR5, while the central cell retains one H3.1

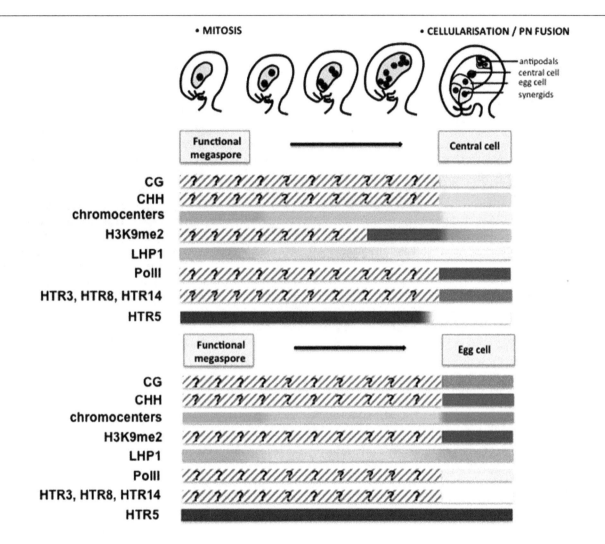

FIGURE 4 | Chromatin dynamics during female gametogenesis. This scheme summarizes mostly cytogenetic and GFP reporter protein analyses suggesting large-scale chromatin dynamics events during female gametophyte development. Although genome-wide, molecular profiling of the chromatin state is currently missing, these data provide, like for **Figure 3**, a conceptual framework for apprehending the extent and potential significance of chromatin dynamics during this developmental stage. Following cellularization, a dimorphic chromatin landscapes are established between the egg cell and the central cell. The central cell chromatin harbors a decondensed chromatin with a low heterochromatin content, correlating with low levels of H3K9me2 and the H3K27me3 reader protein LHP1, but is enriched in active PolII (Ser2 phosphorylated PolII) allowing for active transcription (Pillot et al., 2010). The notable absence of DNA methyltransferases and the presence of the DNA glycosylase DEMETER catalyzing DNA methylation suggest a hypomethylated genome. In contrast, the egg cell harbors heterochromatin foci, though not as prominently as in somatic nuclei and high levels of H3K9me2 and LHP1, but undetectable levels of PolII, suggesting a repressed transcriptional state. Somatic histone variants are depleted from both gametes, with only HTR3, HTR8 and HTR14 retained in the central cell and HTR5 in the egg cell. The model for dynamic changes of CG and CHH methylation is speculative, and is inferred from the analysis of DNA methylation in the endosperm and embryo (Hsieh et al., 2009; Ibarra et al., 2012), as well as the differential expression of DNA methyltransferases between the central cell and egg cell (Jullien et al., 2012). The epigenetic dimorphism concerning heterochromatin content, H3K9me2 and LHP1 seems established just after cellularization.

(HTR3) and two H3.3 variants (HTR8 and HTR14; Ingouff et al., 2010; **Figure 4**). It was considered that the absence of H3.1 in the egg cell may be caused by the arrested cell cycle before S-phase, as H3.1 incorporation is linked with DNA synthesis (Ingouff et al., 2010; Stroud et al., 2012). The specific eviction of core histone H2B in the egg cell, rather than in the central cell, further underlines dimorphic chromatin composition between the gametes (Pillot et al., 2010).

In addition, compared to that in the egg cell where low levels of maintenance DNA methyltransferases including MET1 and CMT3, and high levels of de novo DNA methyltransferases (DRM1/2) are detected, the central cell keeps barely detectable levels of MET1 and CMT3 and low levels of DRM1/2 (Jullien et al., 2012), where *MET1* was proposed to be repressed in the central cell via a Retinoblastoma pathway (Jullien et al., 2008). Furthermore, the DNA demethylase DME is specifically expressed in the central cell, but not in the egg cell prior to fertilization (Choi et al., 2002). Differential expression of those enzymes suggests that the central cell has a globally hypomethylated genome compared to the egg cell (**Figure 4**). While this model is often taken for granted largely due to inferences made from DNA methylome profiling data in the fertilization products at a relatively late stage of seed development (Gehring et al., 2009; Hsieh et al., 2009; Zemach et al., 2010), probing the genome for effective DNA methylation, in sequence context, using cytogenetic and molecular profiling approaches remain necessary to confirm the quantitative and qualitative distinction between the female gametes. In addition, the possibility remains that some loci may be preferentially demethylated after fertilization rather than in the central cell (Jahnke and Scholten, 2009). While instances of hypomethylated genes in the central cell could be described for a few loci in isolated maize gametes (Gutierrez-Marcos et al., 2006; Jahnke and Scholten, 2009), genome-wide profiling of the DNA methylomes, and histone modifications, specifically in the egg and central cells remains currently an immense challenge, due to the extreme difficulty in isolating those cells at a large scale.

FUNCTIONS OF CHROMATIN DYNAMICS DURING GAMETOGENESIS
Derepression of gametic-specific genes

Both the female and male gamete transcriptomes are characterized by a set of specific expressed genes that are otherwise silent in somatic tissues (Wuest et al., 2010; Russell et al., 2012). A few examples report on a contribution of chromatin-mediated repression in this process: for instance, some male-gamete-specific genes were found to be actively repressed by H3K27me1 and H3K27me3 in the sporophyte (Hoffmann and Palmgren, 2013). Thus, chromatin dynamics occurring during gametogenesis and achieving cell-specific epigenetic landscapes (see above) may create a favorable environment for the derepression of those gamete-specific genes. To investigate this hypothesis, it would be of interest to monitor the precise timing of gamete-specific gene expression in relation to the chromatin dynamics events reported above.

Companion cell-dependent TE silencing in the gametes to preserve genome integrity

The problem of maintaining genome integrity in the germline has been exposed in section TE silencing during sporogenesis. In mice, the requirement of a TE control in the germline is restricted to **PGC** development and meiosis (Bao and Yan, 2012), since the meiotic product directly produces the mature gamete. In plants, however, the mitotic developmental phase of the gametophyte, following meiosis, imposes the necessity to prolong a control over TE activity until the mature gametes.

Unlike the sperm cells, the vegetative cell does not contribute to the next generation. Yet, this companion cell seems to influence the epigenetic setup of the sperm cells. The current model involves TE-derived 21nt siRNAs produced by the vegetative cell (following passive and active DNA demethylation) that act in trans on the sperm cells' chromatin to reinforce TE silencing via RNA-directed DNA methylation (RdDM; Slotkin et al., 2009; Ibarra et al., 2012). The efficient silencing of a GFP reporter gene in the sperm cells by expressing the corresponding artificial microRNA in the vegetative cell under the LAT52 promoter supports the model of small RNA transfer from the

companion cell to the male gametes (Slotkin et al., 2009; Feng et al., 2013). However, in another study using a promoter specifically activated in the vegetative cell and not at earlier stage of microspore development (unlike LAT52) this trans silencing experiment could not be reproduced suggesting that TEs siRNAs in the sperm cells may be inherited from the microspore (Grant-Downton et al., 2013). Thus, although further analysis is needed, the model prevails that the companion cell provides a process of genome integrity maintenance in sperm cells that are transcriptionally silent and thus unable to provide the effectors for TE silencing.

Likewise in sperm cells, a control over TE activity in the egg cell would be meaningful. It has been proposed that, similar to the vegetative cell toward the sperm cells, the central cell may play a role in reinforcing TE silencing in the egg cell. This model is inferred from the observation that when the endosperm is derived from a central cell lacking the activity of the DEMETER DNA glycosylase, hypermethylation of TEs is observed, suggesting that those loci are normally demethylated (by DME) and may, likewise the vegetative cell produces TE-derived siRNA. Similarly, trans-silencing of a reporter gene was successfully achieved in the egg by expressing the corresponding amiRNA in the central cell (Ibarra et al., 2012), comforting the idea that siRNA transpose from the central cell to the egg cell to maintain genome integrity in the female germline too (reviewed in Feng et al., 2013).

Setting epigenetic asymmetry for genomic imprinting

Genomic imprinting refers to epigenetic regulations leading to unequal expression of both parental alleles in a diploid cell, thereby conveying possible parent-of-origin-specific effects at the molecular, cellular, tissue, or organismal level. In plants, imprinting occurs in both the embryo and endosperm (reviewed in Dickinson and Scholten, 2013; Gehring, 2013). Genetic studies indicated that imprinting regulation involves differentially methylated regions (DMRs) but also PRC2-mediated histone modifications and likely other, yet unknown, epigenetic mechanisms (Gehring, 2013). The mechanisms of imprinting regulation are extensively reviewed elsewhere (e.g., Raissig et al., 2011; Köhler et al., 2012; Dickinson and Scholten, 2013; Gehring, 2013) and will not be treated in detail here. However, it is relevant to outline the basic principle that imprinting regulation relies on an asymmetric epigenetic setup between the parental alleles that has to be established prior to fertilization. So far, the current model suggests that the parental alleles are, by default, set in an epigenetically repressed state inherited from the somatic cells while a gender-specific erasure of a, e.g., silencing mark enables priming expression after fertilization. For instance, maternally expressed genes (MEG) active in the endosperm are demethylated in the central cell via both active and passive mechanisms (DME-mediated DNA demethylation and lack of DNA methylation maintenance by MET1, respectively, Jullien and Berger, 2010), while their paternal counterpart are hypermethylated in the sperm cells. This is the case for instance for FLOWERING WAGENINGEN (FWA), FERTILIZATION INDEPENDENT SEEDS2 (FIS2), MEDEA (MEA; see reviews cited above), although the latter can be maternally activated in a DME and DNA-methylation-independent manner (Wohrmann et al., 2012). Interestingly, the absence of DME in conjunction with the presence of *de novo* DNA methyltransferases in the egg cell, together with genetic studies on embryo-*MEG* regulation, suggests that the establishment of imprints to be inherited to the embryo relies on distinct mechanisms (Dickinson and Scholten, 2013; Raissig et al., 2013a). The wide range of chromatin changes, including male- or female-gamete-specific resetting of the histone H3 repertoire, and possibly of other histone variants, may offer alternative means to asymmetrically mark the parental alleles of imprinted loci.

Interestingly, gametogenesis is the sole developmental window considered so far for establishing the epigenetic setup of imprinted loci. However, sporogenesis, more particularly SMC differentiation that undergoes massive reprogramming of its chromatin landscape (see Chromatin Dynamics During Sporogenesis) offers another window of opportunity to establish parental imprints. Erasure of DNA methylation at MEG loci for instance may be achieved in the MMC following the eviction of H1 and H2AZ (She et al., 2013) known for their interplay with DNA methylation (Wierzbicki and Jerzmanowski, 2005; Rea et al., 2012; Zemach et al., 2013).

Pre-patterning the post-fertilization fates

The distinct chromatin states established in the egg cell and the central cell after cellularization of the female gametophyte reflect distinct epigenetic and transcriptional status. An interesting explanation for the transcriptionally quiescent state of the egg cell may be a role for establishing totipotency in the zygote likewise in animals. In animals, the zygote is transcriptionally inactive for a duration that varies depending on the species; this transient status (preceding zygotic genome activation) is thought to be necessary to epigenetically reprogram the zygotic genome toward a totipotent ability, through priming developmental regulator genes for expression (Seydoux and Braun, 2006; Surani et al., 2007). The case of plants may mirror that of the animals whereby chromatin dynamics in the egg may pattern the transcriptionally quiescent chromatin of the future zygote. In stark contrast to the egg, the central cell is epigenetically relaxed toward a highly permissive state and *de facto* transcriptionally active (Pillot et al., 2010), a state largely inherited in the endosperm following fertilization. Thus again here, chromatin dynamics in the central cell is likely a pre-patterning event of its post-fertilization fate. What are the critical epigenetic remodeling events that contribute to the identity of the gametes themselves and their post-fertilization products is however still unknown. Clearly, transcriptional quiescence is not enough to define a totipotent state, since an artificially induced transcriptionally silent state in the central cell results in abortion of its post-fertilization program (the endosperm fails to develop; Pillot et al., 2010). Conversely, mutant zygotes deficient in RdDM-mediated gene silencing are transcriptionally active, yet developmentally competent to form a viable embryo (Autran et al., 2011). Being able to profile the epigenetic landscape, genome-wide and at single-gene resolution is critically required

to decipher the targets and role of chromatin dynamics in the gametes.

CHROMATIN DYNAMICS FOLLOWING DOUBLE FERTILIZATION
DIMORPHIC CHROMATIN LANDSCAPES ESTABLISHED IN TWO FERTILIZATION PRODUCTS

Embryogenesis is a long developmental process progressing along consecutive phases of proliferation, morphogenesis, organogenesis, and maturation. Our knowledge is too scarce to draw a developmental atlas of chromatin dynamics events during those phases; large-scale processes have been mostly reported both immediately following fertilization, on which we will focus below, or at maturation stages (van Zanten et al., 2011).

Soon after fertilization, rapid exchanges of gametic H3.3 histone variants occur in the zygote and endosperm and a somatic pattern of H3.3 variant composition is reestablished in the zygote (Ingouff et al., 2007, 2010). This suggests a limited inheritance of H3-based epigenetic information from the gametes to the fertilization products. Yet, the modest resolution of microscopic investigations does not allow excluding inheritance for discrete loci or small chromosomal segments. Clearly, however, the transcriptional states of the fertilized products are largely inherited from their female gametic progenitor: the zygote seems transcriptionally quiescent with barely detectable PolII activity while the endosperm harbors a transcriptionally active chromatin state as shown by abundant levels of engaged RNA PolII (Pillot et al., 2010). Additionally, the dimorphic pattern of H3K9me2 (high in the zygote, low in the endosperm) is also similar to that in the female gametes suggesting inheritance of at least some levels of chromatin organization. The functional requirement of the DNA methyltransferase CMT3 further suggests a connective interplay between H3K9 and DNA methylation in establishing this dimorphism (Pillot et al., 2010).

Furthermore, developmental progression of the fertilization products entails additional chromatin dynamics as inferred by the molecular profiles of DNA methylation patterns in the embryo and endosperm in well-developed seeds (6–8 days after pollination). Particularly, the maternal genome of the endosperm undergoes DME-mediated global DNA demethylation, while, comparatively, DNA methylation levels are higher in embryo in all sequence context (Hsieh et al., 2009). This dimorphism is consistent with the antagonist abundance of DNA methyltransferases including MET1, DRM2, and CMT3 in the embryo and endosperm, respectively (Hsieh et al., 2009; Jullien et al., 2012). The detection of those DNA methyltransferases in the embryo proper at early stage suggests the hypothesis that reprogramming of the DNA methylation landscape already occurs soon after fertilization, likewise in mice (Seisenberger et al., 2013).

Overall, while several evidence suggest dynamic reprogramming of the chromatin states (histone and DNA modifications) in the fertilization products, the data remain uneven with distinct developmental stages investigated (e.g., profiles at early developmental stages are missing) and at different resolution levels (molecular profiles versus microscopic detection of immunostaining signals or fluorescently tagged chromatin modifiers). Clearly, temporally resolved profiles of histone and DNA methylation patterns in the developing embryo and endosperm starting from soon after fertilization is necessary to elucidate the epigenetic landscape and its dynamics. But for this, daunting technical difficulties remain to be solved to enable massive, and tissue specific nuclei isolation suited for epigenome profiling, particularly from the embryo and endosperm – that are embedded within the maternal seed – at very early stages.

FUNCTIONS OF CHROMATIN DYNAMICS IN THE FERTILIZATION PRODUCTS

Reprogramming toward totipotency acquisition in the zygote?

Fertilization unites two differentiated and very specialized cells, the egg cell and the sperm cell. To enable embryo development-with the establishment of novel cell types and organ symmetries toward the basic body plan of the future plant – the newly formed zygote must be alleviated from the gametic programs and acquire totipotency. With analogy to animals, it is tempting to speculate that epigenetic reprogramming may occur in the plant zygote toward setting the future transcriptional program. The abrupt replacement of histone H3 variants in the zygote (Ingouff et al., 2010) may enable a rapid resetting of histone modifications toward this goal. However, we are currently lacking detailed molecular profiles in the gametes and the zygote to draw any meaningful comparison. Yet, strikingly, the plant zygote remains transcriptionally relatively quiescent (Pillot et al., 2010), a situation reminiscent to that of animal zygotes and which is necessary to totipotency acquisition (reviewed in Seydoux and Braun, 2006). Future efforts in elucidating and manipulating epigenome dynamics in the plant zygote are necessary, yet extremely challenging, to conclude about possible evolutionary convergent scenarios between the two kingdoms in the role of epigenetic reprogramming in totipotency acquisition.

Genomic imprinting

Imprinting in the embryo suggests the existence of mechanisms – yet to be discovered – enabling resistance at specific loci against the proposed genome-wide reprogramming of DNA methylation and histone modification landscapes. Imprinted loci seem to have, however, a shorter lifetime in the plant embryo than in animals. In contrast to animals where imprinting persists in adult tissues, no imprinted expression has been detected to date in the seedling which strongly suggests an erasure process of imprints at late stage of embryo development (Raissig et al., 2013a). The suggested, active remethylation of the embryo genome via the DNA methyltransferases DRM1/DRM2 (Jullien and Berger, 2010) may be important in this process (Jullien et al., 2012). Clearly, temporally resolved, DNA methylome profiles of the embryonic genome are awaited to comprehend the timing, the extent and the nature of loci affected by DNA methylation reprogramming following fertilization. While still challenging to perform, recent methodological progress in *Arabidopsis* embryo isolation and bisulfite sequencing from small input fractions offers the realistic possibility in a near future (Schmidt et al., 2012; Raissig et al., 2013b). In addition, the active maintenance of asymmetric histone modifications set by the PRC2 complex is necessary to perpetuate imprinting at several embryo

imprinted loci (Raissig et al., 2013a), yet is in apparent contradiction with the global eviction of maternal H3 variants in the zygote (Ingouff et al., 2010). Thus clearly, chromatin dynamics cannot be described only globally but has to be resolved at the gene-specific level to understand its role in imprinting regulation in the embryo.

TE control for genome integrity across generation

The maternal genome of *Arabidopsis* endosperm undergoes extensive CG demethylation at TE loci, which at least partially requires *DME* activity (Hsieh et al., 2009). Similar to the situation described in the companion cells of the pollen, it has been proposed that TEs from the central cell, and possibly the endosperm as well, may produce specific siRNAs that reinforce TE silencing in the zygote, thereby dooming those genomic elements potentially harmful for the genome integrity of the ensuing generations (Mosher et al., 2009; Ibarra et al., 2012). Although the mobility of siRNA from the endosperm to the embryo remains to be confirmed, it was reported that demethylation of endosperm maternal genome is accompanied by CHH hypermethylation of TEs in the embryo (Hsieh et al., 2009). An endosperm-driven control of genome integrity surveillance in the embryo is likely a conserved mechanism across flowering plants, with evidence reported in both eudicots and monocots (Mosher et al., 2009; Zemach et al., 2010).

CONCLUSIONS AND FUTURE PROSPECTIVES

To date, a broad range of genetic and molecular regulators have been identified that contribute to cell specification processes during sexual reproduction in flowering plants. Yet, with the increasing body of evidence that these processes are accompanied by large-scale chromatin dynamics events, an exciting area is opening; further efforts are needed to comprehend a yet, underestimated level of control mediated by chromatin dynamics likely potentiating the (re)programming of genome expression during those processes. Exciting findings in the past decades uncovered dynamic events of chromatin modifications, DNA methylation, nucleosome remodeling, and small RNA regulation that take place throughout sexual reproduction in flowering plants, particularly during cell fate specification (**Figure 5**). The possible functions of these events range from epigenetic reprogramming of the genome toward pluri- or totipotency, maintenance of genome integrity, regulation of imprinting but may also functions in immediate cellular tasks at meiosis, mitosis cellularization and patterning in the gametophyte and embryo. In the absence of cell-specific epigenome profiles, however, the impact of chromatin dynamics on epigenetic reprogramming remains largely speculative. Establishing a dogma still requires efforts to overcome the daunting obstacles that obstruct cell-specific epigenome profiling in the reproductive lineage, particularly in the model plant *Arabidopsis thaliana*. For these experiments, the choice of other model plants (e.g., model crops) where the gametes are more amenable to mechanical isolation may be judicious. The development of cell-specific nuclei isolation approaches (Deal and Henikoff, 2010) may prove a real asset in these efforts, though it still requires improvement for optimization (Wuest et al., 2013). Alternatively, probing the genome at the microscopic scale for its chromatin composition and organization, at high-resolution, at the single-cell level and in

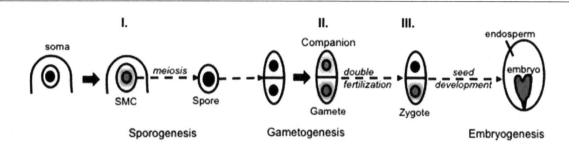

FIGURE 5 | Three main waves of chromatin dynamics during plant reproduction (Model). Sexual plant reproduction can be seen as a three-step process involving sporogenesis, gametogenesis, and embryogenesis taking place in floral organs. Sporogenesis initiates with the specification of spore mother cells (SMCs) within the sporangium tissues. SMCs are primed toward meiosis while undergoing a somatic-to-reproductive cellular fate transition that generates a pluripotent spore. The spore develops a (male or female) multicellular gametophyte generating distinct cell types: the companion (or accessory) cells and the gametic cells (a schematically reduced form is shown, for more details see **Figure 1**).

Fertilization enables the formation of a totipotent zygote, generating in turn the plant embryo. The acquisition of the SMC fate, the gametic fate and the totipotent zygotic fate is associated with three main waves of chromatin dynamics (I.–III., colored nuclei) comprising large-scale reorganization of the chromatin structure, composition and organization, hence reshaping the epigenetic landscape (as reviewed in the text). Whereas some of those events clearly contribute to cell fate establishment (e.g., I., see the text), the challenge of future investigations is to elucidate the functional role of chromatin dynamics in defining the cells' potency versus operating cell fate establishment during sexual reproduction.

a quantitative manner, has proven a valid and fruitful approach (She et al., 2014). It enabled describing unsuspected chromatin dynamics events during SMC and female gamete specification and, in combination with genetic analyses, revealed a functional link with the acquisition of developmental competences (Pillot et al., 2010; She et al., 2013). The completion of such analyses on the male reproductive lineage, in several (model or non-model plants) will be instrumental in determining whether cell specification during reproduction relies on robust, reiterative chromatin dynamics events across developmental phases and genders, and whether an evolutionary conserved scenario exists across eudicots and monocots.

ACKNOWLEDGMENTS

The authors thank Anja Herrmann (University of Zürich) for critical reading of the manuscript, the reviewers for valuable suggestions, and Ueli Grossniklaus (University of Zürich) for insightful discussions and financial support. Our research in this field is supported by the University of Zürich and grants from the Swiss National Foundation to Ueli Grossniklaus (31003AB-126006) and Célia Baroux (31003A_130722).

REFERENCES

Agger, K., Cloos, P. A., Christensen, J., Pasini, D., Rose, S., Rappsilber, J., et al. (2007). UTX and JMJD3 are histone H3K27 demethylases involved in HOX gene regulation and development. *Nature* 449, 731–734. doi: 10.1038/nature06145

Andrey, P., Kieu, K., Kress, C., Lehmann, G., Tirichine, L., Liu, Z., et al. (2010). Statistical analysis of 3D images detects regular spatial distributions of centromeres and chromocenters in animal and plant nuclei. *PLoS Comput. Biol.* 6:e1000853. doi: 10.1371/journal.pcbi.1000853

Armstrong, S. J., and Jones, G. H. (2003). Meiotic cytology and chromosome behaviour in wild-type *Arabidopsis thaliana*. *J. Exp. Bot.* 54, 1–10. doi: 10.1093/jxb/erg034

Autran, D., Baroux, C., Raissig, M. T., Lenormand, T., Wittig, M., Grob, S., et al. (2011). Maternal epigenetic pathways control parental contributions to *Arabidopsis* early embryogenesis. *Cell* 145, 707–719. doi: 10.1016/j.cell.2011.04.014

Bao, J., and Yan, W. (2012). Male germline control of transposable elements. *Biol. Reprod.* 86;162, 161–114. doi: 10.1095/biolreprod.111.095463

Baroux, C., Raissig, M. T., and Grossniklaus, U. (2011). Epigenetic regulation and reprogramming during gamete formation in plants. *Curr. Opin. Genet. Dev* 21, 124–133. doi: 10.1016/j.gde.2011.01.017

Berger, F., and Twell, D. (2011). Germline specification and function in plants. *Annu. Rev. Plant Biol.* 62, 461–484. doi: 10.1146/annurev-arplant-042110-103824

Berr, A., Mccallum, E. J., Menard, R., Meyer, D., Fuchs, J., Dong, A., et al. (2010). *Arabidopsis* SET DOMAIN GROUP2 is required for H3K4 trimethylation and is crucial for both sporophyte and gametophyte development. *Plant Cell* 22, 3232–3248. doi: 10.1105/tpc.110.079962

Borde, V., Robine, N., Lin, W., Bonfils, S., Geli, V., and Nicolas, A. (2009). Histone H3 lysine 4 trimethylation marks meiotic recombination initiation sites. *EMBO J.* 28, 99–111. doi: 10.1038/emboj.2008.257

Borges, F., Gomes, G., Gardner, R., Moreno, N., Mccormick, S., Feijo, J. A., et al. (2008). Comparative transcriptomics of *Arabidopsis* sperm cells. *Plant Physiol.* 148, 1168–1181. doi: 10.1104/pp.108.125229

Bryant, J. M., Govin, J., Zhang, L., Donahue, G., Pugh, B. F., and Berger, S. L. (2012). The linker histone plays a dual role during gametogenesis in *Saccharomyces cerevisiae*. *Mol. Cell. Biol.* 32, 2771–2783. doi: 10.1128/MCB.00282-12

Calarco, J. P., Borges, F., Donoghue, M. T., Van Ex, F., Jullien, P. E., Lopes, T., et al. (2012). Reprogramming of DNA methylation in pollen guides epigenetic inheritance via small RNA. *Cell* 151, 194–205. doi: 10.1016/j.cell.2012.09.001

Chan, S. W., Henderson, I. R., and Jacobsen, S. E. (2005). Gardening the genome: DNA methylation in *Arabidopsis thaliana*. *Nat. Rev. Genet.* 6, 351–360. doi: 10.1038/nrg1601

Choi, Y., Gehring, M., Johnson, L., Hannon, M., Harada, J. J., Goldberg, R. B., et al. (2002). DEMETER, a DNA glycosylase domain protein, is required for endosperm gene imprinting and seed viability in *Arabidopsis*. *Cell* 110, 33–42. doi: 10.1016/S0092-8674(02)00807-3

Christophorou, M. A., Castelo-Branco, G., Halley-Stott, R. P., Oliveira, C. S., Loos, R., Radzisheuskaya, A., et al. (2014). Citrullination regulates pluripotency and histone H1 binding to chromatin. *Nature* 507, 104–108. doi: 10.1038/nature12942

Contreras, A., Hale, T. K., Stenoien, D. L., Rosen, J. M., Mancini, M. A., and Herrera, R. E. (2003). The dynamic mobility of histone H1 Is regulated by cyclin/CDK phosphorylation. *Mol. Cell. Biol.* 23, 8626–8636. doi: 10.1128/mcb.23.23.8626-8636.2003

Cooper, D. C. (1937). Macrosporogenesis and embryo-sac development in euchlaena mexicana and zea mays. *J. Agric. Res.* 55, 539–551.

Costa, L. M., Marshall, E., Tesfaye, M., Silverstein, K. A., Mori, M., Umetsu, Y., et al. (2014). Central cell-derived peptides regulate early embryo patterning in flowering plants. *Science* 344, 168–172. doi: 10.1126/science.1243005

Deal, R. B., and Henikoff, S. (2010). A simple method for gene expression and chromatin profiling of individual cell types within a tissue. *Dev. Cell* 18, 1030–1040. doi: 10.1016/j.devcel.2010.05.013

Dickinson, H., and Scholten, S. (2013). And baby makes three: genomic imprinting in plant embryos. *PLoS Genet.* 9:e1003981. doi: 10.1371/journal.pgen.1003981

Drews, G. N., and Koltunow, A. M. (2011). The female gametophyte. *Arabidopsis Book* 9:e0155. doi: 10.1199/tab.0155

Feng, X., Zilberman, D., and Dickinson, H. (2013). A conversation across generations: soma-germ cell crosstalk in plants. *Dev. Cell* 24, 215–225. doi: 10.1016/j.devcel.2013.01.014

Fransz, P., and de Jong J. H. (2002). Chromatin dynamics in plants. *Curr. Opin. Plant Biol.* 5, 560–567. doi: 10.1016/S1369-5266(02)00298-4

Fransz, P., Ten Hoopen, R., and Tessadori, F. (2006). Composition and formation of heterochromatin in *Arabidopsis thaliana*. *Chromosome Res.* 14, 71–82. doi: 10.1007/s10577-005-1022-5

Gehring, M. (2013). Genomic imprinting: insights from plants. *Annu. Rev. Genet.* 47, 187–208. doi: 10.1146/annurev-genet-110711-155527

Gehring, M., Bubb, K. L., and Henikoff, S. (2009). Extensive demethylation of repetitive elements during seed development underlies gene imprinting. *Science* 324, 1447–1451. doi: 10.1126/science.1171609

Grant-Downton, R., Kourmpetli, S., Hafidh, S., Khatab, H., Le Trionnaire, G., Dickinson, H., et al. (2013). Artificial microRNAs reveal cell-specific differences in small RNA activity in pollen. *Curr. Biol.* 23, R599–R601. doi: 10.1016/j.cub.2013.05.055

Greer, E., Martin, A. C., Pendle, A., Colas, I., Jones, A. M., Moore, G., et al. (2012). The Ph1 locus suppresses Cdk2-type activity during premeiosis and meiosis in wheat. *Plant Cell* 24, 152–162. doi: 10.1105/tpc.111.094771

Grob, S., Schmid, M. W., Luedtke, N. W., Wicker, T., and Grossniklaus, U. (2013). Characterization of chromosomal architecture in *Arabidopsis* by chromosome conformation capture. *Genome Res.* 14:R129.

Gutierrez-Marcos, J. F., Costa, L. M., Dal Pra, M., Scholten, S., Kranz, E., Perez, P., et al. (2006). Epigenetic asymmetry of imprinted genes in plant gametes. *Nat. Genet.* 38, 876–878. doi: 10.1038/ng1828

Hackett, J. A., Zylicz, J. J., and Surani, M. A. (2012). Parallel mechanisms of epigenetic reprogramming in the germline. *Trends Genet.* 28, 164–174. doi: 10.1016/j.tig.2012.01.005

Hajkova, P. (2011). Epigenetic reprogramming in the germline: towards the ground state of the epigenome. *Philos. Trans. R. Soc. Lond. B Biol. Sci.* 366, 2266–2273. doi: 10.1098/rstb.2011.0042

Hajkova, P., Ancelin, K., Waldmann, T., Lacoste, N., Lange, U. C., Cesari, F., et al. (2008). Chromatin dynamics during epigenetic reprogramming in the mouse germ line. *Nature* 452, 877–881. doi: 10.1038/nature06714

Hajkova, P., Erhardtb, S., Lanec, N., Haafd, T., El-Maarrie, O., Reikc, W., et al. (2002). Epigenetic reprogramming in mouse primordial germ cells. *Mech. Dev.* 117, 15–23. doi: 10.1016/S0925-4773(02)00181-8

Hoffmann, R. D., and Palmgren, M. G. (2013). Epigenetic repression of male gametophyte-specific genes in the *Arabidopsis* sporophyte. *Mol. Plant* 6, 1176–1186. doi: 10.1093/mp/sst100

Houben, A., Kumke, K., Nagaki, K., and Hause, G. (2011). CENH3 distribution and differential chromatin modifications during pollen development in rye (*Secale cereale L.*). *Chromosome Res.* 19, 471–480. doi: 10.1007/s10577-011-9207-6

Hsieh, T. F., Ibarra, C. A., Silva, P., Zemach, A., Eshed-Williams, L., Fischer, R. L., et al. (2009). Genome-wide demethylation of *Arabidopsis* endosperm. *Science* 324, 1451–1454. doi: 10.1126/science.1172417

Ibarra, C. A., Feng, X., Schoft, V. K., Hsieh, T. F., Uzawa, R., Rodrigues, J. A., et al. (2012). Active DNA demethylation in plant companion cells reinforces transposon methylation in gametes. *Science* 337, 1360–1364. doi: 10.1126/science.1224839

Ingouf, M., Hamamura, Y., Gourgues, M., Higashiyama, T., and Berger, F. (2007). Distinct dynamics of HISTONE3 variants between the two fertilization products in plants. *Curr. Biol.* 17, 1032–1037. doi: 10.1016/j.cub.2007.05.019

Ingouf, M., Rademacher, S., Holec, S., Soljic, L., Xin, N., Readshaw, A., et al. (2010). Zygotic resetting of the HISTONE 3 variant repertoire participates in epigenetic reprogramming in *Arabidopsis*. *Curr. Biol.* 20, 2137–2143. doi: 10.1016/j.cub.2010.11.012

Ito, H., Gaubert, H., Bucher, E., Mirouze, M., Vaillant, I., and Paszkowski, J. (2011). An siRNA pathway prevents transgenerational retrotransposition in plants subjected to stress. *Nature* 472, 115–119. doi: 10.1038/nature09861

Jahnke, S., and Scholten, S. (2009). Epigenetic resetting of a gene imprinted in plant embryos. *Curr. Biol.* 19, 1677–1681. doi: 10.1016/j.cub.2009.08.053

Jenuwein, T., and Allis, C. D. (2001). Translating the histone code. *Science* 293, 1074–1080. doi: 10.1126/science.1063127

Jullien, P. E., and Berger, F. (2010). DNA methylation reprogramming during plant sexual reproduction? *Trends Genet.* 26, 394–399. doi: 10.1016/j.tig.2010.06.001

Jullien, P. E., Mosquna, A., Ingouff, M., Sakata, T., Ohad, N., and Berger, F. (2008). Retinoblastoma and its binding partner MSI1 control imprinting in *Arabidopsis*. *PLoS Biol.* 6:e194. doi: 10.1371/journal.pbio.0060194

Jullien, P. E., Susaki, D., Yelagandula, R., Higashiyama, T., and Berger, F. (2012). DNA methylation dynamics during sexual reproduction in *Arabidopsis thaliana*. *Curr. Biol.* 22, 1825–1830. doi: 10.1016/j.cub.2012.07.061

Kniewel, R., and Keeney, S. (2009). Histone methylation sets the stage for meiotic DNA breaks. *EMBO J.* 28, 81–83. doi: 10.1038/emboj.2008.277

Köhler, C., Wolff, P., and Spillane, C. (2012). Epigenetic mechanisms underlying genomic imprinting in plants. *Annu. Rev. Plant Biol.* 63, 331–352. doi: 10.1146/annurev-arplant-042811-105514

Kumar, S. V., and Wigge, P. A. (2010). H2A.Z-containing nucleosomes mediate the thermosensory response in *Arabidopsis*. *Cell* 140, 136–147. doi: 10.1016/j.cell.2009.11.006

Law, J. A., and Jacobsen, S. E. (2010). Establishing, maintaining and modifying DNA methylation patterns in plants and animals. *Nat. Rev. Genet.* 11, 204–220. doi: 10.1038/nrg2719

Lu, F., Cui, X., Zhang, S., Jenuwein, T., and Cao, X. (2011). *Arabidopsis* REF6 is a histone H3 lysine 27 demethylase. *Nat. Genet.* 43, 715–719. doi: 10.1038/ng.854

Maheshwari, P. (1950). *An Introduction to the Embryology of Angiosperms.* New York: McGraw-Hill.

Mansour, A. A., Gafni, O., Weinberger, L., Zviran, A., Ayyash, M., Rais, Y., et al. (2012). The H3K27 demethylase Utx regulates somatic and germ cell epigenetic reprogramming. *Nature* 488, 409–413. doi: 10.1038/nature11272

Martinho, R. G., Kunwar, P. S., Casanova, J., and Lehmann, R. (2004). A noncoding RNA is required for the repression of RNApolII-dependent transcription in primordial germ cells. *Curr. Biol.* 14, 159–165. doi: 10.1016/j.cub.2003.12.036

Melamed-Bessudo, C., and Levy, A. A. (2012). Deficiency in DNA methylation increases meiotic crossover rates in euchromatic but not in heterochromatic regions in *Arabidopsis*. *Proc. Natl. Acad. Sci. U.S.A.* 109, E981–E988. doi: 10.1073/pnas.1120742109

Mirouze, M., Lieberman-Lazarovich, M., Aversano, R., Bucher, E., Nicolet, J., Reinders, J., et al. (2012). Loss of DNA methylation affects the recombination landscape in *Arabidopsis*. *Proc. Natl. Acad. Sci. U.S.A.* 109, 5880–5889. doi: 10.1073/pnas.1120841109

Morales-Ruiz, T., Ortega-Galisteo, A. P., Ponferrada-Marin, M. I., Martinez-Macias, M. I., Ariza, R. R., and Roldan-Arjona, T. (2006). DEMETER and REPRESSOR OF SILENCING 1 encode 5-methylcytosine DNA glycosylases. *Proc. Natl. Acad. Sci. U.S.A.* 103, 6853–6858. doi: 10.1073/pnas.0601109103

Mosher, R. A., Melnyk, C. W., Kelly, K. A., Dunn, R. M., Studholme, D. J., and Baulcombe, D. C. (2009). Uniparental expression of PolIV-dependent siRNAs in developing endosperm of *Arabidopsis*. *Nature* 460, 283–286. doi: 10.1038/nature08084

Nakamura, A., Shirae-Kurabayashi, M., and Hanyu-Nakamura, K. (2010). Repression of early zygotic transcription in the germline. *Curr. Opin. Cell Biol.* 22, 709–714. doi: 10.1016/j.ceb.2010.08.012

Nonomura, K., Morohoshi, A., Nakano, M., Eiguchi, M., Miyao, A., Hirochika, H., et al. (2007). A germ cell specific gene of the ARGONAUTE family is essential for the progression of premeiotic mitosis and meiosis during sporogenesis in rice. *Plant Cell* 19, 2583–2594. doi: 10.1105/tpc.107.053199

Ohno, R., Nakayama, M., Naruse, C., Okashita, N., Takano, O., Tachibana, M., et al. (2013). A replication-dependent passive mechanism modulates DNA demethylation in mouse primordial germ cells. *Development* 140, 2892–2903. doi: 10.1242/dev.093229

Olmedo-Monfil, V., Durán-Figueroa, N., Arteaga-Vázquez, M., Demesa-Arévalo, E., Autran, D., Grimanelli, D., et al. (2010). Control of female gamete formation by a small RNA pathway in *Arabidopsis*. *Nature* 464, 628–632. doi: 10.1038/nature08828

Pecinka, A., Schubert, V., Meister, A., Kreth, G., Klatte, M., Lysak, M. A., et al. (2004). Chromosome territory arrangement and homologous pairing in nuclei of *Arabidopsis thaliana* are predominantly random except for NOR-bearing chromosomes. *Chromosoma* 113, 258–269. doi: 10.1007/s00412-004-0316-2

Pillot, M., Baroux, C., Vazquez, M. A., Autran, D., Leblanc, O., Vielle-Calzada, J. P., et al. (2010). Embryo and endosperm inherit distinct chromatin and transcriptional states from the female gametes in *Arabidopsis*. *Plant Cell* 22, 307–320. doi: 10.1105/tpc.109.071647

Probst, A. V., Fransz, P. F., Paszkowski, J., and Scheid, O. M. (2003). Two means of transcriptional reactivation within heterochromatin. *Plant J.* 33, 743–749. doi: 10.1046/j.1365-313X.2003.01667.x

Qin, Y., Zhao, L., Skaggs, M. I., Andreuzza, S., Tsukamoto, T., Panoli, A., et al. (2014). ACTIN-RELATED PROTEIN6 regulates female meiosis by modulating meiotic gene expression in *Arabidopsis*. *Plant Cell* 26, 1612–1628. doi: 10.1105/tpc.113.120576

Rabiger, D. S., and Drews, G. N. (2013). MYB64 and MYB119 are required for cellularization and differentiation during female gametogenesis in *Arabidopsis thaliana*. *PLoS Genet.* 9:e1003783. doi: 10.1371/journal.pgen.1003783

Raissig, M. T., Baroux, C., and Grossniklaus, U. (2011). Regulation and flexibility of genomic imprinting during seed development. *Plant Cell* 23, 16–26. doi: 10.1105/tpc.110.081018

Raissig, M. T., Bemer, M., Baroux, C., and Grossniklaus, U. (2013a). Genomic imprinting in the *Arabidopsis* embryo is partly regulated by PRC2. *PLoS Genet.* 9:e1003862. doi: 10.1371/journal.pgen.1003862

Raissig, M. T., Gagliardini, V., Jaenisch, J., Grossniklaus, U., and Baroux, C. (2013b). Efficient and rapid isolation of early-stage embryos from *Arabidopsis thaliana* seeds. *J. Vis. Exp.* 7:76. doi: 10.3791/50371

Ravi, M., Shibata, F., Ramahi, J. S., Nagaki, K., Chen, C., Murata, M., et al. (2011). Meiosis-specific loading of the centromere-specific histone CENH3 in *Arabidopsis thaliana*. *PLoS Genet.* 7:e1002121. doi: 10.1371/journal.pgen.1002121

Rea, M., Zheng, W., Chen, M., Braud, C., Bhangu, D., Rognan, T. N., et al. (2012). Histone H1 affects gene imprinting and DNA methylation in *Arabidopsis*. *Plant J.* 71, 776–786. doi: 10.1111/j.1365-313X.2012.05028.x

Reinders, J., Mirouze, M., Nicolet, J., and Paszkowski, J. (2013). Parent-of-origin control of transgenerational retrotransposon proliferation in *Arabidopsis*. *EMBO Rep.* 14, 823–828. doi: 10.1038/embor.2013.95

Roudier, F., Ahmed, I., Berard, C., Sarazin, A., Mary-Huard, T., Cortijo, S., et al. (2011). Integrative epigenomic mapping defines four main chromatin states in *Arabidopsis*. *EMBO J.* 30, 1928–1938. doi: 10.1038/emboj.2011.103

Russell, S. D., Gou, X., Wong, C. E., Wang, X., Yuan, T., Wei, X., et al. (2012). Genomic profiling of rice sperm cell transcripts reveals conserved and distinct elements in the flowering plant male germ lineage. *New Phytol.* 195, 560–573. doi: 10.1111/j.1469-8137.2012.04199.x

Sasaki, Y., Yasuda, H., Ohba, Y., and Harada, H. (1990). Isolation and characterization of a novel nuclear protein from pollen mother cells of lily. *Plant Physiol.* 94, 1467–1471. doi: 10.1104/pp.94.3.1467

Schmidt, A., Schmid, M. W., and Grossniklaus, U. (2012). Analysis of plant germline development by high-throughput RNA profiling: technical advances and new insights. *Plant J.* 70, 18–29. doi: 10.1111/j.1365-313X.2012.04897.x

Schmidt, A., Wuest, S. E., Vijverberg, K., Baroux, C., Kleen, D., and Grossniklaus, U. (2011). Transcriptome analysis of the *Arabidopsis* megaspore mother cell uncovers the importance of RNA helicases for plant germline development. *PLoS Biol.* 9:e1001155. doi: 10.1371/journal.pbio.1001155

Schoft, V. K., Chumak, N., Mosiolek, M., Slusarz, L., Komnenovic, V., Brownfield, L., et al. (2009). Induction of RNA-directed DNA methylation upon decondensation of constitutive heterochromatin. *EMBO Rep.* 10, 1015–1021. doi: 10.1038/embor.2009.152

Schubert, V., Lermontova, I., and Schubert, I. (2014). Loading of the centromeric histone H3 variant during meiosis-how does it differ from mitosis? *Chromosoma* doi: 10.1007/s00412-014-0466-9 [Epub ahead of print].

Schulz, P., and Jensen, W. A. (1981). Pre-fertilization in *Capsella*: ultrastructure and ultrachemical localization of acid phosphatase in female meiocytes. *Protoplasma* 107, 27–45. doi: 10.1007/BF01275605

Seisenberger, S., Peat, J. R., Hore, T. A., Santos, F., Dean, W., and Reik, W. (2013). Reprogramming DNA methylation in the mammalian life cycle: building and breaking epigenetic barriers. *Philos. Trans. R. Soc. Lond. B Biol. Sci.* 368:20110330. doi: 10.1098/rstb.2011.0330

Seydoux, G., and Braun, R. E. (2006). Pathway to totipotency: lessons from germ cells. *Cell* 127, 891–904. doi: 10.1016/j.cell.2006.11.016

She, W., Grimanelli, D., and Baroux, C. (2014). An efficient method for quantitative, single-cell analysis of chromatin modification and nuclear architecture in whole-mount ovules in *Arabidopsis*. *J. Vis. Exp.* 88, e51530. doi: 10.3791/51530

She, W., Grimanelli, D., Rutowicz, K., Whitehead, M. W., Puzio, M., Kotlinski, M., et al. (2013). Chromatin reprogramming during the somatic-to-reproductive cell fate transition in plants. *Development* 140, 4008–4019. doi: 10.1242/dev.095034

Singh, M., Goel, S., Meeley, R. B., Dantec, C., Parrinello, H., Michaud, C., et al. (2011). Production of viable gametes without meiosis in maize deficient for an ARGONAUTE protein. *Plant Cell* 23, 443–458. doi: 10.1105/tpc.110.079020

Slotkin, R. K., Vaughn, M., Borges, F., Tanurdzic, M., Becker, J. D., Feijo, J. A., et al. (2009). Epigenetic reprogramming and small RNA silencing of transposable elements in pollen. *Cell* 136, 461–472. doi: 10.1016/j.cell.2008.12.038

Sniezko, R. (2006). "Meiosis in plants," in *Plant Cell Biology*, eds W. V. Dashek, P. Harrison (New Hampshire: Science Publisher), 227–258.

Sprunck, S., and Gross-Hardt, R. (2011). Nuclear behavior, cell polarity, and cell specification in the female gametophyte. *Sex. Plant Reprod.* 24, 123–136. doi: 10.1007/s00497-011-0161-4

Stroud, H., Otero, S., Desvoyes, B., Ramírez-Parra, E., Jacobsen, S. E., and Gutierrez, C. (2012). Genome-wide analysis of histone H3.1 and H3.3 variants in *Arabidopsis thaliana*. *Proc. Natl. Acad. Sci. U.S.A.* doi: 10.1073/pnas.1203145109

Surani, M. A., Hayashi, K., and Hajkova, P. (2007). Genetic and epigenetic regulators of pluripotency. *Cell* 128, 747–762. doi: 10.1016/j.cell.2007.02.010

Tachibana, M., Nozaki, M., Takeda, N., and Shinkai, Y. (2007). Functional dynamics of H3K9 methylation during meiotic prophase progression. *EMBO J.* 26, 3346–3359. doi: 10.1038/sj.emboj.7601767

Tanaka, I., Ono, K., and Fukuda, T. (1998). The developmental fate of angiosperm pollen is associated with a preferential decrease in the level of histone H1 in the vegetative nucleus. *Planta* 206, 561–569. doi: 10.1007/s004250050433

Tariq, M., and Paszkowski, J. (2004). DNA and histone methylation in plants. *Trends Genet.* 20, 244–251. doi: 10.1016/j.tig.2004.04.005

Ueda, K., Kinoshita, Y., Xu, Z., Ide, N., Ono, M., Akahori, Y., et al. (2000). Unusual core histones specifically expressed in male gametic cells of *Lilium longiflorum*. *Chromosoma* 108, 491–500. doi: 10.1007/s004120050401

van Zanten, M., Koini, M. A., Geyer, R., Liu, Y., Brambilla, V., Bartels, D., et al. (2011). Seed maturation in *Arabidopsis thaliana* is characterized by nuclear size reduction and increased chromatin condensation. *Proc. Natl. Acad. Sci. U.S.A.* 108, 20219–20224. doi: 10.1073/pnas.1117726108

Vaucheret, H. (2008). Plant ARGONAUTES. *Trends Plant Sci.* 13, 350–358. doi: 10.1016/j.tplants.2008.04.007

Wendrich, J. R., and Weijers, D. (2013). The *Arabidopsis* embryo as a miniature morphogenesis model. *New Phytol.* 199, 14–25. doi: 10.1111/nph.12267

Wierzbicki, A. T., and Jerzmanowski, A. (2005). Suppression of histone H1 genes in *Arabidopsis* results in heritable developmental defects and stochastic changes in DNA methylation. *Genetics* 169, 997–1008. doi: 10.1534/genetics.104.031997

Wohrmann, H. J., Gagliardini, V., Raissig, M. T., Wehrle, W., Arand, J., Schmidt, A., et al. (2012). Identification of a DNA methylation-independent imprinting control region at the *Arabidopsis* MEDEA locus. *Genes Dev.* 26, 1837–1850. doi: 10.1101/gad.195123.112

Wuest, S. E., Schmid, M. W., and Grossniklaus, U. (2013). Cell-specific expression profiling of rare cell types as exemplified by its impact on our understanding of female gametophyte development. *Curr. Opin. Plant Biol.* 16, 41–49. doi: 10.1016/j.pbi.2012.12.001

Wuest, S. E., Vijverberg, K., Schmidt, A., Weiss, M., Gheyselinck, J., Lohr, M., et al. (2010). *Arabidopsis* female gametophyte gene expression map reveals similarities between plant and animal gametes. *Curr. Biol.* 20, 506–512. doi: 10.1016/j.cub.2010.01.051

Xu, C., Tian, J., and Mo, B. (2013). siRNA-mediated DNA methylation and H3K9 dimethylation in plants. *Protein Cell* doi: 10.1007/s13238-013-3052-7 [Epub ahead of print].

Xu, H., Swoboda, I., Bhalla, P., and Singh, M. B. (1999). Male gametic cell-specific expression of H2A and H3 histone genes. *Plant Mol. Biol.* 39, 607–614. doi: 10.1023/A:1006162120037

Yamaji, M., Seki, Y., Kurimoto, K., Yabuta, Y., Yuasa, M., Shigeta, M., et al. (2008). Critical function of Prdm14 for the establishment of the germ cell lineage in mice. *Nat. Genet.* 40, 1016–1022. doi: 10.1038/ng.186

Yang, H., Lu, P., Wang, Y., and Ma, H. (2011). The transcriptome landscape of *Arabidopsis* male meiocytes from high-throughput sequencing: the complexity and evolution of the meiotic process. *Plant J.* 65, 503–516. doi: 10.1111/j.1365-313X.2010.04439.x

Yoshida, S., Barbier De Reuille, P., Lane, B., Bassel, G. W., Prusinkiewicz, P., Smith, R. S., et al. (2014). Genetic control of plant development by overriding a geometric division rule. *Dev. Cell* 29, 75–87. doi: 10.1016/j.devcel.2014.02.002

Zemach, A., Kim, M. Y., Hsieh, P.-H., Coleman-Derr, D., Eshed-Williams, L., Thao, K., et al. (2013). The *Arabidopsis* nucleosome remodeler DDM1 allows DNA methyltransferases to access H1-containing heterochromatin. *Cell* 153, 193–205. doi: 10.1016/j.cell.2013.02.033

Zemach, A., Kim, M. Y., Silva, P., Rodrigues, J. A., Dotson, B., Brooks, M. D., et al. (2010). Local DNA hypomethylation activates genes in rice endosperm. *Proc. Natl. Acad. Sci. U.S.A.* 107, 18729–18734. doi: 10.1073/pnas.1009695107

Zhang, D., Luo, X., and Zhu, L. (2011). Cytological analysis and genetic control of rice anther development. *J. Genet. Genomics* 38, 379–390. doi: 10.1016/j.jgg.2011.08.001

25

Thiol-based redox regulation in sexual plant reproduction: new insights and perspectives

Jose A. Traverso[1], Amada Pulido[2], María I. Rodríguez-García[1] and Juan D. Alché[1]*

[1] *Estación Experimental del Zaidín, Consejo Superior de Investigaciones Científicas, Granada, Spain*
[2] *Departamento de Fisiología Vegetal, Universidad de Granada, Granada, Spain*

Edited by:
Jean-Philippe Reichheld, Centre National de la Recherche Scientifique, France

Reviewed by:
John Hancock, University of the West of England, UK
María de la Cruz G. García, Universidad de Sevilla, Spain

*Correspondence:
Juan D. Alché, Plant Reproductive Biology Group, Department of Biochemistry, Cell and Molecular Biology of Plants, Estación Experimental del Zaidín, Spanish Council for Scientific Research, Profesor Albareda 1, 18008 Granada, Spain
e-mail: juandedios.alche@ eez.csic.es

The success of sexual reproduction in plants involves (i) the proper formation of the plant gametophytes (pollen and embryo sac) containing the gametes, (ii) the accomplishment of specific interactions between pollen grains and the stigma, which subsequently lead to (iii) the fusion of the gametes and eventually to (iv) the seed setting. Owing to the lack of mobility, plants have developed specific regulatory mechanisms to control all developmental events underlying the sexual plant reproduction according to environmental challenges. Over the last decade, redox regulation and signaling have come into sight as crucial mechanisms able to manage critical stages during sexual plant reproduction. This regulation involves a complex redox network which includes reactive oxygen species (ROS), reactive nitrogen species (RNS), glutathione and other classic buffer molecules or antioxidant proteins, and some thiol/disulphide-containing proteins belonging to the thioredoxin superfamily, like glutaredoxins (GRXs) or thioredoxins (TRXs). These proteins participate as critical elements not only in the switch between the mitotic to the meiotic cycle but also at further developmental stages of microsporogenesis. They are also implicated in the regulation of pollen rejection as the result of self-incompatibility. In addition, they display precise space-temporal patterns of expression and are present in specific localizations like the stigmatic papillae or the mature pollen, although their functions and subcellular localizations are not clear yet. In this review we summarize insights and perspectives about the presence of thiol/disulphide-containing proteins in plant reproduction, taking into account the general context of the cell redox network.

Keywords: redox regulation, sexual plant reproduction, thioredoxin, glutaredoxin, pollen, pistil, pollen-pistil interaction, stigma

INTRODUCTION

Sexual Plant Reproduction involves complex biochemical pathways under a tight genetic control, which lead to drastic architectural changes at developmental and cellular levels (Franklin-Tong, 1999). These changes begin with the generation of the haploid gametes in specialized structures know as the mega- and microgametophytes: the embryo sac in the pistils and the pollen grain in the anthers. Afterwards, mature pollen grains land over a receptive stigma, hydrate and germinate, emerging a pollen tube, which enlarges at its apical region at exceptionally high growing rates. This pollen tube penetrates throughout the style toward the embryo sac in order to deliver the male gametes, which fertilize the egg cell and the polar nuclei generating an embryo and the endosperm, respectively, and ultimately the offspring (Sprunck et al., 2013).

Plants, like other eukaryotes, have evolved dedicating enormous resources and efforts to guarantee sexual reproduction. Among others, they have developed molecular mechanisms, which allow a tight regulation of all developmental events underlying the process. These mechanisms not only have contributed to assure the emergence of genetically-improved progenies, but also allowing plants to tune this process according to challenging environmental conditions, which has let flowering plants evolve dominating almost every terrestrial ecosystem (Hiscock, 2011).

Redox regulation has recently emerged as a crucial mechanism able to manage significant stages during sexual plant reproduction where ROS, nitric oxide (NO) and other classical antioxidant protein and molecules are critically involved (Feijó et al., 2004; Prado et al., 2004; Dresselhaus and Franklin-Tong, 2013). In addition, several isoforms of plant redoxins like TRXs or GRXs seem to be specifically associated with reproductive tissues according to their precise space-temporal expression patterns (**Table 1**), although no clear functions have usually been assigned in this context.

TRXs and GRXs are redox proteins whose activity depends on conserved cysteine (Cys) residues present in their active centers (Meyer et al., 2012). These Cys residues are usually maintained in their reduced state within the cell (Foyer and Noctor, 2005a). TRXs and GRXs carry out oxide-reductive reactions on essential Cys residues of a variety of plant proteins. The *Arabidopsis* genome contains about 40 genes encoding TRXs or TRX-related proteins grouped in different clusters and subclusters according to several aspects like sequence similarity, subcellular localization or intron position, which have been described as putatively involved in a plethora of plant roles (Buchanan and Balmer, 2005; Meyer et al., 2012). GRXs share important similarities with the TRX family like their protein structures (both protein types belong to the TRX superfamily) and the fact that higher plants also possess

Table 1 | Described redoxins and other related proteins involved in sexual plant reproduction.

	Protein / Gene	Accession number	Organism	Localization associated with sexual plant reproduction	Potential role	Reference(s)
TRXs	AtTRXh1	At3g51030	*A. thaliana*	Gene expressed in style Protein detected in pollen tube	Unknown	Reichheld et al., 2002; Ge et al., 2011
	AtTRXh4	At1g19730	*A. thaliana*	Gene expressed in pollen and pollen tube	Unknown	Reichheld et al., 2002
	AtTRXh5	At1g45145	*A. thaliana*	Protein detected in pollen tube	Unknown	Ge et al., 2011
	PsTRXh1	AJ310990	*P. sativum*	Gene expression and protein localized in pollen and stigma	Unknown	Traverso et al., 2007
	TRX h (subgr. 1)	CI000057:1 (Contig)	*C. sativus* (Saffron)	Gene highly expressed in stigma	Unknown	D'Agostino et al., 2007
	THL-1	AF273844	*B. rappa*	Protein in pollen coat and stigma	Unknown (pollen) and Self-Incompatibility response (stigma)	Toriyama et al., 1998; Cabrillac et al., 2001; Ivanov and Gaude, 2009a
	TRX h (subgr. 3)	AF159388	mono and dicot	Conserved genes expressed in pollen	Unknown	Juttner et al., 2000
	NaTRXh	AAY42864	*N. alata*	Protein secreted into the extracellular matrix (stylar transmitting tract)	Self-Incompatibility?	Juárez-Díaz et al., 2006
	Protein S	X81992	*P. coerulescens*	Gene expressed in mature pollen	Self-Incompatibility (grass model)	Li et al., 1997
	ACHT3	At2g33270	*A. thaliana*	Gene highly expressed in pollen	Unknown	Becker et al., 2003; Lee and Lee, 2003
	PsTRXf and PsTRXm	X63537 and X76269	*P. sativum*	Gene expressed in pollen grains, anthers, style and ovules	Unknown; Photosynthesis regulation in style?	de Dios Barajas-López et al., 2007
	AtNTRC	At2g41680	*A. thaliana*	Gene expressed in style	Unknown	Kirchsteiger et al., 2012
GRXs	CC-type GRX	GRX; patent US2009/0038028A1	Maize	Gene expressed in anthers	Anther development	Chaubal et al., 2003; Timofejeva et al., 2013
	CC-type GRX (ROXY1)	At3g02000	*A. thaliana*	Gene expressed in anthers	Petal and anther development	Xing et al., 2005
	CC-type GRX (ROXY2)	At5g14070	*A. thaliana*	Gene expressed in anthers	Anther development (gametogenesis?)	Xing and Zachgo, 2008
	CC-type GRX	LOC_Os07g05630	rice	Gene and protein expression in anthers	Male gametogenesis and anther development	Hong et al., 2012
	AtGrxC2	At5g40370	*A. thaliana*	Extracellular protein secreted from germinated pollen tube	Unknown	Ge et al., 2011
	GRX	G4XH75_9POAL	Triticale (*Triticum* spp x *Secale* spp)	Extracellular protein released from pollen coat upon pollen hydration	Unknown	Zaidi et al., 2012
OTHERS	CBSX1	At4g36910	*A. thaliana*	Gene and protein expressed in anthers	Anther dehiscence	Yoo et al., 2011
	CP12-1	At2g47400	*A. thaliana*	Gene expressed in style	Unknown; Photosynthesis regulation in style?	Singh et al., 2008
	CP12-2	At3g62410	*A. thaliana*	Gene expressed in style	Unknown; Photosynthesis regulation in style?	Singh et al., 2008
	PRX (TPx2)	At1g70950	*A. thaliana*	Gene highly expressed in pollen	Unknown	Lee and Lee, 2003

(Continued)

Table 1 | Continued

Protein / Gene	Accession number	Organism	Localization associated with sexual plant reproduction	Potential role	Reference(s)
AtPRXII-C	At1g65970	*A. thaliana*	Gene expressed in mature pollen	Unknown	Bréhélin et al., 2003
AtPRXII-D	At1g60740	*A. thaliana*	Gene expressed in pollen and pollen tube	Unknown	Bréhélin et al., 2003
AtPRDII-E	At3g52960	*A. thaliana*	Gene expressed in immature anthers and ovules	Unknown	Bréhélin et al., 2003

a larger number of GRX genes per genome when compared to other organisms (about 50 genes encoding for GRXs or GRX-related proteins in *Arabidopsis*) (Couturier et al., 2009; Meyer et al., 2012). The physiological roles of plant GRXs are less-known than those of TRXs, although they have also been involved in a variety of functions (Meyer et al., 2012).

Oxidized TRXs and GRXs are physiologically reduced by dedicated systems according to their subcellular localizations (for an extended review see Meyer et al., 2012). Cytosolic and mitochondrial TRXs are mainly reduced by NADPH in a reaction catalyzed by the enzyme NADPH-TRX-Reductase (NTR) (Arner and Holmgren, 2000; Laloi et al., 2001), while the isoforms from plastids are reduced by ferredoxin via the enzyme Ferredoxin-TRX-Reductase (FTR) (Shürmann and Jacquot, 2000; Balmer et al., 2006) and the GRXs are generally described as reduced by reduced glutathione (GSH). However, alternative crosstalk between both TRX and GRX systems has also been demonstrated in addition to these classical schemes, which reveals a high plasticity of the thiol-based redox regulation in plants (Gelhaye et al., 2003; Balmer et al., 2006; Reichheld et al., 2007; Bandyopadhyay et al., 2008; Marty et al., 2009).

TRXs and GRXs are described as the main protein families responsible for the redox status of protein Cys residues within the cell. These Cys are particularly susceptible to oxidations by reactive species, this fact being usually identified by researchers as a regulatory mechanism of the protein activity, as well as a protective or redox signaling mechanism (Couturier et al., 2013). These thiol-based regulations have been interpreted as a sensing mechanism of the cellular redox state, which acts between stress perception and plant response against environmental challenges (Foyer and Noctor, 2005a).

In this review we summarize, discuss and hypothesize about the occurrence of these thiol/disulfide containing proteins in reproductive tissues, pointing out an increased importance of the thiol-based redox regulation and signaling mechanisms in sexual plant reproduction.

REDOX REGULATION BY ROS/RNS IN SEXUAL PLANT REPRODUCTION

Reactive species are produced in living cells as an unavoidable consequence of their own metabolism (Foyer and Noctor, 2005a). Apart from their activity damaging different macromolecules, they have been shown to act as secondary messengers, reason why the concept of oxidative stress has been re-evaluated (Foyer and Noctor, 2005b). Taking into account that protein Cys residues are particularly affected by reactive species, we review in this chapter the most important results involving ROS and RNS at different stages of the sexual plant reproduction.

Reactive species have been shown to increase in a rapid and transient manner by specific molecular mechanisms for a proper plant growth and development, including sexual plant reproduction (Foreman et al., 2003; Potocký et al., 2007). ROS and NO have been involved in redox signaling taking place previously and during pollen-pistil interactions (**Figure 1**; Sharma and Bhatla, 2013). A high production of H_2O_2, exclusively confined to receptive stigmatic papillae was suggested to serve as a redox signal promoting pollen-pistil interaction and/or germination, as well as a defense mechanism against microbe attack (**Figure 1A**; McInnis et al., 2006a; Wilson et al., 2009; Zafra et al., 2010). In parallel to ROS production, a Stigma-Specific Peroxidase (SSP) was shown to be active only during a short period of stigma receptivity in *Senecio squalidus* (**Figures 1A,B**; McInnis et al., 2005, 2006b). *In vitro* experiments also revealed that mature pollen grains produce a high level of NO, which inhibits ROS production in the stigmatic papillae (**Figure 1B**; McInnis et al., 2006a; Bright et al., 2009; Zafra et al., 2010).

In addition, physiological mechanisms of ROS generation have been indicated to be required for normal pollen tube development (**Figures 1B,C**; Cardenas et al., 2006; Potocký et al., 2007, 2012). ROS production has been described to be mainly originated by the activity of specific isoforms of NADPH oxidases (NOX) localized at the tip of tobacco pollen tube (**Figure 1B**), whose activity was found to be increased by Ca^{2+} (Potocký et al., 2007). This represents a conserved mechanism of polarized cell tip growth (Bushart and Roux, 2007; Konrad et al., 2011).

In cucumber germinated pollen, ROS and NO production was specifically tip-localized during the initial germination steps, although was extended along the pollen tubes and the pollen grain later during germination (**Figure 1B**) (Šírová et al., 2011).

Both redox chemical agents and external conditions have shown to alter the production of these reactive species (Šírová et al., 2011). The addition of a reducer like ascorbate abolishes this production probably due to its capacity to effectively scavenge the intracellular ROS. The treatment with NO donors inhibits pollen germination and growth, and the addition of NO scavengers increases pollen germination rates (Prado et al., 2004; Šírová et al., 2011). NO seems also to be involved in the inhibition of germination and tube growth after pollen exposure to UV-B (Feng et al., 2000; He et al., 2007; Wang et al., 2010a,b). Curiously, NO exerts just the opposite effect in gymnosperm pollens, since it was shown that Pine pollen tube growth rate is accelerated by NO donors, whereas NO scavengers affect contrary (Wang et al., 2009a,b).

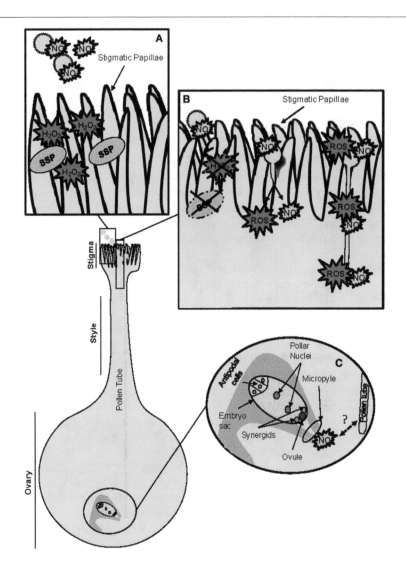

FIGURE 1 | ROS/NO are implicated in plant reproduction signaling.
Model gynoecium of an Angiosperm. **(A)** Receptive stigma at high
magnification, before pollen-pistil interaction. Pollen grains are active NO
producers, while the receptive stigma papillae generate large quantities
of H_2O_2 and display enhanced peroxidase activity, in some cases in
the form of stigma-specific peroxidase (SSP). **(B)** Stigma and pollen
grains during diverse phases of their interaction. Both the enhanced
peroxidase activity and the high levels of H_2O_2 become significantly
reduced after pollen landing on the stigma and pollen germination, likely
through pollen-produced NO signaling. The germinating pollen grains and
the elongating pollen tubes produce ROS and NO, particularly at their
tips. **(C)** The embryo sac is reached by the pollen tube tip trough the
micropylar end. NO produced in micropyle could be involved in pollen
tube guidance.

Although the mechanisms controlling pollen tube guidance and pollen-pistil interaction are still unknown (Boavida
et al., 2005; Higashiyama and Hamamura, 2008; Márton and
Dresselhaus, 2008) there is also evidence about the involvement
of NO as a key molecule at this regard (Prado et al., 2004).
Prado et al. (2008) showed that a Ca^{2+} specific response to NO
induces pollen tube re-direction toward the ovule. Prado's work
also describes the detection of NO production in the micropyle,
where it was suggested to participate in pollen tube guidance to
the ovules (**Figure 1C**; Prado et al., 2008). The generation of reactive oxygen species has also been involved in microsporogenesis,
usually throughout programmed cell death (Jiang et al., 2007;
Wan et al., 2007). In addition, it was recently shown that the male

germ cell fate critically depends on H_2O_2 levels of the precursor cells (Kelliher and Walbot, 2012). Moreover, the molecular
models developed in order to explain self-incompatibility (SI) in
plants, usually include important roles for ROS or NO (McClure
and Franklin-Tong, 2006; McInnis et al., 2006a,b; Wilkins et al.,
2011).

Several of the detailed molecular mechanisms through which
ROS and NO exert these functions are beginning to be outlined,
and some of them involve thiol modifications. At this regard,
Cys residues in proteins are particularly affected by these reactive species, and Cys-based signaling by ROS and/or RNS is a
well-described feature affecting an increasing number of proteins, some of them from plants (Couturier et al., 2013; Corpas

and Barroso, 2013). In bacteria, fungi or mammals, Cys modification by ROS and/or RNS has been described to affect DNA binding properties of some transcription factors (D'Autreaux and Toledano, 2007). In plant cells, the Cys-based actions of RNS (S-nitrosylation) and ROS (oxidation) on NPR1 and TGA1 proteins regulate plant systemic defense (Tada et al., 2008; Lindermayr et al., 2010). Protein S-nitrosylation is produced by the interaction of specific Cys residues with NO generated by different types of RNS, and this modification is emerging as a crucial regulatory mechanism involved in several aspect of plant physiology (Corpas et al., 2013). As an example, the activity of NADPH oxidases can be controlled by the S-nitrosylation of a C-terminal Cys (Yun et al., 2011). NOX proteins are highly involved in pollen tube growth and the subject of further regulation mechanisms (Potocký et al., 2007, 2012).

S-nitrosoglutathione (GSNO), an abundant molecule in plant tissues (Airaki et al., 2011) originated by S-nitrosylation of reduced glutathione (GSH) (Broniowska et al., 2013) is considered a reservoir, vehicle and biological donor of NO in plant cells (Corpas et al., 2013). Protein S-nitrosylation by GSNO (S-transnitrosation) also seems to be a feasible physiological mechanism for post-translational modification of proteins (Begara-Morales et al., 2013), however, not yet sufficiently described in plant reproductive tissues.

According to all these data, further experiments must be carried out to identify key proteins involved in the regulation of sexual plant reproduction via ROS or NO-mediated Cys oxidations. The specific presence of redoxins at these stages (Table 1; Figure 3) together with the importance of these reactive species suggests a more critical thiol-based regulation of several stages of the sexual plant reproduction than initially thought.

SPECIFIC REDOXINS INVOLVED IN ANTHER DEVELOPMENT AND MALE GAMETOGENESIS

Successful sexual reproduction depends on the proper formation of specialized complex structures in the flower: anthers and pistils. Initially, a group of somatic cells must switch from the mitotic to the meiotic pathway to generate the haploid gametes. All processes are developed according to both environmental and developmental signals (Bhatt et al., 2001). Later, anther dehiscence will produce the release of mature pollen grains. Anther development and male gametogenesis processes are known to be critically influenced by the redox activity of specific thiol-based redox proteins. CC-type GRXs and the redox chloroplastidial system including CBSX (single cystathionine β-synthase domain–containing proteins)/TRX/PRX proteins play important roles in redox homeostasis and development in male reproductive tissues (Figure 2) (Wang et al., 2009a,b; Yoo et al., 2011).

CC-type GRXs, (also named ROXY proteins), are conserved plant-specific GRXs involved in anther and male gamete differentiation and flower development (Figure 2A; Xing et al., 2006; Wang et al., 2009a,b). The first evidence about such involvement was described by Chaubal et al. (2003) during the characterization of the maize mutant mscal. In this mutant, all anther cell layers were transformed into non-differentiated vegetative tissues. This phenotype was associated later with the lack of a GRX (Xing et al., 2011), and recently corroborated during the screening of

male sterile lines in maize (Timofejeva et al., 2013). Culture of the mscal mutant under hypoxia conditions (low oxygen / H_2O_2) allows a rescue of the differentiation of the germinal line in the mutant flowers (Kelliher and Walbot, 2012).

However, probably the most important data concerning the role of ROXY proteins in anther development was provided by studies based on A. thaliana. Initially, the redox activity of the GRX ROXY1 was identified as a major regulator of early petal organ initiation and further steps of floral morphogenesis (Xing et al., 2005). Afterwards, the functionally redundant GRXs ROXY 1 and 2 were described to perform essential redox-dependent activities in early steps of anther and tapetum differentiation (Figure 2A; see anther structure in Figure 2B) by affecting the expression of a large variety of anther genes supporting critical roles (Xing and Zachgo, 2008). During anther development, they act via the redox activation of TGA9 and 10 transcription factors, probably among other protein targets (Murmu et al., 2010). Arabidopsis ROXY proteins were also suggested to be involved in male gametogenesis (Xing and Zachgo, 2008). In fact, this involvement has been recently evidenced in monocots (Hong et al., 2012). These authors have shown that the rice MIL1 gene encodes for a CC-type GRX which is not only involved in the differentiation of the surrounding somatic layer of the anthers, but also in the switch of microsporocytes from mitosis to meiosis (Figure 2A). According to these results, pollen mother cells contain specific meiosis-initiation machinery in which this nuclear GRX (MIL1) plays preponderant roles, probably acting also via TGA-type transcription factors. In this context, the results from Kelliher and Walbot (2012), demonstrating that changes in the redox status critically control the male germ lineage fate in maize, suggest a master or integrator role of these types of GRXs in the redox regulation associated with anther and gamete differentiation.

In the anthers, the chloroplast redox system comprising CBSXs, TRXs and peroxiredoxins (PRXs) is involved in anther dehiscence and therefore pollen release via the control of H_2O_2 (Ok et al., 2012), which ultimately allows connecting plant nutritional information and pollen release (Figure 2B). CBSX are redox proteins characterized by sharing only a single pair of Cystathione β-Synthase domains (CBS) in their structures that belong to the CBS-containing protein (CDCPs) superfamily. Arabidopsis genome contains six genes encoding CBSX proteins (CBSX1-6), which have recently been described as cellular sensors involved in the control of plant redox homeostasis and development (Yoo et al., 2011). They act interacting and increasing the activity of TRXs by sensing cellular changes of adenosine nucleotides. CBSX1 is a member of this family in Arabidopsis, which is preferentially expressed in the chloroplast of the anther. This protein is able to interact and increase the activity of all four types of plastidial TRXs (f, m x and y). This augmentation is favored by the presence of AMP, but not by ADP or ATP (Figure 2B; Yoo et al., 2011; Ok et al., 2012). The overexpression of CBSX1 or CBSX2 in Arabidopsis transgenic plants yields plants showing a severe sterility as a consequence of the inhibition of their anther dehiscence, which prevents the liberation of mature pollen grains (Yoo et al., 2011; Jung et al., 2013). This sterility is due to a decrease of H_2O_2 in the anthers, which causes

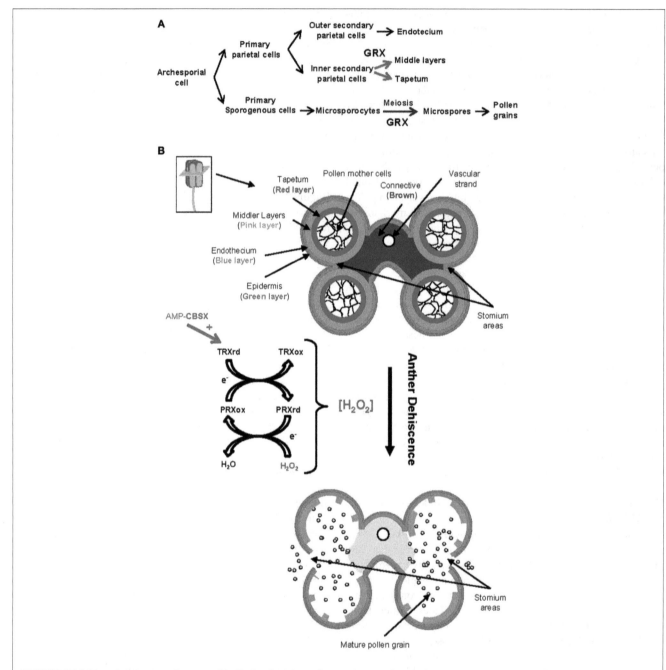

FIGURE 2 | Thiol-based redox proteins are critically involved in male gametogenesis, anther development and dehiscence. (A) Ontogenic development of the male germline. CC-type GRXs are essential for the switch from mitosis to meiosis in the microsporocytes (blue arrow), which ultimately originate the haploid pollen grains. They also participate in the development of the anther layers surrounding the microsporocytes (red arrows). **(B)** Diagram representing the histological structure of an anther (transversal section) and major changes occurring after dehiscence. H_2O_2 is required for cell wall lignification, which induces a thickening of the endotecium leading to anther dehiscence. CBSXs can regulate the level of H_2O_2 via their ability to active TRXs, which ultimately reduce PRXs in the plastids. This activity is enhanced by the presence of AMP, thus connecting the nutrition state with anther development.

a lignin deficiency that originates a failure in the secondary wall thickening of the endotecium layer, and subsequently a very narrow crevice in the stomium area (region of the anther where dehiscence occurs and pollen grains leave the anthers; **Figure 2B**). Male sterility caused for this same reasons (a limitation of H_2O_2) have previously been reported (Karlsson et al., 2005; Villarreal et al., 2009). According to these authors, CBSX1 regulates the level of H_2O_2 via the activation of plastid TRXs, which reduce and activate peroxiredoxins (PRX) directly detoxifying this radical (**Figure 2B**). In our opinion, an evaluation of the roles of other enzymes or non-enzymatic systems known to be involved in the homeostasis of H_2O_2 would be of great interest at this regard.

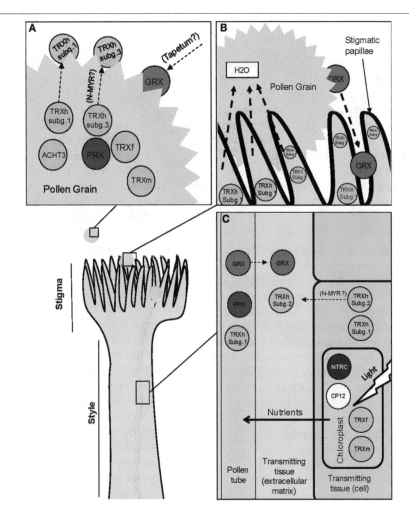

FIGURE 3 | Redoxins and related proteins are critical for pollen-pistil interactions. Illustration of the stigma and style during pollen-pistil interaction. **(A)** Representation of a pollen grain and the pollen coat. The presence of redoxins has been described both within the pollen grain and in the pollen coat. It is particularly remarkable the specific and conserved occurrence of h-type TRXs in the pollen grains, which can be secreted to the pollen coat. We hypothesize that N-terminal lipidations may play a role in this mechanism of secretion. Also, some redoxins synthesized in the tapetum may become integrated into the pollen coat after tapetum degeneration **(B)** Image representing the initial stages of the pollen-stigma interaction. Upon pollen arrival, pollen starts to hydrate, rapidly releasing GRXs present in the pollen coat. Also, the stigmatic papillae are rich in subgroup 1 TRXs. These TRXs have been involved in SI processes in *Brassica* although their occurrence has also been described for self-compatible species. **(C)** Representation of a pollen tube growing throughout the transmitting tissues of the style. Different types of redoxins are present in the pollen tube, some of them being secreted to the extracellular matrix of the transmitting tissue. Other redoxins are mainly expressed in stylar cells and may also be secreted to the extracellular matrix. Several redoxins related to photosynthesis are specifically expressed in the cells of the stylar tissues, which suggest that high photosynthetic rates are probably supporting pollen tube growth.

For example, the NADPH-TRX-Reductase C (NTRC) protein has been shown to be the main reducer of type-2 PRXs in chloroplasts (Kirchsteiger et al., 2009; Pulido et al., 2010). Apoplastic type III peroxidases (POXs), which are involved in cell wall polymerization via H_2O_2 regulation, should be assessed into their participation in this process (Hiraga et al., 2001). It is also well known the role of the ascorbic acid as main redox buffer in apoplast compartments (Foyer and Noctor, 2005a). Finally, overexpression of ROXY GRXs has also been shown to alter the level of H_2O_2 (Wang et al., 2009a,b).

Plant genomes contain higher number of genes encoding redoxins (like TRXs or GRXs) than other species, allowing them assigning specific isoforms to precise plant metabolic functions.

This includes anther development and male gametogenesis. Remarkably, a similar case is found in mammals, where male germ cells are endowed with three testis-specific thioredoxins named Sptrx-1, Sptrx-2, and Sptrx-3, which are specially involved in spermatogenesis (Jiménez et al., 2004; Miranda-Vizuete et al., 2004).

UNEXPECTED AND SPECIFIC OCCURRENCE OF REDOXINS IN PLANT REPRODUCTIVE TISSUES SUGGEST KEY FUNCTIONS IN POLLEN-PISTIL INTERACTIONS

Once a mature and dehydrated pollen grain lands on the appropriate stigma, it rapidly hydrates and incorporates nutrients from the stigma exudates (wet stigmas) or the stigma papillae (dry stigmas). This fact transforms the pollen grain into a polarized cell, which organizes its cytoplasm and cytoskeleton to support the extension of a tube within minutes after hydration (Edlund et al., 2004). This tube must grow to enter the transmitting tract along the style and finally reach the embryo sac where it will deliver the sperm nuclei that will participate of the double fertilization. At these stages, it is required a continuous interchange of both physical and chemical signals between partners (pollen, stigma, ovules...), which has to take place in a tight time frame (Dresselhaus and Franklin-Tong, 2013). Redox regulation and signaling by reactive species (**Figure 1**; see part 2) and probably by some redoxins (**Table 1**; **Figure 3**), might be critically involved in these signal exchanges.

Redoxins in pollen grain

TRXs *h* from subgroup 1 (Gelhaye et al., 2004), and GRXs are among the most abundant redoxins present in the pollen grain and the pollen coat (**Figure 3**). THL-1 is an h-type TRX that was immunodetected as a *B. rapa* extracellular pollen protein (Toriyama et al., 1998). However, although THL-1 is a well-known stigmatic protein involved in SI (Cabrillac et al., 2001), there is no clear data about its function in the pollen coat. The *Arabidopsis* h-type TRXh4 was shown to be expressed in the pollen grain and the pollen tube (Reichheld et al., 2002). Their counterparts AtTRXh1 and AtTRXh5, and the GRX AtGrxC2 proteins were detected in the pollen tube but not in the mature pollen grain. AtGrxC2 was also identified as a secreted protein (**Figures 3A,C**; Ge et al., 2011). Molecular and proteomic approaches also identified a GRX from triticale as a major pollen protein rapidly released upon pollen hydration on the stigma papillae (**Figure 3B**; Zaidi et al., 2012). This GRX contains a Gly in position 2, predicted to be co-translationally modified by N-myristoylation (N-MYR), a type of lipidation assisting protein anchoring to membranes, which therefore could account for secretion (**Figure 3B**; Denny et al., 2000; Utsumi et al., 2005; Martinez et al., 2008). The occurrence of this N-terminal modification has been recently evidenced *in vitro* for a similar GRX, AtGrxC1 from *Arabidopsis* (At5g63030) (Traverso, Pers. Commun.). It must be mentioned here that some of the proteins of the pollen coat are originated in the tapetum, and then incorporated into the pollen coat after the degeneration of this layer (**Figure 2B**), a mechanism that could be also suggested for the GRX released after hydration (**Figures 2A,B**; Zaidi et al., 2012).

The h-type TRXs belonging to the subgroup 3 (Gelhaye et al., 2004) were initially identified as a highly conserved group of pollen-expressed TRXs from both mono and dicots, featured by the presence of a N-terminal extension which contains conserved Gly and Cys residues in positions 2 and 4, respectively (Juttner et al., 2000). Curiously, these N-terminal extensions have been recently identified as a substrate for N-terminal myristoylation (N-MYR) as well as N-terminal palmitoylation (N-PAL). This last is another type of lipidation, usually identified in plasma membrane proteins (Traverso et al., 2008, 2013). No clear information is available concerning the specific roles or the subcellular localizations of these TRXs in pollen, although a member of this subgroup in *A. thaliana* (AtTRXh9) was shown to move from cell to cell via its N-terminal extension (Meng et al., 2010). According

to these results, we hypothesized that these lipidations can be involved in the release of this subgroup of TRXs to the extracellular matrix, where others TRXs and redoxins have already been identified (**Figure 3A**; Ge et al., 2011; Zaidi et al., 2012).

Other types of redoxins have been identified as highly expressed in pollen. Two independent transcriptomic analyses (Becker et al., 2003; Lee and Lee, 2003) have shown ACHT3 TRX to be highly expressed in *Arabidopsis* pollen. However, no functional data are associated with this presence. In addition, PRXs are also among the redoxins displaying specific localization in pollen grains (**Figure 3A**). Transcriptomic analysis of *Arabidopsis* revealed two PRXs among the 50 most expressed genes in pollen. One of them (TPx2) showed increased expression under cold treatment (Lee and Lee, 2003). In addition, some type II PRXs from *Arabidopsis* showed specific expression patterns associated with male reproductive tissues (**Figures 3A,C**) (Bréhélin et al., 2003). The cytosolic AtPRXII-C is almost exclusively expressed in mature pollen, whereas its counterpart AtPRXII-D is detected in mature pollen, germinating pollen and pollen tubes, where both proteins could be reduced by AtTRXh4 (Reichheld et al., 2002; Bréhélin et al., 2003). Finally, the plastid-addressed AtPRDII-E has been described as mainly expressed in immature anthers and ovules (Bréhélin et al., 2003).

Considering that no clear functional data are associated with this specific occurrence in pollen grains, further work is necessary to better understand the precise redox mechanisms underlying pollen function. It is well-known that when pollen grains reach the appropriated stigma, they release a number of proteins from pollen coat, which together with the proteins released from pistil surrounding tissues, seem to play important roles during pollen-pistil interaction, adhesion, germination or pollen tube growth as well as providing protection against pathogen attack (Andersson and Lidholm, 2003; Grote et al., 2008; Zaidi et al., 2012). In a different context, redoxins from the pollen coat have been attributed with some allergic potential (Toriyama et al., 1998). However, there is no evidence supporting this fact, with the exception of the description of a 1-Cys PRXs and a h-type TRX as respiratory wheat flours allergens from maize (Fasoli et al., 2009; Pahr et al., 2012).

Occurrence of GR/GRX and NTR/TRX systems during pollen germination and pollen tube growth

Several works based on the characterization of *A. thaliana* mutant lines have evidenced that GRXs and TRXs are specifically involved in pollen germination and pollen tube growth. The *Arabidopsis* double mutant *ntra ntrb*, lacking NTR activity to reduce cytosolic h-type TRXs, showed a reduced fitness due to defects in pollen functions (Reichheld et al., 2007). Under this genetic background, GRXs are also able of directly reduce h-type TRXs although in lesser extension, thus revealing a more complex *in vivo* interplay between the TRX and glutathione pathways (Reichheld et al., 2007). In fact, the additional disruption of the glutathione reductase 1 gene (*GR1*) under this double mutant background (triple mutant *ntra ntrb gr1*) led to a pollen lethal phenotype (Marty et al., 2009). Noteworthy, the characterization of the single mutant *gr1* demonstrated that a residual reduction of GSSG could be directly attributed to h-type TRXs, reason why this NTR/TRX system was described as a functional backup for the activity of GR1 (Marty et al., 2009).

The characterizations of the double (*ntra ntrb*) and triple mutant (*ntra ntrb gr1*) headed to other important conclusions. Defects in pollen grains were not associated with gametogenesis, since mature pollen grains inside the anthers were viable in both mutants (Reichheld et al., 2007; Marty et al., 2009). Then it is important to elucidate, in our opinion, in which subsequent process these redox systems are critically involved: (i) pollen germination, (ii) pollen tube growth, (iii) polarity or guidance, or (iv) pollen tube–embryo sac interactions and fertilization.

A second conclusion is that the lack of both redox systems drastically affected male gametophyte functions, contrary to what occurred in the female haploid gametophyte. The unpaired NTR-TRX System (NTS) in the double *ntra ntrb* mutant yielded reduced fitness in pollen grains, probably derived from a limited reduction of h-type TRXs, although the diploid sporophyte or the female gametophyte did not show such drastic phenotypes (Reichheld et al., 2007). In a similar way, the *ntra ntrb gr1* triple mutant produced a lethal phenotype in pollen, while the female embryo sac was unaffected (Marty et al., 2009). This strong difference is probably associated with the exceptional burst metabolism occurring during pollen germination, that produces important redox imbalances and would justify the exceptional requirement of both thiol-dependent redox GR/GRX and NTR/TRX systems. This is in agreement with the results compiled in our present review, showing how the h-type TRXs or the GRXs are specifically found in pollen grain, pollen tube or pollen coat.

Within this context, it must be noted the importance of GSH and auxins for pollen functionality. Zechmann et al. (2011) have shown that GSH availability is essential for pollen germination and early elongation steps of the pollen tube, since its depletion triggers disturbances in the auxin metabolism, which led to inhibition of pollen germination. Considering that NTR/TRX and GSH pathways are involved in auxin homeostasis (Bashandy et al., 2010, 2011), we suggest that both redox systems can act somehow linking the GSH cellular status and auxin downstream signals during and/or after pollen germination.

Redoxins in female reproductive tissues

TRXs h have been shown to be specifically expressed in the stigmatic papillae of different plants (**Figure 3B**). The PsTRXh1 from *P. sativum* has been immunolocalized in the receptive stigmatic papillae and the mature pollen grain of this plant (Traverso et al., 2007). THL-1 from *B. rapa*, a TRXs h from the same subgroup, which has been involved in SI (Cabrillac et al., 2001) also shows a similar pattern of expression in flower tissues than PsTRXh1. Considering that *P. sativum* is a self-compatible species, the role associated with this dual localization in pea flower is unlikely implicated in SI. Besides, an h-type TRX from subgroup I was included within the 50 most highly expressed genes characterized in a transcriptomic analysis from saffron stigmas (D'Agostino et al., 2007), and another TRX, h-type *Arabidopsis* h1 was distinctively detected in the style (Reichheld et al., 2002).

Classical chloroplastidial TRXs or TRX-related proteins have also been associated with pistils (**Figure 3C**). For example, expression of Pea TRXs m and f types was specifically detected in pollen

grains, tapetum, style and ovules (de Dios Barajas-López et al., 2007). Other plastidial TRX-related proteins like the NTRC or CP12 redox proteins are also present in the style (**Figure 3C**). CP12 are small, dithiol-based redox-sensitive proteins which together with the plastidial TRX f, regulate the activity of the Calvin cycle in response to rapid light changes (Howard et al., 2008). The redox state of CP12 has been shown to be regulated by TRXs (Marri et al., 2009). The *Arabidopsis* genome contains three genes encoding CP12 proteins (CP12-1, 2 and 3), which could be involved in plant reproduction since CP12-1 and 2 are specifically expressed in the style (Singh et al., 2008). No clear roles are associated with these expressions, although CP12-antisense lines of tobacco and *Arabidopsis* display a complex phenotype including reduced fertility (Singh et al., 2008). AtNTRC is also a redox plastidial protein characterized as the principal reducer of 2-Cys PRXs (Kirchsteiger et al., 2009; Pulido et al., 2010). This protein allows the use of NADPH to maintain the redox status of the chloroplast (Spínola et al., 2008), as well as acting as a general molecular switch able to convert NADPH into redox signal in non-photosynthetic plastids (Kirchsteiger et al., 2012). Remarkably, this protein also shows a high level of expression in the style (Kirchsteiger et al., 2012).

Pollen tube growth throughout the style depends on nutrients supported by female tissues (Lind et al., 1996; Taylor and Hepler, 1997; Wu et al., 2001). The presence of photosynthesis regulatory proteins like those described here might be derived from the necessity of extra nutrients from the stylar photosynthetic cell to feed pollen tubes. However, we cannot exclude other possibilities like their involvement in the control of the redox imbalance derived from the burst metabolism of the growing pollen tube. Under this context, the redox state of two key enzymes of the Benson-Calvin cycle (fructose-1,6-bisphosphatase and phosphoribulokinase) could be regulated by S-nitrosylation, since they have been previously identified as S-nitrosylated proteins (Lindermayr et al., 2005; Begara-Morales et al., 2013).

SELF-INCOMPATIBILITY AND h-type TRXs

Self-incompatibility (SI) is a mechanism adopted by flowering plants to prevent self-fertilization as well as to avoid crossings with genetically-related plants, promoting outcrossing and subsequently, genetic diversity in the offspring (Iwano and Takayama, 2012). Therefore, this mechanism has been critically relevant on the dominant position reached by flowering plants in the biosphere during evolution (Gaude and Cabrillac, 2001). Genes involved in SI display high allelic variability, and are referred as haplotypes in the literature. SI systems are based on the discrimination between male-specificity and female-specificity determinants (S-determinants) localized in the surface of pollen grains and the stigmatic papillae respectively, both encoded at the S-locus. Therefore, SI depends on a specific interaction between male and female S-determinants derived from the same or different S-haplotypes (depending on the plant family. For a review see Iwano and Takayama, 2012). Although these determinants differ among plant species, the redox activity of the TRXs h has been described as critically involved within most of the proposed SI molecular models.

In the plant *Phalaris coerulescens*, the SI model system is based on two multiallelic not-linked loci, called S and Z (Hayman and Richter, 1992; Li et al., 1997; Klaas et al., 2011). Locus S encodes a monocot-conserved protein exclusively expressed in mature pollen. Its N-terminal domain is highly variable and contains the allelic information, whereas the C-terminal end contains a conserved TRX h-type motif responsible of the catalytic activity and its disruption yields the lost of the SI (Li et al., 1994, 1995). Although no enough data are available for assembling a clear model, the redox activity of the TRX domain has been shown to be essential for the SI response in *P. coerulescens* (Li et al., 1996).

In the classical model of *Brassica*, several genes are involved in SI, all of them linked to the S-locus. Two of these genes encode for the female and male determinants, the S-locus Receptor Kinase (SRK) expressed in the stigma-papilla cells and its ligand, the S-locus Cys rich protein (SCR) localized in the pollen coat (Schopfer et al., 1999; Takasaki et al., 2000). Pollen rejection is achieved when SRK recognizes SCR coming from the same S-allele. SRK is a transmembrane protein with a serine threonine kinase activity in the cytoplasmic region, which triggers the downstream biochemical pathways of the SI response (Giranton et al., 2000). In a classical model suggested by Cabrillac et al. (2001), this kinase activity is reversibly blocked by the direct interaction with a h-type TRX (THL-1) in the subcortical cytoplasmic side of stigma papillae cells, and only the allele-specific interaction between SCR and SRK produces the release of the TRX, therefore releasing the kinase activity (Cabrillac et al., 2001). However, according to Ivanov and Gaude (2009a), SRK is mostly intracellular, and the scarce part present in different domains of the plasma membrane can interact with SCR. After ligand recognition, the receptor-ligand complex is internalized in endosomes enriched in the negative regulator TRX h (Ivanov and Gaude, 2009a). In fact, these authors proposed a new model, the SI "domain" model, where the role of the TRX is mainly relegated to the endomembrane system (Ivanov and Gaude, 2009b). h-type TRXs are the biggest cluster of plant TRXs, encoded by almost eleven TRXs in the genome of *A. thaliana*, which have been usually described as soluble proteins due to the lack of transit peptides (Florencio et al., 1988). However, some TRXs h have recently been described to be modified by lipidation (myristoylation and palmitoylation) at their N-terminal extensions. These modifications result in their localization to the endomembrane system or to the plasma membrane (Traverso et al., 2008, 2013) and are in agreement with the endomembrane localizations of the TRXs h shown by Ivanov and Gaude (2009a). Independently from the models suggested, the role of the TRX h in *Brassica* SI mechanism was demonstrated, since antisense transgenic lines for THL-1 displayed a limited level of SI only (Haffani et al., 2004). Furthermore, it has been shown that this role depends on both the redox activity of the TRX, and the occurrence of a specific sequence in their active centers -WCPPC- (Mazzurco et al., 2001).

The SI model mechanism described for *Solanaceae, Rosaceae, and Plantaginaceae* families is based on the activity of secreted S-RNases. In all three families, compatibility is also controlled by a polymorphic S-locus containing at least two genes. S-RNases determine the specificity of pollen rejection in the pistil, and S-locus F-box proteins fulfill this function in pollen. In *N. alata*, Juárez-Díaz et al. (2006) suggested the participation of a h-type TRX (NaTRXh), which interacts with the S-RNase in the extracellular matrix of the stylar transmitting tract (**Figure 3C**). Although these authors suggested that NaTRXh was secreted due to some information present in its N-terminal extension, the secretion mechanism remained unexplained. Remarkably, NaTRXh has also been characterized as a N-myristoylated membrane-associated protein. Therefore, this characteristic could be involved in its secretion (**Figure 3C**; Traverso et al., 2013).

ROS and RNS have also been shown to be involved in SI mechanisms. In the *Papaver* model of SI (where no TRXs have been directly implicated), SI response seems to be mediated by ROS/NO redox signaling. Allele-specific interaction was shown to induce a rapid and transitory increase of ROS and NO in *Papaver rhoeas* pollen tubes. In this model, SI is triggered by an increase of intracellular Ca^{2+} in the pollen tube, which ultimately originates actin reorganization and programmed cell death resulting in the destruction of the self-pollen. ROS/NO seem to act mediating the signal between calcium and PCD (Wilkins et al., 2011). In the S-RNase-based model of SI response, the S-RNase specifically disrupts tip-localized ROS of incompatible pollen tubes via arresting ROS formation in mitochondria and cell walls of *Pyrus pyrifolia* (Wang et al., 2010a,b).

All these SI mechanisms usually originate the rejection of the allele-incompatible pollen grains by affecting the pollen tube growth throughout the stigma and the style. Although independent SI molecular mechanisms are known, which indicates that SI has clearly evolved independently several times during plant evolution (Takayama and Isogai, 2005), the involvement of the TRXs (and probably other redoxins) as well as the signaling mediated by ROS or NO have been evidenced in different SI models proposed. Further experiments are needed to clarify the role of redoxins in all these models within the general context of redox regulation and signaling, which includes reactive species-mediated signaling.

CONCLUDING REMARKS

In this review, we have listed and discussed the most remarkable evidences suggesting that molecular and cellular events involved in sexual plant reproduction are critically influenced by plant-conserved and specific thiol-based redox mechanisms, which include ROS and NO together with several isoforms of TRXs, GRXs and other redox-related proteins. These plant reproductive-specific redox mechanisms likely appeared as a consequence of the high number of isoforms of redoxins available, which emerged during plant evolution and somehow resulted in beneficial characteristic for plant physiology. However, and comparatively with non-reproductive plant tissues and organs, much is still unknown about the nature, presence, localization and molecular features of these proteins.

ACKNOWLEDGMENTS

This work was partially supported by European Regional Development Fund (ERDF) through the projects BFU2011-22779, P2010-AGR6274, P2010-CVI5767 and P2011-CVI-7487. Jose A. Traverso thanks Spanish CSIC funding within the frame of JAE-DOC program.

REFERENCES

Airaki, M., Sánchez-Moreno, L., Leterrier, M., Barroso, J. B., Palma, J. M., and Corpas, F. J. (2011). Detection and quantification of S-nitrosoglutathione (GSNO) in pepper (Capsicum annuum L.) plant organs by LC-ES/MS. Plant Cell Physiol. 52, 2006–2015. doi: 10.1093/pcp/pcr133

Andersson, K., and Lidholm, J. (2003). Characteristics and immunobiology of grass pollen allergens. Int. Arch. Allergy Immunol. 130, 87–107. doi: 10.1159/0000 69013

Arner, E. S. J., and Holmgren, A. (2000). Physiological functions of thioredoxin and thioredoxin reductase. Eur. J. Biochem. 267, 6102–6109. doi: 10.1046/j.1432-1327.2000.01701.x

Balmer, Y., Vensel, W. H., Cai, N., Manieri, W., Schü rmann, P., Hurkman, W. J., et al. (2006). A complete ferredoxin/thioredoxin system regulates fundamental processes in amyloplasts. Proc. Natl. Acad. Sci. U.S.A. 103, 2988–2993. doi: 10.1073/pnas.0511040103

Bandyopadhyay, S., Gama, F., Molina-Navarro, M. M., Gualberto, J. M., Claxton, R., Naik, S. G., et al. (2008). Chloroplast monothiol glutaredoxins as scaffold proteins for the assembly and delivery of [2Fe-2S] clusters. EMBO J. 27, 1122–1133. doi: 10.1038/emboj.2008.50

Bashandy, T., Guillcminot, J., Vernoux, T., Caparros-Ruiz, D., Ljung, K., Meyer, Y., et al. (2010). Interplay between the NADP-linked thioredoxin and glutathione systems in Arabidopsis auxin signaling. Plant Cell 22, 376–391. doi: 10.1105/tpc.109.071225

Bashandy, T., Meyer, Y., and Reichheld, J. (2011). Redox regulation of auxin signaling and plant development in Arabidopsis. Plant Signal. Behav. 6, 117–119. doi: 10.4161/psb.6.1.14203

Becker, J. D., Boavida, L. C., Carneiro, J., Haury, M., and Feijó, J. A. (2003). Transcriptional profiling of Arabidopsis tissues reveals the unique characteristics of the pollen transcriptome. Plant Physiol. 133, 713–725. doi: 10.1104/pp.103.028241

Begara-Morales, J. C., López-Jaramillo, F. J., Sánchez-Calvo, B., Carreras, A., Ortega-Muñoz, M., Santoyo-González, F., et al. (2013). Vinyl sulfone silica: application of an open preactivated support to the study of transnitrosylation of plant proteins by S-nitrosoglutathione. BMC Plant Biol. 12, 13–61. doi: 10.1186/1471-2229-13-61.

Bhatt, A. M., Canales, C., and Dickinson, H. G. (2001). Plant meiosis: the means to 1N. Trends Plant. Sci. 6, 114–121. doi: 10.1016/S1360-1385(00)01861-6

Boavida, L., Becker, J. D., Vieira, A. M., and Feijó, J. A. (2005). Gametophyte interaction and sexual reproduction: how plants make a zygote. Int. J. Dev. Biol. 49, 615–632. doi: 10.1387/ijdb.052023lb

Bréhélin, C., Meyer, E. H., de Souris, J. P., Bonnard, G., and Meyer, Y. (2003). Resemblance and dissemblance of Arabidopsis type II peroxiredoxins: similar sequences for divergent gene expression, protein localization, and activity. Plant Physiol. 132, 2045–2057. doi: 10.1104/pp.103.022533

Bright, J., Hiscock, S. J., James, P. E., and Hancock, J. T. (2009). Pollen generates nitric oxide and nitrite: a possible link to pollen-induced allergic responses. Plant Physiol. Biochem. 47, 49–55. doi: 10.1016/j.plaphy.2008. 09.005

Broniowska, K. A., Diers, A. R., and Hogg, N. (2013). S-Nitrosoglutathione. Biochim. Biophys. Acta 1830, 3173–3181. doi: 10.1016/j.bbagen.2013.02.004

Buchanan, B. B., and Balmer, Y. (2005). Redox regulation: a broadening horizon. Annu. Rev. Plant Biol. 56, 187–220. doi: 10.1146/annurev.arplant.56.032604. 144246

Bushart, T. J., and Roux, S. J. (2007). Conserved features of germination and polarized cell growth: a few insights from a pollen-fern spore comparison. Ann. Bot. 99, 9–17. doi: 10.1093/aob/mcl159

Cabrillac, D., Cock, J. M., Dumas, C., and Gaude, T. (2001). The S-locus receptor kinase is inhibited by thioredoxins and activated by pollen coat proteins. Nature 410, 220–223. doi: 10.1038/35065626

Cardenas, L., McKenna, S. T., Kunkel, J. G., and Hepler, P. K. (2006). NAD(P)H oscillates in pollen tubes and is correlated with tip growth. Plant Physiol. 142, 1460–1468. doi: 10.1104/pp.106.087882

Chaubal, R., Anderson, J. R., Trimnell, M. R., Fox, T. W., Albertsen, M. C., and Bedinger, P. (2003). The transformation of anthers in the msca1 mutant of maize. Planta 216, 778–788. doi: 10.1007/s00425-002-0929-8

Corpas, F. J., Alché, J. D., and Barroso, J. B. (2013). Current overview of S-nitrosoglutathione (GSNO) in higher plants. Front. Plant Sci. 4:126. doi: 10. 3309/fpls.2013.00126

Corpas, F. J., and Barroso, J. B. (2013). Nitro-oxidative stress vs oxidative or nitrosative stress in higher plants. New Phytol. 199, 633–635. doi: 10.1111/nph. 12380

Couturier, J., Chibani, K., Jacquot, J. P., and Rouhier, N. (2013). Cysteine-based redox regulation and signaling in plants. Front. Plant Sci. 4:105. doi: 10.3389/fpls.2013.00105

Couturier, J., Jacquot, J. P., and Rouhier, N. (2009). Evolution and diversity of glutaredoxins in photosynthetic organisms. Cell Mol. Life Sci. 66, 2539–2557. doi: 10.1007/s00018-009-0054-y

D'Autreaux, B., and Toledano, M. B. (2007). ROS as signalling molecules: mechanisms that generate specificity in ROS homeostasis. Nat. Rev. Mol. Cell Biol. 8, 813–824. doi:10.1038/nrm2256

D'Agostino, N., Pizzichini, D., Chiusano, M. L., and Giuliano, G. (2007). An EST database from saffron stigmas. BMC Plant Biology 7:53. doi:10.1186/1471-2229-7-53

de Dios Barajas-López, J., Serrato, A. J., Olmedilla, A., Chueca, A., and Sahrawy, M. (2007). Localization in roots and flowers of pea chloroplastic thioredoxin f and thioredoxin m proteins reveals new roles in nonphotosynthetic organs. Plant Physiol. 145, 946–960. doi: 10.1104/pp.107.105593

Denny, P. W., Gokool, S., Russell, D. G., Field, M. C., and Smith, D. F. (2000). Acylation-dependent protein export in Leishmania. J. Biol. Chem. 275, 11017–11025. doi: 10.1074/jbc.275.15.11017

Dresselhaus, T., and Franklin-Tong, N. (2013). Male-female crosstalk during pollen germination, tube growth and guidance, and double fertilization. Mol. Plant 6, 1018–1036. doi: 10.1093/mp/sst061

Edlund, A. F., Swanson, R., and Preuss, D. (2004). Pollen and stigma structure and function: the role of diversity in pollination. Plant Cell 16, S84–97. doi: 10.1105/tpc.015800

Fasoli, E., Pastorello, E. A., Farioli, L., Scibilia, J., Aldini, G., Carini, M., et al. (2009). Searching for allergens in maize kernels via proteomic tools. J. Proteomics 13, 501–510. doi: 10.1016/j.jprot.2009.01.013

Feijó, J. A., Costa, S. S., Prado, A. M., Becker, J. D., and Certal, A. C. (2004). Signalling by tips. Curr. Opin. Plant Biol. 7, 589–598. doi: 10.1016/j.pbi.2004. 07.014

Feng, H., Lizhe, A., Tan, L., Hou, Z., and Wang, X. (2000). Effect of enhanced ultraviolet-B radiation on pollen germination and tube growth of 19 taxa in vitro. Environ. Exp. Bot. 43, 45–53. doi: 10.1016/S0098-8472(99)00042-8

Florencio, F. J., Yee, B. C., Johnson, T. C., and Buchanan, B. B. (1988). An NADP/thioredoxin system in leaves: purification and characterization of NADP-thioredoxin reductase and thioredoxin h from spinach. Arch. Biochem. Biophys. 266, 496–507. doi: 10.1016/0003-9861(88)90282-2

Foreman, J., Demidchik, V., Bothwell, J. H., Mylona, P., Miedema, H., Torres, M. A., et al. (2003). Reactive oxygen species produced by NADPH oxidase regulate plant cell growth. Nature 27, 442–446. doi: 10.1038/nature01485

Foyer, C. H., and Noctor, G. (2005a). Redox homeostasis and antioxidant signaling. A metabolic interface between stress perception and physiological responses. Plant Cell 17, 1866–1875. doi: 10.1105/tpc.105.033589

Foyer, C. H., and Noctor, G. (2005b). Oxidant and antioxidant signalling in plants: A reevaluation of the concept of oxidative stress in a physiological context. Plant Cell Environ. 28, 1056–1071. doi: 10.1111/j.1365-3040.2005.01327.x

Franklin-Tong, V. E. (1999). Signaling and the modulation of pollen tube growth. Plant Cell 11, 727–738.

Gaude, T., and Cabrillac, D. (2001). Self-incompatibility in flowering plants: the Brassica model. C. R. Acad. Sci. III 324, 537–542. doi: 10.1016/S0764-4469(01)01323-3

Ge, W., Song, Y., Zhang, C., Zhang, Y., Burlingame, A. L., and Guo, Y. (2011). Proteomic analyses of apoplastic proteins from germinating Arabidopsis thaliana pollen. Biochim. Biophys. Acta 1814, 1964–1973. doi: 10.1016/j.bbapap. 2011.07.013

Gelhaye, E., Rouhier, N., and Jacquot, J. P. (2003). Evidence for a subgroup of thioredoxin h that requires GSH/Grx for its reduction. FEBS Lett. 555, 443–448. doi: 10.1016/S0014-5793(03)01301-2

Gelhaye, E., Rouhier, N., and Jacquot, J. P. (2004). The thioredoxin h system of higher plants. Plant Physiol. Biochem. 42, 265–271. doi: 10.1016/j.plaphy.2004. 03.002

Giranton, J. L., Dumas, C., Cock, J. M., and Gaude, T. (2000). The integral membrane S-locus receptor kinase of Brassica has serine/threonine kinase activity in a membranous environment and spontaneously forms oligomers in planta. Proc. Natl. Acad. Sci. U.S.A. 97, 3759–3764. doi: 10.1073/pnas.97.7.3759

Grote, M., Westritschnig, K., and Valenta, R. (2008). Immunogold electron microscopic localization of the 2 EF-hand calcium-binding pollen allergen Phl p 7 and its homologues in pollens of grasses, weeds and trees. *Int. Arch. Allergy Immunol.* 146, 113–121. doi: 10.1159/000113514

Haffani, Y. Z., Gaude, T., Cock, J. M., and Goring, D. R. (2004). Antisense suppression of thioredoxin h mRNA in Brassica napus cv. Westar pistils causes a low level constitutive pollen rejection response. *Plant Mol. Biol.* 55, 619–630. doi: 10.1007/s11103-004-1126-x

Hayman, D. L., and Richter, J. (1992). Mutations affecting self-incompatibility in Phalaris-Coerulescens Desf (Poaceae). *Heredity* 6, 8495–8503.

He, J. M., Bai, X. L., Wang, R. B., Cao, B., and She, X. P. (2007). The involvement of nitric oxide in ultraviolet-B-inhibited pollen germination and tube growth of Paulownia tomentosa *in vitro. Physiol. Plant.* 131, 273–282.

Higashiyama, T., and Hamamura, Y. (2008). Gametophytic pollen tube guidance. *Sex. Plant Reprod.* 21, 17–26. doi: 10.1007/s00497-007-0064-6

Hiraga, S., Sasaki, K., Ito, H., Ohashi, Y., and Matsui, H. (2001). A large family of class III plant peroxidases. *Plant Cell Physiol.* 42, 462–468. doi: 10.1093/pcp/pce061

Hiscock, S. J. (2011). Sexual plant reproduction. *Ann. Bot.* 108, 585–587. doi: 10.1093/aob/mcr217

Hong, L., Tang, D., Zhu, K., Wang, K., Li, M., and Cheng, Z. (2012). Somatic and reproductive cell development in rice anther is regulated by a putative glutaredoxin. *Plant Cell* 24, 577–588. doi: 10.1105/tpc.111.093740

Howard, T. P., Metodiev, M., Lloyd, J. C., and Raines, C. A. (2008). Thioredoxin-mediated reversible dissociation of a stromal multiprotein complex in response to changes in light availability. *Proc. Natl. Acad. Sci. U.S.A.* 105, 4056–4061. doi: 10.1073/pnas.0710518105

Ivanov, R., and Gaude, T. (2009a). Endocytosis and endosomal regulation of the S-receptor kinase during the self-incompatibility response in Brassica oleracea. *Plant Cell* 21, 2107–2017. doi: 10.1105/tpc.108.063479

Ivanov, R., and Gaude, T. (2009b). Brassica self-incompatibility: a glimpse below the surface. *Plant Signal. Behav.* 4, 996–998. doi: 10.4161/psb.4.10.9714

Iwano, M., and Takayama, S. (2012). Self/non-self discrimination in angiosperm self-incompatibility. *Curr. Opin. Plant Biol.* 15, 78–83. doi: 10.1016/j.pbi.2011.09.003

Jiang, P., Zhang, X., Zhu, Y., Zhu, W., Xie, H., and Wang, X. (2007). Metabolism of reactive oxygen species in cotton cytoplasmic male sterility and its restoration. *Plant Cell Rep.* 26, 1627–1634. doi: 10.1007/s00299-007-0351-6

Jiménez, A., Zu, W., Rawe, V. Y., Pelto-Huikko, M., Flickinger, C. J., Sutovsky, P., et al. (2004). Spermatocyte/spermatid-specific thioredoxin-3, a novel Golgi apparatus-associated thioredoxin, is a specific marker of aberrant spermatogenesis. *J. Biol. Chem.* 279, 34971–34982. doi: 10.1074/jbc.M404192200

Juárez-Díaz, J. A., McClure, B., Vázquez-Santana, S., Guevara-García, A., León-Mejía, P., Márquez-Guzmán, J., et al. (2006). A novel thioredoxin h is secreted in Nicotiana alata and reduces S-RNase in vitro. *J. Biol. Chem.* 281, 3418–3424. doi: 10.1074/jbc.M511687200

Jung, K. W., Kim, Y. Y., Yoo, K. S., Ok, S. H., Cui, M. H., Jeong, B. C., et al. (2013). A Cystathionine-β-synthase domain-containing protein, CBSX2, regulates endothecial secondary cell wall thickening in anther development. *Plant Cell Physiol.* 54, 195–208. doi: 10.1093/pcp/pcs166

Juttner, J., Olde, D., Langridge, P., and Baumann, U. (2000). Cloning and expression of a distinct subclass of plant thioredoxins. *Eur. J. Biochem.* 267, 7109–7117. doi: 10.1046/j.1432-1327.2000.01811.x

Karlsson, M., Melzer, M., Prokhorenko, I., Johansson, T., and Wingsle, G. (2005). Hydrogen peroxide and expression of hipI-superoxide dismutase are associated with the development of secondary cell walls in Zinnia elegans. *J. Exp. Bot.* 56, 2085–2093. doi: 10.1093/jxb/eri207

Kelliher, T., and Walbot, V. (2012). Hypoxia triggers meiotic fate acquisition in maize. *Science* 337, 345–348. doi: 10.1126/science.1220080

Kirchsteiger, K., Ferrández, J., Pascual, M. B., González, M., and Cejudo, F. J. (2012). NADPH thioredoxin reductase C is localized in plastids of photosynthetic and nonphotosynthetic tissues and is involved in lateral root formation in *Arabidopsis. Plant Cell* 24, 534–548. doi: 10.1105/tpc.111.092304

Kirchsteiger, K., Pulido, P., González, M., and Cejudo, F. J. (2009). NADPH Thioredoxin reductase C controls the redox status of chloroplast 2-Cys peroxiredoxins in *Arabidopsis thaliana. Mol. Plan.* 2, 298–307. doi: 10.1093/mp/ssn082

Klaas, M., Yang, B., Bosch, M., Thorogood, D., Manzanares, C., Armstead, I. P., et al. (2011). Progress towards elucidating the mechanisms of self-incompatibility in the grasses: further insights from studies in Lolium. *Ann. Bot.* 108, 677–685. doi: 10.1093/aob/mcr186

Konrad, K. R., Wudick, M. M., and Feijó, J. A. (2011). Calcium regulation of tip growth: new genes for old mechanisms. *Curr. Op. Plant Biol.* 14, 721–730. doi: 10.1016/j.pbi.2011.09.005

Laloi, C., Rayapuram, N., Chartier, Y., Grienenberger, J. M., Bonnard, G., and Meyer, Y. (2001). Identification and characterization of a mitochondrial thioredoxin system in plants. *Proc. Natl. Acad. Sci. U.S.A.* 98, 14144–14149. doi: 10.1073/pnas.241340898

Lee, J. Y., and Lee, D. H. (2003). Use of serial analysis of gene expression technology to reveal changes in gene expression in *Arabidopsis* pollen undergoing cold stress. *Plant Physiol.* 132, 517–529. doi: 10.1104/pp.103.020511

Li, X. M., Paech, N., Nield, J., Hayman, D., and Langridge, P. (1997). Self-incompatibility in the grasses: evolutionary relationship of the S gene from *Phalaris coerulescens* to homologous sequences in other grasses. *Plant Mol. Biol.* 34, 223–232. doi: 10.1023/A:1005802327900

Li, X., Nield, J., Hayman, D., and Langridge, P. (1994). Cloning a putative self-incompatibility gene from the pollen of the grass *Phalaris coerulescens. Plant Cell* 6, 1923–1932. doi: 10.2307/3869918

Li, X., Nield, J., Hayman, D., and Langridge, P. (1995). Thioredoxin activity in the C terminus of Phalaris S protein. *Plant J.* 8, 133–138. doi: 10.1046/j.1365-313X.1995.08010133.x

Li, X., Nield, J., Hayman, D., and Langridge, P. (1996). A self-fertile mutant of Phalaris produces an S protein with reduced thioredoxin activity. *Plant J.* 10, 505–513. doi: 10.1046/j.1365-313X.1996.10030505.x

Lind, J. L., Bönig, I., Clarke, A. E., and Anderson, M. A. (1996). A style-specific 120-kDa glycoprotein enters pollen tubes of Nicotiana alata *in vivo. Sex. Plant Reprod.* 9, 75–86. doi: 10.1007/BF02153054

Lindermayr, C., Saalbach, G., and Durner, J. (2005). Proteomic indentification of S-nitrosylated proteins in *Arabidopsis. Plant Physiol.* 2005, 137, 921–930. doi: 10.1104/pp.104.058719

Lindermayr, C., Sell, S., Muller, B., Leister, D., and Durner, J. (2010). Redox regulation of the NPR1-TGA1 system of *Arabidopsis thaliana* by nitric oxide. *Plant Cell* 22, 2894–2907. doi: 10.1105/tpc.109.066464

Marri, L., Zaffagnini, M., Collin, V., Issakidis-Bourguet, E., Lemaire, S. D., Pupillo, P., et al. (2009). Prompt and easy activation by specific thioredoxins of Calvin cycle enzymes of *Arabidopsis thaliana* associated in the GAPDH/CP12/PRK supramolecular complex. *Mol. Plant.* 2, 259–269. doi: 10.1093/mp/ssn061

Martinez, A., Traverso, J. A., Valot, B., Ferro, M., Espagne, C. Ephritikhine, G., et al. (2008). Extent of N-terminal modifications in cytosolic proteins from eukaryotes. *Proteomics* 8, 2809–2831. doi: 10.1002/pmic.200701191

Márton, M. L., and Dresselhaus, T. (2008). A comparison of early molecular fertilization mechanisms in animals and flowering plants. *Sex. Plant Reprod.* 21, 37–52. doi: 10.1007/s00497-007-0062-8

Marty, L., Siala, W., Schwarzländer, M., Fricker, M. D., Wirtz, M., Sweetlove, L. J., et al. (2009). The NADPH-dependent thioredoxin system constitutes a functional backup for cytosolic glutathione reductase in *Arabidopsis. Proc. Natl. Acad. Sci. U.S.A.* 106, 9109–9114. doi: 10.1073/pnas.0900206106

Mazzurco, M., Sulaman, W., Elina, H., Cock, J. M., and Goring, D. R. (2001). Further analysis of the interactions between the Brassica S receptor kinase and three interacting proteins (ARC1, THL1 and THL2) in the yeast two-hybrid system. *Plant Mol. Biol.* 45, 365–376. doi: 10.1023/A:1006412329934

McClure, B. A., and Franklin-Tong, V. (2006). Gametophytic self-incompatibility: understanding the cellular mechanisms involved in "self" pollen tube inhibition. *Planta* 224, 233–245. doi: 10.1007/s00425-006-0284-2

McInnis, S. M., Costa, L. M., Gutiérrez-Marcos, J. F., Henderson, C. A., and Hiscock, S. J. (2005). Isolation and characterization of a polymer phic stigma-specific class III peroxidase gene from *Senecio squalidus* L. (Asteraceae). *Plant. Mol. Biol.* 57, 659–677. doi: 10.1007/s11103-005-1426-9

McInnis, S. M., Desikan, R., Hancock, J. T., and Hiscock, S. J. (2006a). Production of reactive oxygen species and reactive nitrogen species by angiosperm stigmas and pollen: potential signalling crosstalk? *New Phytol.* 172, 221–228. doi: 10.1111/j.1469-8137.2006.01875.x

McInnis, S. M., Emery, D. C., Porter, R., Desikan, R., Hancock, J. T., and Hiscock, S. J. (2006b). The role of stigma peroxidases in flowering plants: insights from further characterization of a stigma-specific peroxidase (SSP) from *Senecio squalidus* (Asteraceae). *J. Exp. Bot.* 57, 1835–1846. doi: 10.1093/jxb/erj182

Meng, L., Wong, J. H., Feldman, L. J., Lemaux, P. G., and Buchanan, B. B. (2010). A membrane-associated thioredoxin required for plant growth moves from cell to cell, suggestive of a role in intercellularcommunication. *Proc. Natl. Acad. Sci. U.S.A.* 107, 3900–3905. doi: 10.1073/pnas.0913759107

Meyer, Y., Belin, C., Delorme-Hinoux, V., Reichheld, J. P., and Riondet, C. (2012). Thioredoxin and glutaredoxin systems in plants: molecular mechanisms, crosstalks, and functional significance. *Antioxid. Redox. Signal.* 17, 1124–1160. doi: 10.1089/ars.2011.4327

Miranda-Vizuete, A., Sadek, C. M., Jiménez, A., Krause, W. J., Sutovsky, P., and Oko, R. (2004). The mammalian testis-specific thioredoxin system. *Antioxid. Redox Signal.* 6, 25–40. doi: 10.1089/152308604771978327

Murmu, J., Bush, M. J., DeLong, C., Li, S., Xu, M., Khan, M., et al. (2010). *Arabidopsis* basic leuine-zipper transcription factors TGA9 and TGA10 interact with floral glutaredoxins ROXY1 and ROXY2 and are redundantly required for anther development. *Plant Physiol.* 154, 1492–1504. doi: 10.1104/pp.110.159111

Ok, S. H., Yoo, K. S., and Shin, J. S. (2012). CBSXs are sensor relay proteins sensing adenosine-containing ligands in *Arabidopsis*. *Plant Signal. Behav.* 7, 664–667. doi: 10.4161/psb.19945

Pahr, S., Constantin, C., Mari, A., Scheiblhofer, S., Thalhamer, J., Ebner, C., et al. (2012). Molecular characterization of wheat allergens specifically recognized by patients suffering fromwheat-induced respiratory allergy. *Clin. Exp. Allergy* 42, 597–609. doi: 10.1111/j.1365-2222.2012.03961.x

Potocký M., Jones, M. A., Bezvoda, R., Smirnoff, N., and Žďárskı, V. (2007). Reactive oxygen species produced by NADPH oxidase are involved in pollen tube growth. *New Phytol.* 174, 742–751. doi: 10.1111/j.1469-8137.2007.02042.x

Potocký, M., Pejchar, P., Gutkowska, M., Jiménez-Quesada, M. J., Potocká, A., Alché, J. D., et al. (2012). NADPH oxidase activity in pollen tubes is affected by calcium ions, signaling phospholipids and Rac/Rop GTPases. *J. Plant Physiol.* 169, 1654–1663. doi: 10.1016/j.jplph.2012.05.014

Prado, A. M., Colaço, R., Moreno, N., Silva, A. C., and Feijó, J. A. (2008). Targeting of Pollen Tubes to Ovules Is Dependent on Nitric Oxide (NO) Signaling. *Mol. Plant* 1, 703–714. doi: 10.1093/mp/ssn034

Prado, A. M., Porterfield, D. M., and Feijó, J. A. (2004). Nitric oxide is involved in growth regulation and re-orientation of pollen tubes. *Development* 131, 2707–2714. doi: 10.1242/dev.01153

Pulido, P., Spínola, M. C., Kirchsteiger, K., Guinea, M., Pascual, M. B., Sahrawy, M., et al. (2010). Functional analysis of the pathways for 2-Cys peroxiredoxin reduction in *Arabidopsis thaliana* chloroplasts. *J. Exp. Bot.* 61, 4043–4054. doi: 10.1093/jxb/erq218

Reichheld, J. P., Mestres-Ortega, D., Laloi, C., and Meyer, Y. (2002). The multigenic family of thioredoxin h in *Arabidopsis thaliana*: specific expression and stress response. *Plant Physiol. Biochem.* 40, 685–690. doi: 10.1016/S0981-9428(02)01406-7

Reichheld, J. P., Khafif, M., Riondet, C., Droux, M., Bonnard, G., and Meyer, Y. (2007). Inactivation of thioredoxin reductases reveals a complex interplay between thioredoxin and glutathione pathways in *Arabidopsis* development. *Plant Cell* 19, 1851–1865. doi: 10.1105/tpc.107.050849

Schopfer, C. R., Nasrallah, M. E., and Nasrallah, J. B. (1999). The male determinant of self-incompatibility in Brassica. *Science* 286, 1697–1700. doi: 10.1126/science.286.5445.1697

Sharma, B., and Bhatla, S. C. (2013). Accumulation and scavenging of reactive oxygen species and nitric oxide correlate with stigma maturation and pollen–stigma interaction in sunflower. *Acta Physiol. Plant.* 35, 2777–2787. doi: 10.1007/s11738-013-1310-1

Shürmann, P., and Jacquot, J. P. (2000). Plant thioredoxin systems revisited. *Annu. Rev. Plant Physiol. Plant Mol. Biol.* 51, 371–400. doi: 10.1146/annurev.arplant.51.1.371

Singh, P., Kaloudas, D., and Raines, C. A. (2008). Expression analysis of the *Arabidopsis* CP12 gene family suggests novel roles for these proteins in roots and floral tissues. *J. Exp. Bot.* 59, 3975–3398. doi: 10.1093/jxb/ern236

Šírová, J., Sedláρová, M., Piterková, J., Luhová, L., and Petρivalskı, M. (2011). The role of nitric oxide in the germination of plant seeds and pollen. *Plant Sci.* 181, 560–572. doi: 10.1016/j.plantsci.2011.03.014

Spínola, M. C., Pérez-Ruiz, J. M., Pulido, P., Kirchsteiger, K., Guinea, M., González, M., et al. (2008). NTRC new ways of using NADPH in the chloroplast. *Physiol. Plant.* 133, 516–524. doi: 10.1111/j.1399-3054.2008.01088.x

Sprunck, S., Rademacher, S., Vogler, F., Gheyselinck, J., Grossniklaus, U., and Dresselhaus, T. (2013). Egg cell–secreted EC1 triggers sperm cell activation during double fertilization. *Science* 338, 1093–1097. doi: 10.1126/science.1223944

Tada, Y., Spoel, S. H., Pajerowska-Mukhtar, K., Mou, Z., Song, J., Wang, C., et al. (2008). Plant immunity requires conformational changes of NPR1 via S-nitrosylation and thioredoxins. *Science* 321, 952–956. (Erratum in *Science* 325:1072). doi: 10.1126/science.1156970

Takasaki, T., Hatakeyama, K., Suzuki, G., Watanabe, M., Isogai, A., and Hinata, K. (2000). The S receptor kinase determines self-incompatibility in Brassica stigma. *Nature* 403, 913–916. doi: 10.1038/35002628

Takayama, S., and Isogai, A. (2005). Self-incompatibility in plants. *Annu. Rev. Plant. Biol.* 56, 467–489. doi: 10.1146/annurev.arplant.56.032604.144249

Taylor, L. P., and Hepler, P. K. (1997). Pollen germination and tube growth. *Annu. Rev. Plant Physiol. Plant Mol. Biol.* 48, 461–491. doi: 10.1146/annurev.arplant.48.1.461

Timofejeva, L., Skibbe, D. S., Lee, S., Golubovskaya, I., Wang, R., Harper, L., et al. (2013). Cytological characterization and allelism testing of anther developmental mutants identified in a screen of maize male sterile lines. *G3 (Bethesda).* 3, 231–249. doi: 10.1534/g3.112.004465

Toriyama, K., Hanaoka, K., Okada, T., and Watanabe, M. (1998). Molecular cloning of a cDNA encoding a pollen extracellular protein as a potential source of a pollen allergen in Brassica rapa. *Febs Lett.* 424, 234–238. doi: 10.1016/S0014-5793(98)00174-4

Traverso, J. A., Meinnel, T., and Giglione, C. (2008). Expanded impact of protein N-myristoylation in plants. *Plant Signal. Behav.* 3, 501–502. doi: 10.4161/psb.3.7.6039

Traverso, J. A., Micalella, C., Martinez, A., Brown, S. C., Satiat-Jeunemaître, B., Meinnel, T., et al. (2013). Roles of N-Terminal fatty acid acylations in membrane compartment partitioning: *Arabidopsis* h-Type thioredoxins as a case study. *Plant Cell* 25, 1056–1077. doi: 10.1105/tpc.112.106849

Traverso, J. A., Vignols, F., Cazalis, R., Pulido, A., Sahrawy, M., Cejudo, F. J., et al. (2007). PsTRXh1 and PsTRXh2 are both pea h-type thioredoxins with antagonistic behavior in redox imbalances. *Plant Physiol.* 143, 300–311. doi: 10.1104/pp.106.089524

Utsumi, T., Ohta, H., Kayano, Y., Sakurai, N., and Ozoe, Y. (2005). The N-terminus of B96Bom, a *Bombyx mori* G-protein-coupled receptor, is N-myristoylated and translocated across the membrane. *FEBS J.* 272, 472–481. doi: 10.1111/j.1742-4658.2004.04487.x

Villarreal, F., Martín, V., Colaneri, A., González-Schain, N., Perales, M., Martín, M., et al. (2009). Ectopic expression of mitochondrial gamma carbonic anhydrase 2 causes male sterility by anther indehiscence. *Plant Mol. Biol.* 70, 471–485. doi: 10.1007/s11103-009-9484-z

Wan, C., Li, S., Wen, L., Kong, J., Wang, K., and Zhu, Y. (2007). Damage of oxidative stress on mitochondria during microspores development in Honglian CMS line of rice. *Plant Cell Rep.* 26, 373–382. doi: 10.1007/s00299-006-0234-2

Wang, C. L., Wu, J., Xu, G. H., Gao, Y. B., Chen, G., and Wu, J. Y. (2010a). S-RNase disrupts tip-localized reactive oxygen species and induces nuclear DNA degradation in incompatible pollen tubes of *Pyrus pyrifolia*. *J. Cell Sci.* 123, 4301–4309. doi: 10.1242/jcs.075077

Wang, S., Xie, B., Yin, L., Duan, L., Li, Z., Egrinya, A., et al. (2010b). Increased UV-B radiation affects the viability, reactive oxygen species accumulation and antioxidant enzyme activities in maize (*Zea mays* L.) Pollen. *Photochem. Photobiol.* 86, 110–116. doi: 10.1111/j.1751-1097.2009.00635.x

Wang, Y., Chen, T., Zhang, C., Hao, H., Liu, P., Zheng, M., et al. (2009a). Nitric oxide modulates the influx of extracellular Ca2+ and actin filament organization during cell wall construction in *Pinus bungeana* pollen tubes. *New Phytol.* 182, 851–862. doi: 10.1111/j.1469-8137.2009.02820.x

Wang, Z., Xing, S., Birkenbihl, R. P., and Zachgo, S. (2009b). Conserved functions of *Arabidopsis* and rice CC-type glutaredoxins in flower development and pathogen response. *Mol. Plant* 2, 323–335. doi: 10.1093/mp/ssn078

Wilkins, K. A., Bancroft, J., Bosch, M., Ings, J., Smirnoff, N., and Franklin-Tong, V. E. (2011). Reactive oxygen species and nitric oxide mediate actin reorganization and programmed cell death in the self-incompatibility response of papaver. *Plant Physiol.* 156, 404–416. doi: 10.1104/pp.110.167510

Wilson, I. D., Hiscock, S. J., James, P. E., and Hancock, J. T. (2009). Nitric oxide and nitrite are likely mediators of pollen interactions. *Plant Signal. Behav.* 4, 416–418. doi: 10.4161/psb.4.5.8270

Wu, H., de Graaf, B., Mariani, C., and Cheung, A. Y. (2001). Hydroxyproline-rich glycoproteins in plant reproductive tissues: structure, functions and regulation. *Cell Mol. Life Sci.* 58, 1418–1429. doi: 10.1007/PL00000785

Xing, S., and Zachgo, S. (2008). ROXY1 and ROXY2, two *Arabidopsis* glutare-doxin genes, are required for anther development. *Plant J.* 53, 790–801. doi: 10.1111/j.1365-313X.2007.03375.x

Xing, S., Lauri, A., and Zachgo, S. (2006). Redox regulation and flower develop-ment: a novel function for glutaredoxins. *Plant Biol.* 8, 547–555. doi: 10.1055/s-2006-924278

Xing, S., Rosso, M. G., and Zachgo, S. (2005). ROXY1, a member of the plant glutaredoxin family, is required for petal development in *Arabidopsis thaliana*. *Development* 132, 1555–1565. doi: 10.1242/dev.01725

Xing, S., Salinas, M., and Huijser, P. (2011). New players unveiled in early anther development. *Plant Signal. Behav.* 6, 934–938. doi: 10.4161/psb.6.7.15668

Yoo, K. S., Ok, S. H., Jeong, B. C., Jung, K. W., Cui, M. H., Hyoung, S., et al. (2011). Single cystathionine β-synthase domain-containing proteins modulate development by regulating the thioredoxin system in *Arabidopsis*. *Plant Cell* 23, 3577–3594. doi: 10.1105/tpc.111.089847

Yun, B. W., Feechan, A., Yin, M., Saidi, N. B., Le Bihan, T., Yu, M., et al. (2011). S-Nitrosylation of NADPH oxidase regulates cell death in plant immunity. *Nature* 478, 264–268. doi: 10.1038/nature10427

Zafra, A., Rodríguez-García, M. I., and Alché, J. D. (2010). Cellular localization of ROS and NO in olive reproductive tissues during flower development. *BMC Plant Biol.* 10:36. doi: 10.1186/1471-2229-10-36

Zaidi, M. A., O'Leary, S., Wu, S., Gleddie, S., Eudes, F., Laroche, A., et al. (2012). A molecular and proteomic investigation of proteins rapidly released from triticale pollen upon hydration. *Plant Mol. Biol.* 79, 101–121. doi: 10.1007/s11103-012-9897-y

Zechmann, B., Koffler, B. E., and Russell, S. D. (2011). Glutathione synthesis is essential for pollen germination *in vitro*. *BMC Plant Biol.* 11:54. doi: 10.1186/1471-2229-11-54

26

Recent advances in understanding of meiosis initiation and the apomictic pathway in plants

Chung-Ju R. Wang[1] and Ching-Chih Tseng[1,2]*

[1] Institute of Plant and Microbial Biology, Academia Sinica, Taipei, Taiwan
[2] Institute of Plant Biology, National Taiwan University, Taipei, Taiwan

Edited by:
Ian Henderson, University of Cambridge, UK

Reviewed by:
Daphné Autran, Institut de Recherche pour le Développement, France
Gabriela Carolina Pagnussat, Universidad Nacional de Mar del Plata, Argentina

*Correspondence:
Chung-Ju R. Wang, Institute of Plant and Microbial Biology, Academia Sinica, Room 120, Section 2, Academia Road, Taipei 11529, Taiwan
e-mail: rwang@gate.sinica.edu.tw

Meiosis, a specialized cell division to produce haploid cells, marks the transition from a sporophytic to a gametophytic generation in the life cycle of plants. In angiosperms, meiosis takes place in sporogenous cells that develop *de novo* from somatic cells in anthers or ovules. A successful transition from the mitotic cycle to the meiotic program in sporogenous cells is crucial for sexual reproduction. By contrast, when meiosis is bypassed or a mitosis-like division occurs to produce unreduced cells, followed by the development of an embryo sac, clonal seeds can be produced by apomixis, an asexual reproduction pathway found in 400 species of flowering plants. An understanding of the regulation of entry into meiosis and molecular mechanisms of apomictic pathway will provide vital insight into reproduction for plant breeding. Recent findings suggest that AM1/SWI1 may be the key gene for entry into meiosis, and increasing evidence has shown that the apomictic pathway is epigenetically controlled. However, the mechanism for the initiation of meiosis during sexual reproduction or for its omission in the apomictic pathway still remains largely unknown. Here we review the current understanding of meiosis initiation and the apomictic pathway and raised several questions that are awaiting further investigation.

Keywords: apomixis, meiosis initiation, RNA-directed DNA methylation, plant reproduction, ameiotic1

INTRODUCTION

Meiosis is an extremely important step in sexual reproduction. It is widely accepted that it evolved from mitosis and shares certain features with mitosis (Maynard Smith, 1978). Yet at least three meiosis-specific events make meiosis a specialized cell division: meiotic recombination and pairing between homologous chromosomes during prophase I, the suppression of sister-chromatid separation during the first meiotic division, and the absence of chromosome replication at the start of the second division (Kleckner, 1996). While these meiosis-specific events have been studied extensively, the mechanisms that switch mitosis into meiosis are still puzzling. In multicellular organisms, meiosis initiation takes place within multicellular organs; consequently, mechanisms that initiate meiosis must integrate developmental cues. In plants, the decision to start meiosis may also be connected with reproductive cell fate specification since plants do not have pre-determined germ lines. Thus, the switch of somatic fate to germinal cell fate and the mitosis–meiosis cell cycle transition occur sequentially during the development of reproductive organs (i.e., anthers and ovules; Ma, 2005). Importantly, these sexual processes can be replaced by the asexual apomictic pathway in which meiosis is bypassed or a mitosis-like division occurs to produce unreduced daughter cells, followed by the development of an embryo without fertilization, apomictic plants can then produce diploid seeds with identical genetic content to their maternal genome. This phenomenon is called apomixis that occurs naturally in some flowering plants (Barcaccia and Albertini, 2013). If apomixis is engineered into crops to produce clonal seeds, its application on agriculture will be broad and profound. Here, we review the current understanding of the cell cycle transition that directs

sporogenous cells to leave the mitotic cell cycle and enter the meiotic program in higher plants and additionally discuss advances in the apomictic pathway.

WHAT HAVE WE LEARNED FROM OTHER MODEL SPECIES ABOUT INITIATION OF MEIOSIS?

The cellular events during meiosis are evolutionarily conserved among species; however, the mechanisms controlling the initiation of meiosis are diverse (Pawlowski et al., 2007). The molecular controls elucidated to date involve signaling pathways, transcriptional and translational regulations of meiotic genes, and cyclin-dependent kinase (CDK) circuits. Although different mechanisms are adopted, the final readout is likely the activation of a specific cyclin–CDK complex to initiate the meiotic S phase. It was also suggested from many studies that the decision to start meiosis is made before the onset of the pre-meiotic S phase (Watanabe et al., 2001). Here, we first summarize the discoveries from several model species, and then discuss recent advances in plants.

The meiosis decision in single-celled yeasts is often cued by environmental conditions. In the budding yeast *Saccharomyces cerevisiae*, starvation induces expression of the *Initiator of Meiosis I* (*IME1*) gene, which encodes a transcription factor responsible for activating early meiotic genes (Chu et al., 1998). One of these target genes, *IME2*, which encodes a Ser/Thr protein kinase, promotes meiotic DNA replication by directly phosphorylating Rfa2, a subunit of replication protein A (Foiani et al., 1996; Clifford et al., 2005). Sic1, an inhibitor of the CDK (CDC28), is also phosphorylated by Ime2p and then leads to its degradation. Subsequently, CDC28, in conjunction with the B-type S phase cyclins, Clb5, and

Clb6, triggers the initiation of the premeiotic S phase (Dirick et al., 1998; Stuart and Wittenberg, 1998).

In fission yeast *S. pombe*, a key transcription factor *STE11*, which is produced in response to environmental conditions, is responsible for early meiotic genes expression (Sugimoto et al., 1991). MEI2, an RRM-type RNA-binding protein, plays a crucial role in promoting entry into meiosis by regulating meiosis-specific mRNAs accumulation. During mitosis, MEI2 is inactivated by PAT1 kinase. Under meiosis-inducing conditions, this repression of MEI2 is released, allowing binding to and stabilization of meiosis-specific mRNAs at the G1 phase (Kitamura et al., 2001). In addition, this process reinforces stabilization also by sequestrating MMI1 protein, which function is to eliminate these meiotic mRNAs (Harigaya et al., 2006). Finally, CDC2 kinase binding with cyclin CIG2 is essential for entry into the pre-meiotic S phase (Borgne et al., 2002). Recently, protein S-palmitoylation, a lipid modification was also found to regulate the entry into meiosis (Zhang et al., 2013).

In mammals, meiosis is initiated at different stages of development in females and males (Bowles and Koopman, 2007). Mouse studies have revealed that retinoic acid (RA) produced during embryonic development can induce meiosis in both sexes. The level of RA is negatively regulated by the *Cyp26b1* enzyme that has RA degradation activity (Bowles et al., 2006; Koubova et al., 2006). *Stimulated by RA 8* (*Stra8*), a vertebrate-specific gene, is then induced by RA and is required for the transition to meiosis (Anderson et al., 2008). *Stra8* plays no role in the mitotic phases of embryonic germ-cell development, but in females it is required for pre-meiotic DNA replication and the subsequent events of meiotic prophase. On the other hand, *Dmrt1* represses *Stra8* transcription in the mitotic phase, thereby preventing meiosis (Matson et al., 2010).

From these studies, the mechanisms that initiate meiosis are very different, and more importantly, the genes involved share no similarity. No doubt different strategies evolved because of the different reproductive requirements of diverse organisms.

THE DECISION OF MITOSIS–MEIOSIS SWITCH IN PLANTS

In plants, meiosis is initiated in sporogenous cells that are differentiated in ovules and anthers (Bhatt et al., 2001). In each ovule, only a single megaspore mother cell (MMC) surrounded by the somatic nucellar cells is differentiated and then undergoes meiosis (**Figure 1**). During anther development, after primary sporogenous cells (i.e., the precursor of pollen mother cells, PMCs) are differentiated, they first undergo several rounds of mitosis to proliferate, and then meiosis occurs synchronously in all the PMCs of each anther (**Figure 1**; Palmer, 1971). Thus, the decision of mitosis–meiosis transition must coordinate with the developmental stages of anthers and ovules. For example, the signal that starts meiosis in an anther must be generated after complete development of the somatic layers of anthers (Kelliher and Walbot, 2011). Interestingly, the signal can also establish the synchronization of the meiotic cell cycle in an anther. On the other hand, only a single MMC in each ovule is specified to enter meiosis, which accompanies the development of ovule in parallel. Thus, the regulatory mechanism of meiosis initiation may be different between female and male in plants because of distinct development of sporogenesis.

The first discovery about meiosis initiation was the isolation of the maize *ameiotic1* (*am1*) mutant by Rhoades (1956). The original *am1* mutant allele does not undergo meiosis; instead mitosis-like divisions take place in well-developed meiocytes in both female and male organs (Golubovskaya et al., 1993). *Am1* encodes a plant specific coiled-coil protein with unknown functions (Pawlowski et al., 2009). All five null mutant alleles display identical phenotypes in male meiocytes in which mitosis replaces meiosis. However, female MMCs in the mutant may either undergo mitosis, or arrest at interphase. Interestingly, the *am1-praI* allele carrying a single amino acid substitution (R358W) can enter meiosis but cells arrest in the leptotene/zygotene stage, resembling the phenotype of the rice *am1* mutant that also carries an amino acid substitution (R360W) in a conserved region (Golubovskaya et al., 1997; Pawlowski et al., 2009; Che et al., 2011). These results suggest that AM1 is required for meiosis initiation and may also regulate meiotic progression. In contrast to maize and rice, mutants in the closest homolog of *Am1* in *Arabidopsis*, *switch1/dyad* (*swi1*), exhibit abnormal meiosis with sister chromatid cohesion defects in male meiocytes, and the mitosis-like division was only observed in female meiocytes (Mercier et al., 2001, 2003). These differences among species may indicate that the AM1-related genes have undergone species-specific diversification.

While the molecular functions of AM1/SWI1 are still unknown, microarray analyses showed that AM1 is required for normal expression of many meiotic genes (Nan et al., 2011). Using Agilent 44K microarrays, the authors compared transcriptomes in 1-mm and 1.5-mm anthers of *am1-489* (null allele) and *am1-praI* (point mutant allele) and their fertile siblings. In 1-mm anthers when meiosis is about to start in the wild-type, 484 genes were missing and 1208 genes were ectopically expressed in *am1-489* anthers. These genes are considered to contribute to the initiation of meiosis or the suppression of mitosis. In 1.5-mm anthers, during prophase I in the wild-type, 3700 transcripts were missing and another 3107 genes were differentially expressed in *am1-489* anthers. Nearly 60% of transcriptome changes, regardless of stage, were genes enriched in PMCs and many putative meiosis-related genes were found among them. However, none of the meiosis-related genes were regulated in an absolute On/Off pattern on the *am1-489* allele, somewhat surprising given that the *am1-489* PMCs perform mitosis instead of abnormal meiosis. These results redefine the role of AM1 in the modulation of transcript accumulation for many meiotic genes rather than simply switching them on or off (Nan et al., 2011).

Recently, microarray analyses on laser-captured germinal and somatic initials from maize 0.3-mm anthers (right after sporogenous cells are differentiated) found about 2500 genes specific or enriched in germinal initials (Kelliher and Walbot, 2014). Surprisingly, more than 100 meiotic genes are expressed in the mitotic amplification period that is long before the onset of meiosis initiation. This finding raises a possibility that precocious expression of meiotic genes permits gradual dilution of mitotic chromatin components, a hypothesis recently proposed for the mouse germline (Hackett et al., 2013). Another possibility is that those PMC

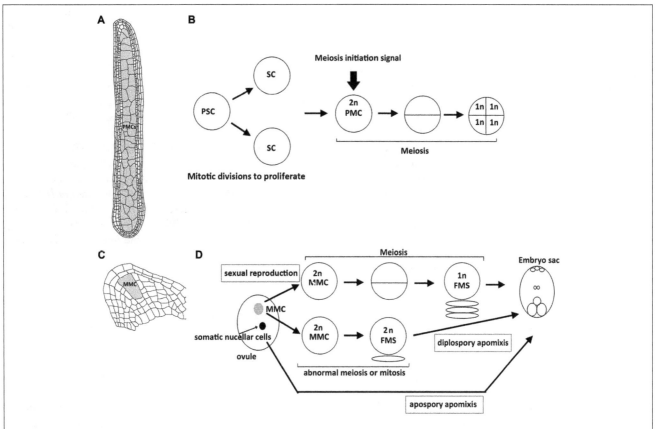

FIGURE 1 | Structure of plant reproductive organs in maize and sequence of events leading to spore or gametophyte formation in anthers and ovules. (A) Longitudinal section of an anther with numerous pollen mother cells (PMCs, shown in gray) that are proliferated from primary sporogenous cells by mitosis, which accompanies the development of surrounding 4 layers of somatic cells. **(B)** After primary sporogenous cells (PSCs) are differentiated, they first undergo mitotic divisions to produce sporogenous cells (SCs) and further develop into PMCs. By the time when the development of surrounding somatic cells (shown in **A**) is complete, unknown meiosis initiation siganl is generated to start meisois synchronously in all PMCs of an anther. Each PMC enters meiosis to produce four haploid spore cells. **(C)** Longitudinal section of an ovule with a single megaspore mother cell (MMC, shown in gray). **(D)** Schematic illustration showing the sequential development of embryo sac through sexual reproduction or apomictic pathways. In sexual reproduction, the single MMC (shown in gray) is differentiated and then enters meiosis to produce a haploid functional megaspore (FMS), and then develops into an embryo sac. In diplospory apomixis, the specified MMC unergoes an abnormal meiosis or mitosis to produce a diploid FMS. In apospory apomixis, somatic nucellar cells develop into embryo sac without meiosis.

precursors are preparing for meiosis at the transcriptional level, and may store some meiotic transcripts for translation at later developmental stages (Zhang et al., 2014). Regardless, this finding suggests that the decision to start meiosis is a series of consecutive steps rather than a single switch. Perhaps, the expression of meiotic genes may be one of the earliest actions, and the following regulatory cascade finally governs the initiation and progression of meiosis. Thus, which transcription factors are responsible for early meiotic gene expression and whether meiotic genes are under translational control are interesting questions for further study. In addition, identification of components in the regulatory cascade will provide better understanding of this process.

Another gene that has been reported to be involved in meiosis initiation is rice *MEL2*, named for its "meiosis arrested at leptotene" phenotype. *MEL2* encodes an RNA-recognition-motif (RRM) protein, and it is required for regulating the premeiotic G1/S-phase transition of male and female germ cells (Nonomura et al., 2011) as most germ cells fail to enter pre-meiotic S phase in *mel2* mutant. A small proportion of PMCs can escape

from the defects and undergo meiosis with a significant delay or continued mitotic cycles. How an RRM protein affects the initiation of meiosis is unclear at the molecular level, but this result implied a possible link between mRNA processing, transport or stability, and entry into meiosis in plants. Studies in yeast have shown that the final trigger to start meiosis is the activation of specific cyclin–CDK complexes to initiate the meiotic S phase. *Arabidopsis* has at least 50 cyclins and only a few of them are specifically expressed in the inflorescence (Bulankova et al., 2013). Mutant analyses revealed that some of these cyclins contribute to distinct meiosis-related processes, but none of cyclin mutants showed meiosis initiation defects, which was attributed to gene redundancy. Thus, it will be interesting to know which, if any, cyclin–CDK complex is responsible for the transition. Besides cyclin–CDK complexes, some meiosis-specific regulators, such as replication factor MUM2 and cohesion protein REC8, are involved at the meiotic S phase although much of the basic replication apparatus is employed (Strich, 2004). Therefore, what is special about the pre-meiotic S phase and which are the specific genes that differ from the mitotic S phase in plants? Understanding of these meiosis-specific components at meiotic S phase will help us to illustrate the molecular mechanisms of meiosis initiation. A proteomics study may offer valuable information on this aspect.

To date, mutants directly affecting meiosis initiation showed similar phenotypes in that some of reproductive cells fail to enter meiosis in either female, male, or both sexes. Although some of these mutants produce unreduced daughter cells by mitosis-like division, there is no evidence that these resulting diploid cells in ovules would undergo the apomictic pathway without fertilization. However, an interesting study demonstrated that in the *Arabidopsis swi1/dyad* mutant, few seeds were produced when pollinated with wild-type pollens. Most of the progeny were triploid, suggesting that unreduced female daughter cells after mitosis-like division are able to develop further and be fertilized by haploid male gametes (Ravi et al., 2008).

CURRENT ADVANCES IN THE APOMICTIC PATHWAY

Apomixis is a type of asexual reproduction through seeds that avoid both meiosis and fertilization. In the apomictic pathway, differentiated MMCs or other somatic cells in ovules that gain germinal cell fate are able to bypass meiosis or undergo an abnormal meiosis to produce unreduced spores that further divide mitotically to form an embryo sac (**Figure 1**; Koltunow, 1993; Carman, 1997). Although apomixis is genetically regulated and occurs naturally in more than 400 species of flowering plants, its implementation at the molecular level is still unclear. Over the past few years, there has been increasing evidence to show that epigenetic control may regulate apomixis. In *Arabidopsis, argonaute 9 (ago9)* mutants exhibit multiple MMCs compared to a single MMC in the wild-type ovule, and additional MMCs in the mutant are able to initiate gametogenesis without undergoing meiosis, resembling apospory (**Figure 1**; Olmedo-Monfil et al., 2010). AGO9 preferentially interacts with 24-nucleotide small interfering RNAs (siRNA) derived from transposable elements to direct homolog-based RNA-dependent DNA methylation (RdDM). Moreover, mutations in *SUPPRESSOR OF GENE SILENCING3* (*SGS3*) and *RNA-DEPENDENT RNA POLYMERASE6* (*RDR6*), two genes required for siRNA biogenesis, also lead to an identical defect to that in *ago9* mutants (Olmedo-Monfil et al., 2010). Similarly, maize AGO104, the homolog of *Arabidopsis* AGO9, is found to regulate reproductive fate despite some differences between maize *ago104* and *Arabidopsis ago9* phenotypes (Singh et al., 2011). The maize *ago104* mutant has a single MMC; however, defective female meiosis with aberrant condensation results in functional female gametes with an unreduced chromosome set, resembling diplospory (**Figure 1**). Additionally, AGO104 is required for heterochromatic CHG and CHH methylation. Consistent with the idea that epigenetics regulates apomixis, mutations of two DNA methyltransferases, DMT102 and DMT103 in maize, also exhibit apomictic development (Garcia-Aguilar et al., 2010). Thus, loss of RdDM seems to direct somatic cells to distinct reproductive cells with an apomictic fate (seen in *Arabidopsis ago9* mutant) or lead to apomixis in correctly specified MMCs (seen in maize *ago104* mutant). Interestingly, both AGO9 in *Arabidopsis* and AGO104 in maize are specifically expressed in surrounding somatic nucellar cells, and not in the reproductive cells, implying that both genes control the apomictic pathway in a non-cell-autonomous manner. siRNAs produced from somatic cells may move to germinal cells to regulate the chromatin state by suppressing transposable elements. Indeed, many transposable elements are silenced in *Arabidopsis* wild-type ovules, in an AGO9-dependent manner (Durán-Figueroa and Vielle-Calzada, 2010). These results suggest a link between siRNA-dependent chromatin remodeling and the apomictic pathway (Garcia-Aguilar et al., 2010; Grimanelli, 2012).

Another gene belonging to the ARGONAUTE family with meiotic phenotypes is rice *MEL1*. It encodes an AGO5 protein that is required for maintaining germ cell identity and normal meiosis progression. Interestingly, the *mel1* mutant also shows defective chromosome condensation with abnormal pericentromere histone modification (Nonomura et al., 2007). Recently, MEL1 has been shown to bind to 21-nucleotide phased small interfering RNA (Komiya et al., 2014). Further investigation is needed to understand the epigenetic regulation of plant reproduction.

Over the past few years, the identification of mutants has shed light on genetic control of epigenetic mechanisms involved in apomixis. However, it is still not clear how the RdDM-dependent process affects cell fate specification, meiosis, and gametophyte development? Why is there a need for transposes-derived siRNA in the germ line? Is it possible that RdDM resets cell fate in the germ line, a role also demonstrated for the animal PIWI pathway (Houwing et al., 2008; Juliano et al., 2011)? Perhaps, identifying the targets of the RdDM pathway at different stages will be essential for further definition of their roles. In addition, what is the relationship between AM1/SWI1-dependent meiosis initiation and the RdDM pathway? It is worth noting that alterations in histone modification were observed in the *swi1* mutant (Boateng et al., 2008), raising the possibility that somehow AM1/SWI1 is involved in chromatin remodeling. Many exciting questions are awaiting further investigation.

CONCLUSION

Understanding the initiation of meiosis and apomixis in plants will be enlightening, and may have many potential applications for plant breeding and in agriculture including developing a strategy for acquiring apomixis in crops, and allowing manipulation of the meiotic cell cycle. It will be crucial to identify more participants in the mitosis–meiosis decision and the apomictic pathway and to explore their molecular functions.

ACKNOWLEDGMENTS

We would like to thank the members of the Wang lab for insightful discussions and critical comments. This work was supported by the Career Development Award from Academia Sinica, Taiwan.

REFERENCES

Anderson, E. L., Baltus, A. E., Roepers-Gajadien, H. L., Hassold, T. J., de Rooij, D. G., van Pelt, A. M. et al. (2008). *Stra8* and its inducer, retinoic acid, regulate meiotic initiation in both spermatogenesis and oogenesis in mice. *Proc. Natl. Acad. Sci. U.S.A.* 105, 14976–14980. doi: 10.1073/pnas.0807297105

Barcaccia, G., and Albertini, E. (2013). Apomixis in plant reproduction: a novel perspective on an old dilemma. *Plant Reprod.* 26, 159–179. doi: 10.1007/s00497-013-0222-y

Bhatt, A. M., Canales, C., and Dickinson, H. G. (2001). Plant meiosis: the means to 1N. *Trends Plant Sci.* 6, 114–121. doi: 10.1016/S1360-1385(00)01861-6

Boateng, K. A., Yang, X., Dong, F., Owen, H. A., and Makaroff, C. A. (2008). SWI1 is required for meiotic chromosome remodeling events. *Mol. Plant.* 4, 620–633. doi: 10.1093/mp/ssn030

Borgne, A., Murakami, H., Ayté, J., and Nurse, P. (2002). The G1/S cyclin Cig2p during meiosis in fission yeast. *Mol. Biol. Cell.* 13, 2080–2090. doi: 10.1091/mbc.01-10-0507

Bowles, J., Knight, D., Smith, C., Wilhelm, D., Richman, J., Mamiya, S., et al. (2006). Retinoid signaling determines germ cell fate in mice. *Science* 312, 596–600. doi: 10.1126/science.1125691

Bowles, J., and Koopman, P. (2007). Retinoic acid, meiosis and germ cell fate in mammals. *Development* 134, 3401–3411. doi: 10.1242/dev.001107

Bulankova, P., Akimcheva, S., Fellner, B., and Riha, K. (2013). Identification of *Arabidopsis* meiotic cyclins reveals functional diversification among plant cyclin genes. *PLoS Genet.* 9:e1003508. doi: 10.1371/journal.pgen.1003508

Carman, J. G. (1997). Asynchronous expression of duplicate genes in angiosperms may cause apomixis, bispory, tetraspory, and polyembryony. *Biol. J. Linn. Soc.* 61, 51–94. doi: 10.1111/j.1095-8312.1997.tb01778.x

Che, L., Tang, D., Wang, K., Wang, M., Zhu, K., Yu, H., et al. (2011). OsAM1 is required for leptotene-zygotene transition in rice. *Cell Res.* 21, 654–665. doi: 10.1038/cr.2011.7

Chu, S., DeRisi, J., Eisen, M., Mulholland, J., Botstein, D., Brown, P. O., et al. (1998). The transcriptional program of sporulation in budding yeast. *Science* 282, 699–705. doi: 10.1126/science.282.5389.699

Clifford, D. M., Stark, K. E., Gardner, K. E., Hoffmann-Benning, S., and Brush, G. S. (2005). Mechanistic insight into the Cdc28-related protein kinase Ime2 through analysis of replication protein A phosphorylation. *Cell Cycle* 12, 1826–1833. doi: 10.4161/cc.4.12.2214

Dirick, L., Goetsch, L., Ammerer, G., and Byers, B. (1998). Regulation of meiotic S phase by Ime2 and a Clb5,6-associated kinase in *Saccharomyces cerevisiae*. *Science* 281, 1854–1857. doi: 10.1126/science.281.5384.1854

Durán-Figueroa, N., and Vielle-Calzada, J. P. (2010). ARGONAUTE9-dependent silencing of transposable elements in pericentromeric regions of *Arabidopsis*. *Plant Signal. Behav.* 5, 1476–1479. doi: 10.1038/nature08828

Foiani, M., Nadjar-Boger, E., Capone, R., Sagee, S., Hashimshoni, T., and Kassir, T. (1996). A meiosis-specific protein kinase, Ime2, is required for the correct timing of DNA replication and for spore formation in yeast meiosis. *Mol. Gen. Genet.* 253, 278–288. doi: 10.1007/s004380050323

Garcia-Aguilar, M., Michaud, C., Leblanc, O., and Grimanelli, D. (2010). Inactivation of a DNA methylation pathway in maize reproductive organs results in apomixis-like phenotypes. *Plant Cell* 22, 3249–3267. doi: 10.1105/tpc.109.072181

Golubovskaya, I., Avalkina, N., and Sheridan, W. F. (1997). New insights into the role of the maize ameiotic1 locus. *Genetics* 147, 1339–1350.

Golubovskaya, I., Grebennikova, Z. K., Avalkina, N. A., and Sheridan, W. F. (1993). The role of the ameiotic1 gene in the initiation of meiosis and in subsequent meiotic events in maize. *Genetics* 135, 1151–1166.

Grimanelli, D. (2012). Epigenetic regulation of reproductive development and the emergence of apomixis in angiosperms. *Curr. Opin. Plant Biol.* 15, 57–62. doi: 10.1016/j.pbi.2011.10.002

Hackett, J. A., Sengupta, R., Zylicz, J. J., Murakami, K., Lee, C., Down, T. A., et al. (2013). Germline DNA demethylation dynamics and imprint erasure through 5-hydroxymethylcytosine. *Science* 339, 448–452. doi: 10.1126/science.1229277

Harigaya, Y., Tanaka, H., Yamanaka, S., Tanaka, K., Watanabe, Y., Tsutsumi, C., et al. (2006). Selective elimination of messenger RNA prevents an incidence of untimely meiosis. *Nature* 442, 45–50. doi: 10.1038/nature04881

Houwing, S., Berezikov, E., and Ketting, R. F. (2008). Zili is required for germ cell differentiation and meiosis in zebrafish. *EMBO J.* 27, 2702–2711. doi: 10.1038/emboj.2008.204

Juliano, C., Wang, J., and Lin, H. (2011). Uniting germline and stem cells: the function of Piwi proteins and the piRNA pathway in diverse organisms. *Annu. Rev. Genet.* 45, 447–469. doi: 10.1146/annurev-genet-110410-132541

Kelliher, T., and Walbot, V. (2011). Emergence and patterning of the five cell types of the Zea mays anther locule. *Dev. Biol.* 350, 32–49. doi: 10.1016/j.ydbio.2010.11.005

Kelliher, T., and Walbot, V. (2014). Maize germinal cell initials accommodate hypoxia and precociously express meiotic genes. *Plant J.* 77, 639–652. doi: 10.1111/tpj.12414

Kitamura, L., Katayama, S., Dhut, S., Sato, M., Watanabe, Y., Yamamoto, M., et al. (2001). Phosphorylation of Mei2 and Ste11 by Pat1 kinase inhibits sexual differentiation via ubiquitin proteolysis and 14-3-3 protein in fission yeast. *Dev. Cell.* 1, 389–399. doi: 10.1016/S1534-5807(01)00037-5

Kleckner, N. (1996). Meiosis: how could it work. *Proc. Natl. Acad. Sci. U.S.A.* 93, 8167–8174. doi: 10.1073/pnas.93.16.8167

Koltunow, A. M. (1993). Apomixis: embryo sacs and embryos formed without meiosis or fertilization in ovules. *Plant Cell* 5, 1425–1437. doi: 10.1105/tpc.5.10.1425

Komiya, R., Ohyanagi, H., Niihama, M., Watanabe, T., Nakano, M., Kurata, N., et al. (2014). Rice germline-specific Argonaute MEL1 protein binds to phasiRNAs generated from more than 700 lincRNAs. *Plant J.* 78, 385–397. doi: 10.1111/tpj.12483

Koubova, J., Menke, D. B., Zhou, Q., Capel, B., Griswold, M. D., and Page, D. C. (2006). Retinoic acid regulates sex-specific timing of meiotic initiation in mice. *Proc. Natl. Acad. Sci. U.S.A.* 103, 2474–2479. doi: 10.1073/pnas.0510813103

Ma, H. (2005). Molecular genetic analyses of microsporogenesis and microgametogenesis in flowering plant. *Annu. Rev. Plant Biol.* 56, 393–434. doi: 10.1146/annurev.arplant.55.031903.141717

Matson, C. K., Murphy, M. W., Griswold, M. D., Yoshida, S., Bardwell, V. J., and Zarkower, D. (2010). The mammalian doublesex homolog DMRT1 is a transcriptional gatekeeper that controls the mitosis versus meiosis decision in male germ cells. *Dev. Cell.* 19, 612–624. doi: 10.1016/j.devcel.2010.09.010

Maynard Smith, J. (1978). *The Evolution of Sex.* Cambridge: Cambridge University Press.

Mercier, R., Armstrong, S. J., Horlow, C., Jackson, N. P., Makaroff, C. A., Vezon, D., et al. (2003). The meiotic protein SWI1 is required for axial element formation and recombination initiation in *Arabidopsis*. *Development* 130, 3309–3318. doi: 10.1242/dev.00550

Mercier, R., Vezon, D., Bullier, E., Motamayor J. C., Sellier A., Lefevre, F., et al. (2001). SWITCH1 (SWI1): a novel protein required for the establishment of sister chromatid cohesion and for bivalent formation at meiosis. *Gene Dev.* 15, 1859–1871. doi: 10.1101/gad.203201

Nan, G. L., Ronceret, A., Wang, R. C., Fernandes, J. F., Cande, W. Z., and Walbot, V. (2011). Global transcriptome analysis of two ameiotic1 alleles in maize anthers: defining steps in meiotic entry and progression through prophase I. *BMC Plant Biol.* 11:120. doi: 10.1186/1471-2229-11-120

Nonomura, K., Eiguchi, M., Nakano, M., Takashima, K., Komeda, N., Fukuchi, S., et al. (2011). A novel RNA-recognition-motif protein is required for premeiotic G1/S-phase transition in rice (*Oryza sativa* L.). *PLoS Genet.* 7:e1001265. doi: 10.1371/journal.pgen.1001265

Nonomura, K., Morohoshi, A., Nakano, M., Eiguchi, M., Miyao, A., Hirochika, H., et al. (2007). A germ cell specific gene of the ARGONAUTE family is essential for the progression of premeiotic mitosis and meiosis during sporogenesis in rice. *Plant Cell* 19, 2583–2594. doi: 10.1105/tpc.107.053199

Olmedo-Monfil, V., Durán-Figueroa, N., Arteaga-Vázquez, M., Demesa-Arévalo, E., Autran, D., Grimanelli, D., et al. (2010). Control of female gamete formation by a small RNA pathway in *Arabidopsis*. *Nature* 464, 628–632. doi: 10.1038/nature08828

Palmer, R. G. (1971). Cytological studies of ameiotic and normal maize with reference to premeiotic pairing. *Chromosoma* 35, 233–246. doi: 10.1007/BF00326276

Pawlowski, W. P., Sheehan, M. J., and Ronceret, A. (2007). In the beginning: the initiation of meiosis. *Bioessays* 29, 511–514. doi: 10.1002/bies.20578

Pawlowski, W. P., Wang, C. R., Golubovskaya, I. N., Szymaniak, J. M., Shi, L., Hamant, O., et al. (2009). Maize AMEIOTIC1 is essential for multiple early meiotic processes and likely required for the initiation of meiosis. *Proc. Natl. Acad. Sci. U.S.A.* 106, 3603–3608. doi: 10.1073/pnas.0810115106

Ravi, M., Marimuthu, M. P., and Siddiqi, I. (2008). Gamete formation without meiosis in *Arabidopsis*. *Nature* 451, 1121–1124. doi: 10.1038/nature06557

Rhoades, M. M. (1956). Genetic control of chromosomal behavior. *Maize Genet. Coop. News L.* 30, 38–42.

Singh, M., Goel, S., Meeley, R. B., Dantec, C., Parrinello, H., Michaud, C., et al. (2011). Production of viable gametes without meiosis in maize deficient for an ARGONAUTE protein. *Plant Cell* 23, 443–458. doi: 10.1105/tpc.110.079020

Strich, R. (2004). Meiotic DNA replication. *Curr. Top. Dev. Biol.* 61, 29–60. doi: 10.1016/S0070-2153(04)61002-61007

Stuart, D., and Wittenberg, C. (1998). CLB5 and CLB6 are required for premeiotic DNA replication and activation of the meiotic S/M checkpoint. *Genes Dev.* 12, 2698–2710. doi: 10.1101/gad.12.17.2698

Sugimoto, A., Iino, Y., Maeda, T., Watanabe, Y., and Yamamoto, M. (1991). Schizosaccharomyces pombe ste11+ encodes a transcription factor with an HMG motif that is a critical regulator of sexual development. *Genes Dev.* 5, 1990–1999. doi: 10.1101/gad.5.11.1990

Watanabe, Y., Yokobayashi, S., Yamamoto, M., and Nurse, P. (2001). Pre-meiotic S phase is linked to reductional chromosome segregation and recombination. *Nature* 409, 359–363. doi: 10.1038/35053103

Zhang, H., Egger, R. L., Kelliher, T., Morrow, D., Fernandes, J., Nan, G. L., et al. (2014). Transcriptomes and proteomes define gene expression progression in pre-meiotic maize anthers. *G3 (Bethesda)* 4, 993–1010. doi: 10.1534/g3.113.009738

Zhang, M. M., Wu, P. J., Kelly, F. D., Nurse, P., and Hang, H. C. (2013). Quantitative control of protein S-palmitoylation regulates meiotic entry in fission yeast. *PLoS Biol.* 11:e1001597. doi: 10.1371/journal.pbio.1001597

Permissions

List of Contributors

Timothy Kelliher
Syngenta Biotechnology Inc., Research Triangle Park, NC, USA

Rachel L. Egger, Han Zhang and Virginia Walbot
Department of Biology, Stanford University, Stanford, CA, USA

Shu-Nong Bai
State Key Laboratory of Protein & Plant Gene Research, Quantitative Biology Center, College of Life Science, Peking University, Beijing, China

Diego Hojsgaard and Elvira Hörandl
Department of Systematics, Biodiversity and Evolution of Plants, Albrecht-von-Haller Institute for Plant Sciences, Georg-August University of Göttingen, Göttingen, Germany

Justin S. Williams
Department of Biochemistry and Molecular Biology, Pennsylvania State University, University Park, PA, USA

Lihua Wu, Shu Li and Penglin Sun
Intercollege Graduate Degree Program in Plant Biology, Pennsylvania State University, University Park, PA, USA

Teh-Hui Kao
Department of Biochemistry and Molecular Biology, Pennsylvania State University, University Park, PA, USA
Intercollege Graduate Degree Program in Plant Biology, Pennsylvania State University, University Park, PA, USA

Claudia Köhler and Clément Lafon-Placette
Department of Plant Biology, Uppsala BioCenter, Linnean Center of Plant Biology, Swedish University of Agricultural Sciences, Uppsala, Sweden

Haibian Yang and Yuling Jiao
State Key Laboratory of Plant Genomics, National Center for Plant Gene Research, Institute of Genetics and Developmental Biology – Chinese Academy of Sciences, Beijing, China

Yingying Han
State Key Laboratory of Plant Genomics, National Center for Plant Gene Research, Institute of Genetics and Developmental Biology – Chinese Academy of Sciences, Beijing, China
University of Chinese Academy of Sciences, Beijing, China

Bo Sun
Temasek Life Sciences Laboratory, 1 Research Link, National University of Singapore, Singapore

Toshiro Ito
Temasek Life Sciences Laboratory, 1 Research Link, National University of Singapore, Singapore
Department of Biological Sciences, National University of Singapore, Singapore

Aiwu Dong
State Key Laboratory of Genetic Engineering, Collaborative Innovation Center of Genetics and Development, International Associated Laboratory of CNRS-Fudan-HUNAU on Plant Epigenome Research, Department of Biochemistry, Institute of Plant Biology, School of Life Sciences, Fudan University, Shanghai, China

Jinlei Shi and Wen-Hui Shen
State Key Laboratory of Genetic Engineering, Collaborative Innovation Center of Genetics and Development, International Associated Laboratory of CNRS-Fudan-HUNAU on Plant Epigenome Research, Department of Biochemistry, Institute of Plant Biology, School of Life Sciences, Fudan University, Shanghai, China
CNRS, Institut de Biologie Moléculaire des Plantes, Université de Strasbourg, Strasbourg, France

Fabien Lombardo and Hitoshi Yoshida
Rice Biotechnology Research Project, Rice Research Division, National Agriculture and Food Research Organization (NARO) Institute of Crop Science, Tsukuba, Ibaraki, Japan

Nadia M. Atallah and Jo Ann Banks
Department of Botany and Plant Pathology, Purdue University, West Lafayette, IN, USA

Xiaolu Qu
Center for Plant Biology, School of Life Sciences, Tsinghua University, Beijing, China

Yuxiang Jiang, Ming Chang, Xiaonan Liu, Ruihui Zhang and Shanjin Huang
Key Laboratory of Plant Molecular Physiology, Institute of Botany – Chinese Academy of Sciences, Beijing, China

Ce Shi, Liyao Zhang and Meng-Xiang Sun
State Key Laboratory of Hybrid Rice, Department of Cell and Developmental Biology, College of Life Sciences, Wuhan University, Wuhan, China

An Luo
State Key Laboratory of Hybrid Rice, Department of Cell and Developmental Biology, College of Life Sciences, Wuhan University, Wuhan, China
College of Life Sciences, Yangtze University, Jingzhou, China

Kelli F. Henry and Robert B. Goldberg
Department of Molecular, Cell and Developmental Biology, University of California, Los Angeles, Los Angeles, CA, USA

Yulin Yang
Department of Pharmacy, Shanghai Tenth People's Hospital, School of Life Sciences and Technology, Tongji University, Shanghai, China

Liangsheng Zhang
Department of Pharmacy, Shanghai Tenth People's Hospital, School of Life Sciences and Technology, Tongji University, Shanghai, China
Advanced Institute of Translational Medicine, Tongji University, Shanghai, China

Jie Cui, Fang Chang and Yingxiang Wang
State Key Laboratory of Genetic Engineering and Collaborative Innovation Center for Genetics and Development, Ministry of Education Key Laboratory of Biodiversity Science and Ecological Engineering and Institute of Biodiversity Sciences, Institute of Plants Biology, Center for Evolutionary Biology, School of Life Sciences, Fudan University, Shanghai, China

Lei Wang and Hong Ma
State Key Laboratory of Genetic Engineering and Collaborative Innovation Center for Genetics and Development, Ministry of Education Key Laboratory of Biodiversity Science and Ecological Engineering and Institute of Biodiversity Sciences, Institute of Plants Biology, Center for Evolutionary Biology, School of Life Sciences, Fudan University, Shanghai, China
Institutes of Biomedical Sciences, Fudan University, Shanghai, China

Hao Li and Fanrui Meng
State Key Laboratory of Systematic and Evolutionary Botany, Institute of Botany, Chinese Academy of Sciences, Beijing, China
University of Chinese Academy of Sciences, Beijing, China

Margarida Rocheta, Teresa Ribeiro and Leonor Morais-Cecílio
Departamento de Recursos Naturais Ambiente e Território, Instituto Superior de Agronomia, Universidade de Lisboa, Lisboa, Portugal

Pilar García-Jiménez and Rafael R. Robaina
Departamento de Biología, Universidad de Las Palmas de Gran Canaria, Las Palmas de Gran Canaria, Spain

Jochen Gohlke
School of Plant Sciences, University of Arizona, Tucson, AZ, USA

Rómulo Sobral, Joana Magalhães and Maria M. R. Costa
Centre for Biodiversity, Functional & Integrative Genomics, Plant Functional Biology Centre, University of Minho, Braga, Portugal

Maria I. Amorim
Departamento de Biologia, Faculdade de Ciências da Universidade do Porto, Porto, Portugal

Miguel Pinheiro and Conceição Egas
Biocant, Parque Tecnológico de Cantanhede, Cantanhede, Portugal

Zhigang Wu, Xuelin Tan and Qiong Luo
Ministry of Education Key Laboratory of Agriculture Biodiversity for Plant Disease Management, Yunnan Agricultural University, Kunming, China

Ding Tang, Hongjun Wang, Yi Shen, Wenqing Shi, Yafei Li and Zhukuan Cheng
State Key Laboratory of Plant Genomics and Center for Plant Gene Research, Institute of Genetics and Developmental Biology, Chinese Academy of Sciences, Beijing, China

Jianhui Ji
State Key Laboratory of Plant Genomics and Center for Plant Gene Research, Institute of Genetics and Developmental Biology, Chinese Academy of Sciences, Beijing, China
School of Life Sciences, Huaiyin Normal University, Huaian, China

Hongyan Shan, Hongzhi Kong and Chunce Guo
State Key Laboratory of Systematic and Evolutionary Botany, Institute of Botany, Chinese Academy of Sciences, Beijing, China

Eva Knoch, Adiphol Dilokpimol and Naomi Geshi
Department of Plant and Environmental Sciences, University of Copenhagen, Copenhagen, Denmark

Xiaojing Xie, Tiansheng Zhu and Shuigeng Zhou
Shanghai Key Lab of Intelligent Information Processing and School of Computer Science, Fudan University, Shanghai, China

Ying Hua Su, Yu Bo Liu, Bo Bai and Xian Sheng Zhang
State Key Laboratory of Crop Biology, College of Life Sciences, Shandong Agricultural University, Taian, China

Rosalia Deeken
Department of Molecular Plant Physiology and Biophysics, Julius-von-Sachs-Institute, University of Wuerzburg, Wuerzburg, Germany

Tatiana V. Matveeva and Ludmila A. Lutova
Department of Genetics and Biotechnology, St. Petersburg State University, St. Petersburg, Russia

Andrea Pitzschke
Department of Applied Genetics and Cell Biology, University of Natural Resources and Applied Life Sciences, Vienna, Austria

Wenjing She and Célia Baroux
Institute of Plant Biology – Zürich-Basel Plant Science Center, University of Zürich, Zürich, Switzerland

Jose A. Traverso, María I. Rodríguez-García and Juan D. Alché
Estación Experimental del Zaidín, Consejo Superior de Investigaciones Científicas, Granada, Spain

Amada Pulido
Departamento de Fisiología Vegetal, Universidad de Granada, Granada, Spain

Chung-Ju R. Wang
Institute of Plant and Microbial Biology, Academia Sinica, Taipei, Taiwan

Ching-Chih Tseng
Institute of Plant Biology, National Taiwan University, Taipei, Taiwan

Index

Printed in the USA
CPSIA information can be obtained
at www.ICGtesting.com
JSHW051408091023
49903JS00006B/325

9 781641 167376